EXAMINING GCSE

INTEGRATED SCIENCE

BARRY STONE · DAVID ANDREWS · ROY WILLIAMS

EXAMINING GCSE

INTEGRATED SCIENCE

BARRY STONE · DAVID ANDREWS · ROY WILLIAMS

HUTCHINSON

London Melbourne Sydney Auckland Johannesburg

Hutchinson Education

An imprint of Century Hutchinson Ltd
62 – 65 Chandos Place, London WC2N 4NW

Century Hutchinson Pty Ltd
PO Box 496, 16 – 22 Church Street, Hawthorn, Victoria 3122, Australia

Century Hutchinson New Zealand Ltd
PO Box 40 – 086, Genfield, Auckland 10, New Zealand

Century Hutchinson South Africa (Pty) Ltd
PO Box 337, Bergvlei, 2012 South Africa

First published 1988

Text set in 10 on 12pt Palatino

Produced by The Pen and Ink Book Co. Ltd.

Printed and bound in Great Britain by
Butler & Tanner Ltd, Frome and London

British Library Cataloguing in Publication Data

Stone, Barry
Examining GCSE integrated science
1. Science – For schools
I. Title II. Andrews, David III. Williams, Roy
500

ISBN 0 09 173088 0

Acknowledgements

The preparation of this book has been a long and at times difficult task. It has been an immense pleasure but it could not have been done without the help and co-operation of many others. In particular I would like to thank Josephine Fageant for her keyboard skills, Joan for her continued support and encouragement over the years, Reg for his enthusiasm of things scientific, Pat Rowlinson and her colleagues at Century Hutchinson together with Ruth Holmes for their help, advice and understanding, Martyn Lanyon at Spring-board Electronics, Prit and Richard for their professional dedication and advice and the numerous industrial, commercial and public organizations who have supplied copious amounts of practical information, without whom this book would not have been possible.

In particular I would like to acknowledge the help, co-operation and support of my colleagues, David Andrews and Roy Williams.

Barry Stone

Many thanks for the help, support and advice of family, friends and colleagues.

David Andrews

With many thanks to Jean, Sue and Wendy for secretarial help and to Bob McDuell for his help and encouragement.

Roy Williams

The authors and publishers are grateful to the following for permission to reproduce photographs and illustrations:

Access 8.18; J Allan Cash Photo Library 3.4, 3.6, 14.29; Alton Towers 3.48; Ardea 6.10, 6.15; Associated Press Ltd 6.18, 10.20, 10.32; Barnaby's Picture Library 2.42, 3.28, 10.1, 10.14, 14.25, 15.1, 15.29; Barr and Stroud Ltd 10.25b; BEC Mobility Ltd 1.5; Robin Birkett 1.7, 1.9, 3.1, 3.36, 8.18, 8.21, 10.16, 11.1, 13.27, 13.35, 14.14, 14.20, 14.38, 15.11, 15.15, 15.19, 15.23, 15.25, 15.27; BP Oil Ltd 14.22, 15.5; British Museum (Natural History) 4.6; British Railways Board 13.20; British Steel 13.21, 13.23; CS Broomfield 4.11b; Camera Press 2.7, 3.5, 3.45, 4.9, 6.3, 9.11, 12.6, 13.7, 14.17, 14.21, 14.31, 14.37, 15.28, 17.10; Central Office of Information 2.20b; CF Cassella Ltd 4.14a, 4.14c, 4.14d; Chubb Fire 14.13; Copper Development Association 14.9; Crosfield Electronics 10.24; Daily Mail 1.2; Distiller's Co Ltd 15.16; Esso 15.6; Fiat Auto (UK) Ltd 17.14; Fidor 1.15; Fotobank International Colour Library Ltd 13.25, 13.26, 14.23; JFP Galvin 4.11c, 4.11d, 4.11e, 4.14b, 4.14f; Great Universal 9.13; Sally and Richard Greenhill 1.1, 2.32, 3.25, 3.34, 10.13, 14.6, 17.11; MI Holmes 4.11a; Hunt and Sykes, *Chemistry*, Longman, Harlow 1984, 13.18; The Hutchison Library 2.32; ICI 2.33, 14.18; JC Jones/JEOL UK Ltd 11.12; Nick Kamen 10.1; Kratos Analytical 16.6; Frank Lane Picture Agency Ltd 6.15; Mansell Collection 4.4; Mencap 2.30; Minivator 9.6; NASA 6.10, 7.6, 14.8; 6.14 first appeared in *New Scientist*, the weekly review of science and technology; North Thames Gas Board 15.1; Oxford Scientific Films 2.3, 2.35, 2.37, 2.41, 2.42, 5.9, 7.1, 10.11, 10.27a, 14.11, 14.16, 14.19, 14.32; Ann Ronan Picture Library 1.17, 6.3; J Sainsbury plc 3.7; Saint Bartholomew's Hospital 10.27b; Sarabec Ltd 10.28a, 10.28b; Science Museum 11.2; Science Photo Library 2.5, 2.21, 2.22, 3.15, 3.24, 4.5, 6.2, 7.1, 7.3, 7.8b, 7.8c, 9.5a, 9.5b, 10.25a, 14.3, 14.4, cover; Shell UK Ltd 15.1; Barry Stone 4.10, 7.10, 8.6, 8.7, 8.8, 10.9, 17.21; Sunday Times 9.7; TeleFocus 10.28c; Telegraph Colour Library 4.8; Alan Thomas 3.23a, 3.43, 4.14e; Times Newspapers Ltd 2.9, 3.42, 9.7; Today 2.18; Topham Picture Library 13.8; Vauxhall Motor Company Ltd 8.2a, 13.28; Vision International 9.4c; War on Want 3.7; C James Webb 2.36, 6.21; Andrew Williams 11.20, 11.27, 11.28, 13.16, 14.35, 16.3, 17.9.

The data in table 2.2 was taken from J Guillebaud, *Contraception – Your Questions Answered*, Churchill Livingstone, Edinburgh, 1986.

The following examination boards gave permission to reproduce examination questions:
London and East Anglian Group (LEAG), Midland Examining Group (MEG), Northern Examining Association (NEA), Southern Examining Group (SEG).
The answers at the back of the book are provided by the authors, and not by the examining boards.

CONTENTS

CONTENTS

INTRODUCTION

A balanced science education can be achieved within the secondary school curriculum in a number of ways. This may be through an integrated, combined, modular or co-ordinated approach to science. This book has been specifically designed to meet the needs of students following such courses.

During the book's preparation much consideration was given to the National Criteria for science and to the various 'balanced' science courses available from the different examining boards. The contents have therefore been very carefully chosen to be suitable for the majority of GCSE science courses for single and double certification.

The text follows closely an integrated science approach to the presentation of science although no attempt has been made to force this where it was not deemed suitable. Teachers less familiar with this style of presentation should find the text logical, easy to folow and easy to use with their students.

One of the aims of any GCSE science course is to enhance and enrich scientific experience through investigative experimentation. In reality this ideal is not always possible because of the very nature of a school's organization. The practical work contained within each unit aims to meet this ideal wherever possible. However, as the additional time often required to offer a completely investigative approach is not always available there are occasions when a more prescriptive method of approach is useful, and this has therefore been used where it has been thought appropriate and benficial.

The text has been designed to lead and direct students to adopt safe working practices.

The social, economic and technological aspects of science have been emphasized throughout the text. Specific examples are given in the 'Application' and 'Time for Thought' sections, many of which have been supplied by industrial, commerical and public organizations.

We hope you enjoy working with the material in the text, but above all, that it encourages you in an interest of science.

Barry Stone
David Andrews
Roy Williams
1988

Barry Stone was formerly Head of Physics at a large comprehensive in Hampshire and was instrumental in the school's transition from the separate sciences to Integrated Science.
David Andrews is a teacher of Biology and Integrated Science at a London Sixth Form College. He has responsibility for Pastoral Guidance.
Roy Williams is Head of Science at Endon High School, Stoke-on-Trent and a member of the Midland Examining Group's Science Panel.

STARTING SCIENCE

STARTING OUT

Get up and go

How many times has the alarm clock gone off and yet some time later you are still in bed – not quite asleep but certainly not awake? More often than not it happens during the week, on days when you go to work or school. Holidays and weekends are different! Then it's much easier to leap out of bed full of beans. It's not unusual for spare time to be used more energetically than work time.

figure 1.1 Full of energy?

What makes things happen?

Energy is the name of the game. Without it nothing would happen. You will probably already have an idea what the word energy means. It's quite difficult to give it a very precise meaning and almost impossible to say exactly what it is. For the moment think that

energy can make things happen

In any situation if you can see that anything is happening or being made to happen then energy is around somewhere.

57 varieties

A tin of beans can make things happen. They are full of energy, according to all the advertisements (but they don't tell you what that is).

You only have to look out of a window or walk down a street to see things happen. They don't all happen in the same way and they are not all caused by the same things. Energy appears in a number of different disguises called **energy forms**. Fortunately we only need to think of a few energy forms, not 57 varieties.

POTENTIAL AND KINETIC ENERGIES

The different energy forms fall roughly into two groups: **potential energy** and **kinetic energy**.

Potential energy (PE)

If anything has energy stored away that can be used later then it has potential energy or PE. Potential means hidden or stored. Two examples of PE would be the fat reserves in the human body and a wound up clock spring.

Kinetic energy (KE)

If anything moves then it is said to have kinetic energy or KE. 'Kinetic' comes from the Greek word for movement. Examples of KE would be moving pictures at the cinema (kinema) and walking, running, jogging etc.

What sort of energy is it?

Each of the two groups of energies can be split up into sections. We shall look at forms shown in table 1.1 individually. Notice that the energy form 'heat' appears in both columns.

potential energies	kinetic energies
mechanical potential energies	kinetic energy in moving objects
chemicals (including fuels)	waves (including sound)
gravitational energy	electricity
magnetism	heat energy
nuclear energy	
heat energy	

table 1.1 One way of organizing energy forms

Mechanical potential energy
Any mechanism or machine that stores energy stores it as potential energy. The spring in a clock or watch is an example of this type of potential energy, as is a stretched rubber band.

Chemical energy
As you might expect, this form of energy is stored in chemicals. Fuels such as petrol and food are good examples of chemical energy. A less obvious chemical energy form is a battery. When a battery is not working the chemicals are stored. Only when it is connected into a circuit do the chemicals react and produce electricity.

Gravitational energy
Anything which is high up has some of this type of energy. The water behind the dam wall of a hydroelectric power station has vast quantities of gravitational energy stored in it. Gravity pulls the water down. As it does so the water rushes through a set of turbines which create electricity. It is because the water rushes through a set of turbines which create electricity. It is because the water is up high that it has energy. Water itself (on the ground, for example) has hardly any energy stored in it at all (see section on heat).

Magnetism
This is an energy form but it is not very common. From your own experiences you will know that a magnet is able to pick up pieces of certain types of metal (iron, steel, cobalt and nickel are best). A magnet can make things happen, so magnetism is an energy form. Magnetism is used, for example, to stop fridge or cupboard doors from opening – it pulls them closed.

Nuclear energy
Certain materials such as uranium decay naturally (see unit 12, Radioactivity). In doing so they give out radiation. This process, called nuclear fission, causes the uranium and the surrounding materials to become extremely hot. The heat generated by nuclear fission in the core of a nuclear reactor is used to produce superheated steam which in turn drives a generator (steam turbine). The energy contained in nuclear radiations from certain materials can be used to treat hospital patients by killing off unwanted cells, e.g. cancer cells.

Heat energy
A hot-water bottle will act as an energy store for some considerable time. If hotter water is used then the heat energy will last longer. A red-hot electric cooker ring needs to be left for some time before it's safe to touch. Time is needed for the heat energy to escape. If you were to touch the ring while it still contained a lot of energy then some of it would pass to you and you would burn yourself. (The heat energy cooks the proteins in human cells and melts the fats, so killing the cells.)

Heat is to do with the particles (molecules) contained in a material. These particles vibrate. Give them more energy and they vibrate more. Take energy away and they vibrate less. Heat is a measure of particle (molecular) vibration.

It would be quite right for you now to say, "But wait a minute, if anything moves it's got kinetic energy!" This links heat energy and kinetic energy, and so heat appears in both columns of table 1.1.

If you had a large object such as a red-hot poker it would be foolish to say that it moves and has kinetic energy. It certainly contains a store of energy (and a lot of it) but you don't normally see pokers on the move. At the same time, if you could look at a few of the molecules inside the poker you would see them in a state of violent vibration – they would be moving. It is the molecules in the poker which have KE.

Kinetic energy in moving objects
Anything you can think of that moves, shivers, shakes, rattles or rolls has KE, including molecules. A moving car has an enormous amount of KE that has to be removed every time it stops. An increase in either the mass or the velocity will increase the amount of KE of a moving object.

Many cars, such as Volvos, are specially constructed to absorb this energy in the event of a crash. The front and rear portions, rather than the passenger compartment, are designed to crumble should the brakes not be sufficient to stop the car. The KE of the car or cars is used to change the shape of the metal (see page 93).

Waves

figure 1.2 The floods in Lynmouth, Devon in 1952 were quite extensive. The boulders swept along by the water were enormous

There is a number of different types of wave (see unit 10, Waves). Each carries energy from place to place. The damage produced by high seas is quite evident after a storm. Sound (shock waves) can also cause damage if they are too loud. When aircraft fly through the 'sound barrier' (i.e. faster than the speed of sound) the sonic boom left behind can easily shatter windows. Speaking to someone is a method of energy transfer. Sound energy (in the form of vibrations) is transferred from your vocal chords via moving air molecules to someone else's eardrum. There is another set of waves known as the electomagnetic spectrum (EMS). Light, X-rays etc, are all members of this set. They all carry energy from place to place. Visible light, particularly red and blue, is absorbed by plants. Without it they would not have the energy to make the food they need. Solar-powered calculators will not work in darkness.

Electricity
This is probably the most useful energy form for modern living. Electricity is easy to transform into any other energy, e.g. light, heat, sound etc. But an electric circuit will not work unless it is complete. Electricity (electrical energy) travels from one terminal of a source, around the circuit, to the other terminal. If the circuit is broken the flow of electricity stops immediately. A battery on its own does not contain electricity. It contains chemicals which, when allowed to react with one another, produce a current of electricity between its terminals. When the chemicals are used up no more electricity will flow.

Heat energy
See earlier section.

THE JOULE IN THE CROWN

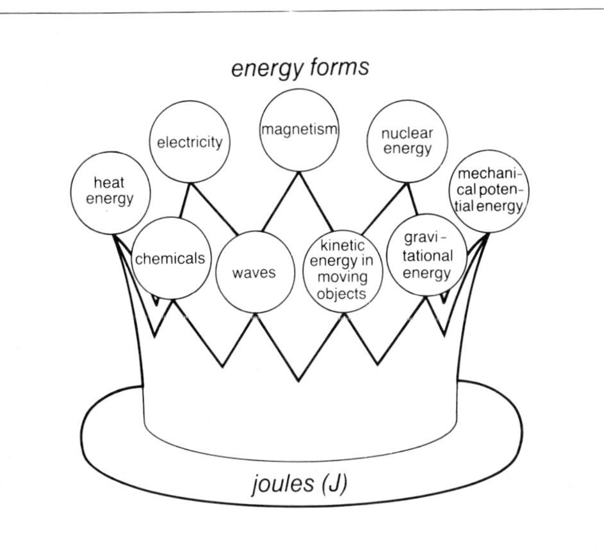

figure 1.3 *The joule in the crown*

No matter what energy form is being considered, energy is always measured in **joules (J)**. Even when energy changes take place, joules are always the unit of measurement.

Transduced before your very eyes

An electric light bulb is an example of an **energy transducer** or energy changer. It takes in electrical energy (measured in joules) and emits energy in the forms of heat and light (both measured in joules). Figure 1.4 shows the energy changes in an ordinary household light bulb as an energy arrow.

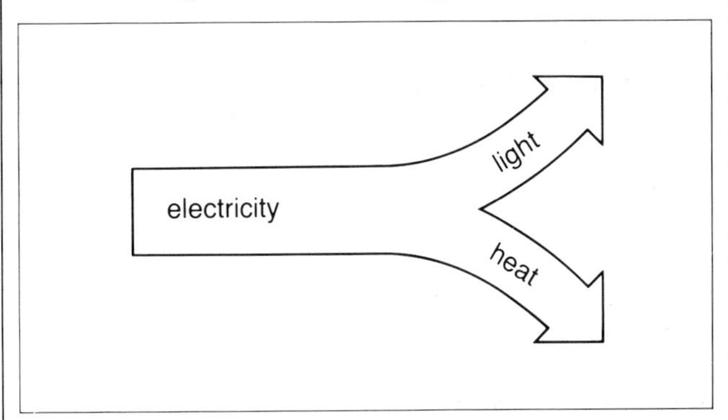

figure 1.4 *A simple energy arrow for a device like a light bulb where electrical energy turns to heat energy and light energy*

Plants are also energy transducers. They take energy from the sun and change it into stored food energy. This is called photosynthesis (see page 82). **Energy arrows** are very useful to show how energy is transduced (changed) from one form into another (or others). It's possible to make up energy arrows for almost any situation where energy is being used. The photograph in figure 1.5 shows an electrically propelled chair. The energy arrow shows the chair's main energy changes.

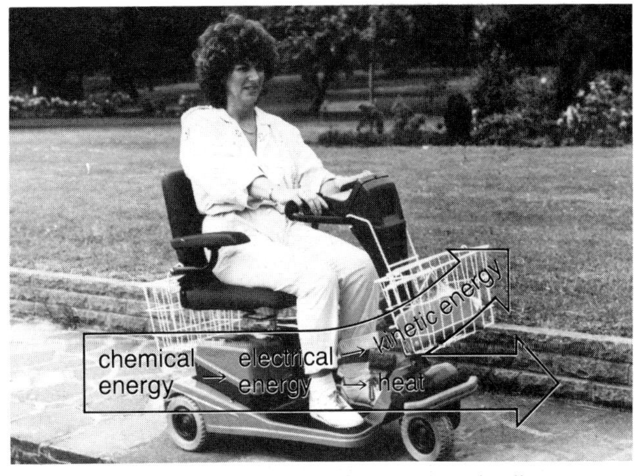

figure 1.5 *An energy arrow for an electrically operated chair*

Most energy changes release some energy in the form of heat. You could compare a bank with an energy changer. When banks exchange foreign currency, for example, they will charge you a commission. You have to pay them for the service. A loss of energy (in the form of heat) is the 'energy service charge' when one form of energy is changed into another.

INVESTIGATION 1.1

ENERGY AND ENERGY CHANGES

Each of the experiments in figure 1.6 shows a set of energy changes taking place. Try each experiment in turn and observe them carefully.

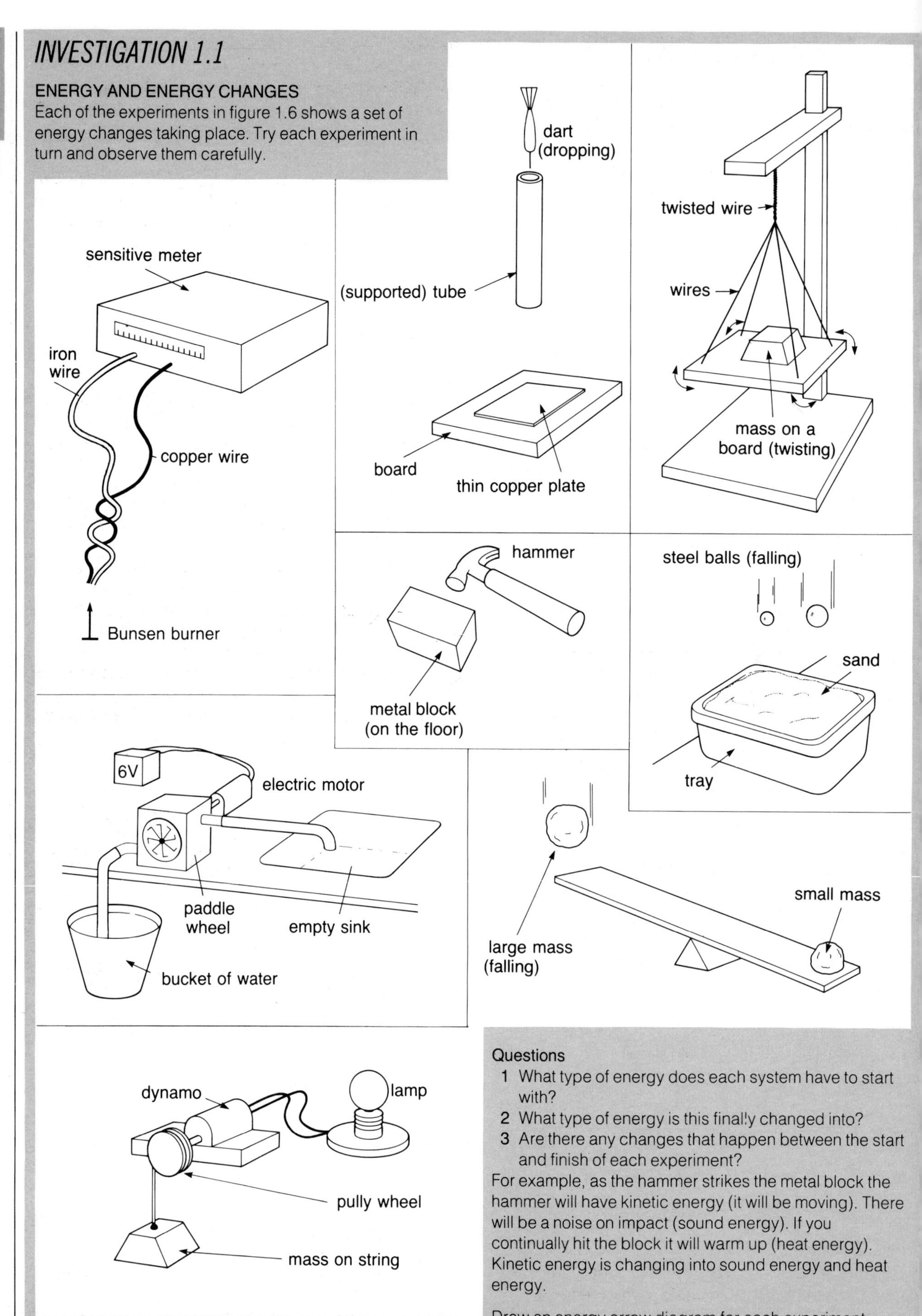

figure 1.6 Experiments involving energy changes

Questions

1 What type of energy does each system have to start with?
2 What type of energy is this finally changed into?
3 Are there any changes that happen between the start and finish of each experiment?

For example, as the hammer strikes the metal block the hammer will have kinetic energy (it will be moving). There will be a noise on impact (sound energy). If you continually hit the block it will warm up (heat energy). Kinetic energy is changing into sound energy and heat energy.

Draw an energy arrow diagram for each experiment showing the main energy changes.

PARTICLES AND MOLECULES

How many times have you heard the words 'particle' and 'molecule' and wondered exactly what they mean? The simplest explanation to start with is 'something small'. Now how many different words are there to describe something small?

Each of the following words means something small or small part, but each is usually only used in a certain way.

splash trace snack morsel bit jot dab fleck speck smut grain crumb drop seed fragment clip shaving

Take the words crumb and trace. Both mean 'small portion' but each has its own separate use. More importantly, neither word means 'the smallest amount'. In the scientific world we tend to use the word 'particle' to mean something small, but not to mean the smallest. (Crumbs can be broken down into even smaller parts, after all!) Having a word for 'small', we now need a word for 'smallest'. From the previous list of words think of 'seed'.

figure 1.7 From crumbs and seeds to particles and molecules

A plant seed is not the smallest thing there is. It would be quite easy to chop one up into smaller parts but once you've done that the seed no longer has an identity. It would no longer by recognizable as a seed. **Molecules** are the smallest single pieces of a material. A molecule of water could be chopped up into smaller parts but it would no longer be recognizable as water. A **particle** may be made up of a number of molecules. The number of molecules contained in a particle may be large or small.

The idea that all matter is made from molecules and particles is known as the **particle nature of matter**.

Molecules: some evidence

Molecules are so small it's impossible to see them with the naked eye or with a microscope. There are several experiments that can be performed in a laboratory that will give an idea of how molecules behave. From this information we can build a model to explain some of the behaviour of matter.

Brownian motion

In the early 1800s Robert Brown, a Scottish botanist, was looking at some pollen grains floating on water. He noticed they were moving. No matter how he altered his experiments he could not stop this movement. It was not until 1836 that another scientist offered an explanation. He suggested that the pollen grains were being bombarded by the water, and that whatever water was made of, it was on the move. This provides important evidence that molecules exist and that they move. This idea became known as Brownian motion.

INVESTIGATION 1.2

BROWNIAN MOTION

Brownian motion in a gas can be observed easily with the apparatus shown in figure 1.8. It is a very useful experiment to do to gain confidence in handling delicate apparatus.

- a The smoke container needs to be filled with smoke from a wax straw. Lit at the top end, smoke (being heavier than air) will fall down the straw and emerge from the bottom.
- b Put the smoke cell, including the smoke container, on the microscope (connected to a power supply) before attempting to pour in the smoke.
- c Fill the container with smoke and place a cover slip over it to stop the smoke escaping.
- d Very carefully adjust the microscope focus to look inside the smoke container. You are looking for specks of smoke (smoke particles) which will appear to be lit up.

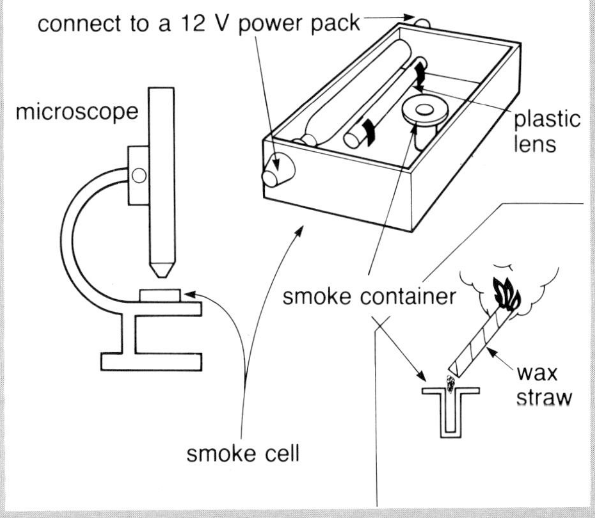

figure 1.8 Using a smoke cell to observe Brownian motion

Questions

1. By observing a large group of smoke particles, carefully describe how they move.
2. Concentrate on a single particle and describe how it moves.
3. Compare your answers to **1** and **2**. What is the same about them? What is different?

The evidence from investigation 1.2 confirms the idea that molecules are in motion. The smoke particles are being continuously bombarded by the molecules in the surrounding air. An individual smoke particle has no particular place to go nor does it have any particular preference for direction. The motion of an individual particle is random – it shows 'random motion'. There is no fixed pattern. You may, however, notice that large groups of particles drift across the container as a convection current in a general direction (e.g. from left to right).

More about molecules: diffusion and osmosis

A simple definition of **diffusion** could be 'the movement of molecules to somewhere where they weren't before'. We can tell molecules are diffusing through the atmosphere if we can smell them a long way away from where they started out. Bacon frying in a pan provides a smell that diffuses quickly. Surprisingly, diffusion takes place in solids as well as in fluids (liquids and gases).

Osmosis is a very special example of diffusion. It is the movement of liquid molecules across or through a barrier. An example of osmosis would be the uptake of water by the roots of a plant. The water moves across the barrier if there is less water on the other side, i.e. if there is less water in the root cells than in the soil.

figure 1.9 *The plant on the right needs a drop of osmosis*

The water enters the plant root cell by crossing the cell wall. The membrane inside the cell wall is an example of a **selectively permeable membrane** which allows certain molecules to pass through it but not others. It acts rather like a filter. A selectively permeable membrane with large 'holes' in it will obviously let larger molecules through. Water and minerals are small enough to pass through the plant's selectively permeable cell membrane. This process is essential to the plant's survival. When dried prunes are soaked in water the same thing happens and the prunes swell up again.

Problem How do you keep foods such as fruit from going bad?
Solution By allowing the foods to dry out (dehydrate) the micro-organisms which cause the decay cannot function. The foods may be soaked in water again before they are eaten.
The diagrams in figure 1.10 show a number of simple experiments that illustrate the movement of molecules. Some take longer to work than others.

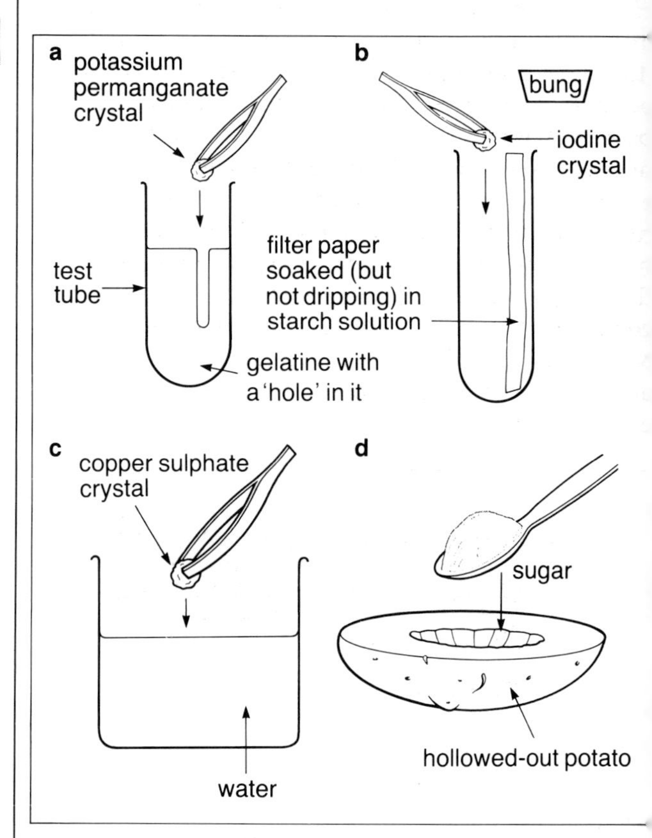

figure 1.10 *Experiments to show diffusion (**a**, **b**, **c**) and osmosis (**d**) at work*

Molecular models

The evidence provided by experiments with Brownian motion, diffusion and osmosis allows us to put forward an idea or **theory** of the way in which molecules might behave or begin to behave. A theory is a suggestion which a scientist makes to explain the results of experiments. Very few theories provide all the answers.

A **model** is an image of something real. We could use a set of basketballs or a group of people as a molecular model. Probably the easiest model to use is a set of small hard spheres, like marbles or snooker balls. Using simple models like this we can go on from single molecules to try and explain some of the things that happen in very large groups of molecules.

Solids, liquids and gases

Jsing the snooker ball or marble model of molecules, he behaviour of the **three states of matter** (**solids, iquids** and **gases**) becomes much easier to ınderstand (see figure 1.11).

'he only real difference between the three states of natter is energy. A gas usually has more energy than liquid which usually has more energy than a solid. 'he molecules in a solid stay in one position. They ıave a pattern which is fixed and cannot change. Iowever, they do vibrate. Given some more energy, nolecules in the solid state will vibrate more violently they heat up!). If they gain enough energy they may e able to break free from their fixed pattern and flow ›ast one another. When molecules have this amount ›f energy the solid **melts** to become a liquid. With ven more energy the molecules will be able to break ree from one another completely. When this ıappens the liquid **boils** and changes into a gas. If he energy is taken from the molecules then they may **ondense** to a liquid or **solidify** back to a solid. Molecules from a liquid may turn to a gas slowly, ılthough it is not being heated. This is called **vaporation**.

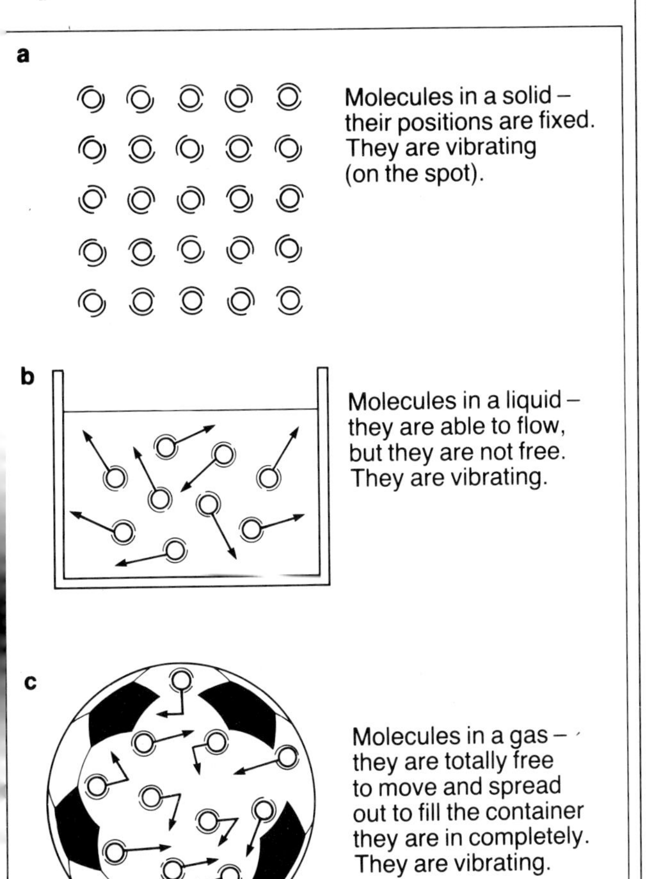

a
Molecules in a solid – their positions are fixed. They are vibrating (on the spot).
b
Molecules in a liquid – they are able to flow, but they are not free. They are vibrating.
c
Molecules in a gas – they are totally free to move and spread out to fill the container they are in completely. They are vibrating.
Arsenal
Rangers

figure 1.11 The three states of matter

Melting (or freezing) and boiling (or condensing) happen at certain fixed temperatures for a given substance. For example, transferring energy to ice will cause it to melt at 0 °C. Transferring energy to water will cause it to boil at 100 °C. There is no temperature change during a change of state, i.e. while ice molecules become water or water molecules become steam.

Cooling curves are graphs drawn to illustrate that the change in state of a material happens at a specific temperature. The two graphs in figure 1.12 are both cooling curves (even though one shows a material being heated and not cooled).

INVESTIGATION 1.3

COOLING CURVES

The top cooling curve in figure 1.12 was probably plotted from a material such as paraffin wax. The bottom curve could represent water.

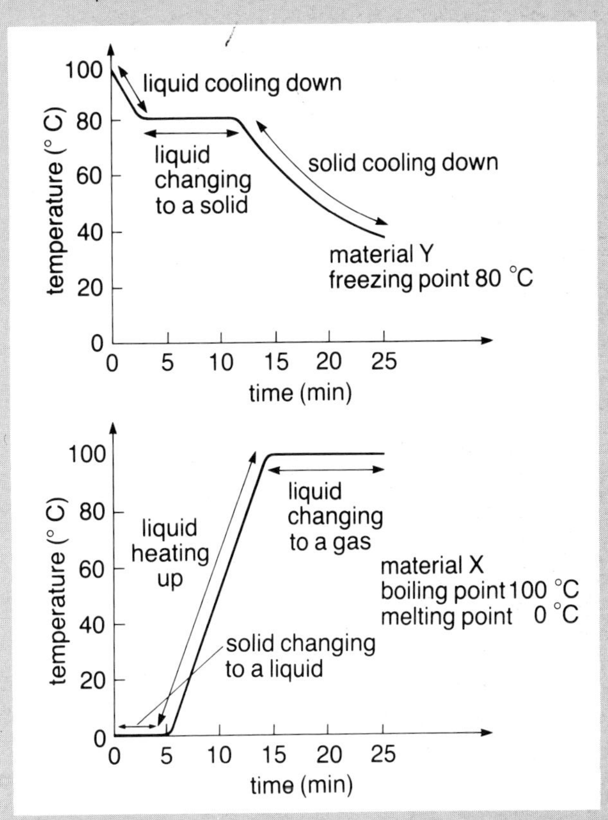

figure 1.12 Cooling curves

Questions

1 What practical design problems would you have to overcome to check the (approximate) shape of those graphs? (e.g. wax is a solid at room temperature and will need to be heated evenly.)
2 Performing these experiments may lead you to other observations about the molecules of a material that fit in with the 'molecular model' idea. What additional evidence do they provide?
3 What design alterations would you suggest to find the boiling point of a liquid such as alcohol?

UNIT 1

Gas pressure

Unlike liquids and solids, gases can be compressed. Figure 1.11c illustrates this idea. A football that has gone a little soft may be pumped up. The volume won't change much but it will become much firmer when there is more air inside. The more air there is inside a football the more often the air molecules (which are on the move) will hit the inside surface. They make the ball feel firmer because they are pushing harder from the inside. The pushing effect caused by the moving gas molecules is known as gas pressure. Gases exert pressure when their molecules hit the inside of the container.

The kinetic theory

What we have learnt about

Brownian motion diffusion osmosis
molecular models molecular movement gas pressure
and changes of state

can be brought together under one heading – **the kinetic theory of matter**. The kinetic theory involves all the laws to do with the movement of molecules.

Compact molecules: mass and density

Normally, when a material changes state, it changes volume too. For example, it is usual for **materials to expand when they are heated**. You would normally expect a material to take up more and more space as it goes from a solid to a gas. The one common exception to this is water. When it freezes, ice takes up more space than the water did originally. A block of ice is lighter than an equal volume of water. Ice floats, but only just. (Most of an iceberg's bulk remains under the water's surface.)

Density is the term or name given to the **mass** (measured in kg) of a certain **volume** (measured in m^3) of material. The 'materials kit' often used in schools contains a number of sets of various materials. A typical set of identical shapes might include glass, Perspex and marble. Although identical in size they have different masses when hung from a spring balance. The increasing order of heaviness or density would be

Perspex glass marble

Marble is the most dense, Perspex the least dense and glass has a density value in between the two. Density is calculated from the following equation.

$$\text{density} = \frac{\text{mass}}{\text{volume}}$$

Density is measured in kg/m^3 (i.e. it is the number of kilograms in every cubic metre of material). Unfortunately, this unit produces some very large numbers which can be awkward to use. For example, lead has a density of 11 300 kg/m^3. It is usually more convenient to use units which give the density value a sensible meaning. The more practical unit for density is g/cm^3. For lead, the density value is 11.3 g/cm^3 (each cm^3 has a mass of 11.3g).

Application 1.1

AEROSOL CANS

In 1929, patent No. 46613 was granted to the Norwegian engineer and inventor, Erik Rotheim. It described the use of a pressure container fitted with a valve suitable for dispensing liquids such as soap and paint.

An **aerosol** requires four essential components: a pressure resistant container, a valve which only allows things to pass *out* of the container (and not *into* it) when it is opened, a product (concentrated) and some form of stored energy. The aerosol shown in figure 1.13 contains a concentrate and a liquid propellant, which is the potential energy store. The propellent is always trying to change into a gas, but usually the pressure is too high and it cannot. When the valve is depressed it opens, and the pressure drops. The propellent turns to a gas and escapes through the valve taking the concentrate with it. Because the concentrate comes out with the propellent gas it can diffuse over a wide area.

An ordinary scent spray works in a similar way, but a scent bottle contains no propellent. Pressing the valve compresses any gas above the concentrate. The compressed gas forces the concentrate through the valve.

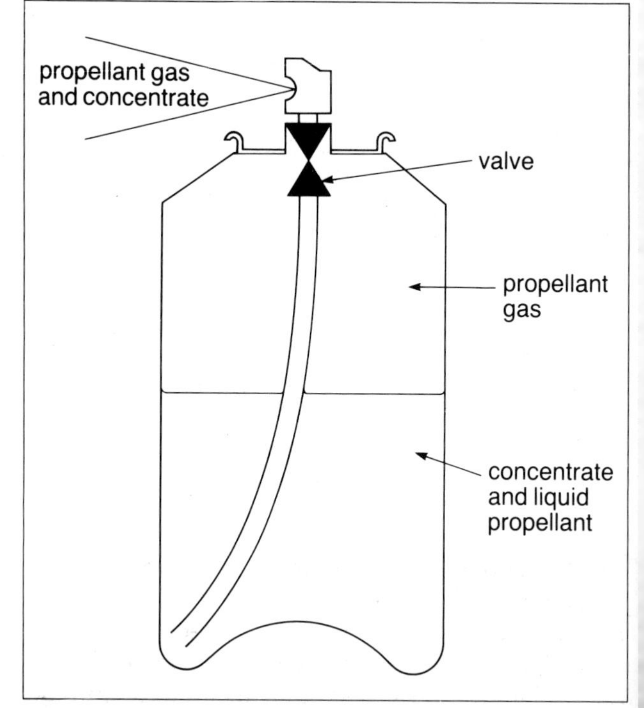

figure 1.13 The kinetic theory of matter at work

Measuring density

To calculate density two practical measurements need to be taken: mass and volume. The mass of an object is easily found with a balance (of suitable scale). Volume can be found in a number of ways. Regular shapes such as cubes, boxes or cylinders are most easily dealt with by measurement with a ruler. The volume is then found by using the appropriate formula. This is not always the best method especially when dealing with smallish objects, because a ruler is not very accurate. You cannot use a ruler with oddly shaped (irregular) objects such as pebbles etc. A **displacement can** and/or a **measuring cylinder** (see figure 1.14) will find the volume of an object accurately.

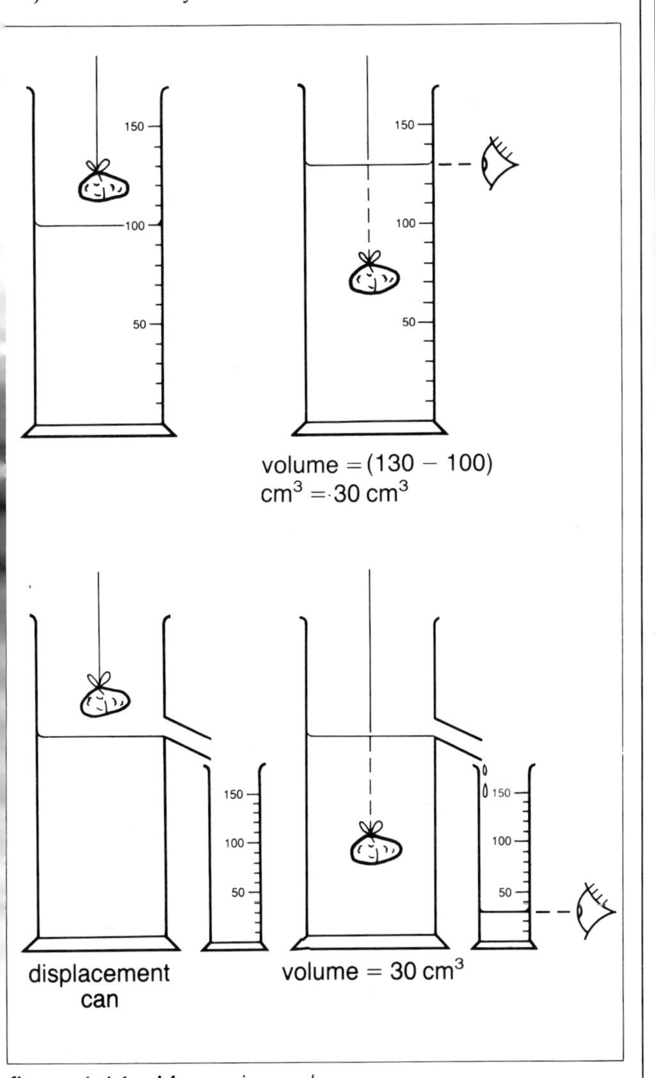

figure 1.14 *Measuring volumes*

The object under investigation is immersed in water. The volume of the water obviously changes by an amount equal to the volume of the object. (This works fine as long as the material is not spoilt by water!) This provides an accurate method for finding the volume of an object as long as you remember to read the measuring cylinder in the correct way.

Application 1.2

DENSITY AROUND THE HOME

As we have already seen, water expands when it freezes, unlike most substances, because ice has a lower density than water. If the temperature inside a house drops below 0 °C, the water in the pipes may freeze. Because ice expands, the pipe may crack. When the temperature rises again and the ice melts, the pipe leaks causing havoc in the house. The best way to avoid this happening is to make sure the pipes are well insulated so the water will not freeze.

Density has a number of other very important applications, from fishing floats to building houses. Two particular applications are in the manufacture of fibre building board and home brewing.

Fibre building boards can be divided into three main groups, depending on how much they have been compressed during manufacture – i.e. how dense they are. The first group is made up of the low-density insulating boards (softboards), insulating board tiles and acoustic boards and tiles. Medium boards of various types form the second density group, while the third group consists of the heavily compressed hardboards (see figure 1.15a).

Hydrometers are floats that sink further into less dense liquids than they do into more dense liquids. Alcohol is less dense than water so the more alcohol there is in a mixture the lower the hydrometer will sink. Home brewers use hydrometers to measure the alcohol content of wines and beers (see figure 1.15b).

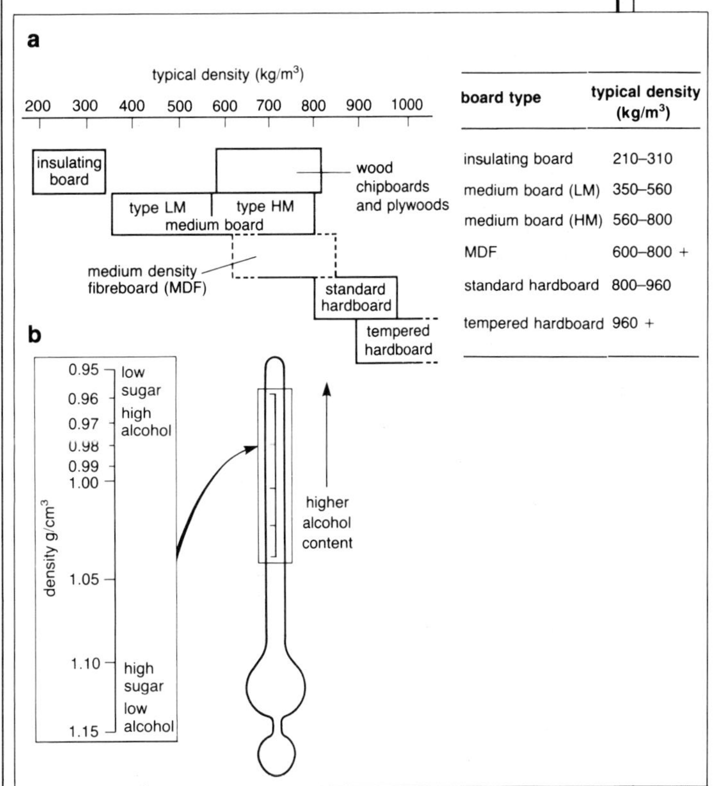

board type	typical density (kg/m^3)
insulating board	210–310
medium board (LM)	350–560
medium board (HM)	560–800
MDF	600–800 +
standard hardboard	800–960
tempered hardboard	960 +

figure 1.15 ***a*** *Density at work in the building industry* ***b*** *A hydrometer used for home brewing*

UNIT 1

IS IT LIVING?

We saw on pages 1-4 how to tell what sort of energy an object has, but how could we tell whether it was alive or not? We take for granted that we can tell if something is living. We have become familiar with the living plants and animals around us. Would it be so easy for someone who was not so familiar with life on earth? Imagine that visitors from a distant planet have just touched down in their spaceship. They observe objects we know as plants and animals (which we call living organisms). They also see a large number of metal objects moving around. We know these objects as motor cars. How could they tell whether motor cars are living? Perhaps they could ask the following questions.

Does it move?

The movement of living organisms usually has a purpose. Animals move away from danger and move towards the things they need such as food and shelter. Plants also move. Sunflowers follow the sun. Other flowers open and close. Some plants move their leaves into better positions for capturing sunlight. Cars also move, but only if a driver operates the controls.

Does it feed?

Animals eat other organisms for their food. Plants can make their own food by using simple substances and the energy from sunlight. A motor car must be filled up with petrol by somebody at intervals.

Does it obtain energy from its food?

Plants and animals can break down food to release the energy it contains. This reaction is similar to burning a fuel and is called **respiration**. Petrol is burnt inside a car engine. This releases energy to move the car.

figure 1.16 *How can an extraterrestrial visitor decide which of these objects are living?*

Does it get rid of its poisonous waste substances?

The reactions going on inside living cells produce a number of poisonous substances. The organism must get rid of these poisons quickly. In other words, living things **excrete** their waste products. Carbon dioxide is a waste gas produced by respiration. It is also one of the gases produced when petrol is burnt. Car engines expel carbon dioxide through the exhaust pipe.

Does it grow?

At some stage in its life an organism will grow in size. This growth occurs because it can take in food and convert it into its own living matter. Living organisms can use this ability to repair damaged cells. Broken bones will mend and cuts will heal. A seed may grow into a tree. A motor car cannot grow in this way and repairs must be carried out by a mechanic.

Does it reproduce?

Living things can produce a new generation of individuals that resemble their parents. The information needed for building a new individual is stored in the form of a code. This code is carried on complex molecules of **DNA** (**deoxyribonucleic acid**). DNA is only found in living organisms. Many identical models of a motor car can be built in factories through the efforts of human workers. On their own, motor cars cannot reproduce themselves!

Does it react to its surroundings?

Living organisms can collect information about their surroundings. Some animals use their sense organs for this. They can then react to this information. When a car stops at a red traffic light it might appear that the car is reacting. However, a closer look would show that it is the driver who made this happen.

Having asked all these questions, our visitor from outer space has a lot to think about. A car takes in fuel and burns it to release energy. This energy is then used to move the car. Waste gases are removed through the exhaust. In these ways a car resembles a living organism. But a car is dependent on humans to be repaired, to reproduce and react to its surroundings. The car is not living because it cannot do these things *on its own*.

Perhaps you can imagine a very advanced robot which could achieve all these feats. In this case we must ask one further question. How is it constructed? Only living organisms are constructed from microscopic building blocks called **cells**.

CELLS

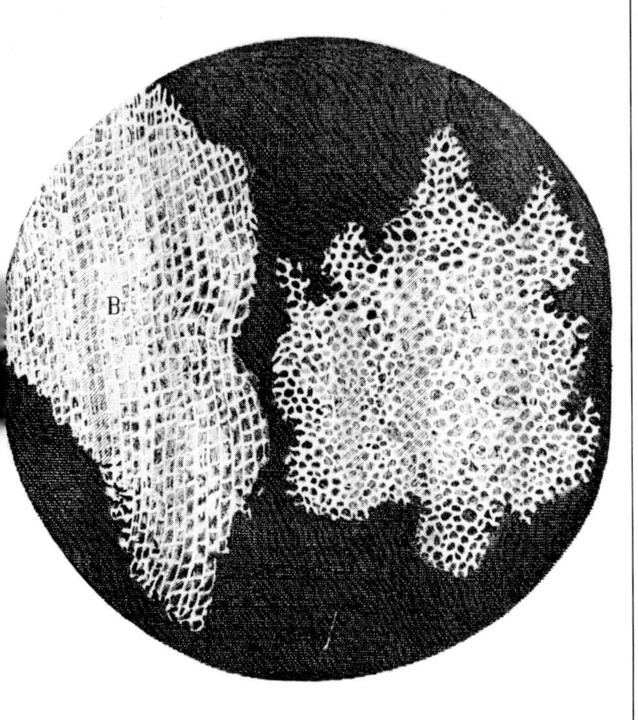

Figure 1.17 These cells from a piece of cork were drawn by Robert Hooke in 1665

Cells were first observed by the scientist Robert Hooke in 1665. He studied a thin slice of cork using his own home-made microscope. He saw the cork was made up of tiny boxes which he called **cells** (see figure 1.17). From observations on a wide variety of plants and animals, scientists concluded that all organisms were made from cells. Even the smallest organisms consist of a single cell. These are known as **unicellular organisms**. All cells consist of **protoplasm**, containing the following structures.

Nucleus

This is a structure that attracts certain dyes and is therefore stained a darker colour than the rest of the cell. It contains thread-like bodies called **chromosomes**. The chromosomes contain instructions for controlling the activities of the cell. Chromosomes are made of DNA and can make copies of themselves so that when the cell divides, identical sets of instructions are passed on.

Cytoplasm

The **cytoplasm** is where the chemical reactions of the cell take place. These chemical reactions may result in the production of substances that are released from the cell (synthesis), or in the breakdown of substances to release energy (respiration). All these reactions are controlled by instructions from the nucleus.

Cell membrane

The cytoplasm is surrounded by a thin **cell membrane** which allows certain substances to enter and leave the cytoplasm. It can also prevent the passage of other substances. The cell membrane is described as **selectively permeable** (see page 6).

Plant cells

Plant cells are different from animal cells. They usually have several structures in addition to the ones described above. They also have a cellulose **cell wall**, a **large vacuole** and a large number of **chloroplasts** and **starch grains**. You can find out the reasons for these differences in investigation 1.4.

INVESTIGATION 1.4

DIFFERENCES BETWEEN PLANT CELLS AND ANIMAL CELLS

Study figure 1.18 and then answer the following questions.

1. Name three structures found in both plant and animal cells.
2. Name four structures found in plant cells but not in animal cells.
3. If the nucleus is removed from a cell it will soon die. Explain why this is.
4. Explain why plant cells are able to make their own food while animal cells cannot.
5. Describe how plant and animal cells differ in the way they store food.

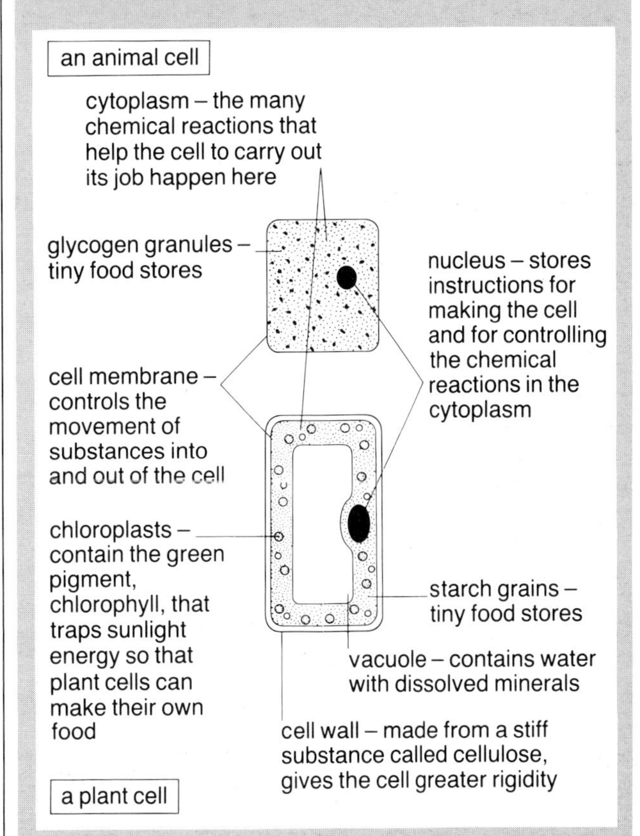

figure 1.18 Plant and animal cells have different structures

The organization of cells

The human body contains several million million cells. There are many different types of cell, each with a particular job to do. When large numbers of cells of the same type are grouped together, they form a **tissue**, e.g. muscle tissue. Different tissues are grouped together to form an **organ**, e.g. the heart. Each tissue contributes to the healthy operation of a particular organ. The stomach is the organ that adds digestive juices to our food and churns it up. It can do this because it is made of muscle tissue that makes the churning movements and glandular tissue that produces digestive juices.

The stomach does not work on its own. Other organs are involved in digesting food, such as the liver, pancreas and small intestine. These organs work together as part of a **system** – the digestive system. Several organ systems work together to keep the body alive, e.g. the nervous system and the blood circulatory system are linked with the digestive system (see figure 1.19).

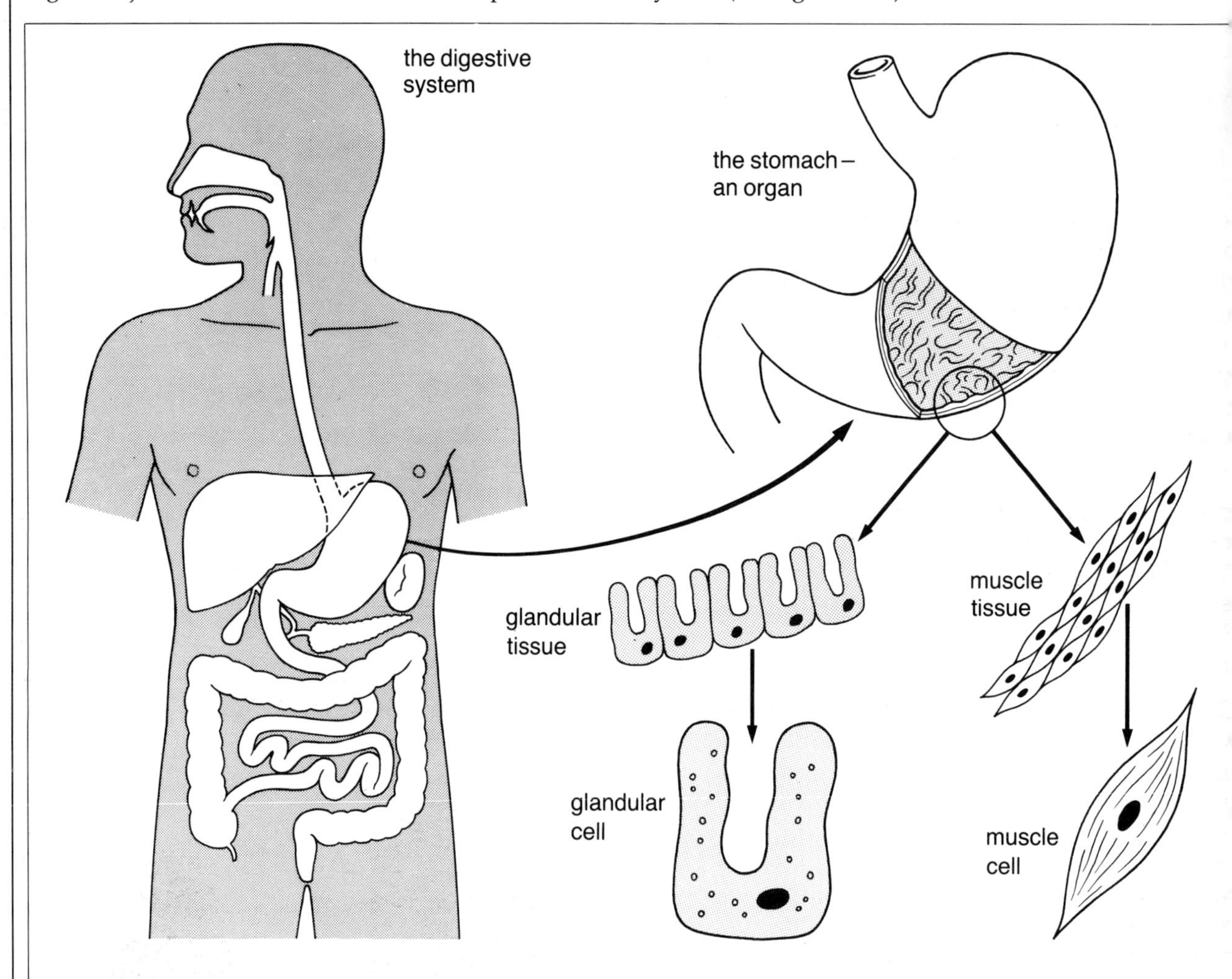

figure 1.19 *The connection between cells, tissues, organs and systems*

INVESTIGATION 1.5

LOOKING AT ONION CELLS

Between the fleshy 'leaves' of an onion there is a very delicate layer only one cell thick. If you are very careful you can remove this layer and observe the cells.

- a Separate two of the 'leaves' from an onion.
- b Use a pair of tweezers to remove a sample of the very thin layer between the leaves.
- c Place your sample on a microscope slide. Spread it out flat using the tweezers.
- d Add a drop of iodine solution and carefully lower a cover slip over the stained cells.
- e Study your slide under a microscope. Draw a group of five onion cells.

Questions

1. What were the main difficulties you came across in carrying out this investigation? How could you avoid them if you repeated the investigation?
2. Compare your diagram with figure 1.18. Use the information in figure 1.18 to label your own diagram.
3. What structures shown in figure 1.18 cannot be observed in the cells you have drawn? Explain why you have been unable to observe them.

THE VARIETY OF LIVING ORGANISMS

A tremendous variety of living organisms live on our planet: about one and a half million different **species**. A species is a group of organisms very similar to one another. Members of one species can only breed with other members of the same species (see pages 38-9). Approximately 1 150 000 of them are animal species and 350 000 are plant species. When a new species is found, it is **classified**. Biologists have a system of classification that is used to sort species into different groups.

To show how this system works we will look at the way in which the human species is classified. The biological name for the species of human beings is *Homo sapiens*. Notice that we are given two names. Our species name is *sapiens*; *Homo* is the name given to a group of different human-like species that are probably closely related. They share many characteristics such as an ability to walk on two legs, and opposable thumbs which can move across the fingers (see figure 1.20). These species make up a group called a **genus**. Other species belonging to the *genus Homo* such as *Homo erectus* may have been our ancestors. They are now all extinct.

The *genus Homo* along with several other genuses are grouped together in the *family Hominidae*. These include our early relatives such as the *genus Australopithecus* (see figure 1.21). The *family Hominidae* is grouped with other families such as the *family Pongidae* (the apes) in the *order Primates*. The Primates include monkeys, apes and ourselves.

Certain orders are grouped together in the *class Mammalia*. This class is recognized by several features including a hairy skin and the ability to give birth to live young (**viviparity**). Five classes belong to an even wider group (or **phylum**): *phylum Chordata*. These classes are fish, amphibians, reptiles, birds and mammals. These animals share an important common feature. They all have a backbone, and are more commonly known as **vertebrates**.

The **invertebrates** (animals without backbones) fall into a number of different phyla. A system of identifying the main groups of animals is illustrated in figure 1.22 (overleaf). This is an example of a **branching key**. You can use figure 1.22 to classify any class of animal. Begin at the left of the diagram. Decide which of the two alternative descriptions fits the appearance of your animal. Repeat the process until you reach the name of the animal group at the far right of the key.

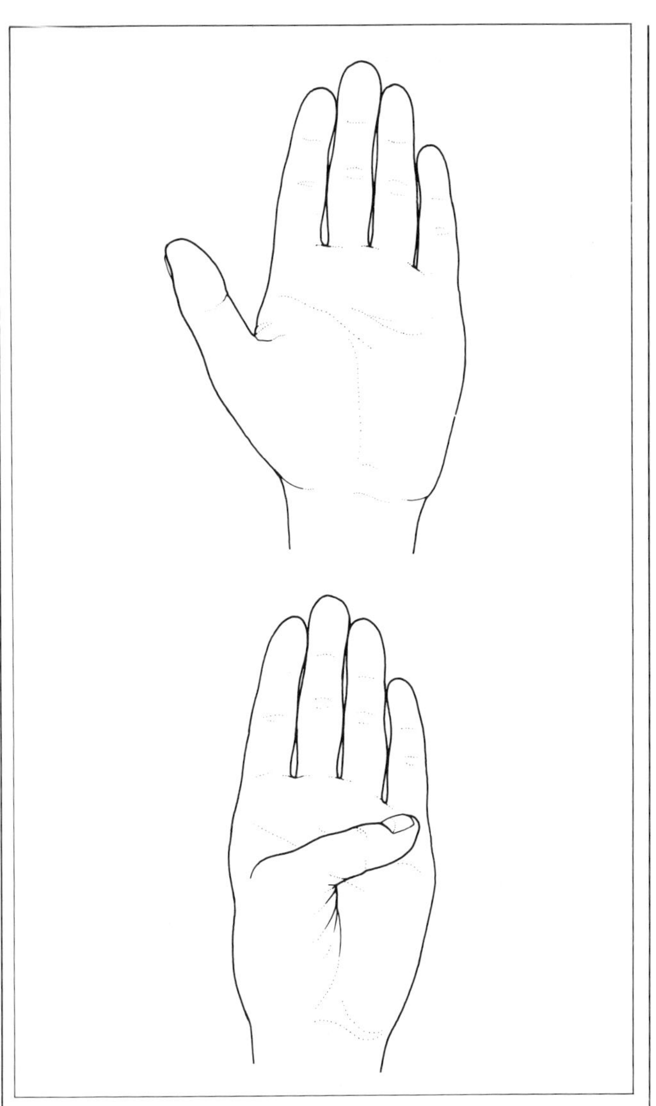

figure 1.20 The opposable thumb

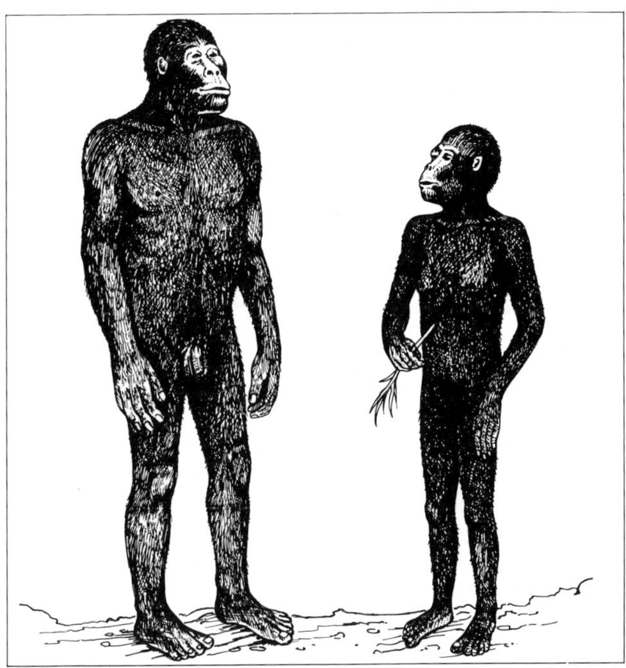

figure 1.21 Australopithecus robustus belongs to the family Hominidae. It became extinct about one million years ago

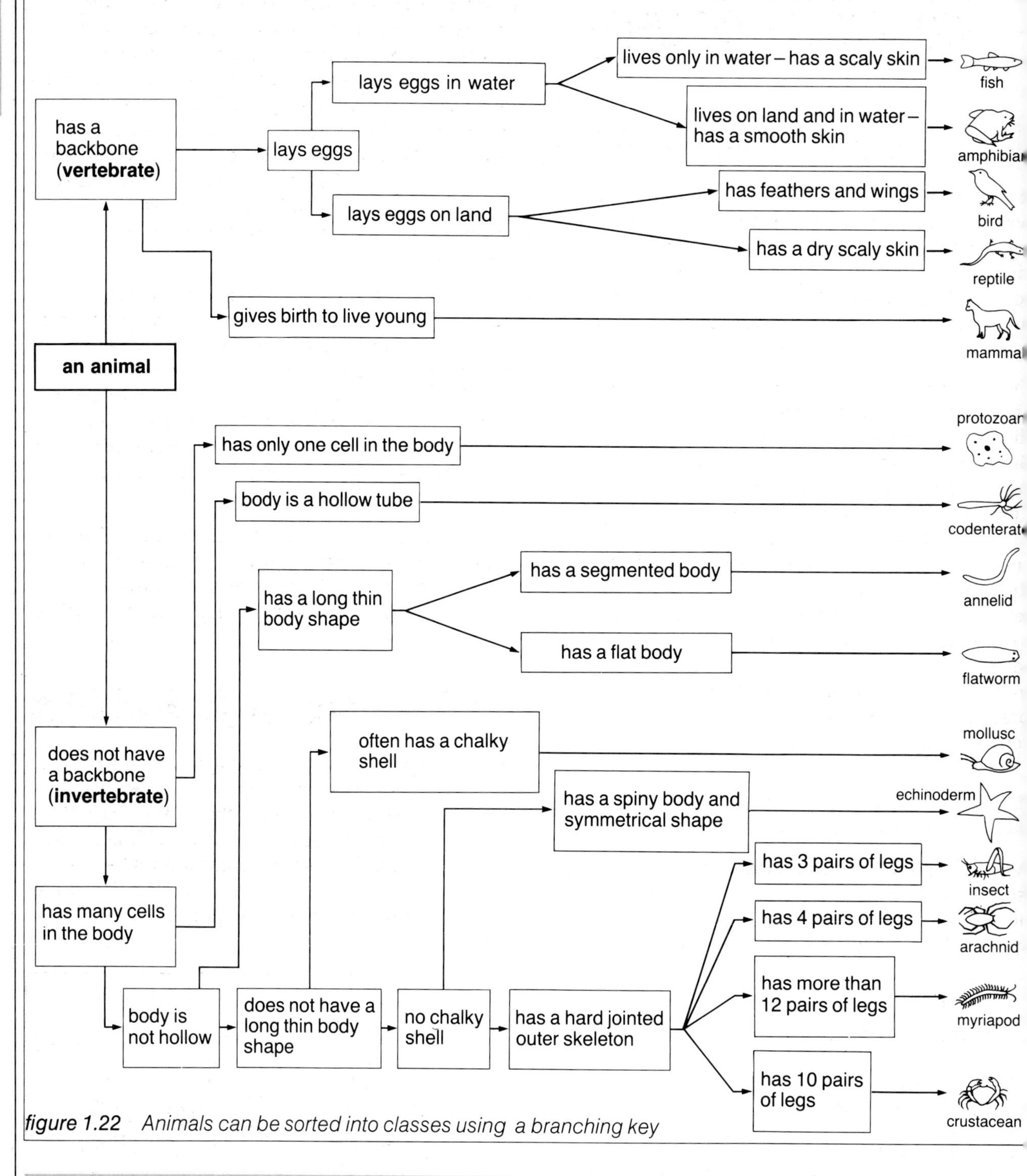

figure 1.22 Animals can be sorted into classes using a branching key

ALL YOUR OWN WORK

1 A tissue is made up of thousands of ______ of the same type. An ______ like the stomach is made from several different types of tissue. The digestive ______ includes many different organs including the stomach and the ______ .
2 What are the main differences between plant and animal cells?
3 Anything that ______ has ______ energy.
4 The unit of ______ forms of energy is the ______ .
5 What are energy arrows and what are they used for?
6 What connection does the kinetic theory have with the three states of matter?
7 a A bottle contains 250 cm^3 of 'French dressing' (a mixture of oil and vinegar used on salads). If the mixture has a mass of 240 g what is its density?
 b When left to stand the oil floats on the vinegar. Suggest a reason for this.
 c Make guesses for suitable values for the densities of oil and vinegar.

CONTINUITY OF LIFE

REPRODUCTION

It would be possible to make an advanced robot that would behave as if it were living. It could take in energy, move around, react to stimuli and make simple decisions. However, it would be impossible to construct a robot which could reproduce itself. The unique ability to reproduce sets living organisms apart from manufactured machines.

Two types of reproduction are possible. The simplest, **asexual reproduction**, involves only one parent. All the offspring produced are identical to the parent. **Sexual reproduction** requires two parents. Each parent contributes a single cell known as a **gamete**. The large female gamete contains a food store and is called the **egg** or **ovum**. The male gamete is much smaller and in animals is called a **sperm**. It has a tail that allows it to move towards the egg. In sexual reproduction, the two gametes fuse together to form a single cell known as the **zygote**. The zygote divides repeatedly and eventually grows into a new individual. We will see how plant and animal breeders use their knowledge of the two types of reproduction in their work.

ASEXUAL REPRODUCTION

Asexual reproduction has certain advantages. An isolated individual can reproduce on its own. There is no need to make contact with another member of the species. This would be useful when stranded on a desert island! In fact asexual reproduction is very common among successful organisms occurring naturally on remote islands. Since the offspring are exact copies of the parent, characteristics that are vital for survival are accurately passed down the generations. However, the lack of variety can pose problems. None of the offspring will survive if there is a sudden change in the environment. A disease can spread through a population and destroy it completely because there will be no resistant varieties.

Single-celled organisms commonly reproduce asexually. They grow to a certain maximum size and then split into two smaller individuals. This allows a very rapid increase in numbers. A single bacterium that divides every twenty minutes could produce a population of 4000 million million in twenty-four hours. This explains how a single bacterium entering the body can cause symptoms of illness in such a short time.

Many plant species can reproduce asexually. Part of the plant becomes detached and develops into a separate individual. This is **vegetative propagation**. Gardeners use vegetative propagation when tending their plants.

SEXUAL REPRODUCTION

When two parents are needed to produce offspring a number of problems have to be solved. The first problem is to find a member of the opposite sex. If the population is spread out thinly over a wide area this may be difficult. This is why many species have a specific breeding site where males and females can be sure of finding each other during the breeding season.

Application 2.1

ASEXUAL REPRODUCTION IN THE GARDEN

Weeds out of control

Vegetative reproduction can sometimes be a nuisance to gardeners. Many weeds reproduce in this way. Brambles are very difficult to control. They reproduce by forming runners. A runner is a branch of the main stem that grows out along the ground. At intervals along the runner, roots grow down into the soil and new stems grow upwards. After several months the parts of the runner between the roots die and a row of new plants is produced.

Taking cuttings

Gardeners who wish to reproduce their favourite plants can do so by taking cuttings. They cut off a healthy young branch just below a node (see figure 2.1). The cut end is then dipped in a special compound (rooting powder) before being placed in the soil. Roots grow from the cut end and a new plant develops.

figure 2.1 *A cutting being taken from a plant*

Another problem is to ensure that the sperm get as near to the egg as possible. Fish and amphibians time their mating very carefully so that the gametes (eggs and sperm) are shed into the water simultaneously. Fertilization happens *outside* the animal's body and is called **external fertilization**. Among the reptiles, birds and mammals the male can deliver sperm into the female's body. This is known as **internal fertilization**. It allows the fertilized egg to be protected during its early stages of development. Birds and reptiles lay eggs with a tough shell to protect the embryo until it hatches. Mammals give birth to live young.

Sexual reproduction has an important advantage compared with asexual reproduction. The offspring of sexual reproduction will vary from one another and from their parents. This variability could make the difference between life and death in certain circumstances. If a fatal disease spreads through the population some of the offspring may be resistant to it. Variation ensures that a few of the offspring will survive if there is a sudden change in the environment.

The male reproductive system

Figure 2.2 shows the structure of the human male reproduction system. The **testes** are a pair of oval structures each consisting of 500 metres of coiled microscopic tube known as **sperm tubules**. The testes manufacture sperm and are held in a sack of skin called the **scrotum**. The scrotum holds the testes below the abdomen in a rather exposed position. They are delicate structures and being struck by a flying object such as a football is very painful. So why are the testes in such an unprotected position? They originally developed inside the abdomen but moved down into the scrotum shortly before birth.

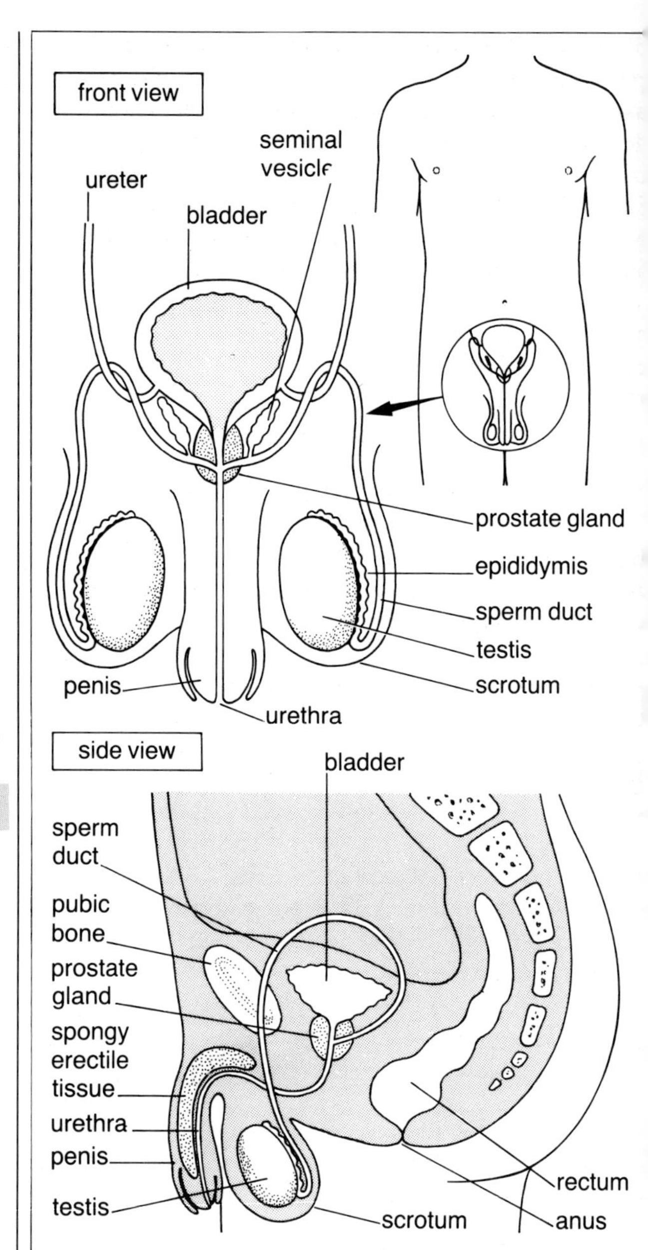

figure 2.2 The human male reproductive system

Although the testes are less protected here, they are kept at a constant temperature about 2 °C cooler than the normal body temperature of 37 °C. At 35 °C the testes are able to produce sperm in large quantities. At a higher temperature sperm production is drastically reduced. Wearing tight trousers can sometimes make a man infertile. The trousers push the testes against the warmth of the body which stops the production of sperm.

Figure 2.3 shows what a cross-section of a sperm tubule looks like when magnified 10 000 times. Sperm begin their life near the edge of the tubule and as they mature they move nearer to the centre. The 'nurse cells' appear to provide food for the developing sperm. The final stage in their development occurs when they grow tails and move to a temporary store next to each testis, called the **epididymis**.

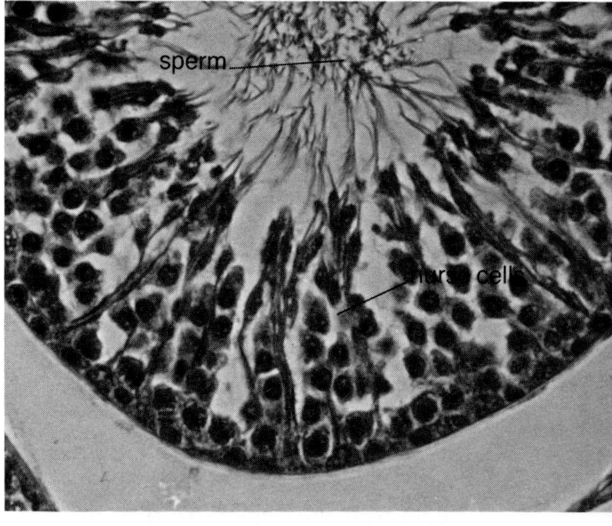

figure 2.3 Photomicrograph of a cross-section of a sperm-producing tubule

A tube called the **sperm duct** leads out from each

ididymis. The two sperm ducts eventually connect with the **urethra**. The urethra is a narrow tube assing down the centre of the **penis**. Several glands ld fluid to the sperm as it passes along the sperm ucts. The **prostrate gland** is about the size and shape a golf ball. It produces fluid that activates the perm, helping them to swim more vigorously. In ome men, particularly when elderly, the prostrate and swells up and has to be removed. The **seminal esicles** also produce large quantities of fluid. The ecretions of these glands when added to the sperm rm the sticky, creamy coloured fluid called **semen**.

The female reproductive system

he structure of the human female reproductive stem is shown in figure 2.4. The two ovaries are val structures about 2 cm long. At birth each ovary of a baby girl contains about 50 000 immature eggs.

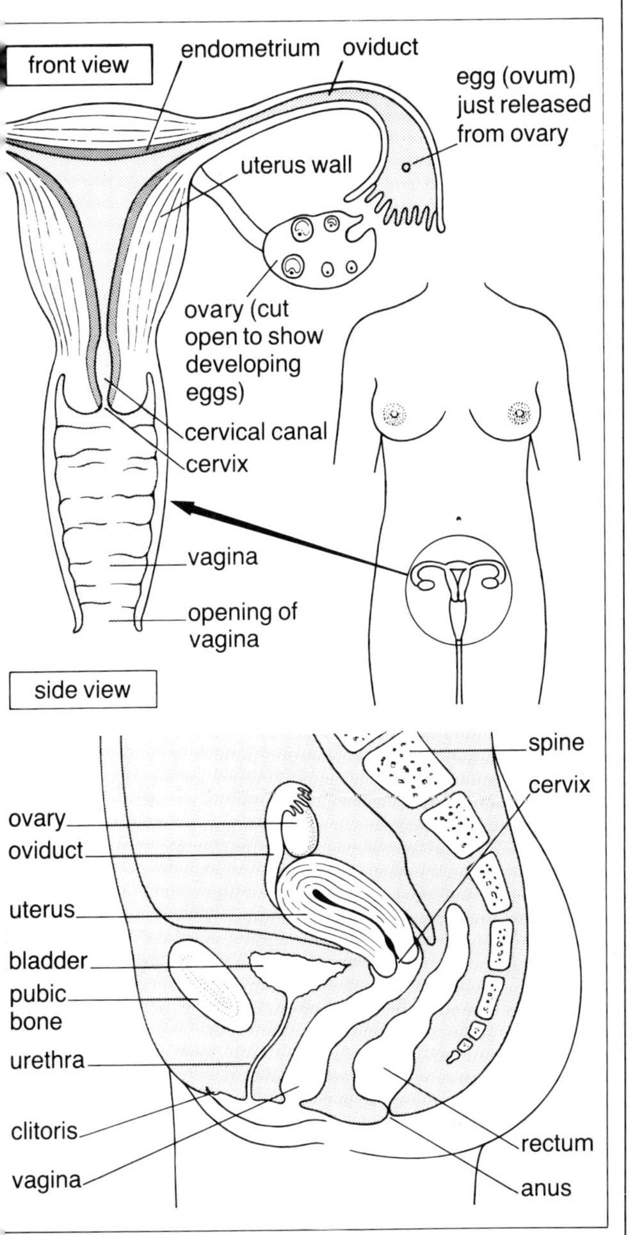

igure 2.4 *The human female reproductive system*

Only a tiny proportion of these eggs will be released during the woman's life. After a girl reaches puberty one egg is released about every 28 days. At about 50 years old, the woman will stop producing eggs, when she reaches the **menopause**. During her life a woman can produce a maximum of about 450 eggs. Inside the ovary, an egg and the cells around it develop into a follicle. This gradually moves towards the ovary wall, and when the egg is ready, it is released through the ovary wall. The release of an egg from the ovary is called **ovulation**. When this happens the egg is drawn into the funnel-shaped entrance to the **oviduct**. If the egg is fertilized it passes down the oviduct to the **uterus**. The uterus is about the same size and shape as a small pear. It is here that the baby will develop if fertilization has occured. The wall of the uterus consists of muscle and elastic fibres which allow it to expand during pregnancy.

The inner lining of the uterus (the **endometrium**) is spongy and well supplied with blood vessels. The bottom of the uterus narrows to a blunt end. This is called the **cervix**. A narrow passage about as wide as a drinking straw, the **cervical canal**, passes through the middle of the cervix. Cancer of the cervix has become more common in recent years and women are advised to have a cervical smear test every five years. This is so cancer cells in the cervix can be detected and treatment given.

The cervix protrudes a short distance into the **vagina**. This is a flexible passage about 10 centimetres long. It expands when the woman becomes sexually aroused.

The entrance to the vagina is protected on either side by folds of skin (**labia**). The **clitoris** is a small projection where the labia come together in front of the vaginal opening. As with the male penis, it swells up with blood and enlarges when the woman is sexually aroused.

Sexual intercourse

'Making love' or 'having sex' are the common names for sexual intercourse. The desire for sex in adults is a biological need similar to hunger or thirst. Sex is enjoyable because it is vital for continuing the species. Without sex no children would be born and the human species would become extinct.

During the first stages of sexual intercourse many changes take place in the male and female. The woman's vaginal wall, labia and clitoris swell up and become sensitive. Glands in the vagina produce a sticky fluid that acts as a lubricant. Spongy tissue in the penis fills up with blood under high pressure and it becomes stiff and erect. This enables the penis to slide into the vagina more easily.

Rhythmic movements of the penis against the woman's vagina and clitoris cause an increase in sexual arousal in both individuals. This leads to a reflex action in the man known as **ejaculation**. The epididymis and sperm ducts contract, forcing semen out of the urethra in several spurts.

The culmination or '**climax**' of sexual intercourse is when the man and woman experience powerful, pleasurable sensations known as **orgasm**. They may not necessarily experience this at the same time.

About a teaspoonful (2 cm^3) of semen containing about 300 million sperm are released each time a man ejaculates. When released into the vagina they swim quickly towards the cervical canal. Ninety seconds after ejaculation many sperm will have passed into the cervical canal. They must swim up through the uterus and into the oviduct. Many sperm die and only about a hundred will eventually make contact with the egg. Only one of these sperm will penetrate the egg membrane and **fertilize** the egg (see figure 2.5) to form a cell that will eventually become an embryo. When the egg is fertilized, **conception** has taken place.

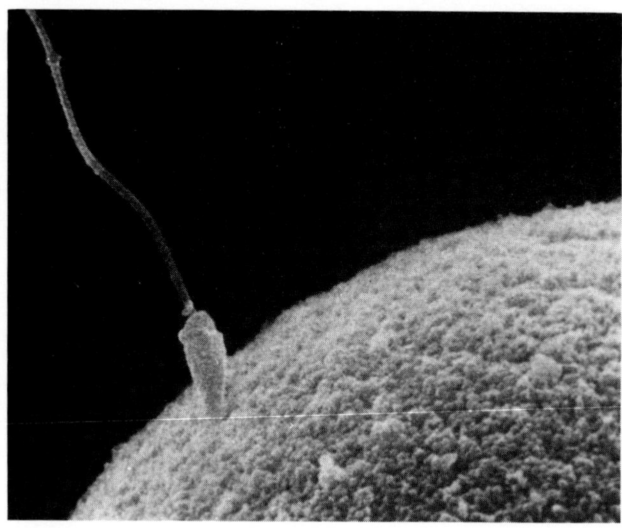

figure 2.5 *Fertilization of a human egg cell*

Development of the fertilized egg

The fertilized egg moves slowly along the oviduct towards the uterus. A few hours after fertilization it divides into two smaller cells. A few hours later each of the two cells divides again to form four cells. These divisions continue until a ball of cells (a **blastocyst**) is formed. The blastocyst is less than 1 mm across and consists of about a hundred cells. Five days after fertilization the blastocyst arrives in the uterus. It begins to sink into the spongy lining of the uterus. The growing individual is now called an embryo. The embryo absorbs oxygen and nutrients from the blood in the uterus lining. A special organ, the **placenta**, develops to carry out this job.

During the following months the embryo cells continue to divide and gradually the organs are formed. The main stages of the embryo's growth are shown in figure 2.6.

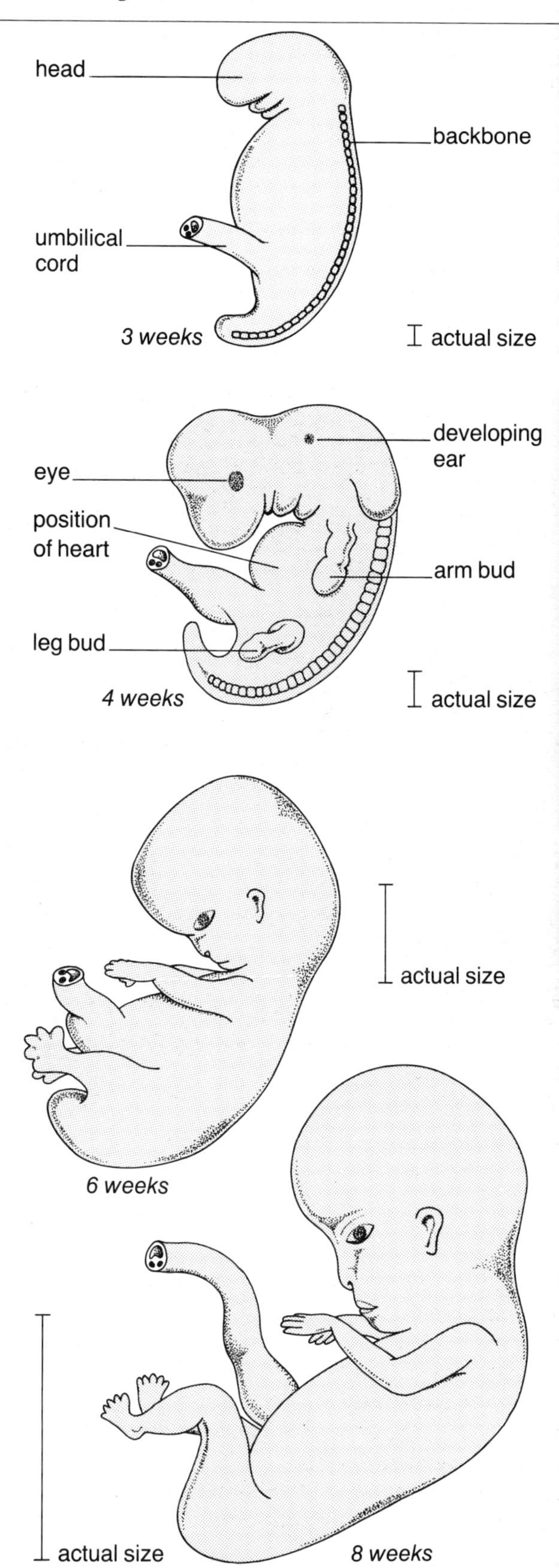

figure 2.6 *The development of the embryo inside the uterus*

Application 2.2

TEST-TUBE BABIES

If a woman's fallopian tubes are blocked, sperm are unable to reach the egg. This blockage can sometimes be cleared by an operation. If this operation is unsuccessful a technique known as *in vitro fertilization* (IVF) can sometimes help.

A woman is given hormones that increase the number of eggs she produces. Using a fine tube passed through the body wall, a doctor collects some of these eggs and places them in a dish containing a nutrient solution. Semen containing active sperm is added to the dish. Fertilization of several eggs will occur. They are allowed to develop for three days before they are inserted into the women's uterus. In most cases at least one of these embryos develops into a baby.

figure 2.7 *Louise Brown, the first 'test-tube baby'*

The placenta – a vital link

The embryo picks up oxygen and nutrients by pumping blood along an **umbilical artery** to the placenta. Here the embryo's blood flows through a fine capillary network. This brings it into close contact with the mother's blood. Waste products diffuse out of the embryo's blood into the mother's blood. Oxygen and nutrients diffuse in the opposite direction. Oxygenated blood flows back to the embryo through the **umbilical vein** (see figure 2.8).

You will notice that blood from the embryo and mother never mix. If mixing occurred it might cause problems due to incompatible blood groups (see

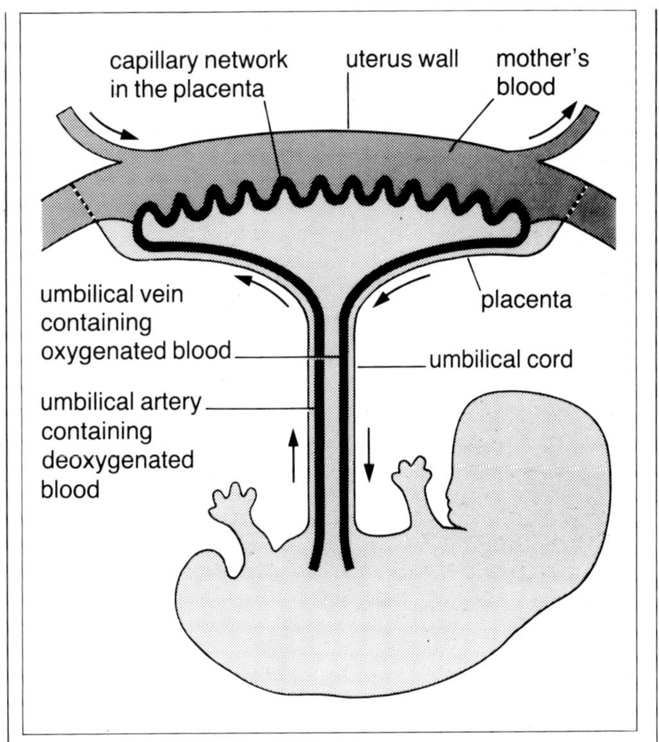

figure 2.8 *The structure of the placenta*

pages 59-60). The placenta acts as a barrier so that bacteria cannot cross from the mother into the embryo. But many other substances *can* cross over and harm the embryo.

Viruses
Viruses are smaller than bacteria and can cross over into the embryo's blood. The rubella (German measles) virus is responsible for causing deafness and blindness in newborn babies. The AIDS virus can also be transmitted in this way.

Nicotine
If the mother smokes cigarettes during pregnancy nicotine will cross over into the embryo. Compared with non-smokers, mothers who smoke tend to have smaller babies that are more likely to die in the first months of life.

Drugs
Pregnant women who are addicted to a drug such as heroin have children who are also born drug addicts. During its development the embryo receives a constant supply of the drug through the placenta. After birth the baby must be gradually weaned off the drug.

Prescribed drugs
Prescribed drugs must be tested very carefully before they are given to pregnant women. In the late 1950s the tranquillizer Thalidomide was prescribed for a large number of pregnant women. This resulted in severe stunting of the growth of the embryos' limbs (see figure 2.9, overleaf).

UNIT 2

figure 2.9 *Terry Wiles is able to move around with the aid of this vehicle specially designed by his father*

Time for thought

Thalidomide, like all prescribed drugs, was first tested on laboratory mice. No side-effects were observed. It was eventually withdrawn from use but only after thousands of malformed children had been born. It was later discovered that the testing had not included investigations on ***pregnant*** *mice.*

- *What might have been the outcome if pregnant mice had been used to test Thalidomide?*
- *In this example, what economic factors do you think influenced the action taken by the company that manufactured Thalidomide?*

Abortion

A **spontaneous abortion**, otherwise known as a miscarriage, happens when the uterus contracts in the early months of pregnancy. The foetus is usually too small to survive. About 75% of all human conceptions are lost in this way. Many miscarriages occur because the foetus is malformed or because the uterus is not ready to hold a foetus.

A **therapeutic abortion** is the deliberate removal of the foetus by a doctor. It is best if the abortion is carried out as early as possible in the pregnancy. There is a greater risk to the mother's mental and physical health if it is left until later.

Time for thought

Suppose you are a prospective parent. You are told that the results of amniocentesis show that your child will be born with Down's syndrome. The doctor says it is your decision whether to keep the child or have it aborted. What will you do? If constant support and professional help throughout the child's life can be provided, how would this affect your decision?

The decision to have an abortion is a very difficult one to make. Women who choose to have an abortion make the decision for a variety of reasons. The foetus may be severely handicapped and the parents cannot face the prospect of caring for the child and the disruption to their lives. The mother may be very young and lack support from the father. The

Application 2.3

AMNIOCENTESIS: LOOKING AT THE EMBRYO'S CHROMOSOMES

The embryo develops inside a bubble of fluid called the **amniotic sac**. This provides a weightless environment in which the embryo can float, supporting its delicate organs as they develop. It also acts as a 'shock absorber' protecting the embryo from bumps as the mother moves around.

A sample of amniotic fluid can be taken from the mother using a long hollow needle attached to a syringe (see figure 2.10). This procedure is called **amniocentesis**. The sample contains many of the embryo's skin cells that have been shed into the fluid. When these cells are examined under the microscope they can provide useful information about the embryo. For example, counting the number of chromosomes in the nucleus will show whether a chromosome mutation has occurred. **Down's syndrome** (see page 32) can be detected by this technique. The risk of Down's syndrome increases with the age of the mother. For this reason most pregnant women over the age of 35 are advised to have the test. Mothers below 35 do not normally have the test as it involves a slight risk of miscarriage.

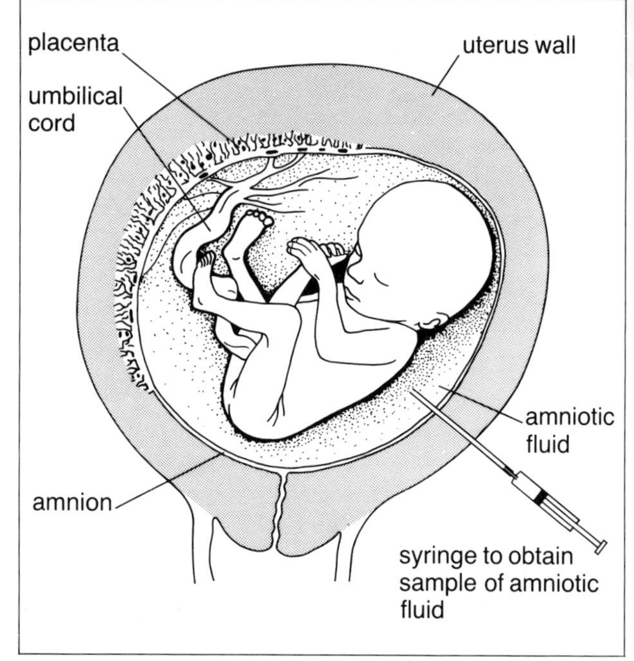

figure 2.10 *During amniocentesis a sample of amniotic fluid is obtained using a long hollow needle*

regnancy may have resulted from a contraceptive ailure and the couple may feel they cannot cope with nancial or emotional demands of an unwanted aby. If couples used contraception sensibly and arefully, many unplanned pregnancies and bortions could be avoided (see page 22–5).

Time for thought

In small groups, discuss the following statements about abortion. How far do you agree with them?

- *Abortion is the same as killing.*
- *If a foetus is abnormal the mother should be encouraged to have an abortion.*
- *Abortion should be easily available to any woman who wants one, whatever the reason.*
- *Abortions are too easily available. They should only be allowed in extreme circumstances.*

Puberty

Puberty is the name given to the dramatic changes in he body that occur during teenage years. During puberty the reproductive organs mature and begin to produce gametes (sperm and eggs). This gives eenagers the ability to produce offspring long before hey are independent enough to look after children. Teenagers reach physical maturity earlier now than hey did a century ago (see investigation 2.1). This ituation will continue to cause difficult problems for eenagers and the society in which they live.

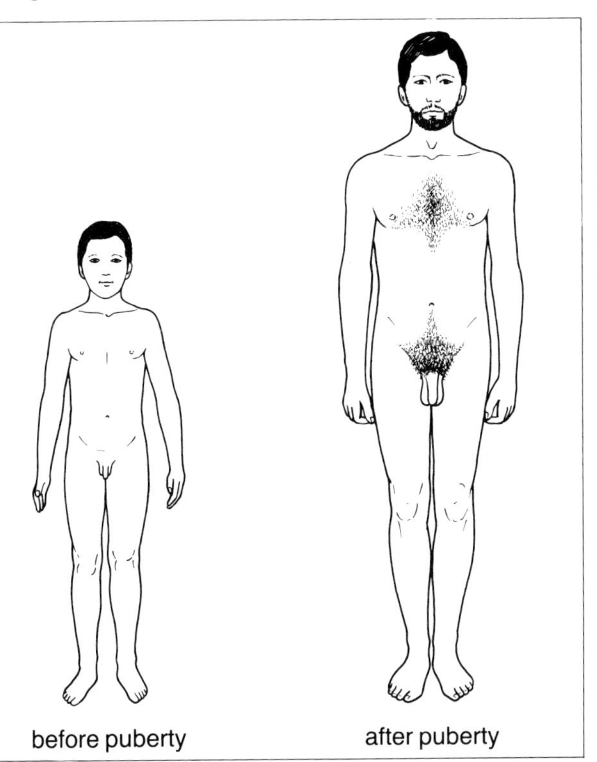

figure 2.11 Changes in the male body during puberty

The changes in the testes and ovaries are the **primary sexual characteristics**. A boy's testes begin to produce sperm. Semen builds up inside the body and is released during sexually exciting dreams ('wet dreams'). A girl's ovaries begin to produce an egg at monthly intervals. The uterus lining, along with a certain amount of blood, is shed each month. This is called **menstruation** or a **period**.

Changes also appear in the appearance of the body. These are the **secondary sexual characteristics**. As figure 2.11 shows, a boy's genitals increase in size and his body hair grows more thickly. Hair begins to grow on his face and his voice begins to deepen due to changes in the shape of the larynx. Changes in a girl's body, shown in figure 2.12, are important for pregnancy and breastfeeding. Her pelvic bones widen and more fatty tissue is laid down to produce a more rounded shape. Her breasts and nipples enlarge and pubic hair begins to develop.

Another important sign that puberty has begun is a sudden increase in growth rate. You can see from figure 2.13 (overleaf) that this growth spurt occurs earlier in girls than in boys. At around the age of 12 girls are, on average, taller than boys. By the time they reach 16 this situation has reversed. It is important to remember that there is a very wide variation in the timing of puberty from one person to another. At the age of 14 some teenagers have yet to begin puberty; while others will have almost completed their development and have an adult physical appearance. It is normal to begin puberty as early as 10 or as late as 16 years old.

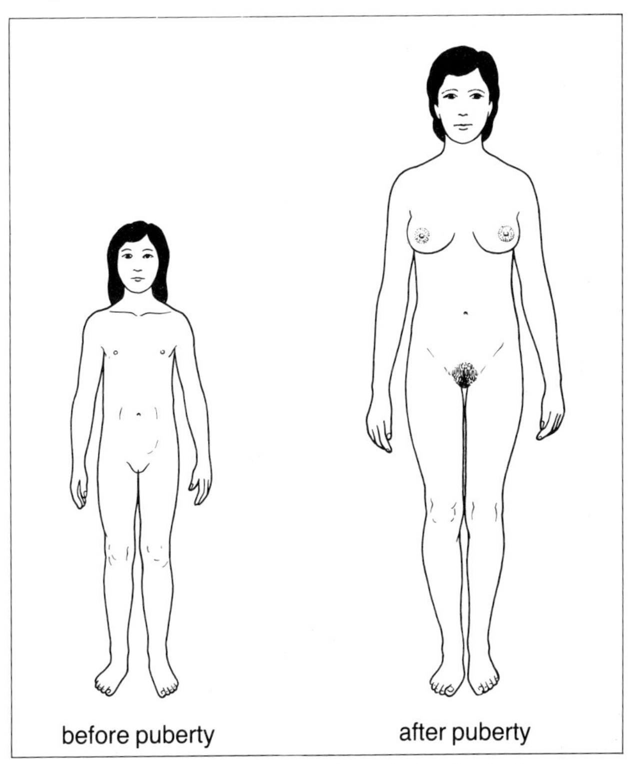

figure 2.12 Changes in the female body during puberty

All these physical changes of puberty are triggered by hormones produced in the **pituitary gland** at the base of the brain. These gonadotrophic hormones circulate in the blood and cause the testes and ovaries to produce sex hormones (see page 263). The sex hormones pass to all parts of the body in the bloodstream and cause the secondary sexual characteristics to develop.

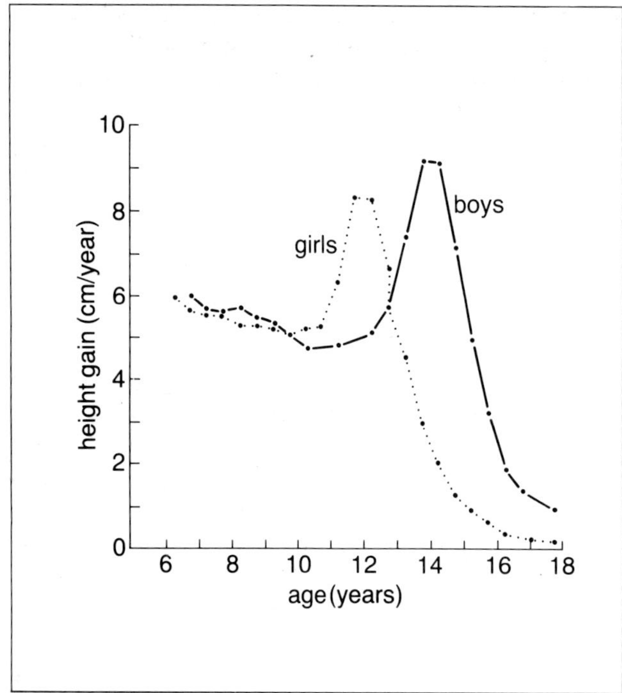

figure 2.13 Graph showing height gain each year during the growing period of girls and boys

CONTRACEPTION

200 years ago women in Britain would commonly give birth to nine or ten children. Nowadays this is unusual. Most families consist of two or three children. The idea of planning for an ideal family size is relatively new. This change has been helped by more reliable methods of contraception. A **contraceptive method** describes any means that a couple can use to avoid an unwanted pregnancy.

Choosing the most suitable contraceptive method should take into account a large number of factors. These include possible health risks and unpleasant side effects; whether the method interferes with intercourse; the method's likelihood of failure (see table 2.1) and the personal feelings and religious beliefs of the couple.

A few types of contraceptive, such as the condom, are available at a pharmacists. The full range of contraceptive methods is available at family planning clinics. At these clinics a woman or a couple can get advice about the most suitable method in their case and obtain free contraceptive supplies.

Withdrawal

Before the development of modern contraception technology, withdrawal was a popular way of preventing pregnancy. It is mentioned in the Bible and in the Qur'an. It involves the man withdrawing

INVESTIGATION 2.1

MATURING AT AN EARLIER AGE

When a girl has her first period, it is known as the **menarche**. It is a clear sign of physical maturity. In several countries accurate records have been made of the average age of menarche. Figure 2.14 shows how the age of menarche has changed in the USA, Great Britain and Norway.

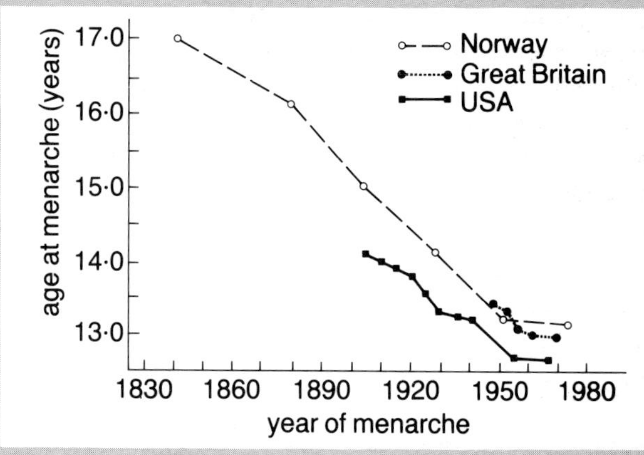

figure 2.14 Age at which girls reached the menarche in the USA, Norway and Great Britain

Questions

1 In which country have records of girls' ages at menarche been kept for the longest time?
2 In Norway what was the average age at menarche
 a in 1850 b in 1950?
3 In 1950 what was the average age at menarche
 a in the USA b in Great Britain?
4 What was the trend in the average age at menarche before 1950? What might explain this trend?
5 What has been the trend in the average age at menarche since 1950?
 Suggest an explanation for this trend.
6 a By how many months did the age at menarche decrease between 1850 and 1950 in Norway?
 b Calculate the *rate of decrease* in the age at menarche during this time. Express your answer in months per decade.
 c If this rate of decrease had continued after 1950, what would be the age at menarche in Norway in 1990?
 d Explain why the age you find from your calculations is very unlikely to agree with the true situation in Norway.
7 Describe and explain the main differences between the changes of age at menarche in the three countries.

ıethod of contraception	failure rate (%)
ṇale sterilization	0 – 0.2
emale sterilization	0 – 0.5
:ombined pill	0.2 – 1
ɔrogesterone only pill ('mini-pill')	0.3 – 5
UD	0.3 – 4
liaphragm	2 – 15
:ondom	2 – 15
vithdrawal	8 – 17
ertility awareness ('safe period')	6 – 25
:ontraceptive sponge	9 – 25
ıo contraception (women under 40)	80 – 90
ıo contraception (women aged 40)	40 – 45

'hese figures show the percentage of women using a particular method of :ontraception who became pregnant over a 12-month period. The failure of a contraceptive method depends on how carefully it is used. This is why a ange of failure rates for each method is given. The lowest figure shows how effective a method can be if it is used correctly and with the greatest care.

ble 2.1 *Failure rate of different contraceptive ethods*

is penis from the woman's vagina before he aculates. It is not very reliable because some sperm ıay be released before ejaculation. It also requires ɔnsiderable self-control. However, in an emergency is better than no contraception at all. Nowadays ɔuples who wish to prevent pregnancy can choose om several much more reliable methods.

Safe period

perm can survive for up to seven days in the uterus nd the egg will die about 24 hours after ovulation if it not fertilized. Couples using the 'safe period' ıethod avoid intercourse for seven days before and 4 hours after ovulation. This is the **fertile period**. It difficult to predict when ovulation will occur. Iowever, there are various methods for discovering which day ovulation happens. One method is to take ccurate daily readings of the woman's body emperature. A slight but definite rise in temperature om one day to the next indicates that ovulation has ccurred.

lthough rather unreliable this 'natural' method of ontraception is chosen by many people who for ersonal or religious reasons object to artificial ontraceptive devices.

The condom

'he earliest condoms date from Roman times when nimal bladders were used to cover the penis during ıtercourse. Nowadays condoms are made from latex ubber. The condom is rolled over the erect penis efore intercourse (see figure 2.15). Sperm are caught ıside the condom and prevented from entering the agina. Condoms are often lubricated with a special elly so that any loss of sensitivity is minimized. ometimes the lubricant contains a spermicide (a hemical that kills sperm).

The condom has a fairly low failure rate if used carefully (see table 2.1). However there is a risk of pregnancy if any sperm escape. This can occur if the condom slips off when the penis is withdrawn or if there is contact between the genitals before a condom has been put on. Even if only 1% of the semen escapes this may contain up to three million sperm.

Condoms are obtainable without a prescription from pharmacists. They have no side-effects. They have the important advantage of offering protection against the spread of sexually transmitted diseases such as gonorrhoea and AIDS.

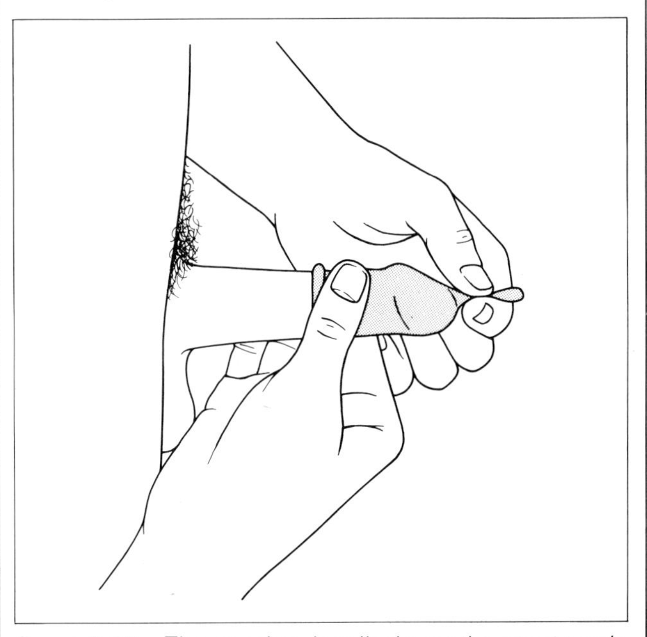

figure 2.15 *The condom is rolled over the erect penis*

Oral contraception – 'the pill'

There are two types of contraceptive pill: the **combined pill** containing two hormones, oestrogen and progesterone, and the so-called **mini-pill** which contains progesterone only. The combined pill was developed during the 1950s and became a very popular method of contraception throughout the world. The most common type of combined pill is taken by the woman once every day for 21 days. She then stops taking it for seven days during which time she has a 'period'. The hormones in the combined pill circulate in the blood and stop the ovary from releasing eggs. This is a very reliable method of preventing pregnancy. It sometimes fails because the woman forgets to take a pill one day and ovulation occurs.

There has been much scientific research into the safety of the combined pill. This research shows that the pill is safe for most women under the age of 35. However, it was discovered that certain groups of pill users were more likely to suffer from diseases of the circulatory system such as vein thrombosis (a clot of

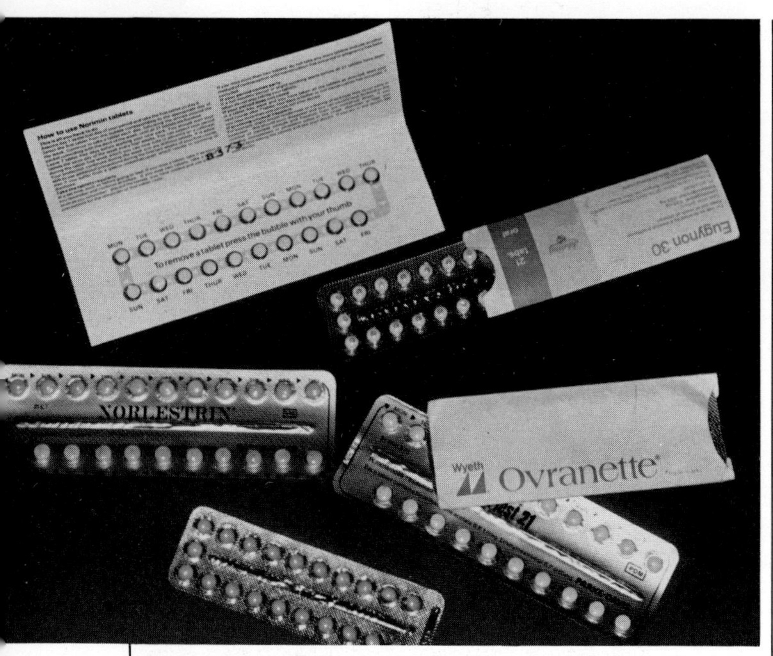

figure 2.16 Different types of contraceptive pill

blood in a vein) (see investigation 2.2). A few women suffer minor side-effects like slight weight gain, headaches and irregular bleeding. These can often be stopped by changing to a different pill.

The mini-pill contains very low doses of progesterone and is taken every day without a break. It appears to work by thickening the mucus in the cervix. This makes it difficult for the sperm to move up into the uterus. In some women the mini-pill also stops ovulation. Although the mini-pill is less reliable in preventing pregnancy, it is often preferable to the combined pill. The lack of oestrogen in the mini-pill means there is less risk of medical complications and side-effects.

The diaphragm

A **diaphragm** (cap) is a small dome of rubber with a rim reinforced with a flexible metal spring. A small amount of spermicide cream or jelly is spread over the diaphragm each time it is used. It is placed into the vagina before intercourse so that it covers the cervix (see figure 2.17). Sperm cannot get past the diaphragm. They remain in the vagina where they are killed by the spermicide and the acidity of the

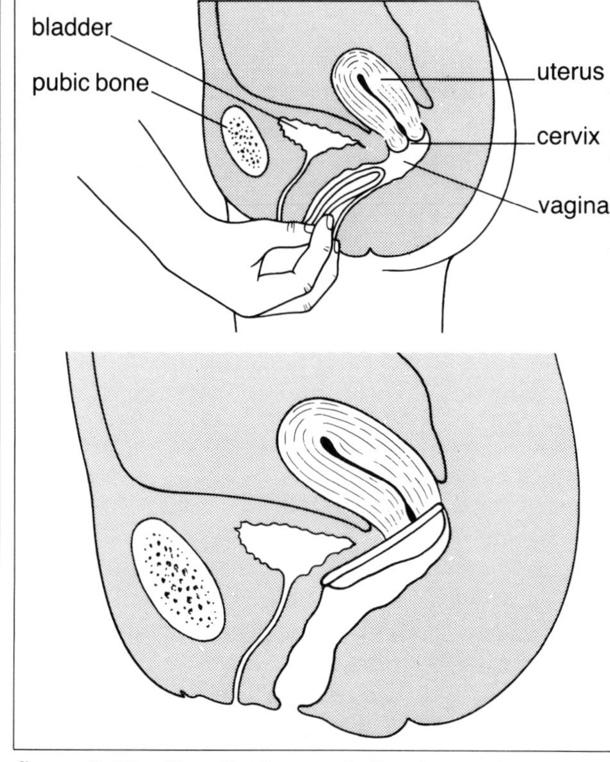

figure 2.17 The diaphragm is fitted over the cervix

INVESTIGATION 2.2

WHAT ARE THE RISKS INVOLVED IN TAKING THE CONTRACEPTIVE PILL?

The risk of an activity can be measured by counting the annual number of deaths it causes. Table 2.2 shows the annual number of deaths per 100 000 people exposed to different types of risk. Among pill users only the deaths due to circulatory disease (e.g. thrombosis) have been taken into account.

Questions

How far do these data support the following statements?

1 Hang-gliding is a very dangerous sport.
2 Developing countries lack medical facilities.
3 As women get older they should consider changing from the pill to another type of contraception.
4 The risk of a woman under 35 dying as a result of her pregnancy is greater than the risk of dying of circulatory disease using the contraceptive pill.
5 Women smokers over 35 face an unacceptable risk of circulatory disease if they use the contraceptive pill.

cause of death	annual number of deaths per 100 000
hang-gliding	200
giving birth in a developing country	200
giving birth in the UK	10
traffic accident – pedestrian	6
traffic accident – driver/passenger	17
accident in the home	4
playing soccer	3
taking the pill: non-smokers, aged under 35	1.3
taking the pill: smokers, aged under 35	10
taking the pill: non-smokers, aged 35-44	15
taking the pill: smokers, aged 35-44	48
taking the pill: non-smokers, aged over 45	41
taking the pill: smokers aged over 45	179

table 2.2 Annual deaths for different activities

gina. The diaphragm is left in place for at least six
urs after intercourse so that there is no chance of
y sperm surviving. Like the condom, the
aphragm has no medical risks and low failure rate.
nlike the condom, the diaphragm comes in a wide
nge of sizes from five to ten centimetres in
ameter. Women who decide to use a diaphragm
ust go to a family planning clinic to be fitted for the
rrect size.

Sponges and spermicides

ne **contraceptive sponge** is made of polyurethane
am which is impregnated with spermicide (see
gure 2.18). The woman moistens it with water and
serts it into the vagina so that it covers the cervix.
ne removes it from the vagina about six hours after
tercourse. It works by killing and absorbing sperm
fore they pass through the cervix. It is less messy
an using a diaphragm and can be bought over the
unter without a prescription. It can be inserted up
24 hours before intercourse. However, its main
sadvantage is its relatively high failure rate. For this
ason it is suitable mainly for older women whose
rtility is declining. Spermicides can also be inserted
rectly into the vagina in cream or foam form. When
plied on their own they are not very reliable but are
metimes used with other contraceptive methods as
back-up.

gure 2.18 *The contraceptive sponge*

Intrauterine devices (IUDs)

n **IUD** or **coil** is a tiny plastic structure that is
serted into the uterus through the cervix by a
ained medical worker. It can remain there for up to
ree years without being replaced. There are many
ifferent shapes of IUD (see figure 2.19). The IUD
oes not prevent fertilization; it works by interfering
ith implantation. The uterus lining will reject a
lastocyst before it can embed itself. Since this
ormally occurs five days after fertilization the IUD is
nacceptable to some women. They view this
ethod as a form of early abortion.

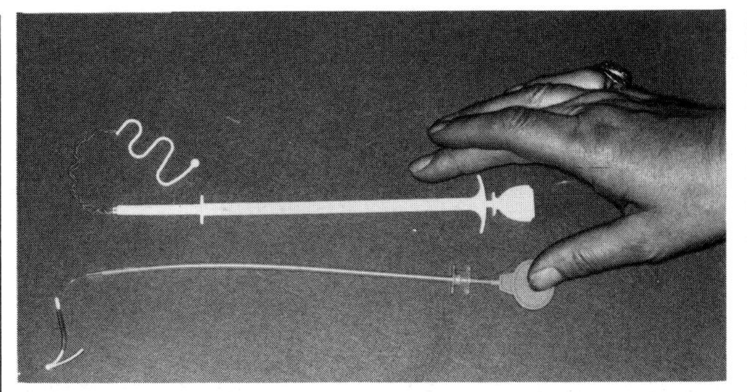

figure 2.19 *Different types of intrauterine device and the applicators used to insert them*

Because it is a foreign body, the IUD may sometimes be expelled by the uterus. If this occurs during a period it may happen without being noticed. This is why tiny threads are attached to the IUD so that the woman can check that the IUD is still in place. Some IUDs such as the 'Copper-7' have a very fine copper wire coiled around the plastic. The copper adds to the effectiveness of the IUD in preventing implantation.

The main advantage of the IUD is that once inserted no further action is required by the couple using it. This avoids the possibility of human error that can lead to the failure of other contraceptive methods such as the pill or condom. However, women who use the IUD have a slightly greater risk of an infection of the uterus and fallopian tubes (pelvic inflammatory disease). Because this can sometimes lead to a blockage of the fallopian tubes and possible infertility, the IUD is more commonly used by women who have already completed their family. It can sometimes cause minor problems like backache and period pains.

Sterilization

Sterilization is an operation that stops the gametes from passing down the tubes of the sex organs. In a man the sperm ducts are cut and tied. This is called a **vasectomy**. Afterwards when he ejaculates, the man's semen will only contain fluid from the prostate gland and seminal vesicles. The sperm cannot pass into the sperm duct and are reabsorbed in the sperm-producing tubules. Since the sperm make up a tiny proportion of the semen the man will not notice any difference when he ejaculates.

Sterilization of a woman involves cutting the fallopian tubes so that sperm no longer reaches the egg. The cut ends are tied. This prevents them from healing together. In rare cases the fallopian tubes grow together and the woman becomes fertile again.

It is extremely difficult to reverse a sterilization operation and rejoin tubes that have been cut. For this reason, it is rarely carried out on men or women under 30.

SEXUALLY TRANSMITTED DISEASE (STD)

A man's urethra and a woman's vagina provide warm and moist conditions ideal for the growth of some bacteria and viruses. During sexual contact it is easy for germs to be transmitted. Infectious diseases passed between sexual partners are called **venereal diseases** or **sexually transmitted diseases** (STDs). Five important STDs are described below.

Gonorrhoea

In the UK about 64 000 cases of gonorrhoea are reported every year. It is caused by a bacterium and can be cured simply by a course of antibiotics. The first signs of the disease appear a few days after intercourse. The individual feels a burning sensation when urinating. Sometimes a yellowish discharge occurs. In women the symptoms are often unnoticed. If not treated, the disease can spread into the uterus, damaging the fallopian tubes and causing infertility.

Syphilis

Syphilis is less common than gonorrhoea: only 4000 cases were reported in 1985. However, syphilis can be very serious. If left untreated it can lead to brain damage and blindness. The disease is caused by a bacterium which produces clear symptoms within a few weeks: a sore appears on the sex organs and is followed by a body rash and fever. If detected at this stage, syphilis can successfully be treated.

Non-specific urethritis (NSU)

NSU refers to a group of STDs that, like gonorrhoea, cause inflammation of the urethra. In men the urethra feels itchy, while in women NSU often goes unnoticed. At least half of the 75 000 reported cases of NSU are caused by a fungus called *Chlamydia*.

AIDS

In 1980 cases of a new disease began to be reported in Africa and the United States. The victims were dying from the combined effects of a variety of infectious diseases and cancers. Their white blood cells did not produce the antibodies which normally provide immunity against infections. This disease quickly spread to many other countries. It was named Aquired Immune Deficiency Syndrome or AIDS.

In 1984 the virus that causes AIDS was discovered (see figure 2.20). It was named Human Immunodeficiency Virus (HIV). HIV invades the white blood cells and can remain there for several years without causing any symptoms. In 1987, it was estimated that 50 000 people in the UK were infected and only about 1000 died of AIDS.

HIV is carried in the blood. It can be transmitted from one person to another in several ways. During sexual intercourse blood vessels lining the penis and vagina come into close contact allowing the virus to pass from one person to another. Anal intercourse can cause the blood vessels of the anus to rupture. This greatly increases the risk of transmission. There is also a great risk if infected blood enters the blood stream. This is why a large number of AIDS victims have been haemophiliacs who received transfusions of infected blood and heroin users who shared the same needle. The virus can also pass from a pregnant woman to her baby, via the placenta.

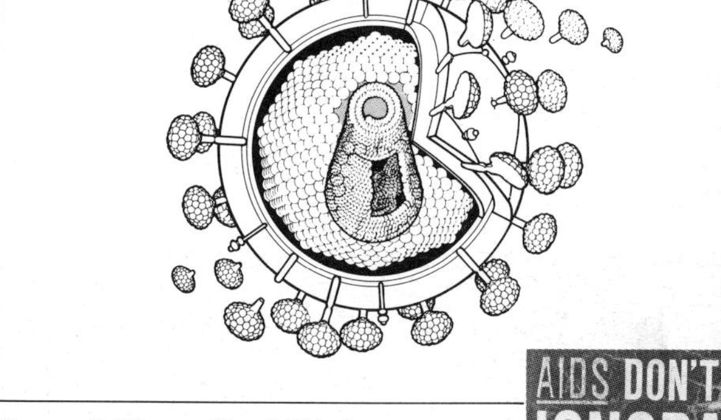

figure 2.20 a *The HIV virus*
b *The campaign against AIDS*

> ### Application 2.4
> **CONTROLLING THE SPREAD OF STDs**
> In the last thirty years the number of cases of STDs has increased dramatically. The spread of AIDS, an STD that is at present incurable, has made people think about the problem more seriously. The following measures are recommended to help stop the spread of sexually transmitted diseases such as AIDS.
>
> 1. A condom used during intercourse provides a secure barrier that stops many bacteria and viruses from passing between sexual partners. The contraceptive pill protects against pregnancy but *not* against STD.
> 2. When the symptoms of an STD first appear the person should cease sexual intercourse until the infection has been fully removed by medical treatment. This will stop the disease spreading to a sexual partner.
> 3. A person who suspects that they may be suffering from an STD should seek medical help. This can be obtained at STD clinics (sometimes called 'special clinics') that are attached to most large hospitals.
> 4. Sexually transmitted diseases spread rapidly if people change their sexual partners frequently. This is sometimes called promiscuity or casual sex. Couples with a stable and faithful sexual relationship are not at risk from STD.

INHERITANCE

How do we get our characteristics?

arents pass on many of their characteristics to their iildren. Some of these characteristics are passed on rough information in the father's sperm and the other's egg. We say that these characteristics are **herited**. Hair, eye and skin colour are examples of herited characteristics. Some inherited iaracteristics can be changed by the influence of the ivironment. For example, a white person's skin olour will change if he or she stays out in strong inlight. Some characteristics are influenced by our ivironment, for example our hobbies or our taste in usic. These are known as **acquired characteristics**.

NA, genes and chromosomes

he information that determines our inherited iaracteristics is stored in complex molecules of a ibstance called **DNA** (**deoxyribonucleic acid**, see gure 2.21). A certain length of DNA is called a **gene**. gene consists of coded instructions for building one pe of protein. Many thousands of genes are needed r building the different proteins in the body. For example, a single gene contains instructions for building the protein haemoglobin which carries oxygen in the red blood cells.

gure 2.21 *The structure of the DNA molecule*

Genes are found along the length of special thread-like structures contained in the nucleus of every cell. These structures are called chromosomes (see page 11). The larger chromosomes are thought to possess 10 000 genes. When the cell divides, the chromosomes shorten and thicken and will take up dye so that they become visible. There is a denser part of the chromosome which does not take up dye called the **centromere**. When stained, magnified and photographed the chromosomes can be sorted into pairs (see figure 2.22). The members of each pair look identical under the microscope and are known as **homologous pairs**. 46 chromosomes arranged in 23 pairs are in the nucleus of every human body cell.

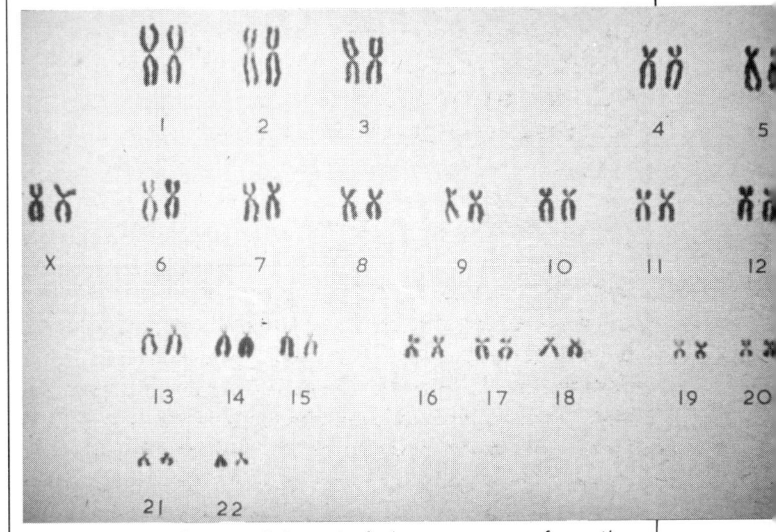

figure 2.22 *A complete set of chromosomes from the body cell of a human female*

Mitosis – copying chromosomes

The instructions for building all the different cells in an organism are contained in the chromosomes of a single cell – the **zygote**. These instructions are genetic information and must be carefully copied every time the cell divides. This ensures that all the cells of the fully grown organism have a complete set of identical instructions. For example, a cell in the brain will contain instructions to become a skin cell, a liver cell, a muscle cell and any other cell in the body. These instructions will be 'switched off'. Only the instructions for becoming a brain cell will be 'switched on'.

When a cell divides each chromosome makes an identical copy of itself. This is known as **replication**. The two copied chromosomes then separate into the two new cells. This type of cell division is known as **mitosis**. Figure 2.23 (overleaf) shows this in more detail.

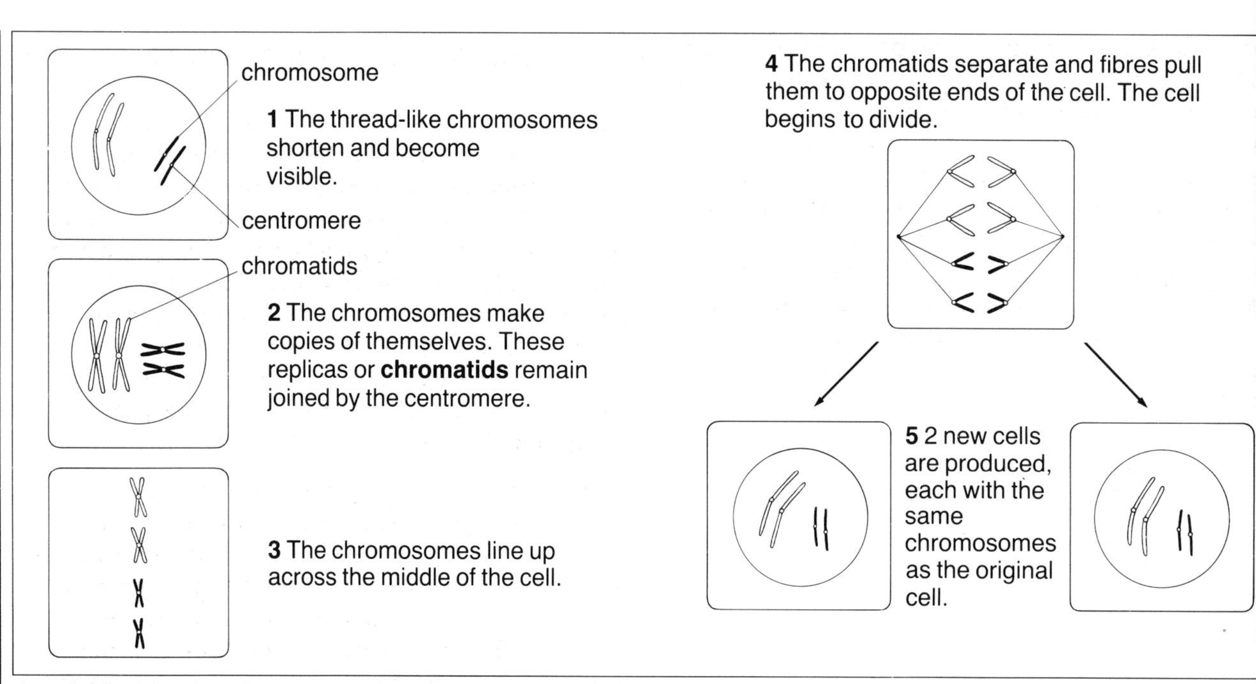

figure 2.23 *The stages of mitosis – producing new cells with identical chromosomes*

Meiosis – halving the diploid number

The number of chromosomes in a normal body cell is known as the **diploid number**. In the human species the diploid number is 46.

All our body cells are produced by division of a single cell, the zygote, which is itself formed by the union of two cells, the sperm and the egg. This means that each gamete must contain *half* the diploid number of chromosomes. Gametes have only 23 chromosomes. This is the **haploid number**. Gametes are produced by a special type of cell division that separates homologous chromosomes into two different cells. This means that only one member of each homologous pair of chromosomes is present in each gamete. This type of cell division is known as **meiosis** and it only occurs in the testis and ovary. The various stages of meiosis are shown in figure 2.24. You will see that meiosis produces gametes with new combinations of genes.

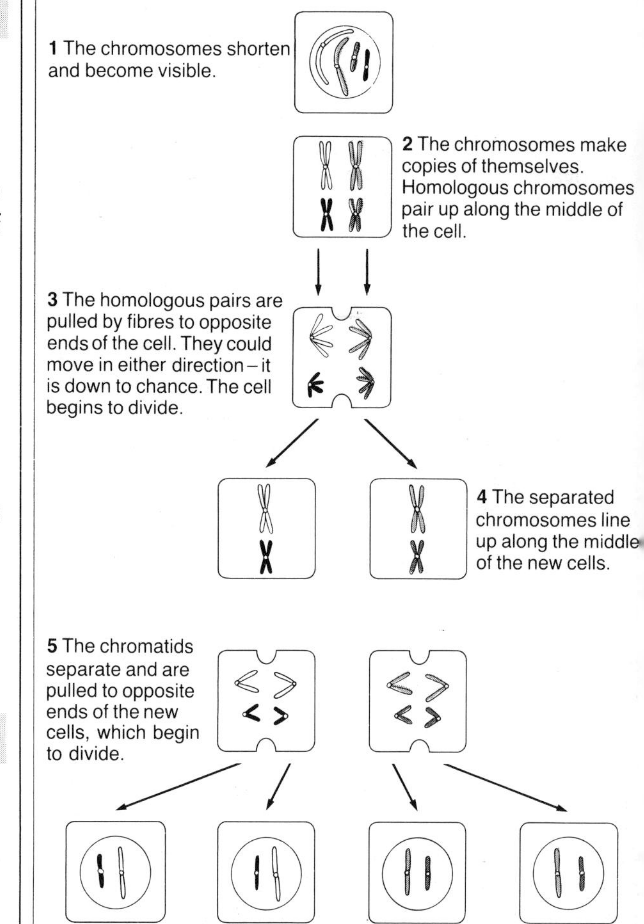

figure 2.24 *The way chromosomes divide and separate during meiosis produces new gene combinations*

Boy or girl?

One pair of chromosomes decides the sex of the individual: the **sex chromosomes**. There are two types of sex chromosome: the **X chromosome** and the **Y chromosome**. The Y chromosome is much shorter and carries fewer genes that the X chromosome. The body cells of a male contain one X and one Y chromosome. Female body cells contain two X chromosomes.

During meiosis, like the other homologous chromosomes, the two sex chromosomes separate and one goes into each gamete. All the ova will ɔntain a single X chromosome, because the female ɔdy cells contain a pair of X chromosomes. ˌowever, half the sperm will contain an X ɪromosome and half will contain a Y chromosome, ecause the male body cells contain one of each. /hen fertilization occurs it is equally likely that an X ɔerm or a Y sperm will combine with the ovum ɔntaining the X chromosome. This leads to an equal robability that the zygote will be XY and develop ɪto a boy, or will be XX and develop into a girl. ˌgure 2.25 summarizes the way the sex ɪromosomes are passed on.

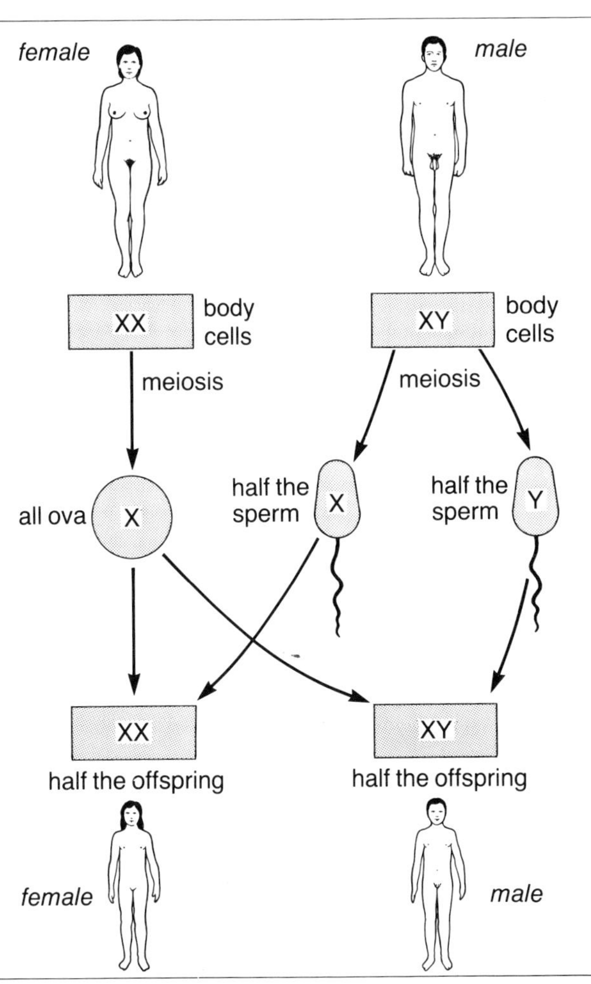

gure 2.25 *The inheritance of X and Y chromosomes ˈecides the sex of a baby*

Genes work in pairs

/e can match up chromosomes into homologous airs because they look alike. The genes carried on omologous chromosomes also match up. The two enes located at a specific point on a homologous hromosome pair both control the same haracteristic. Such genes are called **alleles** and their ɔcation on the chromosome is known as their **locus**. 'he alleles may be *identical* in which case the person is aid to be **homozygous** at that locus. If the two alleles re *different* then the individual is **heterozygous**.

How do genes express themselves?

Let us look at two alleles that control the formation of pigment in the eye. Suppose one of the alleles is a gene for blue eye pigment and the other allele is a gene for brown eye pigment. What colour will this person's eyes be? Will they be a mixture of the two colours – a browny-blue colour? In fact the person's eyes will be brown. This is because the gene for brown eyes overrides the blue-eye gene. We say that the brown-eye gene is **dominant**. The gene for blue eyes is called **recessive** because it is not expressed when combined with a gene for brown eyes.

In genetics we can represent genes by initial letters. A dominant gene is shown by a capital letter and a recessive gene by a lower case letter. In our example the gene for brown eyes is **B** and the gene for blue eyes is **b**. The particular combination of genes possessed by an individual is called the person's **genotype**. The genotype of the person in our example is **Bb**. The genotype consists of instructions that determine the characteristics of the individual. This is the person's **phenotype**. In the example the genotype **Bb** produces the phenotype brown eyes.

The genotypes and phenotypes of the three possible combinations of genes for brown eyes and blue eyes are shown in figure 2.26.

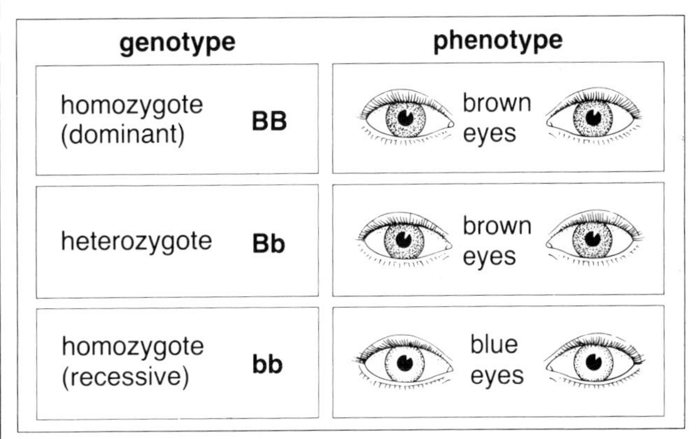

genotype		phenotype
homozygote (dominant)	BB	brown eyes
heterozygote	Bb	brown eyes
homozygote (recessive)	bb	blue eyes

figure 2.26 *The genotypes and phenotypes of the combinations of blue-eyed and brown-eyed genes*

Passing on genes

To understand how genes are passed on we must first look at what happens in the reproductive organs during meiosis. Let us take a brown-eyed man whose genotype is **Bb** and a woman with the same genotype. During meiosis, in the man's testes the homologous chromosomes carrying these alleles will separate and pass into different sperm (see stage 5 on figure 2.24). In the woman's ovary the two alleles carried on homologous chromosomes will separate into different eggs. So half the sperm and half the eggs will contain the **B** gene and the others will contain the **b** gene.

UNIT 2

If this man and woman have a child, we can work out the probabilities of the various combinations of sperm and egg. The four possible combinations are all equally likely to occur. Three of the combinations produce brown-eyed offspring. Only one of the four produces offspring with blue eyes. We can say that the chance of producing a blue-eyed child is ¼ and the chance of producing a brown-eyed child is ¾.

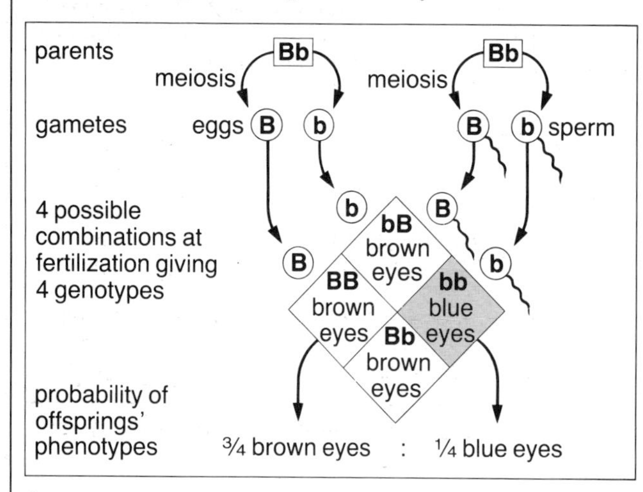

figure 2.27 How genes are passed on

This pattern of inheritance applies to all characteristics controlled by a single pair of genes. The ability to roll the tongue so its edges come together over the top, forming an O shape, is another human characteristic inherited in this way (see investigation 2.3).

INVESTIGATION 2.4

SKIPPING A GENERATION

You may have certain characteristics that your parents do not have but your grandparents do. Characteristics like this 'skip a generation'. This investigation looks at the way a recessive gene can be responsible for this.

Alice (A) has wavy hair but both parents, Brian (B) and Carol (C), have straight hair. The complete family tree can be shown in the form of a diagram as in figure 2.28.

Questions

1 If **w** represents the recessive gene for wavy hair and **W** is the dominant gene for straight hair, what is
  a Alice's genotype
  b the genotype of both Alice's grandmothers?
2 Alice gets one gene from each of her parents, Brian and Carol. What are Brian and Carol's genotypes?
3 Explain how the wavy hair 'skips a generation'.
4 Alice's brother, Derek (D), could be one of two possible genotypes. What are they? Explain your answer.

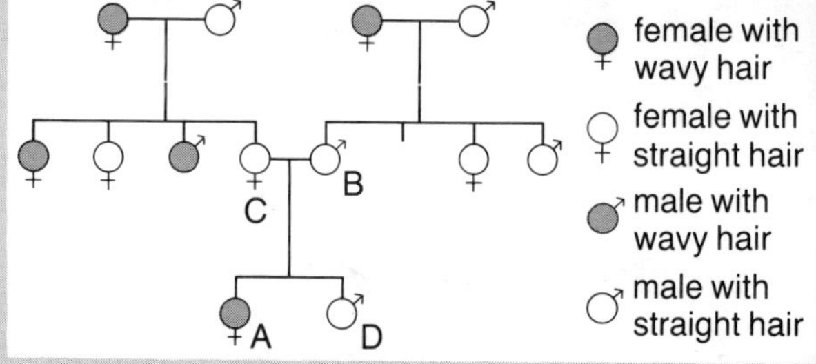

figure 2.28 The complete family tree

INVESTIGATION 2.3

BREEDING TONGUE-ROLLERS

Carrying out breeding experiments on humans involves many problems. Breeding experiments usually require large numbers of offspring. This investigation uses coin-tossing to represent the breeding of a very large human 'family' with as many as 100 offspring!

a Working in pairs, take two coins and attach small sticky labels to both sides of each coin.
b On one side of each coin write the capital letter **R**; on the other side write the lower case letter **r**. **R** represents the dominant tongue-rolling gene and **r** represents the recessive gene for inability to roll the tongue.
c Take a coin each and decide which will represent the sperm and which will represent the egg. Each member of the pair should toss their own coin. The two letters that land uppermost represent the genotype of the offspring.
d Each time the coins are tossed, record the genotype and enter it on a table like table 2.3.
e When your 'family' has at least sixty offspring, add up the totals for each genotype.

Questions

1 How many of the offspring in your family are capable of tongue-rolling?

sperm	egg	tally	total
R	R	𝍸 𝍸 𝍸	15
R	r	𝍸 𝍸 III	13
r	R	𝍸 𝍸 𝍸 II	17
r	r	𝍸 𝍸 𝍸 I	16

table 2.3 Results of tossing a pair of coins

2 How many offspring cannot roll their tongues?
3 In table 2.3 the ratio of rollers to non-rollers is as follows:

$$15 + 13 + 17 : 16$$
$$= 45 : 16$$
$$= \frac{45}{16} : \frac{16}{16}$$
$$= 2.8 : 1$$

In the same way, work out the ratio of rollers to non-rollers in the offspring you have produced.
4 Compare your results with the others in your class. Find an average ratio for the class.
5 In this investigation, why is coin-tossing a suitable way of representing meiosis and fertilization?
6 What does this investigation tell you about the need for large numbers of offspring in breeding exeriments?

Co-dominant genes

n the cases considered so far a dominant allele hides' the effect of the recessive allele so that only one gene is expressed in the phenotype. Sometimes, however, neither of the two alleles is dominant; *both* alleles are expressed in the phenotype. When this occurs the alleles are said to be **co-dominant**. The alleles that determine the ABO blood group system illustrate co-dominance. There are three alleles, **A**, **B** and **O**, only two of which can be present. The **A** and alleles are both dominant to the **O** allele. However, and **B** alleles are co-dominant.

genotype allele 1	allele 2	blood group	explanation
A	B	AB	A is co-dominant with B
A	A	A	homozygous for allele A
A	O	A	A is dominant to O
B	O	B	B is dominant to O
B	B	B	homozygous for allele B
O	O	O	homozygous for allele O

able 2.4 *Inheritance of blood groups*

Genetic mistakes: mutations

When chromosomes are copied during cell division here is always the possibility of a mistake being made. These mistakes are known as **mutations**. They an affect either a single gene or a whole hromosome.

Many factors are known to increase the frequency of mutations (**mutation rate**). Certain chemicals called mutagens will increase the mutation rate. Mutagens re found in cigarette smoke and in the gases released y burning certain types of plastic. Ionizing radiation lso causes an increase in the mutation rate (see unit 2).

Gene mutations

Gene mutations are often harmful. Genetic diseases may be caused by a gene mutation. Haemophilia is a genetic disease that stops the blood from clotting when a blood vessel is cut. It is caused by a single mutant gene carried on the X chromosome (see figure .29). The albino gene is a mutation that prevents the ormation of the skin pigment **melanin**. Melanin is a dark coloured pigment that protects us from the sun's powerful ultraviolet rays. An albino lacks this pigment (see page 33).

Some mutations are advantageous and help the individual to survive in certain environments. These beneficial mutations include the sickle gene that lters the shape of red blood cells. It can help people o survive an attack of malaria (see application 2.8).

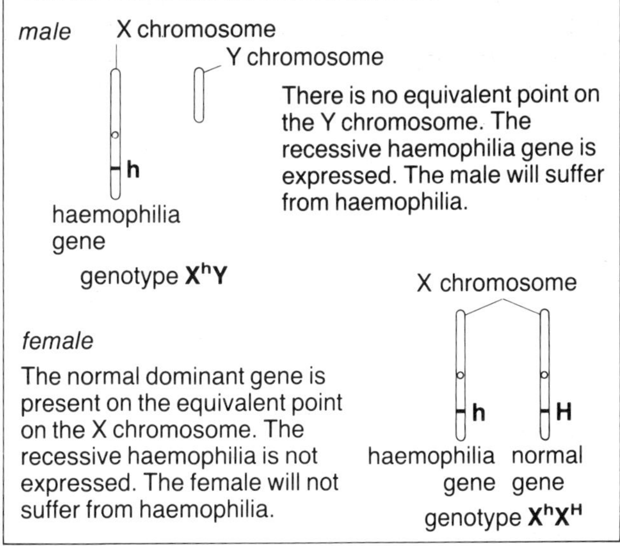

The haemophilia gene is carried on the X chromosome. This is what happens when a *single* haemophilia gene occurs in a male and a female. You will notice that the effect is different.
male
X chromosome
Y chromosome
h
haemophilia gene
There is no equivalent point on the Y chromosome. The recessive haemophilia gene is expressed. The male will suffer from haemophilia.
genotype X^hY
X chromosome
female
The normal dominant gene is present on the equivalent point on the X chromosome. The recessive haemophilia is not expressed. The female will not suffer from haemophilia.
h
H
haemophilia gene
normal gene
genotype X^hX^H

figure 2.29 *Males are more at risk from X-linked genes*

Some mutations have neither a harmful nor a beneficial effect on a person's health or chances of survival. These are called neutral mutations.

Harmful genes

Each one of us carries at least four different harmful genes. Luckily most of these genes are recessive and they do not affect our health. However we *can* pass on these genes to our children.

If a child inherits the same harmful recessive gene from both its parents it will be affected by a **genetic disease**. Almost three thousand genetic diseases have been identified and one in 20 children admitted to hospital in the UK are suffering from a genetic disease. A knowledge of genetics is used by **genetic counsellors** to help people who may be affected by a genetic disease (see application 2.5, overleaf).

PKU (**phenylketonuria**) is a genetic disease that affects one in 10 000 children. Before this disease was understood, babies with PKU suffered from brain damage soon after they were born. Scientists discovered that this was caused by a high level of a particular amino acid in the baby's body. The enzyme that normally helps to break down this amino acid was missing. The PKU sufferer had faulty genes that could not construct the vital enzyme.

Nowadays a tiny blood sample is taken from a newborn baby's foot to discover whether the child has PKU. If the test is positive the child must be given a very restricted diet. The diet must contain none of the amino acid that could build up to cause brain damage.

Chromosome mutations

Chromosomes sometimes become altered during

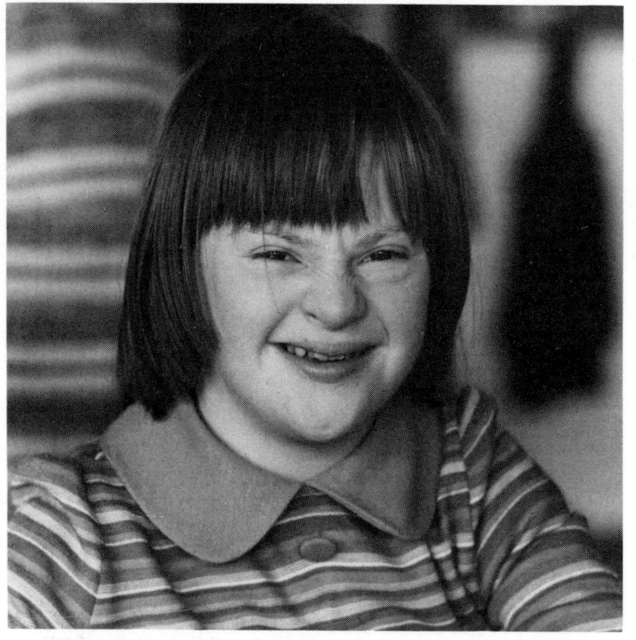

figure 2.30 Down's syndrome children are often affectionate and happy people

meiosis. A section of a chromosome may break off. This chromosome piece may be lost or it may attach itself to another chromosome. Sometimes the addition or loss of a whole chromosome occurs. This happens when a pair of homologous chromosomes fail to separate during meiosis. They both move into the same gamete. If this gamete is fertilized the zygote will have an extra chromosome. An extra chromosome number 21 produces **Down's syndrom** (previously called mongolism). Although this chromosome is one of the smallest, its extra presenc has a severe effect on the development of the child. People with Down's syndrome have a low mental ag and characteristic facial features (see figure 2.30). Down's syndrome children are often extremely affectionate and friendly. With intensive training they can develop many useful skills. Nowadays the presence of a chromosome mutation like this can be detected in the early stages of pregnancy using **amniocentesis** (see application 2.3).

VARIATION

How do individuals vary?

There are two types of variation between individuals of the same species: **continuous variation** and **discontinuous variation**. A characteristic that shows discontinuous variation can be used to separate individuals into distinct groups. Blood groups are a good example of discontinuous variation. We can so individuals into four main blood groups: A, B, AB or O. No one falls between two groups, say with a mixture of groups B and O.

Application 2.5

THE GENETIC COUNSELLOR

When someone is worried that a genetic disease might be passed down to their children, they can discuss their situation with a **genetic counsellor**. The following case study illustrates the type of problem a genetic counsellor might have to deal with.

Problem Jill was 22 years old and engaged to be married. During her childhood her father had died of a disease that affected his behaviour and muscle co-ordination. He had got worse over several years. Jill was worried about this and discussed the problem with a genetic counsellor. She was told that this disease is called **Huntington's chorea**. It is a genetic disease caused by a rare dominant gene. This meant Jill had a 50% chance of developing the disease herself (see figure 2.31). If she was carrying the gene there would be a 50% chance of passing on the disease to her children.

Solution The genetic counsellor helped Jill and her boyfriend to accept the situation. They decided to carry on with their marriage plans but since the risk of having affected children was so high Jill decided that she would be sterilized. Jill and her husband gradually adjusted themselves to the prospect of their married life without children.

Huntington's chorea is very unusual, partly because it is carried by a dominant gene, and partly because people do not know that they suffer from it until late in life. This means that they can have children before they know they are affected by it. Most genetic diseases are defects in metabolism, like PKU, which are obvious from childhood, and the genetic counsellor would then be involved in discussing whether the parents should have any more children, and the life expectancy and treatment of the affected child.

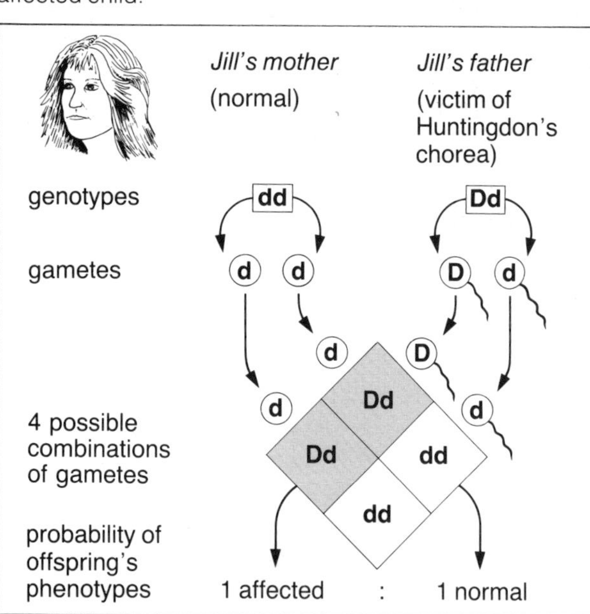

figure 2.31 Huntington's chorea – a genetic disease caused by a dominant gene

gure 2.32 **a** *An albino is good example of iscontinuous variation*

) Differences in height show ***continuous variation***

)iscontinuous variation results from differences in ;enetic instructions. The environment affects liscontinuous variation very little. For example, the omplete lack of skin pigment (albinism – see page 31) esults from a difference in a single gene. No amount of sunshine will give an albino a tanned skin.

Continuous variation applies to features that cannot be used to sort individuals into clearly defined groups. Body measurements such as height or weight are examples of continuous variation. The weight of individuals in a large population will vary from the extremely light to the very overweight with many stages in between.

Continuous variations are likely to be affected by a large number of genes, each contributing a small amount to the characteristic. The environment is also likely to have some affect. A person's weight is influenced by genes that control metabolic rate as well as by the amount of exercise taken and the energy value of their diet.

What influences variation – looking at twins

Which factor has the greatest influence on variation: environment or genes? This question has intrigued scientists for a long time. One method of investigating this problem is to study identical twins. Such twins have identical genes because they both come from a single fertilized egg. The egg split into two after it had been fertilized, and two identical embryos were formed. Any differences between twins must be due to the influence of the environment. Studies of identical twins who were separated from each other shortly after birth and reared in different environments provide evidence about the effect of the environment. If these twins show a similar feature, for example an artistic or sporting talent, it suggests that inheritance was probably the main influence.

INVESTIGATION 2.5

STUDYING DIFFERENCES BETWEEN TWINS

Table 2.5 shows the average difference in certain characteristics between pairs of identical twins reared in different circumstances.

difference in characterstic	50 pairs of twins reared together	50 pairs of twins reared apart
height (cm)	1.7	1.8
weight (kg)	1.9	4.5

table 2.5 *Differences between identical twins brought up together and separately*

Questions

1. a What was the average difference in height between identical twins reared together?
   b Explain why this difference occurs.
2. How does rearing twins in separate families affect the height difference?
3. What does this evidence suggest about the influence of genes on height?
4. Is weight influenced by genes to the same extent as height? Explain your answer.

Application 2.6

BIOTECHNOLOGY – GENETIC ENGINEERING

Genes from different cells come together during fertilization when the sperm nucleus enters the egg cell. In nature this is a unique event. The rest of the time genes remain inside the cell. They cannot transfer to another cell. In the laboratory scientists can take a gene from one cell and transfer it to another cell. For example, it is possible to remove the gene for making insulin from a human pancreas cell. This gene can then be inserted into a bacterium cell. When the bacterium multiplies the millions of offspring can all produce insulin. This method should prove to be an inexpensive way of obtaining large quantities of insulin for the treatment of diabetes. At present insulin has to be extracted from the pancreas of slaughtered animals.

The transfer of genes from one organism to another is known as **genetic engineering**. Growing large numbers of micro-organisms in large fermenters and collecting the product of the transplanted genes is an example of **biotechnology** (see figure 2.33). Interferon, an anti-cancer drug, and Protopin, a human growth hormone, are two other products that are manufactured by genetically engineered bacteria.

figure 2.33 This fermenter contains millions of genetically engineered bacteria. They produce 'Pruteen', a protein used as animal feed

Time for thought

At present, scientists have had most success in transplanting genes from humans into bacteria. In the future it may be possible to transplant genes from any organism into human cells.

- *What would be the dangers of this advance?*
- *Do you think that Parliament should pass laws banning certain types of gene transplantation?*

WHAT IS EVOLUTION?

There are over 200 million different species of plants and animals on the Earth. Humans have looked for an explanation of how this variety of life has occurred. Religions such as Christianity and Islam put forward the belief that all living things were created by God or Allah at a particular moment in the Earth's history. According to this religious view they have remained unchanged since that time. About 150 years ago many scientists became dissatisfied with this explanation. They put forward an alternative – the **theory of evolution**.

According to the theory of evolution, organisms change gradually from one generation to the next. Over millions of generations one species can change to produce a large number of different species. This process began about 4000 million years ago when single-celled organisms first appeared on Earth. These very simple organisms gradually changed, or **evolved**, into all the species presently living on Earth and a lot more which are now extinct. We say that all living organisms share a **common ancestor**.

Natural selection – an explanation of evolution

On 27 December 1831 the HMS Beagle set sail from Plymouth for a five-year voyage around the world (see figure 2.34). On board was a young scientist named Charles Darwin. Darwin's observations on this voyage helped him to develop a theory that could explain how evolution had occurred – the theory of **natural selection**.

Darwin's theory may be outlined in this way. Organisms are capable of producing large numbers of offspring. However, many of these offspring die before they reach adulthood. They die because they are unable to overcome many difficulties: these include the dangers of being eaten by predators, the risks of suffering disease or injury and the problems of finding enough food. Darwin said that there was a **struggle for survival**. However, he made it clear that this struggle was not necessarily a conflict between individuals. Organisms also struggle against a harsh environment. Scientists now describe these difficulties as **selection pressures**.

As there is variation between individuals of the same species, some are better equipped than others to overcome these pressures. For example, the antelope with the most powerful leg muscles will be the one that is most likely to escape from a fast-moving predator. Features like this that help the individual to survive are known as **adaptations**. Now we know that many adaptations are passed down to offspring in genes.

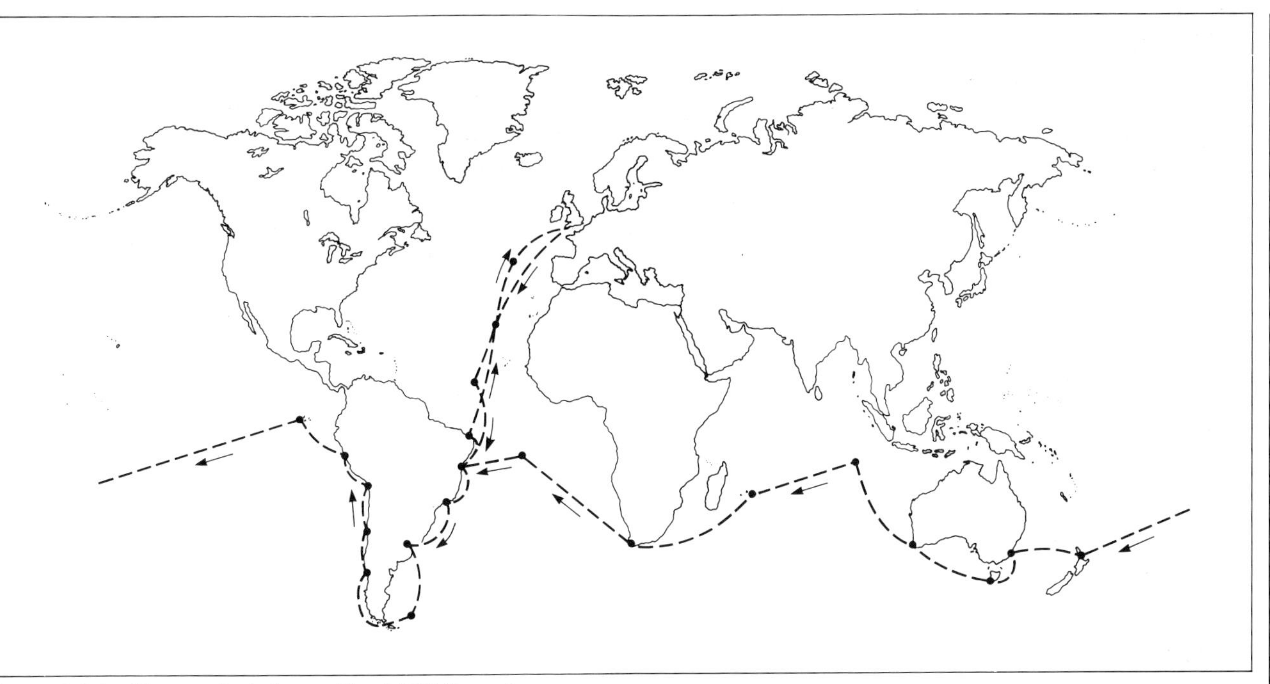

'gure 2.34 *The route followed by HMS Beagle*

)arwin argued that the organisms that survived to ·vercome the difficulties of the environment leave nore offspring. These offspring inherit their parents' daptations. Organisms that were not so well dapted leave fewer or no offspring. Therefore, after nany generations the population consists of rganisms better adapted to the environment. If the environment changes, natural selection chooses those variations that give an advantage in the new conditions (see applications 2.7 and 2.8). It is therefore better if there is variation within a species, because there is more chance of some individuals being suited to an environmental change.

Application 2.7

SELECTION AT WORK – THE PEPPERED MOTH

The peppered moth is so called because of its speckled grey colour. Before 1850 all peppered moths had this appearance. Many Victorians were keen naturalists and anyone with the rare black form of the moth in his or her collection was greatly envied. However, after 1850 the rare black form became much more common, especially around industrial cities. By 1895, 95% of peppered moths in the Manchester area were black. In the countryside, the black form was still rare.

The black form had been a rare mutant. It did not usually survive for long. Sharp-sighted birds easily spotted it against the light-coloured lichen-covered trees. The grey form was well camouflaged against the lichen and did not get eaten. Why did this change after 1850? What allowed the black form to increase?

During the nineteenth century heavy industry developed rapidly. The smoke from factory chimneys blackened the trunks of trees near industrial towns. The grey form showed up clearly on the blackened bark. Birds could quickly spot them. The black form now had the advantage of camouflage. It was less likely to be eaten by birds. It lived longer and produced more offspring than the grey form. As the generations passed the proportion of the black form in the peppered moth population increased.

figure 2.35 **a** *The two forms of the peppered moth against a lichen-covered tree*

b *The two forms of the peppered moth against a blackened tree trunk*

This change is an example of an **adaptation**. An adaptation is any feature of an organism that helps it to survive. Mutation produced the adaptation of a darker wing colour and selection ensured its survival wherever it was advantageous. This helped the peppered moth to survive in the new environment.

Application 2.8

SELECTION AT WORK: SICKLE CELL ANAEMIA

Several inherited diseases cause damage to red blood cells. **Sickle cell anaemia** is the commonest of these diseases. The harmful gene alters the haemoglobin molecule so that it is less efficient at carrying oxygen. The red blood cells become curved and elongated like a sickle.

The sickle gene is recessive. A homozygote for this gene suffers from sickle cell anaemia. Sufferers go through periodic crises and need intensive treatment throughout their lives. A heterozygote with one normal and one sickle gene is said to have **sickle cell trait** but will suffer no ill effects. There are said to be about 200 000 people with sickle cell trait in the UK.

Scientists were curious to find out why this disease was particularly common in central Africa and in people of African descent living in other parts of the world. It was discovered that people with sickle cell trait gained protection against malaria. It was therefore an advantage to be a carrier of the sickle gene. Without this gene children easily became infected with malaria and often died. This is an example of natural selection in a human population. The sickle gene is a mutation that in normal circumstances would be harmful. However, it gives the person an advantage when an area becomes infected with malaria. Individuals with one sickle gene survive longer and leave more children. In other words, the sickle gene improves the **reproductive fitness** of the individual who carries it.

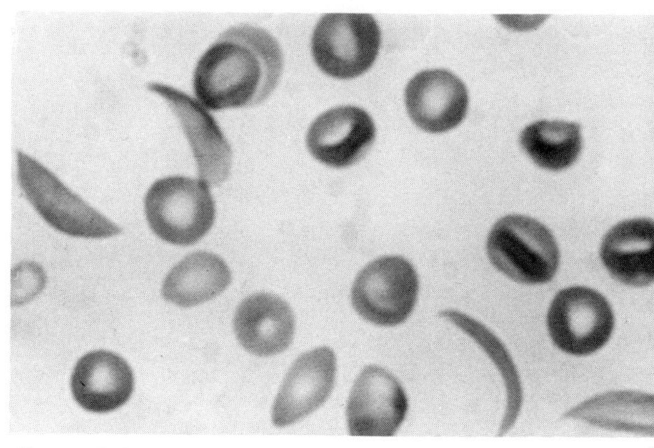

figure 2.36 *Normal and sickle red blood cells*

Controlling evolution – selective breeding

Throughout history, humans have taken many species from the wild and controlled their evolution. Cultivated cereals such as rice and wheat were originally wild grasses. Domesticated animals like the cow and horse were originally wild species. The many varieties of domesticated dog are all distantly related to the wolf.

Changes like these have been brought about by **artificial selection**, otherwise known as **selective breeding**. The breeder examines the variation of a natural population. Some individuals are selected because they have the features that the breeder is looking for. The breeder must be sure that these features are caused by genes and not the environment. The selected individuals are then bred together. For example, grasses with sturdy stems and large starchy seeds would have been selected for cultivation as cereals. Plant breeders today continue to look for wild varieties that could be suitable for cultivation (see application 2.10).

figure 2.37 *These pedigree dogs have all been produced by selective breeding of a single species – the wolf*

Application 2.9

SELECTIVE BREEDING FOR CALMER COWS

In India the cow is a sacred animal that is allowed to roam freely. Indian scientists decided to breed these cows so that they could produce a high yield of milk. Unfortunately milking was not very efficient because the cows were nervous creatures. They were not relaxed enough to be milked unless a calf was suckling on one nipple. This reduced the amount of milk that could be taken.

After several years of selective breeding a less excitable breed of cattle was produced. These cows were content to be milked entirely by hand. Helped by changes to the cows' diet, the breeders obtained a cattle breed with a greatly improved milk yield as well as greater docility.

Application 2.10

IMPROVING THE POTATO

There are over 160 different species of wild potato. The selective breeding of cultivated potatoes has concentrated on just eight species. Like other major crops the potato is cultivated in large fields of a single variety. This system, known as **monoculture**, means that the farmer can harvest the crop more easily. Monoculture has an important disadvantage: a single variety is very susceptible to diseases that can spread quickly and destroy an entire crop. The wild species of potato provide a valuable 'reservoir' of genetic variation. Some of these wild species include potatoes that are resistant to different types of crop disease. The conservation of these species in their natural environment is essential for the future breeding of disease-resistant potatoes.

Application 2.11

HYBRID CORN

It is possible to breed together two different varieties, each with valuable characteristics. This is known as **cross-breeding**. The offspring are called **hybrids** and possess the characteristics of both varieties. This can produce crops that have a combination of qualities such as a high yield *and* resistance to many diseases. In the 1930s the introduction of hybrid corn in the United States helped farmers to increase their yield by 50%.

HOW DO WE KNOW THAT EVOLUTION HAS OCCURRED?

The process of evolution takes millions of years so no-one has lived long enough to observe it. We do have evidence, however, that suggests strongly that evolution has occurred.

The fossil record

Fossils are the remains of dead organisms that are preserved in the rocks of the Earth's crust. We can discover the age of the fossil by measuring the radioactivity in the rock (see page 189). If the fossils are arranged in order of age a pattern can be seen. The oldest fossils consist only of very simple organisms such as the invertebrates. The more advanced organisms such as mammals are found only relatively recently (see figure 2.38). This suggests that the simple organisms evolved into more advanced organisms.

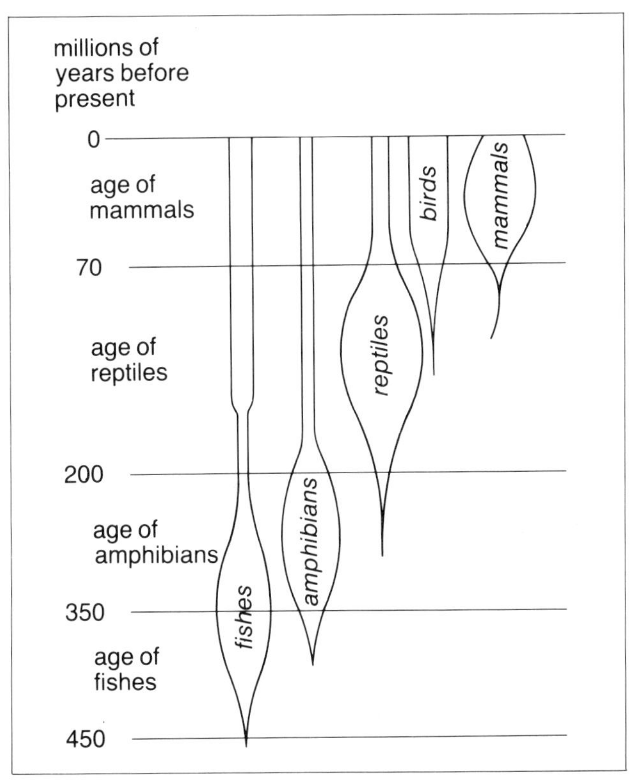

figure 2.38 The appearance of different groups of vertebrates in the fossil record. The width of each band represents how many fossils there were in each group in a particular period

Comparing structures

Figure 2.39 (overleaf)shows the structure of the forelimbs of various groups of vertebrates. Although they appear very different, a closer examination shows they are all based on a similar pattern. This basic structure is known as the **pentadactyl limb** (see figure 2.40, overleaf). This suggests that the vertebrates evolved from a common ancestor. The basic limb structures became modified by natural selection to cope with different requirements. The horse's digits (fingers or toes) became fused together to form sturdy hooves. The bat's digits became elongated to support the thin folds of skin that allow it to fly.

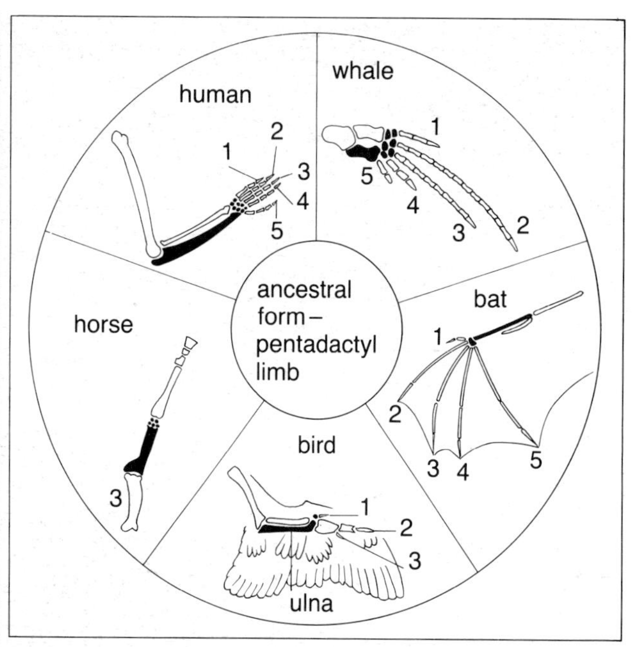

figure 2.39 *The modification of the pentadactyl limb shows adaptation to different environments*

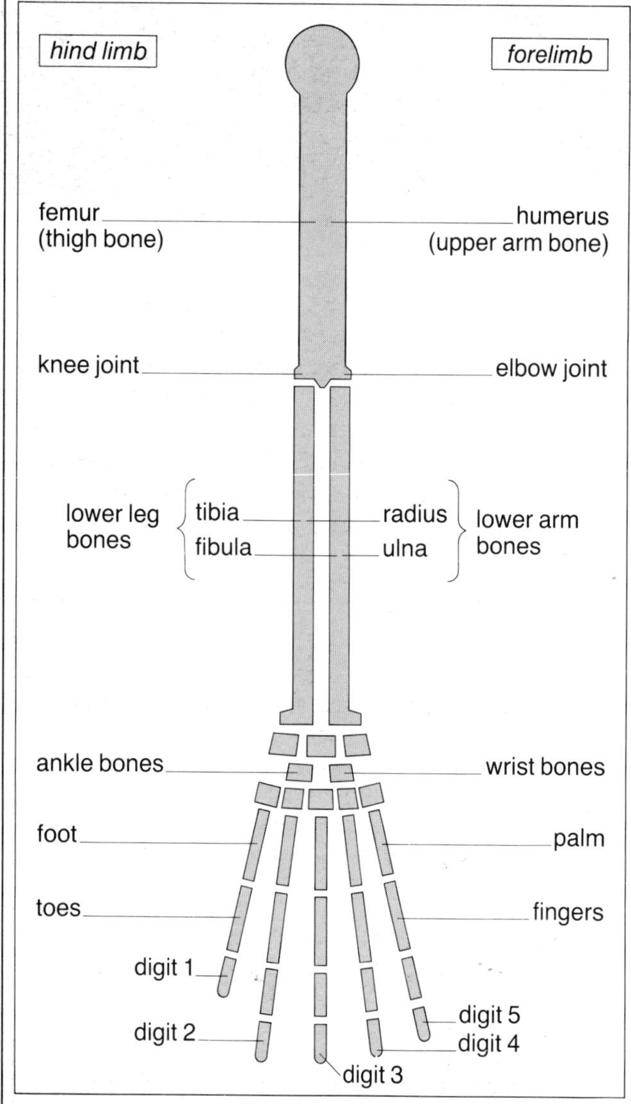

figure 2.40 *The basic structure of the pentadactyl limb and its adaptation in humans*

What is a species?

Are the two individuals in figure 2.41 members of the same species? They look quite different. You might think that they come from different species. However, one species may have individuals that vary greatly from one another. There is a test we can use – if we observe them mating we can be almost certain that they belong to the same species. To be absolutely sure we would need to wait until the offspring of this mating were mature. If *these* offspring can reproduce successfully we can be certain that their parents came from the same species.

figure 2.41 *A peacock and a peahen – male and female of the same species*

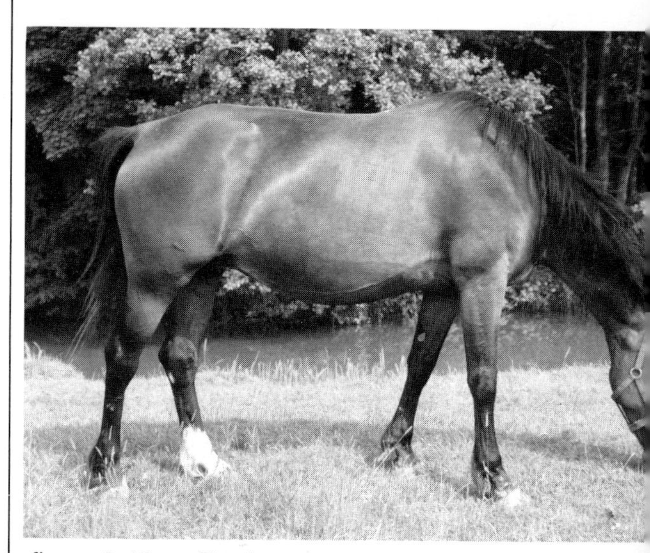

figure 2.42 **a** *The horse*

b The donkey

c The mule, an infertile hybrid produced by the interbreeding of a horse and a donkey

The horse and the donkey can mate to produce the mule. Although it is a hardworking pack animal, the mule is infertile. This tells us that the horse and donkey are different species. We can say that a species is a group of individuals that breed with each other to produce fertile offspring.

How are new species formed?

Natural selection causes a species to change. This change results in the species being better adapted to its environment. Suppose a species becomes separated into two groups by a barrier. The two groups will change and become adapted to their different environments. Over many generations two very different populations will be produced. If the barrier is removed the populations will now be so different that they will not breed with each other. If they do interbreed the hybrid offspring may be infertile like the mule.

The barrier that separates two populations can be a physical one like an ocean or a mountain range. These are **geographical barriers**. A less obvious barrier occurs when two populations occupy different habitats. This is known as an **ecological barrier**. The evolution of the human species involved our ancestors moving into a new habitat (see application 2.12). Differences in the behaviour of two populations may also form a barrier. The way animals behave before mating (courtship behaviour) is particularly important. Different courtship behaviour allows individuals to recognize their own species so that they do not breed with an individual from another species. This is an example of a **behavioural barrier**.

Application 2.12

EVOLUTION OF THE HUMAN SPECIES

Our ancestors may have been ape-like creatures living in the sheltered environment of tropical rain forests. It is thought that certain individuals spent more time on the ground than in trees. These creatures perhaps inhabited the edge of the forests and began to gather food from the open grassland. This less sheltered habitat was potentially more dangerous. It was inhabited by fast-moving carnivores that put our ancestors under great pressure. This encouraged the selection of many adaptations. Walking upright allowed the arms to be used for making tools. A larger brain led to the development of a complex spoken language. This enabled communication to be improved, which paved the way for the greater cooperation and the development of social groups that stayed together over many generations.

INVESTIGATION 2.6

HUMAN ANCESTORS

Questions

1 Look at figure 2.43. Describe how *Homo sapiens* differs from our ancestor *Australopithecus afarensis* in terms of **a** height **b** shape of head **c** use of tools.

2 *Homo erectus* and *Australopithecus robustus* were both living 1.5 million years ago. What evidence from figure 2.43 could suggest why *Homo erectus* survived while *Australopithecus robustus* became extinct?

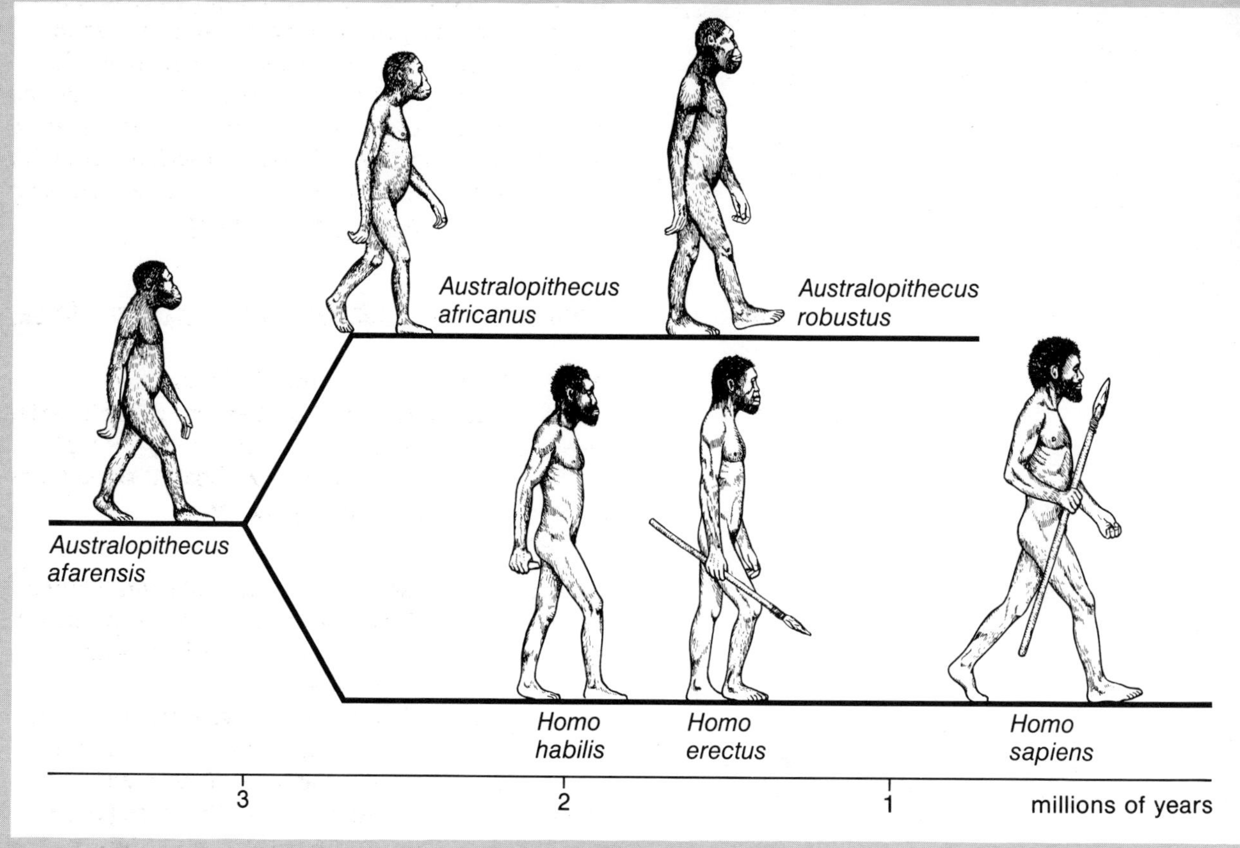

figure 2.43

ALL YOUR OWN WORK

1 In sexual reproduction two gametes fuse together to form a ________. The ovum is the female ________. The ________ is the male gamete. The male gamete has a ________ so that it can swim towards the female gamete.

2 A mistake in copying a gene is called a ________. An example of a chromosome ________ is ________ syndrome.

3 What is meant by co-dominance?

4 Explain how a genetic counsellor can help someone who believes he or she is carrying a harmful gene.

5 Figure 2.44 shows a family tree showing people with red hair and those with brown hair.

a i How many generations are shown?

ii How many females are shown?

iii Which hair colour is dominant? Give a reason for your answer.

b Using the symbols **B** for brown hair and **R** for red hair, explain

i which two genes for hair colour person P must have

ii which two genes for hair colour person Q must have.

[SEG specimen question]

P

Q

□ male with brown hair

○ female with brown hair

▨ male with red hair

⊘ female with red hair

figure 2.44

HEALTHY BODY SYSTEMS

FOOD AND HEALTH

The food we eat is important for our health. In Britain we eat too much of the wrong kinds of food. This is partly responsible for the high levels of certain illnesses, particularly heart disease and diseases of the digestive system. The health of people in poorer countries suffers because they cannot get enough of the right kinds of food. Lack of protein in particular is one of the major problems facing people in these countries.

A healthy balanced diet can be achieved by eating sufficient amounts of each of the following types of food: protein, fat, carbohydrate, fibre, vitamins, minerals and water. These food types are described as nutrients. We will examine each of these nutrients in more detail and look at their importance for our health.

Protein – food for growth

Protein is needed to build new body cells. Children and teenagers require protein so that they can grow. Even adults who have stopped growing need protein because it is used to repair and replace cells. Without protein we would not be able to replace cells like red blood cells that have a limited life. Cells that have worn out or are damaged could not be replaced.

Our bodies are constructed from thousands of different proteins, each with a different job to do. Some proteins form an important part of the body's structure. For example, keratin is a tough hardwearing protein that makes up our skin, hair and nails. Other proteins called enzymes control the chemical reactions in our cells.

figure 3.1 These two plates contain the same amount of protein

It is important that we get the right amount of protein in our diet. As a rough guide, we need to eat each day 1 g of protein for every kilogram of our body mass. An adult whose mass was 60 kg (about 9½ stones) would need a daily protein intake of 60 g. This quantity of protein could be gained by eating a leg of roast chicken and a portion of baked beans (see figure 3.1). Many of us in the UK eat more protein than we need. In poorer developing countries, the opposite is true. Plant foods such as rice and potatoes do not contain as high a proportion of protein as animal produce such as meat or cheese. A person would need to eat 600 g of rice to achieve the recommended protein intake. A shortage of protein in the diet leads to a deficiency disease called **kwashiorkor**. Growth is stunted, the muscles waste away and fluid builds up in the abdomen causing it to swell up.

Saturated and unsaturated fats – choose with care

We could not survive without fat in our diet. However, in certain countries the amount of fat eaten needs to be reduced. In Great Britain, for example, the huge numbers of deaths from heart disease have been linked with the large quantity of fat we consume.

Fat provides a concentrated source of energy which can be stored easily in the body. We have special fat storage cells that form a layer beneath the skin surface. This **subcutaneous fat** is particularly useful in cold climates. Eskimos have a very thick layer of subcutaneous fat that gives them extra insulation in the freezing Arctic conditions. A fat layer also surrounds certain organs such as the kidneys and eyeballs. It cushions them against severe knocks and reduces the chance of damage. Fat is needed to build new cell membranes. Finally, we need a certain amount of fat to provide us with the fat soluble vitamins A and D.

There are two types of fat: **saturated** and **unsaturated**. Unsaturated fats contain a lot of double bonds in their chemical structure which make them more reactive. Many scientists believe that saturated fat (which does not have double bonds) is the main culprit in the diet responsible for heart disease. Red meat, cheese, butter and cream have large quantities of saturated fat. It seems that if we reduced our intake of saturated fat and switched to unsaturated fat, we could improve our health (see tables 3.1 and 3.2, overleaf).

Carbohydrates

Carbohydrates include two very different food types: sugar and starch. Their chemical structure is shown in figure 3.2 (overleaf). Sugar has a limited value in the diet and can cause health problems. Starchy

UNIT 3

meat	percentage fat
fried streaky bacon	45%
grilled streaky bacon	36%
grilled lamb chop	29%
pork pie	27%
luncheon meat	27%
roast lamb (shoulder)	26%
fried pork sausages	25%
roast leg of pork	20%
fried beefburgers	17%
stewed steak	7%
casseroled chicken	7%
fried lambs kidneys	6%
tinned ham	5%
fish	
smoked mackerel	16%
fried fishfingers	13%
cod fried in batter	10%
steamed plaice	2%
cheese	
cream cheeses	50%
stilton	40%
chedder	34%
processed cheese	25%
edam	23%
cheese spread	23%
cottage cheese	4%
milk, butter, oils	
oil (all kinds)	100%
lard	99%
butter	82%
margarine (all kinds)	80%
double cream	50%
dairy ice-cream	7%
gold-top milk	5%
silver-top milk	4%
yoghurt	1%
skimmed milk	less than 1%

table 3.1 *The fat in food*

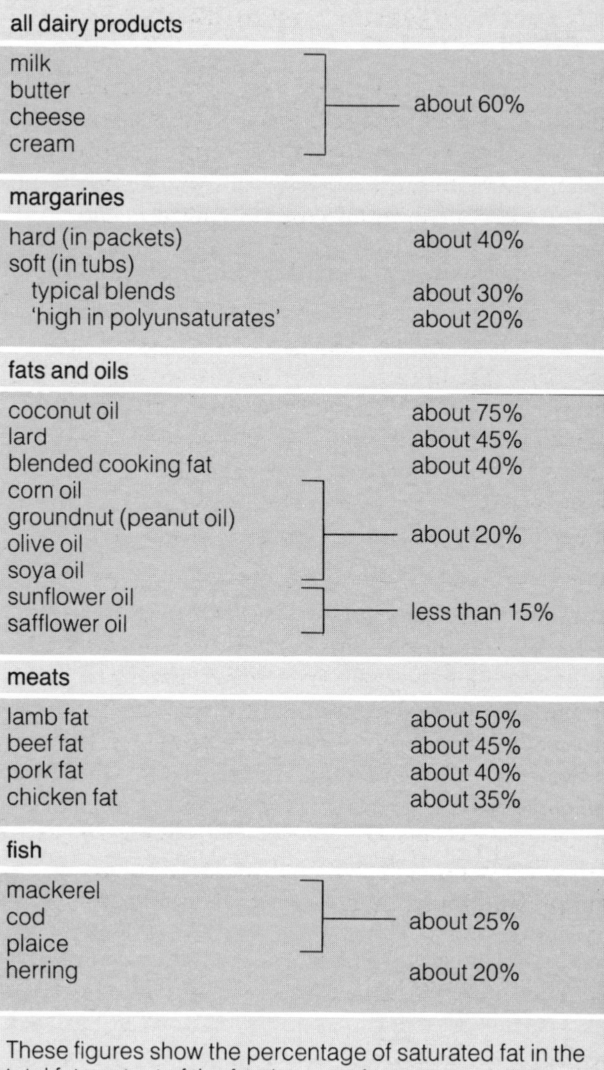

all dairy products	
milk butter cheese cream	about 60%
margarines	
hard (in packets)	about 40%
soft (in tubs)	
typical blends	about 30%
'high in polyunsaturates'	about 20%
fats and oils	
coconut oil	about 75%
lard	about 45%
blended cooking fat	about 40%
corn oil groundnut (peanut oil) olive oil soya oil	about 20%
sunflower oil safflower oil	less than 15%
meats	
lamb fat	about 50%
beef fat	about 45%
pork fat	about 40%
chicken fat	about 35%
fish	
mackerel cod plaice	about 25%
herring	about 20%

These figures show the percentage of saturated fat in the total fat content of the food we eat, for example full cream milk (gold-top) is 5% fat. Two-thirds of that fat is saturated.

table 3.2 *The saturated fat in fatty foods*

foods, sometimes called complex carbohydrates, such as rice, bread, pasta and potatoes, are important in the diet because they can provide a good source of fibre.

Carbohydrates give us energy. Starchy foods are preferable to sugars as an energy source. Sugary foods such as sweets, cakes and jams can easily stick between the teeth. Bacteria grow on them and produce acid, causing tooth decay. Someone who includes a lot of sugar in their diet is more likely to be very overweight (obese) and suffer health problems as a result (see page 144).

In many parts of the world one type of starchy carbohydrate forms the main part of the diet, for example, rice, yams or cassava. However, such plant food is low in protein and can only provide a certain number of the essential amino acids. Including a wide variety of different types of plant foods in the diet can avoid this problem.

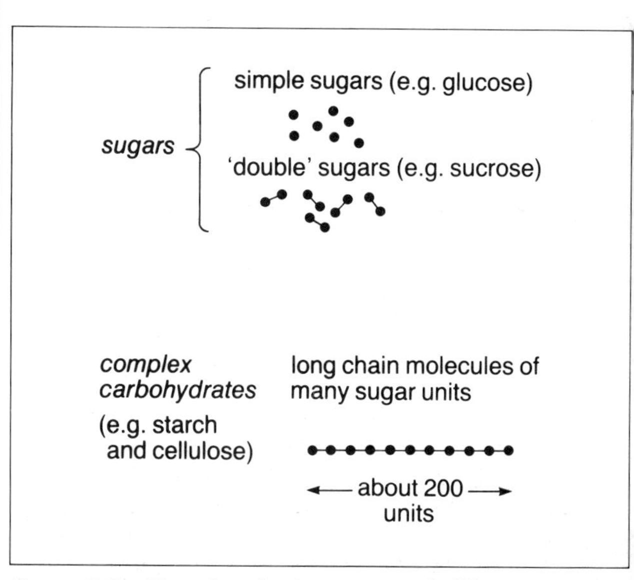

figure 3.2 *The chemical structure of different carbohydrates*

Filling up with fibre

Fibre is only found in plant food. Food from animals such as meat and dairy produce contains no fibre at all. The fibre in plants forms the tough cellulose cell walls. The highest quantities of fibre are found in the husks of cereal grains and the skin of fruits and vegetables. This is why brown rice and wholemeal flour have a higher amount of fibre than white flour and white rice. Baked jacket potatoes are a good source of fibre. Chips or mashed potatoes contain very little. Traditional African and Asian diets have a high fibre content because they consist of large quantities of unrefined plant food. Digestive diseases such as bowel cancer are rarely seen among these populations. This suggests strongly that a lack of fibre can cause diseases of the digestive system.

Why does a lack of fibre cause these problems? Although fibre does not give us energy or nutrients its bulk is very important. It gives the intestine wall muscles something solid to push against. This speeds up the passage of food through the body. Toxic waste products that are sometimes made by certain bacteria in the intestine are removed more rapidly. The extra bulk absorbs water, making the faeces softer and thus preventing constipation. Fibre is also an important part of a slimming diet. It helps to fill the stomach and prevent hunger pangs without adding calories.

Water

Water is essential for our survival. We would die within a few days if we did not have water, yet we can survive for several weeks without food. Water forms about 65% of our total mass. Our blood is about 90% water. Water in cells is the solvent that allows vital chemical reactions to take place.

INVESTIGATION 3.1

HOW TO FIND OUT WHETHER A FOOD CONTAINS PROTEIN

a Using a pestle and mortar, crush up a small quantity of food with a little water.
b Transfer this food sample to a boiling tube.
c Add 2 cm^3 of diluted sodium hydroxide solution and shake well.
d Add a few drops of dilute copper sulphate solution. If the solution turns a pale purple colour this indicates that a significant amount of protein is present.
e Repeat this test for a wide variety of foods. You should include egg white, crushed peanuts, grated cheese, bread, chocolate, pastry, rice and a variety of fruits.

INVESTIGATION 3.2

HOW TO FIND OUT WHETHER A FOOD CONTAINS A SIMPLE SUGAR

The most common simple sugar is glucose. Fructose (fruit sugar) and lactose (milk sugar) are also simple sugars.

a Using a pestle and mortar crush up a small quantity of food with a little water. Pour about 2 cm^3 of the food into a boiling tube.
b Add about 2 cm^3 of Benedict's solution and shake well.
c Place the boiling tube in a beaker of boiling water and leave for a few minutes.

If a green colour forms, it shows that a small amount of simple sugar is present. If an orange/red precipitate forms, it shows that a large amount of simple sugar is present.

INVESTIGATION 3.3

HOW TO FIND OUT WHETHER A FOOD CONTAINS FAT

a Add a small quantity of crushed-up food to 2 cm^3 of ethanol in a test tube.
b Shake the test tube well and then leave it to stand so that any solid food settles to the bottom.
c Pour off the clean ethanol into another test tube containing 2 cm^3 of water.
d If the food contains fat the water will become cloudy due to the formation of an **emulsion**.

INVESTIGATION 3.4

HOW TO FIND OUT WHETHER A FOOD CONTAINS STARCH

a Take a spatula measure of starch powder and place in a petri dish. Add a drop of iodine solution and note any colour changes you observe.
b Repeat this test with a spatula measure of calcium carbonate. Note the colour of the iodine solution.
c Use iodine solution to test a wide range of foods to see if they contain starch. Gently crush a small amount of the food sample using a pestle and mortar and add 1 cm^3 of distilled water before adding a few drops of iodine solution.

Questions

1 How could you use iodine solution to find out whether a food contained starch?
2 Explain why you needed to check the colour of iodine solution when added to a white powder that did *not* contain starch.

vitamin	foods containing the vitamin	why the vitamin is needed	deficiency disease	other information
A retinol	milk, butter, carrots, green vegetables, liver	to make the pigment needed for good vision in dim light, keeps the cells lining the respiratory system healthy	poor vision at night, infections of the respiratory system	fat soluble so can be stored in the liver
B_1 thiamine	yeast extract, wholemeal bread, peas and beans	forms a chemical (co-enzyme) needed in cell respiration	beri-beri – the nervous system is damaged and the muscles weaken; eventually paralysis occurs	polished (not brown) rice and white flour lack the B vitamins
B_2 riboflavin	green leaf vegetables, fish, eggs		pellagra – mouth ulcers, skin disorders	water soluble
C ascorbic acid	citrus fruits: oranges, lemons, limes; green vegetables	forms the 'chemical cement' that binds cells together	scurvy – the gums bleed easily and wounds heal more slowly	destroyed by cooking and water soluble
D calciferol	animal fats and oils; egg yolk; milk; dairy products	helps calcium and phophorus to be absorbed by the bones	rickets – the bones become soft and break more easily	can be made by skin cells if exposed to sunlight, is fat soluble so can be stored in the liver
K	cabbage and spinach	helps the blood to clot when a blood vessel is cut	the blood clots more slowly	small quantities are made by bacteria in the intestine

table 3.3 *Vitamins and why we need them*

Vitamins

Vitamins are complex chemical substances that our cells cannot make themselves. We have to include them in our diet. They are only needed in tiny quantities but without them many chemical reactions in our cells could not take place. Someone whose diet lacks a particular vitamin will suffer from a deficiency disease (see table 3.3).

Certain foods, such as milk, contain several different vitamins. Other foods, like rice, contain only one vitamin. Some foods, such as margarine and breakfast cereals, have vitamins added to them when they are being made. We can make sure that we have enough vitamins by eating a variety of different foods including fresh fruit and vegetables.

Certain vitamins are vital in the diet of some species but not others. Vitamin C is not required in the diet of most animals as their cells can make it. Primates (monkeys and apes), including our own species, have lost this ability; perhaps because the diet includes fruit rich in vitamin C.

Minerals

Minerals are simple chemicals needed for many of the body's activities. We only need small amounts of them to function properly, as with vitamins. Table 3.4 shows the problems caused if we do not have enough of certain minerals.

Application 3.1

DIET DURING PREGNANCY

During pregnancy a good diet is essential to the healthy development of the foetus. The foetus depends for its food entirely on what the mother eats. Although it is sometimes said that 'the mother must eat for two', this does not mean that she needs to double her food intake. Only a moderate increase in certain foods is required. The amount of fat and carbohydrate need not be increased, or the mother might put on too much weight during pregnancy. The extra fat would be an added burden that might cause back pain. The amount of protein should be increased. The foetus will use it for growth. The amount of calcium in the mother's diet should also be greater. It is needed by the foetus for the development of strong bones and teeth. The increase in calcium and protein can be achieved by drinking an extra pint of milk each day.

During pregnancy, a woman makes an extra two litres of blood. To do this her body requires a high level of iron for making haemoglobin. Plenty of green vegetables and liver will provide this. The doctor may prescribe iron tablets if the mother's haemoglobin level is low. A balanced diet with plenty of fresh fruit and vegetables will ensure that enough vitamins and minerals are available for the developing foetus.

Food labelling

The label attached to processed food like that in figure 3.3 can provide a great deal of useful information. The food manufacturer is required by law to include the following information about the food:

a the name and address of the company that packages the food. This gives the customer the chance of returning the food if dissatisfied for some reason

b an accurate and truthful description of the contents of the food package

c the amount of food contained in the package. This is usually given as the mass of the food

d a list of all the ingredients contained in the food, listed in order of mass. This list must include any chemicals (additives) added to the food for a variety of reasons

15 oz 425 gram
SAINSBURY'S
CREAM OF
TOMATO
SOUP
Directions: Pour into a saucepan and gently heat, stirring all the time. Do not boil or overheat as this will impair the flavour
Ingredients: Tomatoes, Water, Sugar, Vegetable oil, Maize starch, Cream, Butter, Salt, Modified starch, Spices, Citric acid

figure 3.3 *A food label provides useful information*

e certain foods will not keep fresh beyond a certain time. Such perishable foods must have a 'best before date' stamped somewhere on the packaging.

INVESTIGATION 3.5

ENERGY FROM FOOD

Fat provides 39 kJ of energy per gram. Carbohydrate provides 17 kJ per gram. When the supply of fat or carbohydrate is low protein can provide 18 kJ per gram. Table 3.5 shows the energy value in kilojoules of fat, carbohydrate and protein in the daily diet of three different groups of people.

	British	Eskimo	rural Kenyan
fat	4200	6100	800
carbohydrate	6800	1000	6600
protein	1700	6400	1700

table 3.5 *Energy values (kJ) of the diets of three groups of people*

Questions

1 Plot this data on a histogram.
2 In which country does the population gain nearly all its energy from carbohydrate?
3 What particular foods captured by Eskimos might explain the high fat and protein content of their diet?
4 Calculate the total energy value of each group's diet.
5 The minimum recommended daily energy intake for an active adult is 9200 kJ. Which group falls below this level?

mineral salt	foods rich in the mineral	why the mineral is needed	deficiency disease	other information
calcium (Ca^{++})	milk, cheese, bread, green vegetables	strengthens teeth and bones, also necessary for muscle contraction	bones and teeth weaken	vitamin D is necessary for the absorption of calcium and phosphorus
phosphates (PO_4^-)	milk, fish, meat, eggs	forms an important part of teeth and bones, also necessary for the chemical reactions of respiration	bones and teeth weaken	
iodine (I)	seafood, seaweed, iodized table salt, most drinking water	forms a vital part of thyroxin – a growth hormine	goitre – the thyroid gland in the neck swells up, growth in children is stunted	goitre used to be common in isolated Alpine communities due to lack of iodine
iron (Fe^{2+})	liver, egg yolk	part of haemoglobin – the red blood pigment that carries oxygen	anaemia – the tissues do not recieve enough oxygen and the person feels tired and weak	
sodium chloride (NaCl)	table salt	forms 1% of the blood plasma – sodium also needed for nerve impulses	very rare	too much salt can lead to high blood pressure
flouride	occurs normally in drinking water in certain parts of the country	strengthens the hard outer layer of the teeth	dental caries (tooth decay) has been found to be common in areas that lack flouride	some water authorities add flouride to drinking water – this is strongly opposed by some people (see page 50)

table 3.4 *Minerals and why we need them*

Application 3.2

'FEED THE WORLD'

About one-third of the world's population go to bed hungry. About half are thought to have less than enough to eat. This food shortage is the result of a number of problems for which people are searching for solutions.

Problem Severe drought lasting several years will drastically reduce the harvest.
Solution Irrigation schemes can ensure that water is stored during periods of rainfall to be used when the drought comes.

figure 3.4 The Aswan Dam in Egypt controls the flow of water from the River Nile

Problem Insect pests, for example locusts and tsetse flies, reduce food supplies. Swarms of locusts can destroy entire fields of crops.
Solution Pests can be controlled by a variety of methods (see page 85).

figure 3.5 A locust swarm

figure 3.6 Modern tractors can replace oxen

Problem The poor quality of soil reduces the yield of food crops.
Solution The quality of the soil can be improved with the *careful* use of fertilizers and modern agricultural techniques.

IT'S NOT ONLY DROUGHTS, FLOODS AND DISEASE THAT ARE CRIPPLING THE THIRD WORLD.

After suffering a succession of natural disasters, many Third World countries now face a problem that is entirely caused by humans. Western banks, having encouraged these hard hit nations to take out huge loans, are now demanding massive interest repayments. (Latin America alone has to repay £23 000 000 000 a year). And while the banks grow rich (British banks reported profits of £3 000 000 000 last year) their debtors remain in poverty.

figure 3.7 Economic problems can produce famines

Problem Many developing countries have borrowed large sums of money from the richer countries. Most of the cereal crop they produce must be sold abroad to pay off the debt.
Solution The richer countries can change the way these debts are repaid. More of the food produced in developing countries can then be used to feed their own population.

FOOD ADDITIVES

Chemicals have been added to food for most of human history to add flavouring or lengthen the time the food could be kept. Sodium chloride (common salt) was one of the earliest additives. When added to meat it would make it last through the winter without spoiling. Nowadays the range of food additives is enormous. It is estimated that 3500 food additives are in use. About 500 of these additives have been given an 'E' number. This shows they have been approved by the EEC. The 'E' number of each additive described below is given in brackets.

Some food additives occur naturally, for example the green colouring chlorophyll (E140) is extracted from plant leaves. Most additives are made artificially in the laboratory.

Flavour enhancers

Some chemicals have very little flavour by themselves but will bring out the flavour of foods to which they are added. Monosodium glutamate (MSG:E621) is a flavour enhancer commonly used in Chinese cooking. It is also added to a wide range of processed foods. Someone who accidently eats too much MSG may feel sick.

Colourings

The bright colours of many foods such as tinned peas, orange squash and fish fingers are due to special chemical colourings. These additives were originally introduced to improve the appearance of the food. Recently manufacturers have brought out new brands that are free from artificial colouring. This is because many people became worried about possible side-effects (see application 3.3).

Emulsifiers and stablizers

Certain foods consist of a number of ingredients that do not normally mix, for example oil and vinegar in salad cream. Emulsifiers such as stearyl tartrate (E483) help the two ingredients to mix together. Stabilizers stop them from separating.

Anti-oxidants

These chemicals are added to many dried foods. They slow down oxidation which might alter the substances that give the food its flavour. They therefore stop foods 'going off'. Biscuits have added anti-oxidants so that they can be kept on shop shelves for long periods.

Preservatives

Chemicals are commonly added to food to prevent the growth of microbes. Benzoic acid (E120) is a common preservative. It also occurs naturally in certain foods such as cranberries. This probably explains why they keep so well.

Flavourings

The distinctive flavours in many manufactured foods are produced by artificial flavourings. Caramel (E150) is a flavouring added to foods such as pickled onions, cola drinks, biscuits and scotch eggs. Many artificial flavourings produced by food chemists are copies of natural flavours. For example a 'cheese flavour' biscuit contains no cheese at all. All the flavour is artificial. If the label says 'cheese flavoured' it means that the biscuit does contain some real cheese.

Application 3.3

HOW SAFE ARE FOOD ADDITIVES?

The safety of food additives can be tested by feeding high doses to animals in the laboratory. If the animals suffer ill effects the additive should be banned. Some additives previously added to food have been banned because they were found to be a risk to people's health. The substance rhodanine B was a luminous pink chemical that used to be added to rock sold at holiday resorts. The government banned it because it was found to cause cancer in experimental animals.

Some individuals may be particularly sensitive to additives that have *not* been banned. The chemical tartrazine (E102), a bright yellow dye, is found in many foods, including certain brands of orange squash. In certain people who are allergic to tartrazine it has been found to cause skin rashes, hay fever and breathing problems. Extreme restlessness and irritability in children, a condition described as **hyperactivity**, has been linked in some cases to a high intake of food additives in the diet.

DIGESTING OUR FOOD

What is digestion?

The food we eat must be broken down into small molecules before it can be absorbed into the bloodstream. Breaking down food is called **digestion**. Digestion is carried out in the long tube going from the mouth to the anus – the alimentary canal (see figure 3.8).

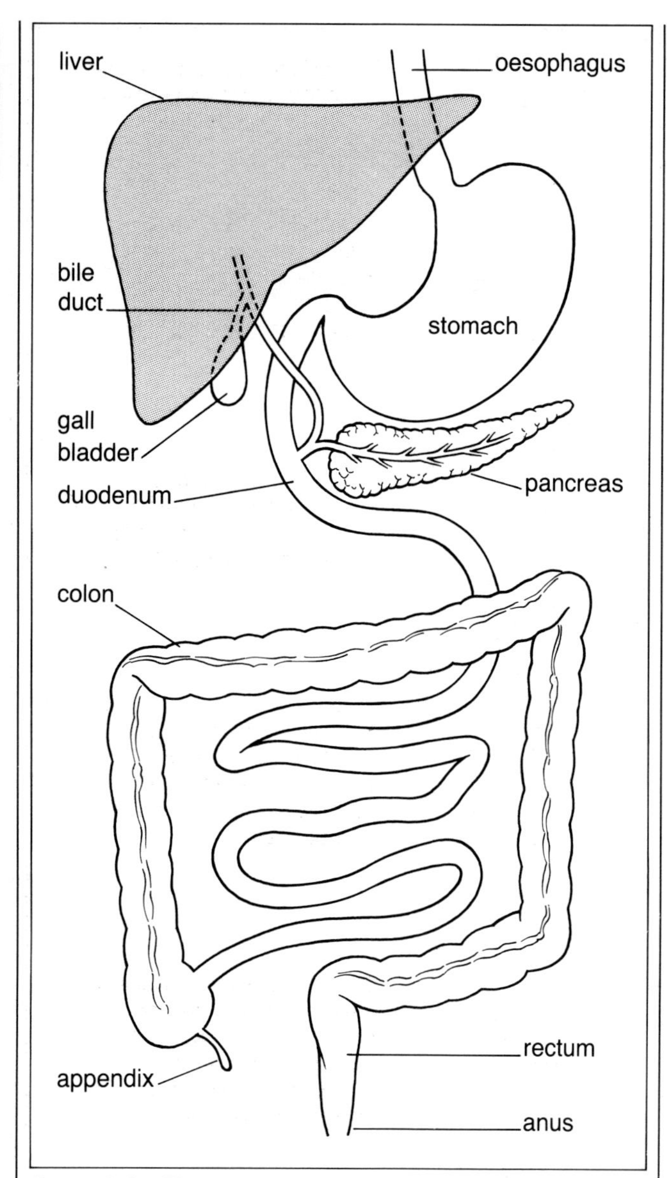

figure 3.8 *The alimentary canal*

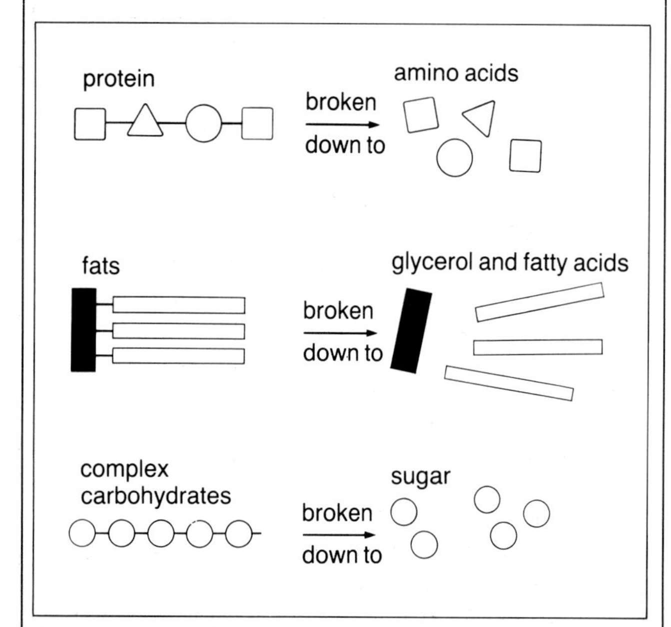

figure 3.9 *During digestion large food molecules are broken down into smaller molecules*

Digestion occurs in several stages. The food must first be broken down physically into smaller pieces. The teeth begin the physical digestion of food. This allows the chemical structure of the food to be attacked. During chemical digestion the large molecules of food are split into smaller molecules (see figure 3.9). The body produces complex substances called digestive enzymes that speed up the breakdown of large food molecules (see page 202). Small food molecules can dissolve while large food molecules are insoluble. Dissolved food can be absorbed into the bloodstream. It can then be carried to where it is needed.

FOOD IN THE MOUTH

Digestion begins in the mouth. Solid food is broken down by the chewing action of the teeth. Chewing also mixes **saliva** with the food. Saliva is made in salivary glands located beneath the tongue. Saliva contains two important substances: **mucus** and **amylase**. Mucus is a slimy fluid that moistens the food and allows it to slip down the oesophagus more easily. Amylase is an enzyme that begins the digestion of starch.

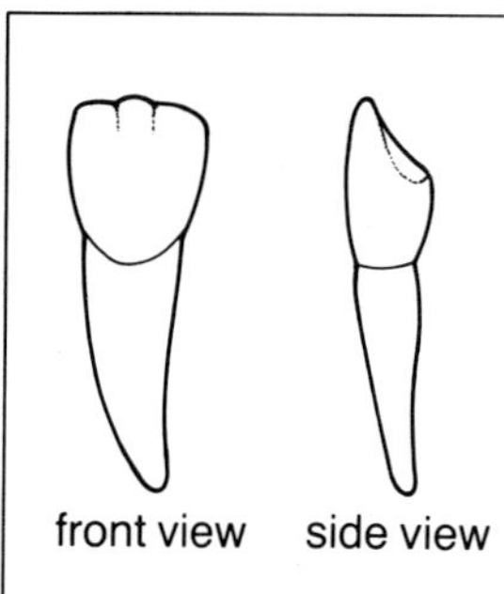

incisor The incisor teeth at the front of the mouth have sharp chisel-shaped edges for biting into food. They are very prominent in animals which eat nuts and bark (e.g. squirrels and rabbits).

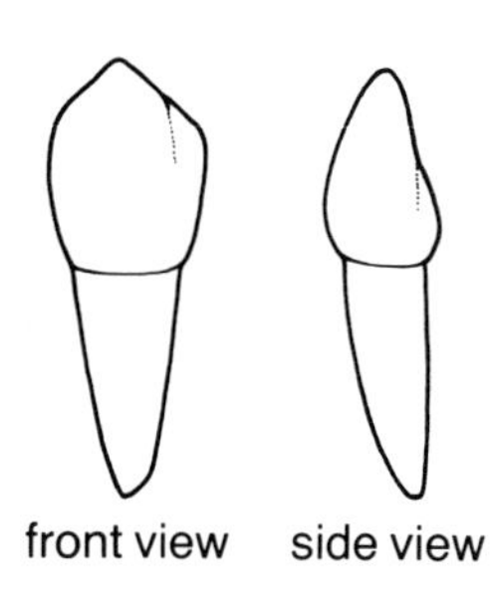

canine Canine teeth are more pointed than other teeth. In flesh-eating mammals such as the tiger, they are very pointed as they are used for killing prey. In humans the canines are not so sharp. They work with the incisors to chop up large pieces of food.

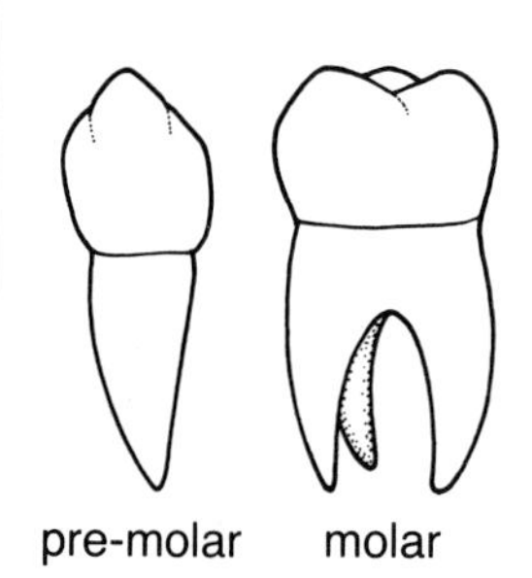

pre-molars and molars The pre-molars and molars are found towards the back of the mouth. They are rather flat with ridges and fissures that can hold food and grind it up. If food becomes trapped in the fissures because of incorrect tooth brushing they can cause a cavity to develop.

figure 3.10 *The four different types of tooth*

Teeth and how to keep them

Our teeth have to be very tough. Grinding up food for about half an hour each day adds up to about 500 days of non-stop grinding in a person's lifetime. Adults have 16 teeth on the top and bottom jaws, making a total of 32. There are four different types of tooth, each with a special shape and a different job to do (see figure 3.10).

Although the teeth look different from the outside, their internal structure is very similar (see figure 3.11). The outer layer of **enamel** gives a strong cutting surface. Beneath the enamel is the **dentine**, a substance that forms the bulk of the tooth. Dentine is tough but it decays more easily than enamel. At the centre of the tooth is the **pulp** which contains nerves and blood capillaries.

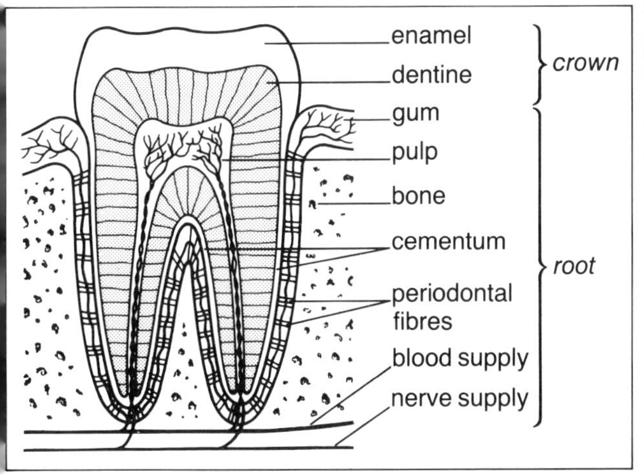

figure 3.11 The structure inside the tooth

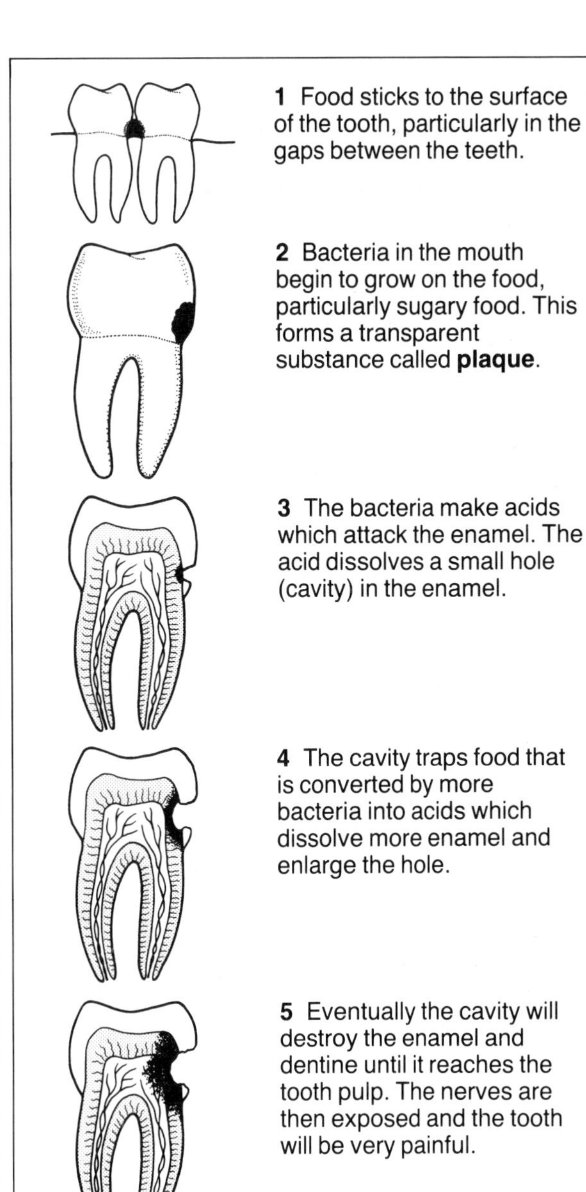

1 Food sticks to the surface of the tooth, particularly in the gaps between the teeth.
2 Bacteria in the mouth begin to grow on the food, particularly sugary food. This forms a transparent substance called **plaque**.
3 The bacteria make acids which attack the enamel. The acid dissolves a small hole (cavity) in the enamel.
4 The cavity traps food that is converted by more bacteria into acids which dissolve more enamel and enlarge the hole.
5 Eventually the cavity will destroy the enamel and dentine until it reaches the tooth pulp. The nerves are then exposed and the tooth will be very painful.

figure 3.12 The stages of tooth decay

What causes tooth decay (dental caries)?

Tooth enamel is one of the toughest materials in the human body. However, it cannot resist attack by acid. Figure 3.12 shows how a tooth becomes decayed.

Healthy gums – hold on to your teeth

Many completely sound teeth have to be extracted simply because the gums become diseased and are too weak to hold the teeth firmly. A very common gum disease is **gingivitis**. This occurs when the gums become inflamed and they may bleed during brushing. If gingivitis is allowed to develop for a long time it may lead to **periodontal disease**. When periodontal disease develops bacteria destroy the fibres holding the root of the tooth firmly in the jawbone. The tooth loosens and may have to be removed by the dentist.

Preventing tooth decay and gum disease

The following measures will help you to achieve healthy teeth and gums.

a Clean the teeth at least twice a day: after breakfast and before going to bed. On at least one of these occasions, clean the teeth thoroughly, concentrating on removing plaque. 'Tartar-control' toothpaste is now available. This prevents the build-up of plaque.

b Avoid sugary food and drinks between meals. If you eat chocolates or sweets, eat them after a meal and then brush your teeth.

c Replace your toothbrush when it wears out (about every four to six months).

d Visit the dentist regularly. Six-monthly visits will ensure that any decay is dealt with quickly before it completely destroys a tooth.

UNIT 3

Fluoride in drinking water

Fluoride is a naturally occurring mineral that is sometimes found in drinking water. It strengthens tooth enamel and this is why it is added to toothpaste. Scientific research in the United States first revealed a link between fluoride and healthy teeth. The teeth of over 7000 children from 21 different cities were checked. The number of filled or decayed teeth was counted and an average figure for each city was calculated. In each city the concentration of fluoride in the drinking water was tested. The fluoride concentrations were then plotted against the number of decayed teeth. The results from this study are shown in figure 3.13.

This data was then used to support a recommendation from dentists that fluoride should be added to drinking water in those areas where the concentration was less than 1 mg per litre, to raise the concentration to this level. In Britain it is estimated that 5½ million people drink water treated in this way. However, some people object strongly to artificial fluoride.

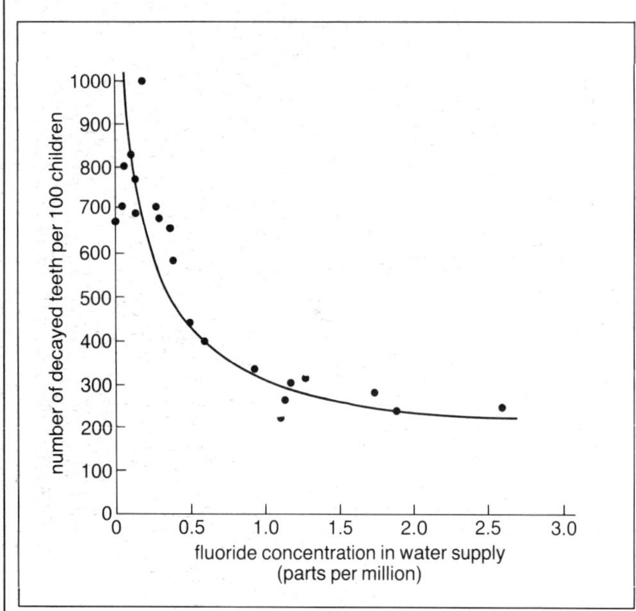

figure 3.13 *The effect of fluoride in drinking water on tooth decay*

Time for thought

The following objections have been raised to the artificial fluoridation of water. What is your reaction to each statement?

- *Fluoride is poisonous. A dose of 2500 mg of fluoride would be enough to kill a person.*
- *If the government is free to add anything it likes to drinking water, where could this lead? People themselves, not the government, should have the right to decide what is added to their drinking water.*
- *The money spent on fluoridation would be better used in persuading people to eat less sugary food and clean their teeth properly.*

INVESTIGATION 3.6

CAN FLOURIDE IN DRINKING WATER IMPROVE CHILDREN'S TEETH?

Two towns, A and B, were used to compare the effect of fluoride concentration on the amount of tooth decay in children. Town A has a fluoride concentration of 0.25 mg per litre of water. Town B has a fluoride concentration of 1.4 mg per litre. The teeth of 500 children aged five were checked and the numbers of decayed, missing or filled (DMF) teeth were counted. Five-year-old children have 20 temporary 'milk teeth', 10 on each jaw. The results on the left and right side of each jaw were added together.

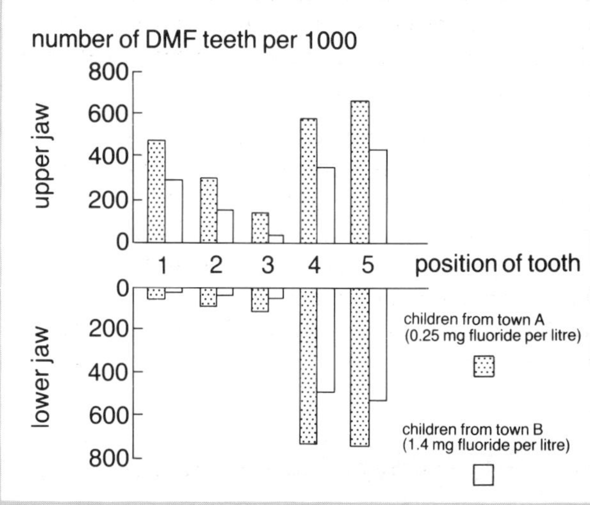

figure 3.14 *Histogram showing the number of DMF teeth per 1000 in two towns*

Questions

1. What percentage of teeth in position 1 of the upper jaw were decayed, missing or filled in
   a town A b town B?
2. What is the difference between the health of children's teeth in towns A and B?
3. What do these results tell you about the link between fluoride and tooth decay?
4. What alternative explanations could you put forward to explain these results?
5. Suggest a possible explanation of the difference in the amount of tooth decay in the upper and lower jaw.
6. Using the data from figure 3.13 explain why 1 mg per litre is the recommended figure for the concentration of fluoride when added to drinking water.

FOOD IN THE STOMACH

The stomach is a muscular bag that will hold food for several hours after a meal. It allows us to eat a large quantity of food at one time. In certain illnesses the stomach has to be completely removed by surgery. This does not seriously affect the person's digestion. It means that instead of having several large meals they must eat small amounts of food throughout the day.

Muscle tissue in the wall of the stomach contracts regularly and churns up the food. When we vomit this muscle tissue contracts violently to force food out of the stomach and back up the oesophagus. Vomiting often occurs if bacteria have entered the stomach with food. Although rather unpleasant, vomiting is a useful way of getting rid of harmful bacteria.

The stomach wall also contains cells that produce two important chemicals: protein-digesting enzymes (proteases) and strong hydrochloric acid (pH 2 – see pages 191-2). This would be strong enough to 'burn' a hole in a carpet! This acid is essential because it provides the correct acidity for the proteases to work. It is also helpful in destroying many bacteria that may have entered the stomach.

Acid in the stomach poses a difficult problem. How can we prevent it from damaging the lining of the stomach? There are special glands in the stomach lining that produce a thick protective layer of mucus. In some people these glands do not work very well. The stomach lining becomes exposed and it is irritated by the acid. This results in a painful stomach ulcer (see application 3.4).

Application 3.4

COPING WITH ULCERS

Ulcers may develop on the lining of the stomach or duodenum. Stress, worry and overwork can sometimes lead to an ulcer. Ulcer patients are advised to avoid drinking coffee or alcohol, smoking cigarettes and eating rich or spicy food. All these things may irritate the ulcer and slow down the healing process.

Doctors can see the inside of the stomach using an instrument called an **endoscope**. This consists of a thin flexible rod made of optical fibre. Optical fibre is a special type of flexible glass that can transmit light along its length (see page 152). The endoscope is inserted through the mouth and pushed gently down the oesophagus into the stomach. A powerful light is shone down the optical fibre rod. This allows the doctor to observe the condition of the stomach lining, and to treat the ulcer. Endoscopes can also be used to investigate the lungs.

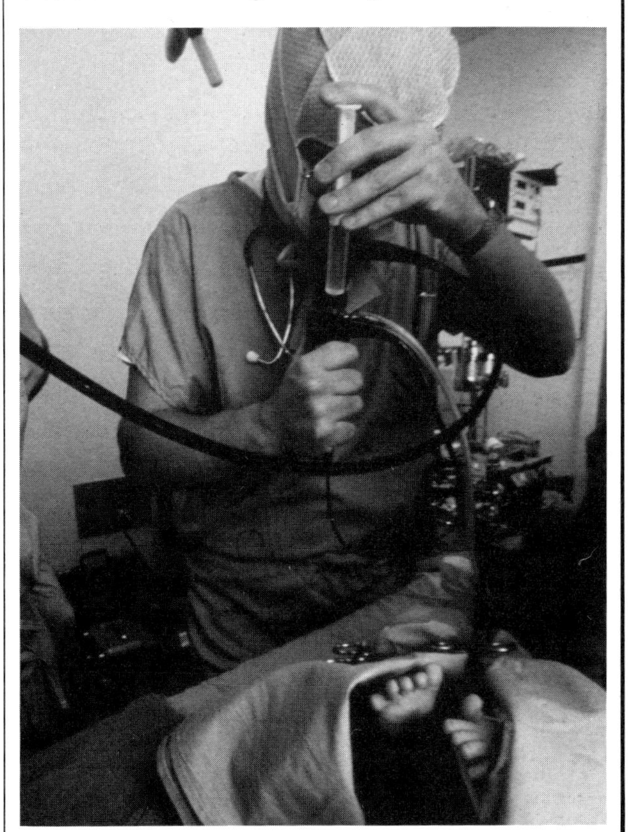

figure 3.15 *A doctor using a fibre optic endoscope to study the inside of the stomach*

FOOD IN THE INTESTINE

A ring of muscle normally closes off the bottom end of the stomach. At intervals this muscle opens to allow food to pass into the **small intestine**. This is a narrow tube about 6 m long. The first 30 cm of the small intestine is known as the **duodenum**. Here the food is mixed with **bile** from the **liver** and enzymes from the **pancreas**. Bile helps digestion by breaking down fat and oil into minute droplets so that enzymes can work faster on them. This is called **emulsification**. Bile also neutralizes the stomach acid. This allows enzymes in the small intestine to work properly.

As the food passes along it is mixed with a wide range of enzymes secreted by glands in the wall of the small intestine (see figure 3.16, overleaf). This completes the digestion of food. The food molecules are now small enough to be absorbed.

Absorption

If the lining of the small intestine were smooth it would not be very efficient at absorbing food (see investigation 3.8). The total surface area is increased by millions of tiny finger-like projections called **villi**. Each villus has its own blood capillaries and a **lacteal** containing **lymph** (see figure 3.16, overleaf; see page 58). Soluble food molecules pass into the blood and lymph. The blood capillaries carry food molecules to the liver. Here they are processed, stored or passed on to other parts of the body in the bloodstream. The lacteals absorb tiny fat droplets. After a fatty meal the lymph has a milky appearance caused by the many fat particles it contains.

The appendix – a dangerous dead end

Food passing from the small intestine into the large intestine meets a T-junction. One turning passes into the **colon**. The other turning leads to a dead end – the

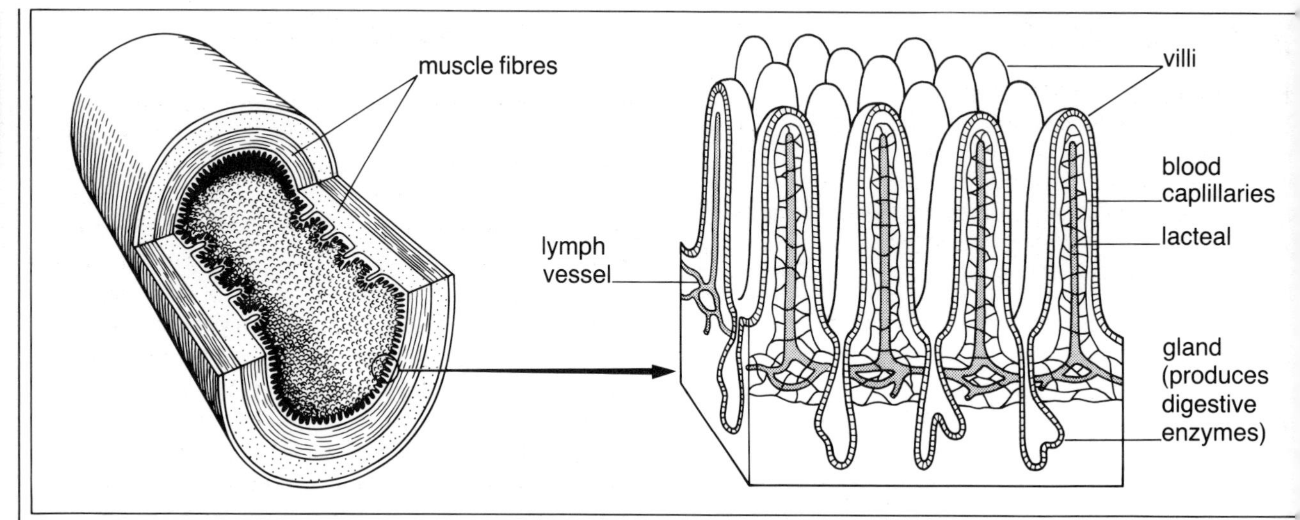

figure 3.16 The inside of the small intestine has many tiny projections called villi, each containing blood capillaries and a lacteal

appendix. Sometimes the appendix becomes infected, causing it to swell up. This is **appendicitis**. When this happens an operation must be carried out quickly to remove the appendix before it bursts, spreading bacteria throughout the abdominal cavity.

The colon

When food enters the colon it is still very liquid. The colon removes water so that the undigested food becomes firmer. At the end of the large intestine the undigested matter is stored in the rectum. It is passed out as solid faeces through the anus.

If it is diseased, the colon sometimes has to be removed by surgery (a **colostomy**). The end of the small intestine is repositioned to pass through a hole in the body wall. Undigested food passes into a bag worn on the outside of the abdomen. The colostomy bag has to be emptied at intervals.

Even in healthy people the colon may not work properly. When someone is suffering from food poisoning or acute anxiety the undigested food is passed rapidly along the colon. The faeces passed out of the body contain a lot of water. This is diarrhoea.

INVESTIGATION 3.7

PASSING THROUGH MEMBRANES: IS IT EASIER FOR GLUCOSE OR STARCH?

Work with a partner for this investigation.

Visking tubing is made from a special type of plastic membrane that allows small molecules to pass through it but not large ones.

a Tie some thread tightly around one end of a length of Visking tubing.

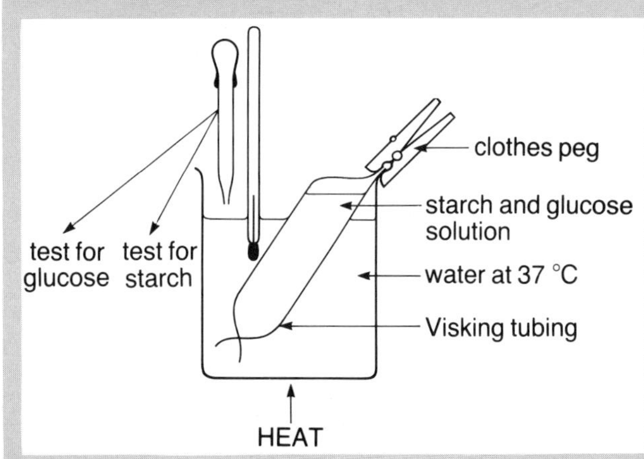

figure 3.17

b While your partner holds the tubing, run a mixture of starch and glucose solution into it from a syringe.
c Rinse the outside of the Visking tubing with tap water to remove any of the starch/glucose mixture.
d Place the filled tubing in a beaker of water, attaching the open end to the lip of the beaker with a clothes peg. Heat the water in the beaker to 37 °C.
e After five minutes remove some water from the beaker with a pipette and carry out tests for glucose and starch. Repeat the tests after 20 and then after 30 minutes. Record your results in table form.

Questions

1 Why was it important to place the Visking tubing in water at 37 °C?
2 Which substance, glucose or starch, was able to pass through the Visking tubing membrane?
3 What does this investigation tell you about the difference in size between starch and glucose molecules?
4 How is Visking tubing similar to the wall of the small intestine?
5 If you were provided with a starch solution and a solution of amylase, how would you use this apparatus to investigate the breakdown of starch molecules to glucose molecules?

INVESTIGATION 3.8

MAKING MODEL INTESTINES

a Use a sheet of paper to make a long tube about 3 cm in diameter. This represents a length of intestine with a smooth lining.
b Fold up another sheet of paper to make a tube with about 10 ridges. Fasten the ends of the sheet with staples so that it can be fitted inside the first tube.

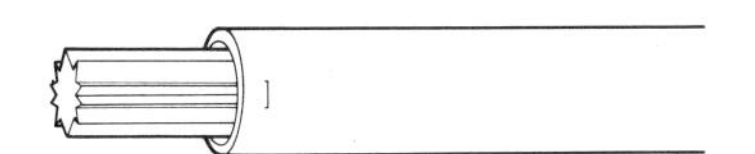

figure 3.18

c Unfold each of your paper tubes. Measure the length of the sides and calculate the area of each sheet.
d You will notice that the folded tube had a large surface area. Calculate the increase in surface area caused by folding using the formula:

$$\text{Increase in surface area} = \frac{\text{area of folded sheet 'intestine'}}{\text{area of smooth 'intestine'}}$$

Questions

1 Which of the two paper tubes is similar to the structure of the human intestine?
2 What structures in the model intestine represent the villi in the human intestine? How does the shape of these structures differ from villi?
3 Use the results of this investigation to explain why a smooth surface is not efficient for absorbing.
4 A certain cattle disease results in the destruction of villi in the small intestine. The lining becomes smooth. What effect is this likely to have?
5 Carry out further investigations to discover how increasing the number of ridges changes the total surface area.

A HEALTHY HEART

The body needs a system for transporting many different substances. For example, oxygen has to be taken from the lungs to all the body cells where it is used up. Carbon dioxide must be carried away from the cells to the lungs where it can be removed from the body. These are just two of the many substances carried around the body in the blood. A network of tubes called **blood vessels** link up all parts of the body so that substances can be transported in the blood to wherever they are needed. The **heart** is the organ that provides the power to drive the blood through the blood vessels. Together the heart and blood vessels form the **circulatory system**. It is vital for our health that we keep the circulatory system in good working order. Later on in this unit you will find out what we can do to help achieve this.

The heart's two pumps

The heart works like a pump driving blood through blood vessels. In fact the heart consists of two pumps working alongside each other. The pump on the right side collects blood that has already passed through the body tissues. This blood contains a lot of carbon dioxide but very little oxygen. It is called **deoxygenated** blood. The right side pumps deoxygenated blood to the lungs. Here the blood becomes recharged with oxygen. It is now **oxygenated** blood. Oxygenated blood returns from the lungs to the left side of the heart. From here it is pumped to all the body's organs, delivering fresh supplies of oxygen.

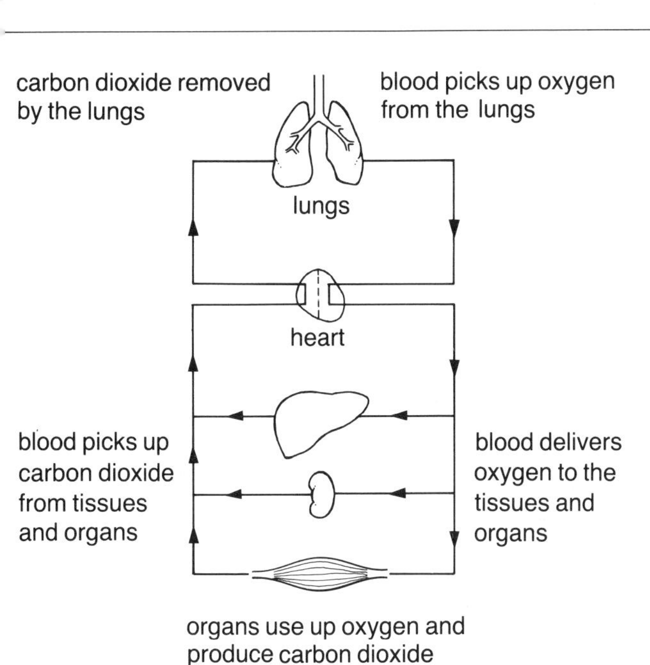

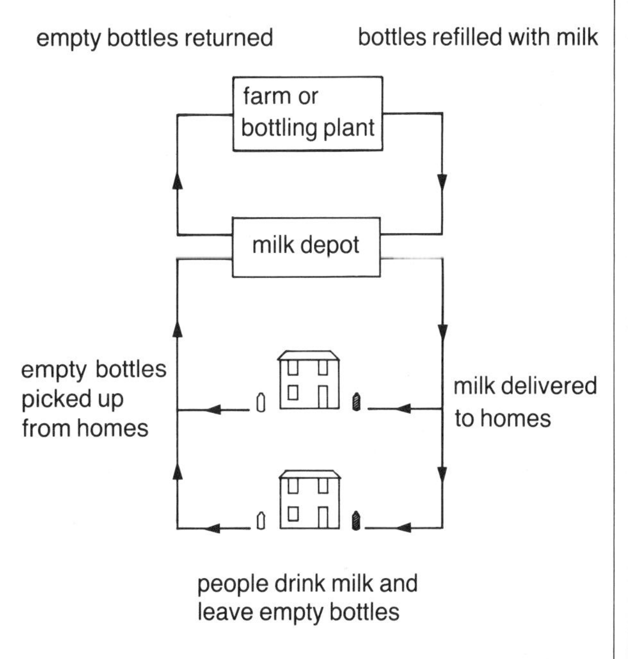

figure 3.19 The two loops of the double circulatory system can be compared with a milk delivery system

The heart's two pumps are linked to two separate transport routes. This forms a double circulatory system. The way this system works can be compared with the delivery of milk (see figure 3.19, previous page). Filled milk bottles are like oxygenated blood. Empty milk bottles represent deoxygenated blood. The 'empties' are first returned to the depot before they are sent to the bottling plant to be refilled. In the same way deoxygenated blood first returns to the heart before it is pumped to the lungs.

Inside the heart

The heart is a hollow muscular bag divided into two. You can see from figure 3.20 that each side of the heart consists of an upper and a lower chamber. The upper chambers are called **atria** and they both receive blood. The left atrium receives blood carried from the lungs by the **pulmonary vein**. The right atrium receives blood returning from the rest of the body in blood vessels called the **venae cavae**. The job of the atria is to collect blood returning to the heart and then pump it through to the lower chambers. The lower chambers are called the **ventricles**. They have thicker more muscular walls than the atria because they have to pump blood out of the heart. The right ventricle pumps blood through the lungs; the left ventricle pumps blood to the other body organs. The left ventricle has a thicker wall than the right ventricle. The right ventricle only has to pump the blood through one set of organs – the lungs. The thicker wall of the left ventricle allows it to produce the greater force needed to pump blood through all the other body organs.

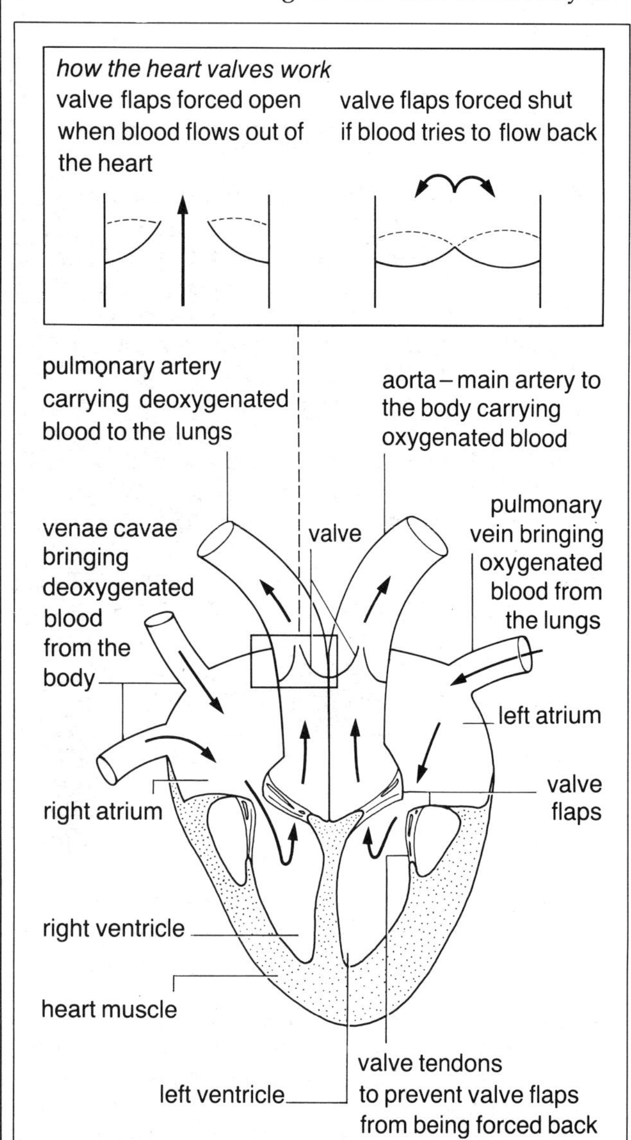

figure 3.20 The structure of the heart

What is a heart attack?

The heart is a very special muscle. We rely on this muscle to contract without tiring about once a second, day and night. This is over 2000 million contractions in a lifetime. Heart tissue, like any other muscle, needs a good supply of oxygen. It gets this supply from tiny coronary arteries. These arteries branch off from the aorta and spread out over the surface of the heart (see figure 3.21).

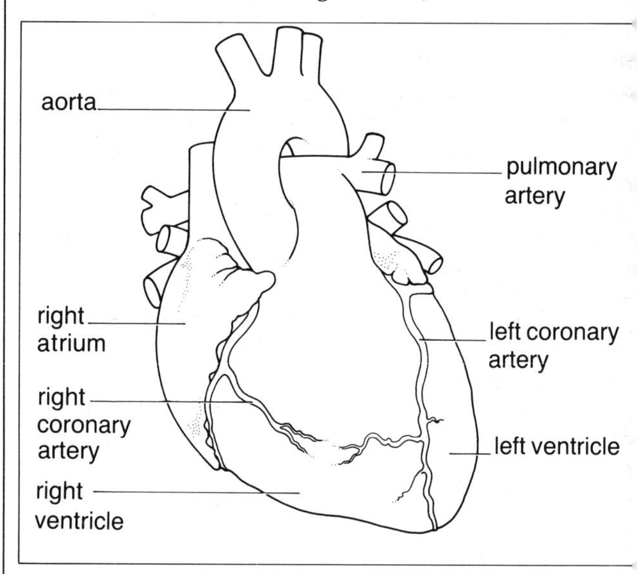

figure 3.21 The coronary arteries

If any of the arteries become blocked the heart muscle will be starved of oxygen. If the blockage is severe the heart will stop beating. This is one cause of a heart attack. Every effort must be made to restart the heart. Without oxygen the brain cells will die within a few minutes.

What is angina?

During a person's lifetime the walls of the coronary arteries can become 'furred up' with a fatty deposit called an **atheroma**. If the atheroma gets too thick the blood flow through the coronary arteries is cut down. This causes a heavy cramp-like pain in the chest whenever the person is exerted. This kind of pain is known as **angina**. The pain will go away after a period of relaxation. Angina by itself is not fatal. However, someone suffering from angina is more likely to have a heart attack.

What causes a heart attack?

If an atheroma has already narrowed one of the coronary arteries, a blood clot is more likely to occur. Blood cannot get to part of the heart and that part may die. The heart may stop beating altogether, or the ventricle may 'tremble' rather than contract. This is a **coronary thrombosis** – a form of heart attack. There is a large number of factors that can affect your chances of suffering from a heart attack. These risk factors are illustrated in figure 3.22.

The nicotine in cigarette smoke increases blood pressure and narrows the coronary arteries. Non-smokers halve the risk of dying from heart disease compared with smokers.
An obese person is carrying extra weight and this places a greater strain on the heart.

Steady, regular exercise strengthens the heart muscle and lowers the risk of heart disease.
If a person eats large quantities of animal fat and dairy products (saturated fat), the risk of heart disease is increased.

Stress speeds up the heart rate and can increase the risk of a heart attack.

figure 3.22 How can we reduce the risk of heart disease?

INVESTIGATION 3.9

IS YOUR HEART RATE AFFECTED BY ACTIVITY?

a Find your own pulse at your wrist using the tips of your middle three fingers. You will feel it by gently pressing an artery against one of your wrist bones (see figure 3.23a). You can use this method to find out how fast your heart is beating.

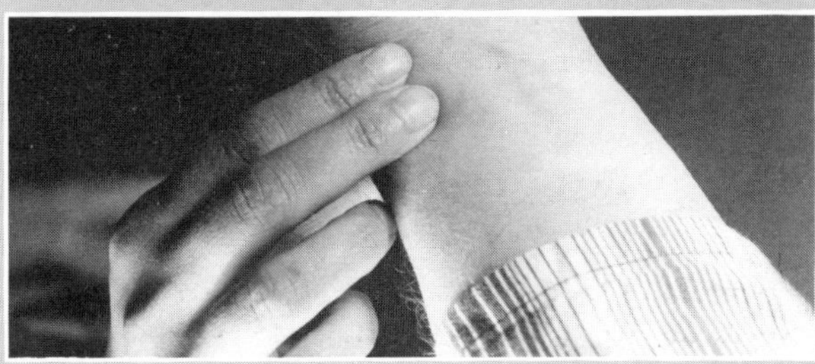

figure 3.23 a Correct hand position for taking a pulse

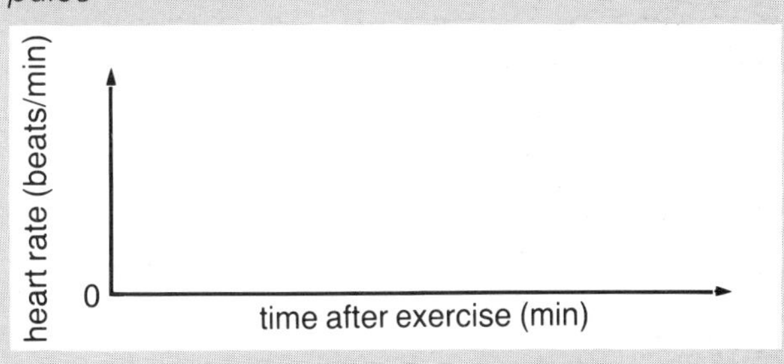

b Copy these axes and plot your results

b Lie down and allow several minutes for your body to relax. Then count your pulse for thirty seconds. Calculate your heart rate in beats per minute. Take three separate counts like this and find the average.

c Repeat step **b** while i sitting ii standing. How does your heart beat change when you are in these different positions?

d Jog for three minutes. Sit down and immediately count your pulse for 30 seconds. Pause for 30 seconds and count your pulse for another 30 seconds.

e Repeat this method of recording until your heart rate returns to the level you previously calculated as normal for sitting down.

f Plot a line graph of your heart rate over the time it takes to return to normal (see figure 3.23b).

g Plot the results of several of your classmates on the same graph.

Questions

1. How long does it take for your heart rate to return to normal?
2. How do you differ from your classmates in this way?
3. What other difference is there between the way your and your classmates' heart rates are affected by exercise?
4. Can you explain these differences?
5. How could you improve this experiment to ensure that everyone carried out the same amount of work when they exercise?
6. Apart from physical activity, what circumstances might affect your heart rate?

Application 3.5

TREATING HEART DISEASE

Problem Some heart problems are due to an irregular heart beat.

Solution A surgeon can attach a special device called a **pacemaker** directly on to the surface of the heart muscle. The pacemaker generates regular and very low voltage electric shocks. These stimulate the heart to contract at regular intervals.

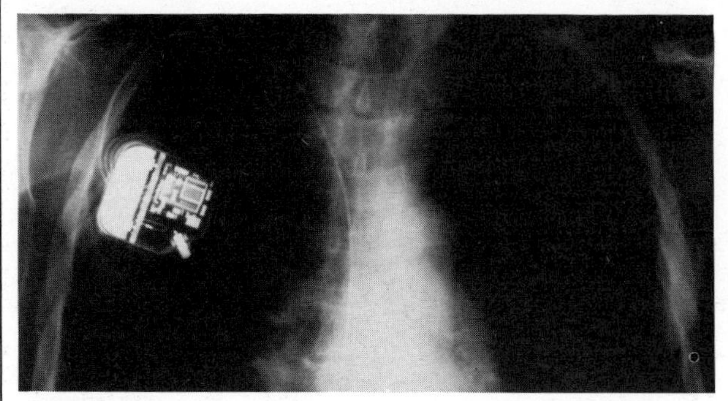

figure 3.24 X-ray of a heart pacemaker implanted in the chest

Problem A heart attack victim has a coronary artery that is in danger of becoming completely blocked by a clot or by atheroma.

Solution Heart surgeons can remove a blood vessel from the patient's leg. They then insert it alongside the blocked artery so that blood passes freely along the new blood vessel. This is known as a **coronary artery by-pass operation**.

TREATING STROKES

Problem A blood vessel in the brain bursts open. This is one cause of **stroke** which is particularly common amongst old people. Some stroke victims find it difficult to move and to express themselves clearly through speaking, because the parts of the brain controlling speech and movement have been damaged.

Solution Stroke victims can be helped in many ways. Speech therapists help them to regain their ability to speak clearly. Special exercises can help to restore control over the movement of their muscles.

figure 3.25 A speech therapist helping a stroke victim

INVESTIGATION 3.10

SMOKING AND YOUR HEART

Susan, Rumana and Evelyn were all locked in a room. One of them started chain smoking; the second fell asleep, but soon woke up again; the third began to do a strenuous workout but soon got bored, stopped exercising and started to read a book. All three girls' heart rates were measured each minute and the results are shown in table 3.6.

time (minutes)	heart rate (beats per minute) Susan	Rumana	Evelyn
0	75	75	75
1	85	75	70
2	95	77	65
3	95	79	60
4	95	81	60
5	95	83	60
6	90	85	60
7	85	87	60
8	80	89	65
9	78	91	70
10	78	93	75
11	78	95	75
12	78	97	75

table 3.6 The heart rates of Susan, Rumana and Evelyn

1. Plot these results on a graph of heart rate (beats/min) against time (mins).
2. Which girl fell asleep? When did she wake up?
3. Which girl did the workout? When did she stop doing it?
4. Which girl was the chain smoker? Explain the reason for your answer.

BLOOD VESSELS

Arteries

Blood is pumped by the ventricles of the heart into **arteries**. Artery walls contain elastic and muscle fibres (see figure 3.27). These fibres allow the wall to stretch when blood passes through. The elastic wall then contracts inwards. This helps the blood to flow more smoothly. You can feel this regular surge of blood as a pulse at a **pressure point** (see figure 3.26). This is where an artery passes just below the skin and can be pressed against a bone. You used one pressure point in investigation 3.9.

If an artery is cut, blood at high pressure will spurt out. Unless first aid is given immediately the victim will lose a large quantity of blood.

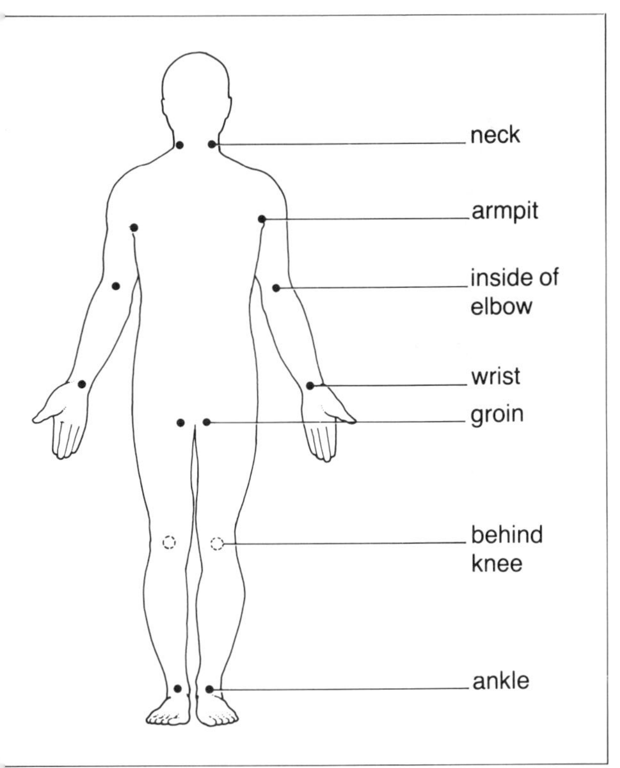

gure 3.26 *The body's pressure points*

Application 3.6

STOPPING THE FLOW OF BLOOD FROM A CUT ARTERY

A cut artery will lose blood at the rate of 750 cm^3 per minute. If someone has a deep cut try to stop the flow of blood by pressing directly on the wound with the palm of your hand. Use a firm consistent pressure until the bleeding stops. If a clean absorbent material is available use this to stop the blood flow. Once the blood has clotted apply a dressing held on firmly by a bandage. If the wound is in the arm or leg, keep the limb raised.

Capillaries

Capillaries are tiny blood vessels forming a network linking arteries with veins (see figure 3.27). They are so narrow that red blood cells have to squeeze through them in single file. The capillary walls are

Application 3.7

VARICOSE VEINS AND FAINTING GUARDSMEN

Blood flowing up the veins in the legs has to defy gravity. The valves are put under considerable pressure to prevent blood flowing downwards.

In some people the pressure on the valves causes them to stretch and they show up as bumps under the skin. This is the condition known as **varicose veins**.

Guardsmen standing to attention for long periods are also putting strain on the valves in their veins. They cannot make the movements that normally squeeze the veins and help blood flow back to the heart. Blood tends to pool in their legs. The brain has a reduced blood supply and they are more likely to faint.

figure 3.28 *Why are these guardsmen in danger of fainting?*

constructed from a single layer of very thin cells. This allows blood plasma to leak out into the spaces between the body cells. Substances dissolved in the plasma such as glucose and amino acids can be delivered directly to the cells.

Veins

Blood returns to the heart in **veins**. Since it has passed through the small, non-elastic capillaries (see figure 3.27), blood now flows steadily without a pulse. Its pressure is much lower. So how does blood get back to the heart? Veins lie between the muscles that move our arms and legs. As we move around these muscles contract and squeeze the veins. Veins contain valves and so blood can only travel in one direction – back towards the heart.

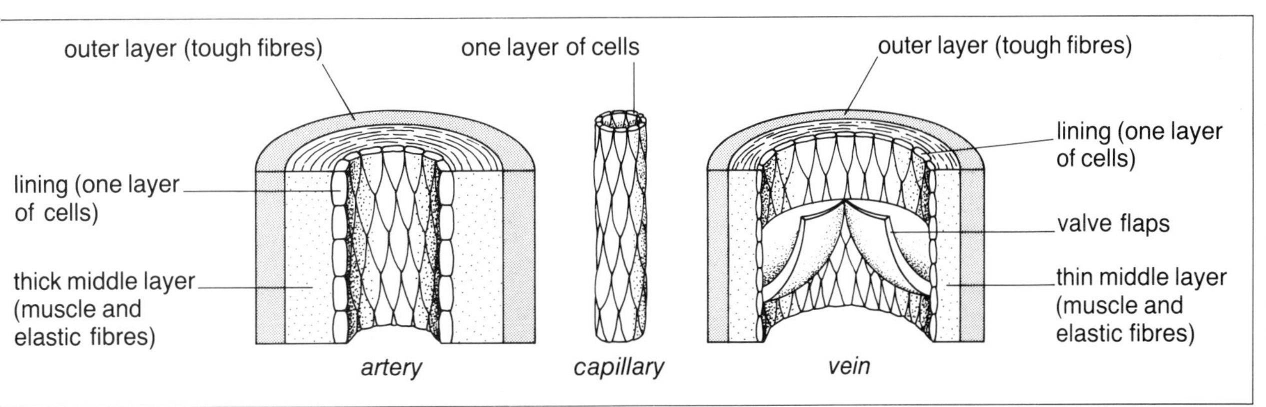

gure 3.27 *Sections through an artery, a capillary and a vein*

Tissue fluid and lymph

The plasma that leaks out from capillaries is called **tissue fluid**. This fluid bathes cells in a solution containing oxygen and nutrients. At the same time waste substances such as carbon dioxide diffuse out of the cells into the tissue fluid. The tissue fluid must be removed rapidly because the waste substances are poisonous. Some of it returns to the capillaries. The rest passes into blind-ended tubes called **lymph vessels** where it forms a fluid called **lymph**. The lymph vessels form a drainage system that remove tissue fluid from all over the body. Lymph vessels also provide a means of collecting fats absorbed in the small intestine (see page 51).

Eventually the lymph vessels join up and empty into veins in the neck. At certain points along the lymph vessels are small swellings called **lymph glands** (see figure 3.29). Lymph glands are located at ideal positions to stop the spread of disease. They are concentrated in the groin, armpits and in the tonsils and adenoids. Lymph glands contain many white blood cells that destroy germs and neutralize toxins. During an infection they may swell up. Inflamed tonsils and adenoids are sometimes removed by surgery if they become too painful.

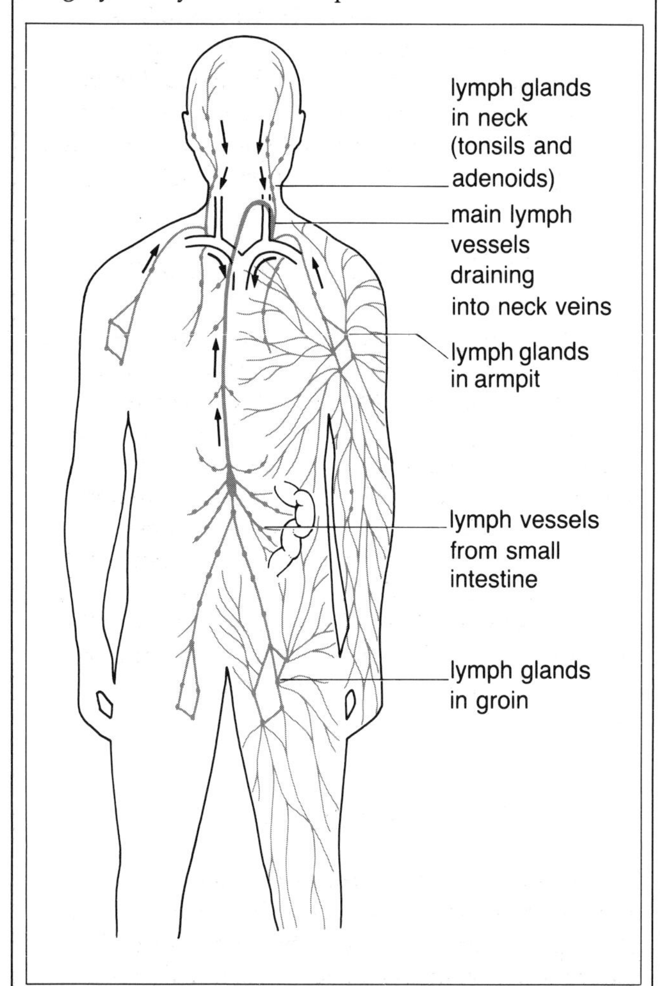

figure 3.29 *The lymphatic system*

BLOOD

Blood plasma

Figure 3.30 shows what happens when blood is taken from the body and left in a test tube for several hours. The blood separates into two distinct layers. The dense bottom layer contains a mixture of blood cells. The less dense upper layer is a pale yellow liquid containing dissolved chemicals and proteins. This liquid is the plasma and it makes up 55% of the total blood volume.

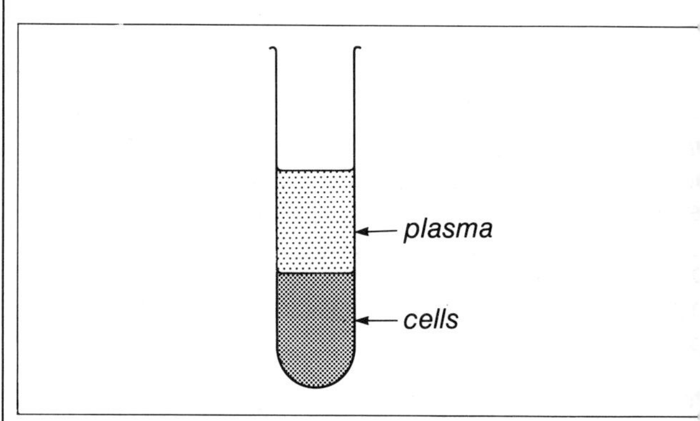

figure 3.30 *Blood separates into two layers*

Red blood cells

A single drop of blood contains 5000 million **red blood cells**. Their job is to pick up oxygen in the lungs and deliver it to the body's cells. They contain a red pigment called **haemoglobin**. Haemoglobin has a powerful attraction for oxygen. In the lungs oxygen combines with haemoglobin to form **oxyhaemoglobin**. When this oxygenated blood passes through our capillaries the oxygen is released. Our cells take in the oxygen and use it up in respiration (see page 142).

Figure 3.31 shows the unusual shape of a red blood cell – it is described as a biconcave disc. It is thinner in the centre and it does not have a nucleus. This makes the cells more flexible so they can squeeze through very narrow capillaries. This shape also increases the surface area so that the red blood cells can absorb oxygen more rapidly.

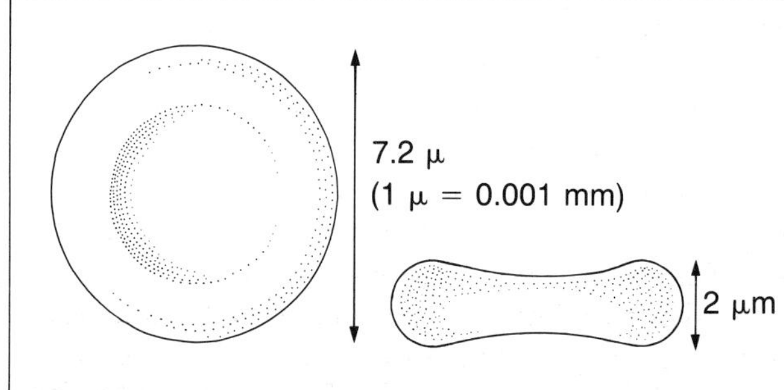

figure 3.31 *A red blood cell*

White blood cells

ere are usually about 700 red blood cells for every hite blood cell. However, when our bodies are ghting infection the number of white blood cells creases. This is because they are important in tacking the bacteria and viruses (**pathogens**) that use infectious diseases. There are two types of hite blood cells – lymphocytes and polymorphs.

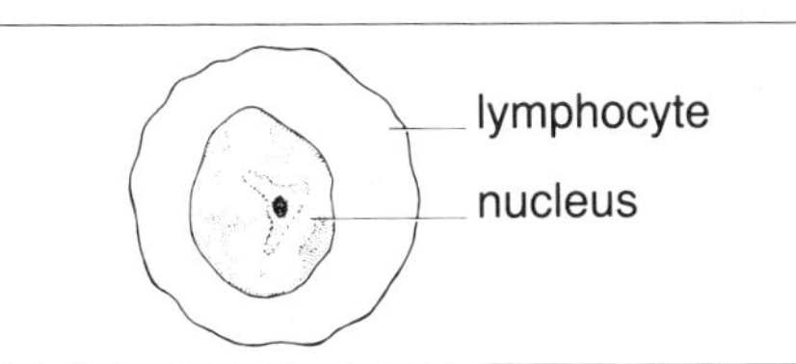

ure 3.32 *Lymphocytes are made in lymph glands. ey produce chemicals (antitoxins) that neutralize e poisons (toxins) produced by pathogens*

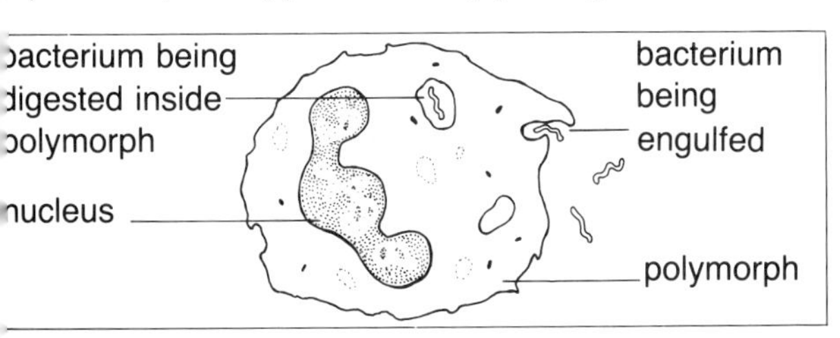

ure 3.33 *Polymorphs are made in the bone arrow. They can change shape and engulf bacteria*

Measuring blood pressure

igh blood pressure (**hypertension**) is very common. may not produce any symptoms but usually quires treatment before it threatens a person's ealth. Blood pressure is measured with a **hygmomanometer** (see figure 3.34). This consists an inflatable cuff which is fastened around the m. When inflated the cuff presses against an artery. he pressure of air in the cuff is measured in illimetres of mercury (mm Hg). A guage connected the cuff shows this pressure.

wo measurements are taken, since the pressure aries during the course of each heart beat. One ading is taken when the ventricles contract. This is hen the blood pressure is at its highest. The person king the blood pressure inflates the cuff until the ulse can no longer be heard. The cuff is then slowly eflated until the pulse just returns. This pressure presents the highest blood pressure. The cuff is en deflated further, until the pulse can again no nger be heard. The pressure in the cuff then rresponds to the lowest blood pressure, when the entricles relax. You will see your blood pressure corded as two figures, for example 120/80 mm Hg. Normal' blood pressure is 100 mm Hg plus your age r the first figure, and between 70 and 90 mm Hg for e second figure.

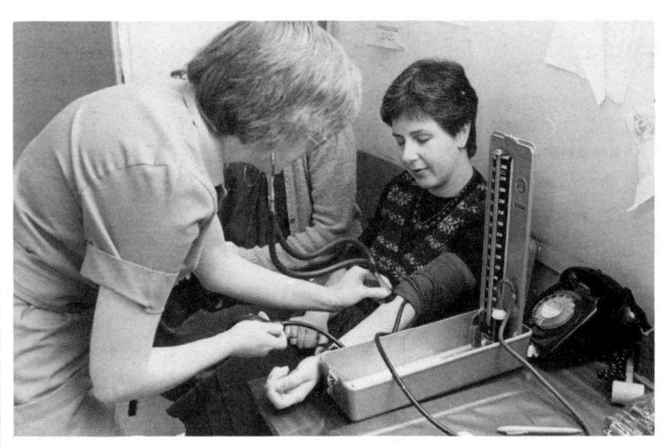

figure 3.34 A sphygmomanometer is used to measure blood pressure

Losing too much blood

People can lose blood for a number of reasons. A blood vessel inside the body may burst open and leak blood into the surrounding tissues. This is an **internal haemorrhage**. Bleeding from a deep wound may be difficult to stop because blood spurts out so quickly. Hospitals keep large supplies of blood in refrigerated blood banks for use in such emergencies. Blood is passed through a tube into a vein in the patient's arm. This is known as a **blood transfusion**. The supplies in hospital blood banks are kept up by healthy people who are willing to be blood **donors**. Without these volunteers many patients would die from a serious loss of blood. Someone who receives a blood transfusion is called a **recipient**.

Problems with blood transfusion

At the beginning of this century doctors first began to realize that it was not always safe to mix blood from two different people. When certain combinations were mixed together they found that the red blood cells struck together in clumps. This is called **agglutination**. It is now known that agglutination occurs when there is a reaction between substances present in the two different bloods. **Antibodies** in the blood plasma can act like glue. They attach themselves to protein molecules called **antigens** that cover the surface of red blood cells (see figure 3.35, overleaf). This causes the red blood cells to stick together.

Blood groups

Your blood group is determined by the type of antigens on your red blood cells. There are two types of antigen – A and B. These antigens are found in different combinations. People with blood group A have A antigens only. People with blood group B

blood group A
red blood cells have antigen A attached to their surfaces
mix together
blood group B
plasma contains anti-A antibodies
anti-A antibodies stick to the antigen A on the surfaces of the red blood cells causing them to clump together

figure 3.35 *What happens when two different blood groups are mixed together*

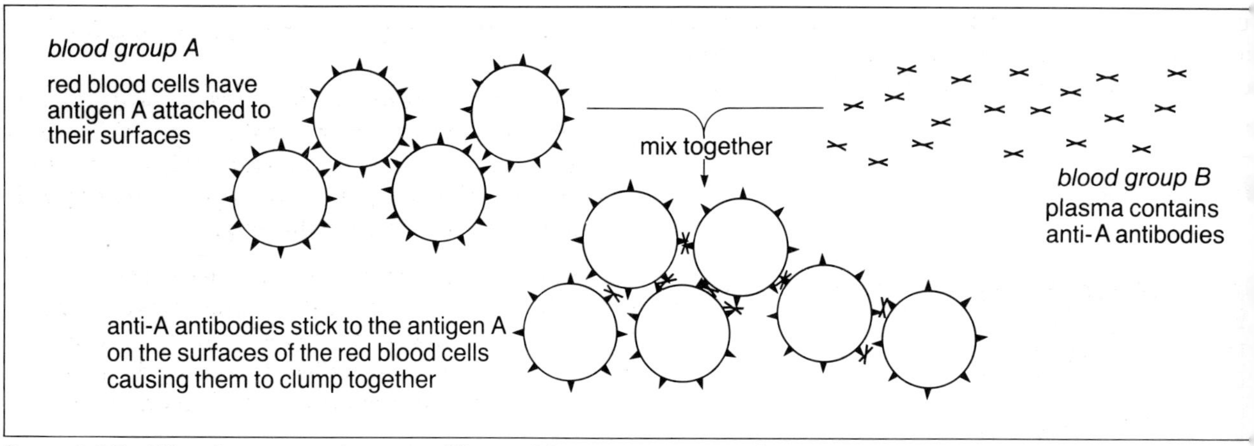

have only B antigens. Blood group AB has both A and B antigens, while blood group O has neither A nor B antigens.

Blood plasma does not contain antibodies which attack its own red blood cell antigens. It *does* contain antibodies which will attack red blood cells belonging to someone with a different blood group. For example, blood group B contains anti-A antibodies which cause the red blood cells of blood group A to agglutinate (see figure 3.35).

Antigens on tissues

Red blood cells are not the only cells that have antigens on their outer surface. Another group of antigens is found on the surface of many other body tissues. These are known as **histocompatability** or **tissue antigens**. A person's tissue antigens are as individual as their signature. When a tissue or organ is transplanted from one person to another the tissue antigens can trigger a reaction in the recipient (see application 3.8). The tissue antigens identify the organ as being different from the recipient's own tissue. The transplanted organ is attacked by the recipient's antibodies and is eventually destroyed. This reaction is known as **rejection**. Rejection is less likely to occur if the donor is a close relative of the recipient. This is because the tissue antigens are controlled by genes inherited from our parents. Rejection will not occur if tissue is taken from another part of the patient's own body.

How do blood clots form?

When we cut ourselves we normally stop bleeding after a while. This is because the blood forms a solid clot when a blood vessel is cut open. The damaged cells of the cut vessel release substances that convert the soluble plasma protein **fibrinogen** into long protein fibres called **fibrin**. Fibrin forms a microscopic net across the cut that traps red blood cells and platelets. This solid mass seals off the cut preventing the loss of blood.

A clot usually forms only when a blood vessel is cut. If a clot forms inside a vessel which has not been cut it is called **thrombosis**. This can be very dangerous especially if it occurs in the brain (perhaps causing a stroke) or in the blood vessels that supply the heart

Application 3.8

TRANSPLANTS

A transplant occurs when a diseased organ is removed from the body and replaced by a healthy organ from another person. In a few cases the diseased organ has been replaced by an artificial organ manufactured by medical engineers. An artificial heart has been constructed and a patient survived for months after the transplant. Artificial organs are still in the early stages of development. The range of organs that can be transplanted is wide – heart, lungs, liver, bone marrow, cornea of the eye, pancreas and kidneys. Some organs and tissues can be removed from the body without damaging the person's health. Certain types of leukaemia can be cured by transplanting bone marrow. The donor builds up new tissue to replace the bone marrow that has been removed. Healthy individuals can donate one of their kidneys without any serious effects. If this is necessary close relatives of the patient are usually asked to be donors. This reduces the problem of organ rejection.

It is more common for organs to be transplanted from the body of someone who has just died in a serious accident such as a road crash. Doctors define death as happening when the brain stops working. This occurs if it has been starved of oxygen for over four minutes. The other body organs can be kept 'alive' by placing the dead person on a heart-lung machine. This machine keeps oxygenated blood flowing around the body until a recipient can be brought to the hospital. The length of time that the organ is outside the body must be kept as short as possible.

After the wearing of seat-belts in cars was made compulsory in 1983, the number of fatal accidents was reduced. One effect of this was a severe shortage of donors, particularly for kidney transplants.

uscle (perhaps causing a coronary thrombosis). The ;elihood of a thrombosis is sometimes increased for omen who take the contraceptive pill (see pages 23-
.

Time for thought

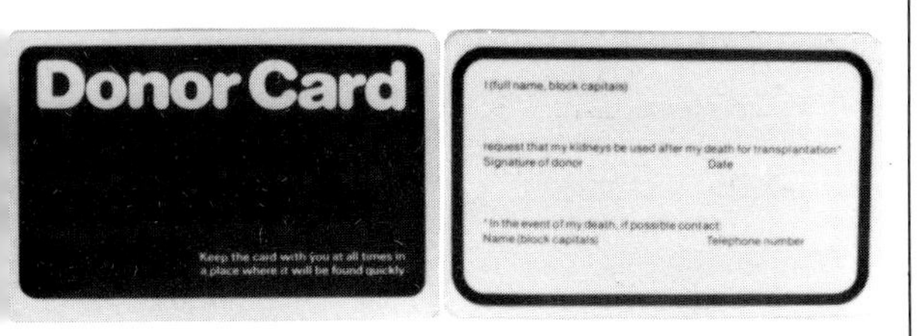

figure 3.36 A donor card tells doctors you would be willing to donate your organs if you died suddenly

At present anyone who wishes to donate their organs if they die suddenly can indicate their wishes by carrying a donor card. It 's thought that about four million people in the UK have signed such a card. However, a survey showed that fewer than one per cent of donors were card holders.

- *What are your views on the donor card?*
- *Would you carry a donor card?*
- *It has been suggested that a census form should ask everyone in the population whether they would agree to be donors and their reply stored on a national computer. If they died suddenly, their wishes could be quickly checked. What are your views on this idea?*
- *In some countries it is assumed that a person is willing to donate their organs unless their relatives object. In the UK, if a person's views are not indicated, permission must be granted by their relatives before organs are removed from their body. What are your views on these two systems?*

HEALTHY LUNGS

Why are the lungs pink, moist and spongy?

he job of the lungs is to absorb oxygen from the air e breathe in. In the lungs oxygen is transferred from ne air to the blood. At the same time carbon dioxide removed from the blood. It passes out of the body hen we breathe out.

you study the lungs of a freshly killed animal you ill notice that they are pink, moist and spongy. All nese features help the lungs to do their job. The pink olour shows that the lungs are well supplied with lood capillaries. This allows the blood to make close ontact with air in the lungs. The spongy feel tells us nat the lungs are filled with millions of tiny air paces (**alveoli**). They increase the surface area and llow a large volume of oxygen to be absorbed into ne blood. The lungs are moist because oxygen is only bsorbed if it is dissolved in water.

A closer look at our breathing apparatus

As we breathe in through our noses the nostrils warm and moisten the air. This helps the lungs to absorb oxygen from it. Hairs in the nostrils help to clean the air. Some of the dust particles and germs are trapped before the air travels on to the lungs.

The air then passes into a tube at the back of the mouth. This tube is called the **trachea** or **windpipe**. When we swallow food, the top of the trachea is covered by a small flap called the **epiglottis**. This stops us from being choked by food entering the windpipe.

At the top of the trachea is a hollow box-shaped structure – the **larynx**. Bands of ligament are stretched across the larynx. These are the vocal cords. When we breathe out through them the cords vibrate and produce sounds. The trachea is kept open by rings of cartilage shaped like the letter 'C'. This shape gives the trachea flexibility while preventing it from collapsing. See figure 3.37.

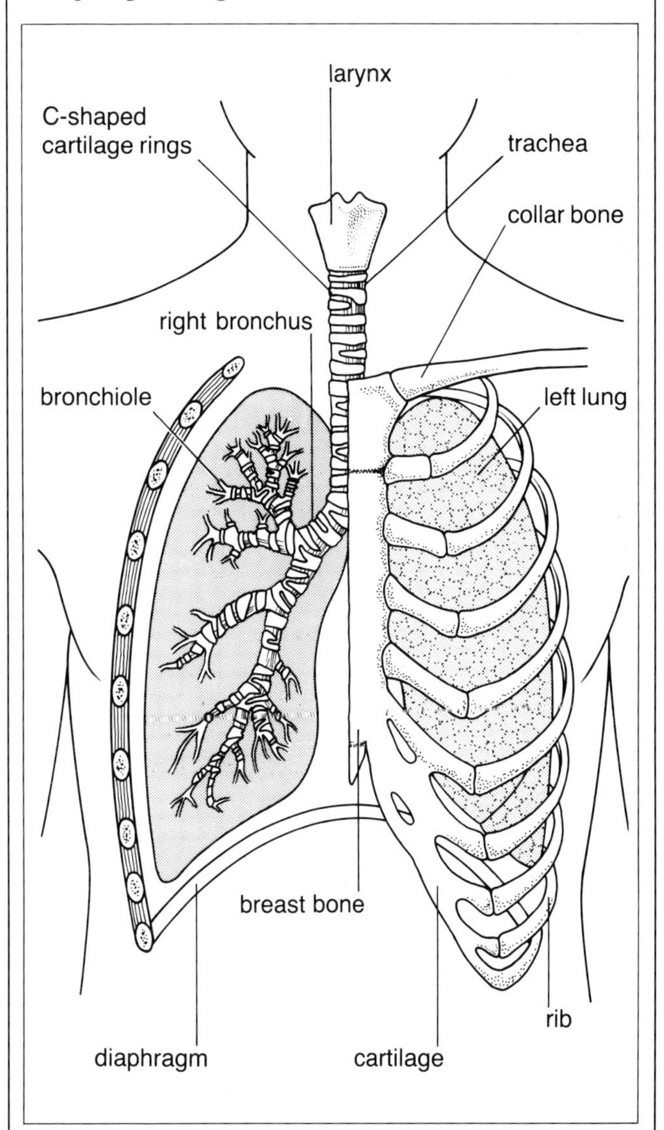

figure 3.37 The breathing organs

INVESTIGATION 3.11

DOES EXERCISE AFFECT YOUR BREATHING RATE?

a Place your hands on the lower half of your rib cage. Feel the in and out breathing movements. Try to breathe normally.
b While sitting, count how many breaths you take in three minutes. What is your average breathing rate per minute when resting?
c Walk at a quick pace for three minutes. Sit down and immediately count your breathing rate for one minute.
d Jog steadily for three minutes. Then count your breathing rate each minute for five minutes after you stop exercising.

Questions

1 How is your breathing rate affected by physical activity?
2 How do you explain the link between a person's breathing rate and physical activity?
3 Compare your results with other class members. What variation is there in breathing rates i when at rest ii after jogging? Suggest explanations for these differences.
4 How long does it take for your breathing rate to return to normal after jogging? Are there differences between class members? Can you explain these differences?

Keeping the lungs clean

The air entering the lungs must be kept clean. If dust and germs get into the alveoli there is a greater risk of lung disease. Some of the cells lining the air tubes produce mucus. Other cells have tiny hairs (cilia) that beat in rhythmic waves. This produces a current of mucus that flows upwards to the back of the throat (see figure 3.38). Dust and germs stick to mucus and are carried away so they do not block up the alveoli.

Chemicals in cigarette smoke interfere with the lungs' self-cleaning mechanism. These chemicals have two serious effects. They slow down the beating of the cilia and increase the amount of mucus produced. The extra mucus collects in the bronchioles. This produces the so-called 'smoker's cough'. Germs can easily get into the alveoli because many cilia are destroyed by the smoke and the smoker is more likel[y] to suffer from lung infections such as bronchitis.

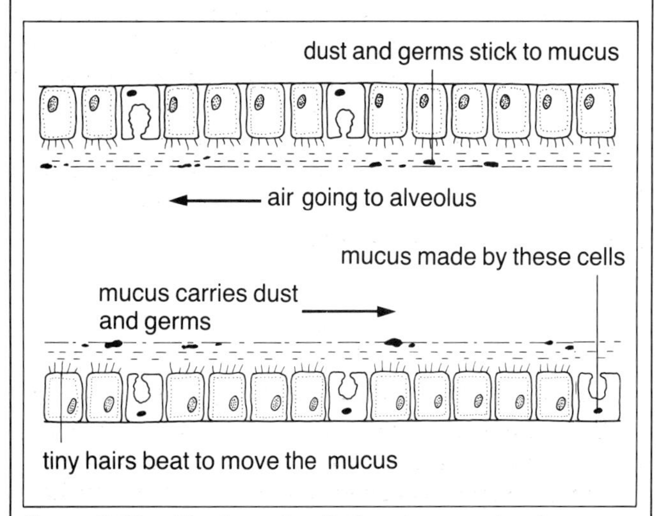

figure 3.38 The lungs' self-cleaning mechanism

INVESTIGATION 3.12

SMOKE + FOG = SMOG

The winter of 1952 was very cold in London and there were dense fogs. People burned a lot of coal on their fires to keep warm. This produced large quantities of smoke which combined with the water droplets of the fog to form **smog**.

Figure 3.39 shows the density of smoke particles measured in milligrams per cubic metre (mg/m^3) in the London atmosphere during the first half of December 1952. The graph also shows the number of deaths that occurred each day during the same period.

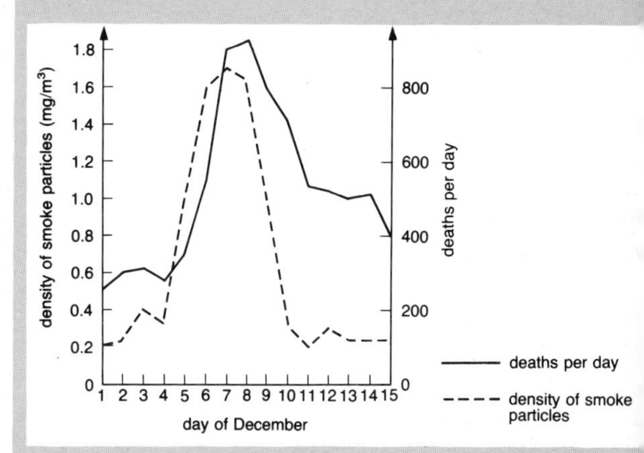

figure 3.39 Graph showing deaths per day and density of smoke particles in London, December 1952

Questions

1 On what day did the density of smoke particles show the greatest increase? Suggest what happened to cause this increase.
2 Between which dates did the number of deaths increase most rapidly?
3 Does the data shown on the graph suggest that smog caused death? Explain your answer.
4 Many of the people who died during the smog were already suffering from lung diseases such as bronchitis. Why were these people particularly at risk?
5 i On which date did the density of smoke particles reach a maximum?
ii On which date did the number of deaths reach a maximum?
6 Suggest why your answers to **5 i** and **ii** are different.
7 After 1952 legislation was introduced to ban the burning of coal fires in many areas. What effect do you think this had on the number of people dying from bronchitis?

f this does not clear up, the structure of the lungs can be permanently damaged. The bronchioles get narrower and this makes breathing more difficult. The alveoli break up to form large air spaces (see figure 3.40). This reduces the available surface area for absorbing oxygen. The person must breathe faster to get the same amount of oxygen into the blood. This illness, known as **emphysema**, will eventually be fatal if the patient does not stop smoking.

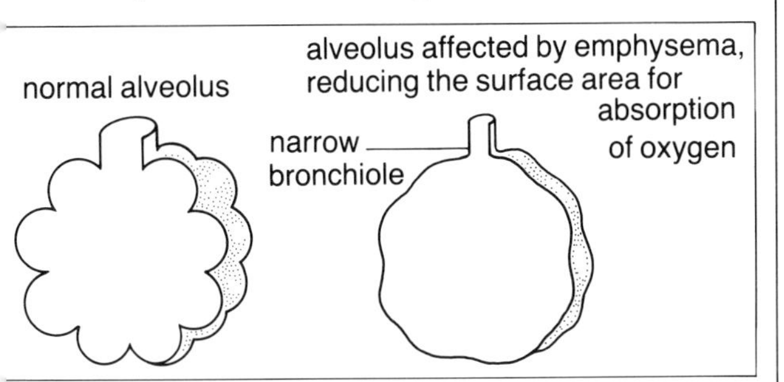

figure 3.40 *How the alveoli and bronchioles are affected by emphysema*

INVESTIGATION 3.13

SMOKING AND LUNG CANCER

Figure 3.41 shows the increase in the risk of dying from lung cancer among smokers, compared with non-smokers.

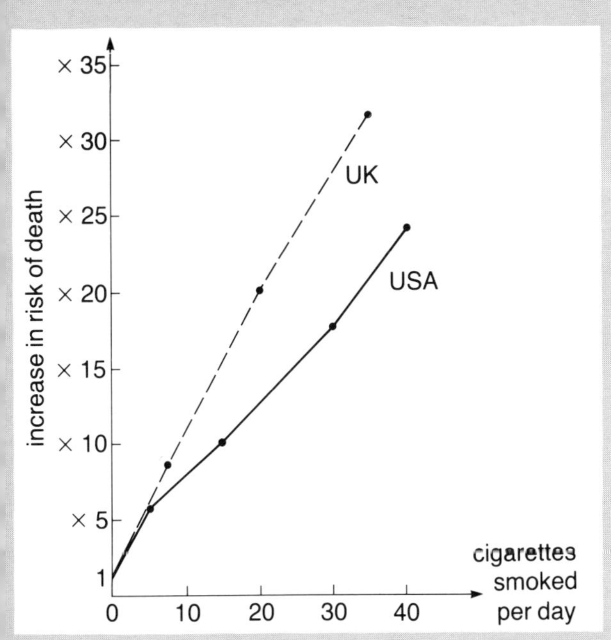

figure 3.41 *Graph showing the increased risk of dying from lung cancer for smokers*

Questions

1. What conclusions can you draw about the link between smoking cigarettes and the risk of dying from lung cancer?
2. What is the increased risk of dying from lung cancer if you smoke five cigarettes a day?
3. Suggest possible reasons for the differences in the results in the USA and the UK.

WHY IS EXERCISE GOOD FOR US?

figure 3.42 *Good for you?*

The runners in figure 3.42 have just finished a marathon. They look completely exhausted. Has it done them any good? On its own the run will not have made them any healthier. However, if it is part of a regular pattern of exercise that will continue throughout their lives, it will improve the health of many of their organs.

Lungs

The lungs of people who exercise regularly can absorb oxygen more efficiently. They will find that their breathing rate returns to normal quickly after exercise. Those who do not exercise regularly will take a longer time to recover.

Heart

Exercise strengthens the heart. When athletes train their hearts get bigger so more blood is pumped with each contraction. When at rest an athlete's heart beats more slowly than that of an average person. A very fit athlete can have a resting heart rate as slow as 40 beats per minute. This compares with 72 beats per minute for an average adult. Even moderate exercise has beneficial effects. It strengthens the heart and heart disease is less likely to occur.

Bones

Regular exercise increases the strength of bones and allows them to grow properly. The first astronauts found that the mass of their bones had fallen during the time they had been weightless in space. This had partly been caused by a lack of exercise.

Joints

Exercise keeps the joints supple. Our joints gradually stiffen as we get older. Exercise slows down this process and allows us to move with greater ease. Yoga exercises can produce dramatic improvements in the suppleness of the joints.

figure 3.43 Yoga improves the suppleness of joints

Muscle tone

Muscle tone is the name given to the slight muscular tension that holds our body in position. Poor muscle tone allows the body to sag into a bad posture. Exercise improves our muscle tone and makes it easier for us to maintain a good posture.

figure 3.44 Exercise improves your posture

Muscle strength

If a particular muscle is not used it will waste away. This is often seen when a broken leg is put in plaster. When the cast is removed after several weeks the muscles are smaller and weaker than those on the unbroken limb. Exercising our body muscles strengthens them and helps us to cope with the normal strain of lifting and carrying heavy objects.

Anyone who takes regular exercise is aware of many other benefits. Exercise can be relaxing. For some people it helps relieve the stress and tension they feel. For others competitive sport is a way of releasing aggressive feelings. Active recreations, like all leisure activities, give us a break from everyday life.

ILLNESS AND HEALTH

Infectious disease

Our body systems are vulnerable to attack from a variety of organisms. These organisms are called **germs** or **pathogens**. After they enter the body cells, pathogens reproduce rapidly, releasing large quantities of poisonous waste products called **toxins**. Toxins carried around the body in the blood cause the symptoms of the disease. Pathogens include species of **virus**, **bacteria**, **protozoa** and **fungi** (see figure 3.46, page 66).

Diseases that can be transmitted from one person to another are called **infectious** diseases. Pathogens can pass into the body through a number of different routes (see figure 3.47, page 66). When we know how a disease is transmitted we can take action to stop it spreading. Let's look at how this is done for a number of different diseases.

Influenza – a virus spread by droplets
Viruses do not have the chemical machinery for reproducing themselves. They cannot survive for long outside living cells. To reproduce they must first invade living cells. The viruses that cause 'flu and colds take over the cells that line the nose and throat. When we cough or sneeze thousands of tiny moisture droplets shoot out into the air. Even breathing out releases droplets into the air. These droplets are swarming with viruses. If other people breathe in these droplets they may suffer an attack of the virus. This method of transmission, known as **droplet infection**, is also responsible for the spread of measles and polio.

The spread of diseases carried by droplet infection can be controlled if sufferers take a few simple

INVESTIGATION 3.14

LEISURE FOR HEALTH

Look at figure 3.45 showing people enjoying different leisure activities. Describe what effect each one is likely to have on the person's physical health and mental well-being.

figure 3.45

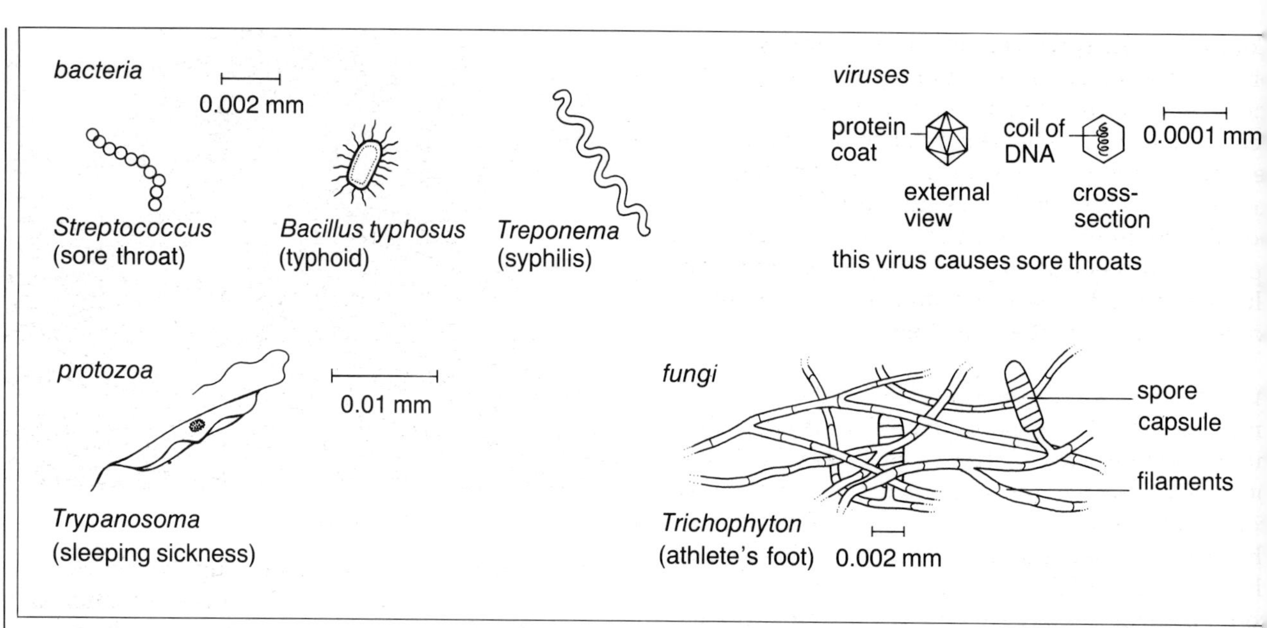

figure 3.46 *The structure of different pathogens*

precautions. They should always cough or sneeze into a tissue. If they avoid contact with other people, especially in confined spaces such as a school classroom, this will slow down the spread of the disease. If rooms are well ventilated contaminated droplets will be carried away quickly.

Cholera – bacteria spread in water

Unlike viruses, **bacteria** can grow and reproduce outside living organisms. Cholera is a sometimes fatal disease caused by bacteria. Although it is rare in Europe and North America it is still common in many parts of the world. The cholera bacteria enter the body through the mouth and infect the digestive system. Cholera causes severe diarrhoea which can lead to death through a loss of water and salt. The bacteria pass out of the body in the faeces. If the disposal of faeces is unhygienic the cholera bacteria can enter the digestive system of other people. In the nineteenth century cholera outbreaks were common in London. It was discovered that human sewage was contaminating the supply of drinking water. The spread of cholera was halted when hygienic sewage disposal was introduced. Nowadays bacteria in drinking water are killed by adding chlorine.

Malaria – a protozoan spread by blood-sucking insects

Malaria is a common disease in tropical countries. It is caused by a species of protozoa called *Plasmodium vivax*. The *Plasmodium* protozoan can infect the blood of humans and cause acute fever. It is carried from one person to another by **mosquitoes**. Mosquitoes are insects with sharp mouthparts that pierce the skin to suck up blood from capillaries just below the surface. The mosquito is an example of a **vector** – an organism that spreads pathogens from one person to another.

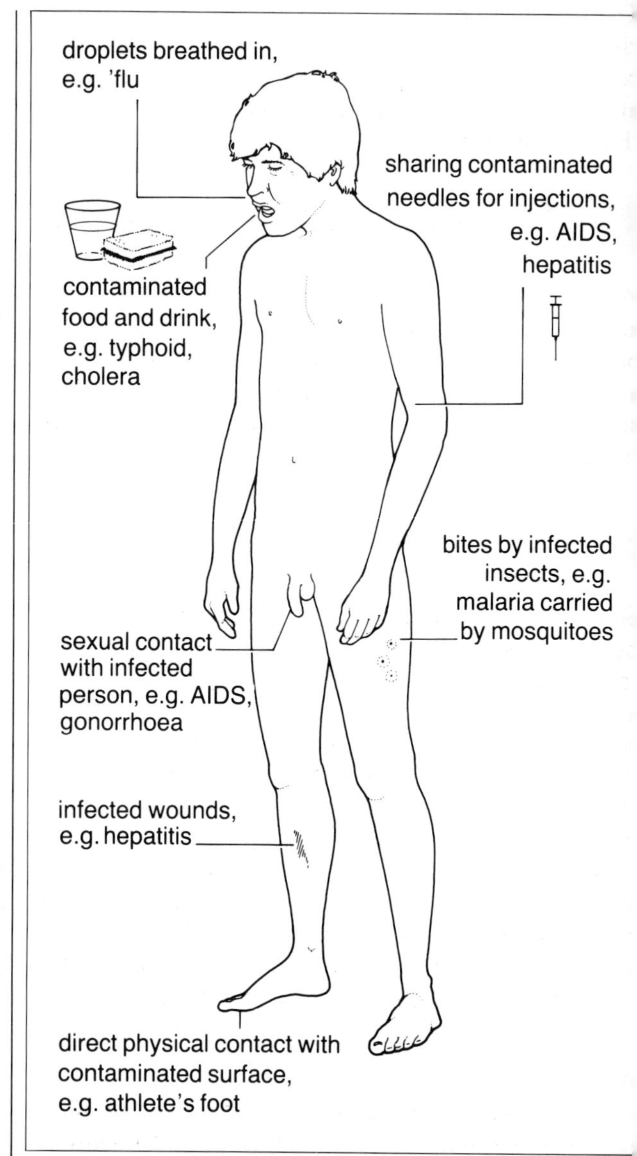

figure 3.47 *The different routes by which pathogens can enter the body*

Controlling the spread of malaria involves two problems: killing mosquitoes and preventing mosquitoes from biting humans. Fine netting over beds and insect repellants spread on the skin may reduce the number of mosquito bites. Spraying mosquitoes with insecticides such as DDT is a possible solution but this can destroy other harmless species and create other problems (see page 85). An alternative solution is to add predatory fish to the water where mosquitoes lay their eggs.

Athlete's foot – a fungus spread by contact
In this country the most common disease caused by fungi is athlete's foot. It causes peeling and irritation of the skin between the toes where the warm, moist conditions are ideal for fungal growth. The fungus that causes athlete's foot consists of tiny filaments. At the tip of some of these filaments are capsules containing spores. When these spores are released they can spread the disease to other people.

This disease is not only found in athletes. Communal changing rooms and shower facilities provide ideal conditions for spreading the spores. Sufferers from athlete's foot should avoid exposing their bare feet in these environments.

STRESS

What does stress do to you? Stress can make people anxious, afraid and unable to think clearly. On some occasions stress can be useful if it helps us to perform a difficult task well. Stress can even be pleasurable: some people enjoy the stressful feelings produced by funfair rides (see figure 3.48).

Drugs

Some people try to deal with stressful feelings by using drugs. For example, a cigarette might be used to 'calm the nerves'. A drug is any substance that affects the normal activity of the body. Drugs are sometimes used to relieve the stress caused by a difficult problem. Although the drug temporarily reduces the feelings of stress, it will not solve the person's problem. Drugs can have one of two different effects: **stimulant** or **depressant**. A stimulant speeds up the activity of the heart and the brain. Depressants have the opposite effect. They slow down body activity and make the person feel more relaxed. We shall look at two of these drugs in more detail.

Alcohol
Alcohol is a legal and socially accepted drug in the UK. In some countries that follow the Muslim faith, such as Saudi Arabia, alcohol is illegal. In certain parts of the United States the sale of alcohol is

figure 3.48 *Stress can be fun*

banned. Alcohol is a depressant drug that slows down the drinkers' reactions. Their co-ordination is clumsier, their vision becomes blurred and their judgement is impaired. They are more likely to have accidents and this is why it is dangerous to drive after drinking alcohol.

In large amounts, alcohol causes uncontrolled behaviour. Violent behaviour such as football hooliganism is often the result of heavy drinking. Nevertheless, in moderation, alcohol can have certain beneficial effects. It is commonly used to help people relax, particularly at social gatherings.

Nicotine
Nicotine is the drug found in tobacco. It speeds up the heart rate and raises the blood pressure even when the body does not *need* this, e.g. for exercise. This is why smoking tobacco increases the likelihood of heart disease. Cigarette smoking involves a number of other health risks resulting from other constituents of the smoke (see page 63). Although tobacco is a legal drug in the UK, its use is prohibited in many public places. However, non-smokers may often find themselves breathing in the smoke from nearby cigarette smokers. This 'passive smoking' has been shown to carry a certain health risk.

Time for thought

The following legislation could be introduced to discourage cigarette smoking.

- *Ban the advertising of cigarettes completely.*
- *Increase the tax on cigarettes so that the cost is doubled.*
- *Ban the sponsorship of sporting events by cigarette companies.*
- *Ban smoking in all enclosed public spaces, e.g. buses, cinemas.*

Do you think these measures would reduce the number of smokers?
What problems might be created if these measures were introduced?

Application 3.9

WHAT IS DRUG ADDICTION?

If a drug is taken regularly over a period of time, larger quantities are gradually needed to achieve the same effect. The size of dose injected by an experienced heroin user would be sufficient to kill someone who had never taken the drug. When the body adjusts to a certain level of the drug **tolerance** has developed and it is difficult to stop taking the drug. The drug taker has become **addicted**. Stopping suddenly causes **withdrawal symptoms**. An alcoholic who stops drinking alcohol may suffer from delirium tremens (DTs) which can involve frightening hallucinations. Someone regularly prescribed tranquillizers such as Valium may suffer from palpitations and trembling when they try to stop taking them. Withdrawal symptoms can be avoided if the quantity of the drug taken is reduced gradually over time.

Teenagers under stress

Thankfully, the death of a teenager is rare. If you study the reasons for death among young people (see table 3.7) you can see that accidents and suicides are the major killers. Stress is often involved in both of these causes of death.

cause of death	death rate (deaths per million population)	
	male	female
infectious diseases	6.7	7.0
cancers	77.9	49.3
endocrine disorders (including diabetes)	13.7	10.7
mental disorders	9.0	2.7
diseases of nervous system	34.8	16.1
diseases of circulatory system	35.6	24.4
diseases of respiratory system	30.7	26.3
diseases of digestive system	5.2	5.4
traffic accidents	310.1	72.7
other accidents and suicides	219.7	63.0
all other causes	28.2	28.0
total	771.6	305.4

table 3.7 *Death rates of people aged 15-24 years in England (1983) and causes of death*

Avoidable accidents

When we are feeling stressed we find it more difficult to concentrate on what we are doing. We can easily make mistakes and cause an accident. Young people are particularly vulnerable to accidents in the following areas:

INVESTIGATION 3.15

SPOT THE HAZARD

Write down the safety hazards you can see in the scenes in figure 3.49. For each hazard explain what should be done to prevent the hazard from causing an accident.

figure 3.49

- *at home* – careless or reckless behaviour can lead to many accidents at home. Make sure you know how to keep your home free from hazards
- *at work* – in many jobs young people learn how to use unfamiliar equipment. Not knowing the correct safety rules can be very dangerous
- *on the roads* – reckless driving of cars and motor bikes claims the lives of about 1500 15–24-year-olds every year. Young inexperienced drivers often make mistakes in stressful traffic conditions
- *at leisure* – leisure activities chosen by teenagers often have an element of risk. This produces feelings of excitement. Sometimes, what starts as 'messing about' can lead to fatal accidents.

STRESS AND MENTAL HEALTH

Stress can sometimes produce disturbing changes in a person's behaviour, feelings and thoughts and he or she becomes mentally ill. In a few cases this may lead to suicide. If the cause of the stress can be found, many cases of mental illness can be treated successfully.

Doctors classify mental illness into two types: **psychoses** and **neuroses**.

Psychosis

A **psychosis** is a severe mental illness where the person loses touch with reality. The most common psychosis is called **schizophrenia**. Schizophrenia covers a range of severe mental illnesses. Schizophrenics are often unaware that they are mentally ill. In many cases they retreat into a world of their own and find it difficult to have normal emotional responses to other people. They may think they can hear voices talking to them and suffer from delusions. For example, a schizophrenic may feel that he or she is being manipulated by outside forces or being plotted against.

About one in 100 of us will suffer from schizophrenia at some point in our lives. Fortunately most cases can be treated successfully so that the person can live a normal life.

Neurosis

A **neurosis** is a less severe type of mental illness. Someone suffering from a neurosis is aware that they are ill. The most common types of neuroses are **phobias**, **depression** and **anxiety states**.

Phobias

A person suffering from a **phobia** has an extreme fear of something. This fear causes them great anxiety and interferes with their life. Phobias can be treated by teaching the sufferers to cope with their anxious feelings. There are many different phobias.

Agoraphobia is a fear of open spaces. An agoraphobic is terrified of going outside his or her home. Animal phobias occur when a person's fear of a particular type of animal can severely disrupt their life. Someone with a spider phobia (**arachnophobia**) may feel unable to enter a room until it has been checked for spiders. **Claustrophobia** is a fear of confined spaces. A claustrophobic will become very anxious when it is necessary to remain inside a confined space such as a car or bus.

Depression

Everyone has felt **depressed** at some time. However, depression can sometimes be so severe that it affects all aspects of a person's life. In this state depressed people feel that they cannot cope with their problems. They may completely lose their interest in looking after themselves. They may lose weight rapidly and have difficulty in sleeping. There are two main types of depression. **Reactive depression** is set off by a specific event such as the death of a close relative or redundancy. **Endogenous depression** results from the individual's personality and is not affected by events in life.

Anxiety states

It is normal to feel **anxious** and worried when faced with stress. When a person becomes over-anxious he or she may suffer physical symptoms such as a rapid heart beat, rapid breathing, shivering and fainting. Tranquillizers are sometimes prescribed to treat anxiety states. Although these drugs reduce the symptoms they can produce dependence (see application 3.9).

STRESS AND PHYSICAL HEALTH

Stress can also affect our physical health. Illnesses such as heart disease (see pages 54-5), obesity (see page 144), anorexia nervosa (see page 144), diabetes (see page 262) and stomach ulcers (see page 51) are all linked with stress. Asthma provides an interesting illustration of how stress combines with other factors to produce illness. During an asthma attack the bronchioles get narrower. The flow of air to the lungs is restricted and the sufferer takes short, rapid breaths. An asthma attack can be brought on by dust or pollen as the sufferer is allergic to small particles entering the lungs. Emotional stress such as anger, anxiety or excitement can also trigger an asthma attack. Some asthmatics sensitive to pollen from flowers have entered a room and suffered an attack, only to discover later that the vase contained plastic flowers.

ALL YOUR OWN WORK

1 Scurvy results from a lack of vitamin ______. If people lack calcium and vitamin D in their diets they may suffer from ______. Poor vision at night can sometimes be corrected by eating more vitamin ______.

2 A pregnant woman should increase the protein in her diet so that her ______ can grow properly. She need not increase the amount of ______ or ______ in her diet.

3 Fluoride prevents tooth decay because it ______ the enamel of the tooth. Adding fluoride to ______ ______ may be a way to reduce tooth decay.

4 Someone who exercises regularly ______ their chances of suffering from heart disease, keeps their joints ______ and increases the ______ of their bones.

5 Name four different types of pathogen. Give an example of a disease caused by each type of pathogen.

6 Explain with examples how a depressant drug differs from a stimulant drug.

7 The following information was seen on a packet of breakfast cereal.

Ingredients: maize, sugar, salt, malt flavouring, niacin, iron, vitamin B6, riboflavin (B2), thiamin (B1), vitamin D3, vitamin B12

typical nutritional composition per 100 g	
energy 1485 kJ	
protein 5.2 g	
vitamins	
niacin	16.0 mg
vit B6	1.8 mg
riboflavin	1.5 mg
thiamin	1.0 mg
vit D3	2.8 mg
vit B12	1.7 mg
iron	6.7 mg

An average of 30 g of this cereal provides at least a quarter of the adult (or one-third of a child's) recommended daily intake of these vitamins.

a Of which vitamin is there most in the cereal?

b Approximately how much energy would be given by a 30 g helping of the cereal?

c If 30 g provides a quarter of the recommended daily intake of vitamins explain why eating 120 g of cereal every day is not a good idea.

d It has been suggested adding milk and sugar to the cereal improves the dietary balance of the meal. Comment on this suggestion.

[SEG specimen question]

each symbol 🯅 represents 5 deaths per year

	people who live in cities	people who live in the country
non-smokers	🯅🯅🯅🯅	🯅
light smokers	🯅🯅🯅🯅🯅🯅 🯅🯅🯅🯅🯅🯅	🯅🯅🯅
heavy smokers	🯅🯅🯅🯅🯅🯅 🯅🯅🯅🯅🯅🯅🯅	🯅🯅🯅🯅 🯅🯅🯅🯅

figure 3.50

8 Figure 3.50 shows information about how many out of every 100 000 people die of lung disease each year. Use it to answer the questions below.

a How does a person's chance of dying from lung disease depend on how much they smoke?

b Are you safe from lung disease if you do not smoke at all?

c How does living in the country instead of a city affect someone's chances of dying from lung disease?

[NEA]

9 a Look at figure 3.51. State two health hazards in the kitchen shown.

b Explain the nature of each health hazard.

[SEG specimen question]

figure 3.51

PLANET EARTH

LONG-TERM CHANGES IN THE EARTH

Structure of the Earth's crust

'he surface of the earth feels solid beneath our feet. It s, in fact, a relatively thin layer of rock, called the **rust**, floating on a liquid layer of molten rock called he **mantle**. On the ocean bed the crust is only 8 km hick. The mantle is nearly 3000 km deep.

'he Earth's crust is divided into several large masses known as **plates**. These plates are moving very slowly at about 1–2 cm per year. The boundaries between these plates and the different directions in which they are moving are shown in figure 4.1. Some of the Earth's plates are moving closer together. When this happens one plate slides beneath the other. This causes changes in the structure of rocks. It plays an important part in the chain of events known as the **rock cycle** which we shall study later. When plates collide the structure of the crust is as affected. The Himalayan mountain range is thought to be due to the rapid movement of the Indian plate northwards into the Eurasian plate. The fact that the height of Mount Everest is increasing by a few centimetres each year supports this view. The strain occurring at plate boundaries also gives rise to earthquakes and volcanoes.

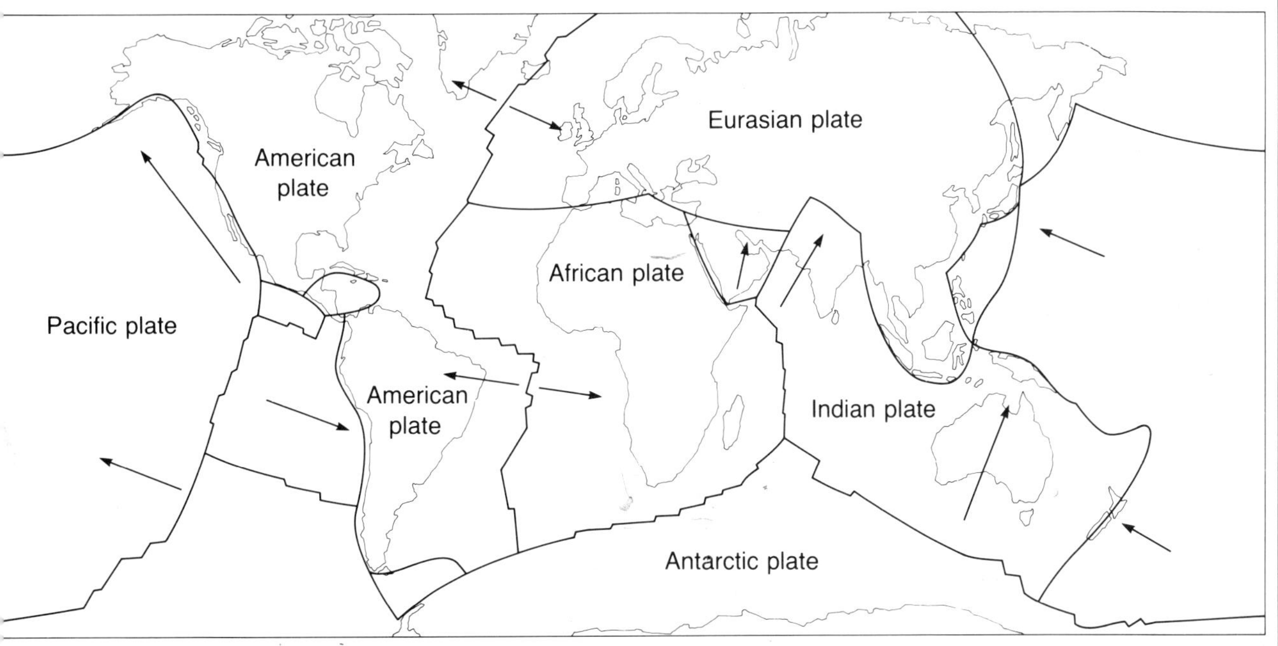

igure 4.1 *The movement of the plates of the Earth's crust*

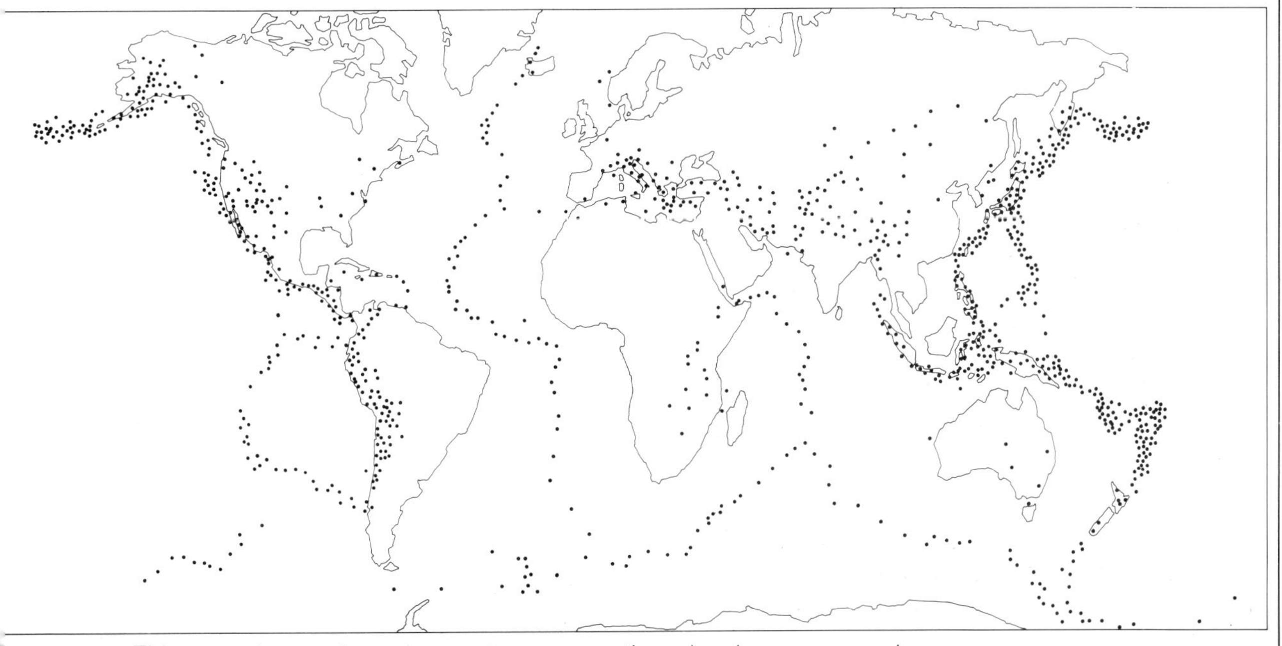

'igure 4.2 *This map shows where the most severe earthquakes have occurred*

Earthquakes

Figure 4.2 shows the location of the most severe earthquakes in recent years. Comparing it with figure 4.1 gives a clue to the causes of earthquakes. An earthquake occurs when there is a sudden movement of a part of the Earth's crust. This is more likely to happen near to the boundary between two plates. The Californian city of San Francisco is located near the boundary of two plates that are gradually sliding past each other in opposite directions. In 1906 a massive earthquake destroyed many of San Francisco's buildings and claimed thousands of lives. Nowadays San Francisco's skyscrapers are constructed with a flexible framework that can absorb the shock of a major earthquake.

Volcanoes

A volcano occurs when molten rock, called **magma**, forces its way to the surface through a weak spot in the Earth's crust (see figure 4.3). This can happen

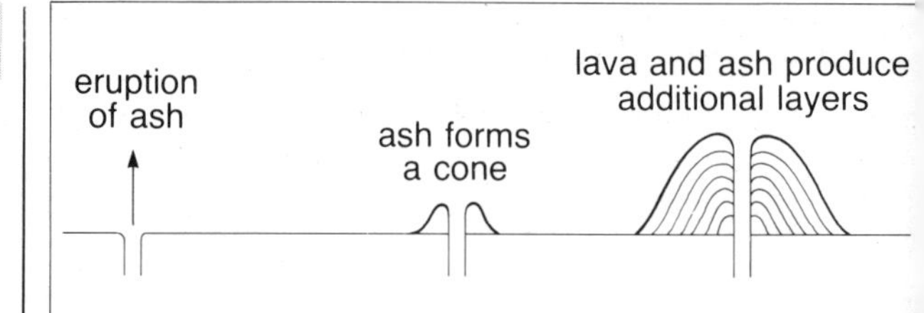

figure 4.3 The formation of a volcano

with dramatic force and a volcanic eruption occurs. When Mount Vesuvius erupted in AD 79 it buried the town of Pompeii and its inhabitants under a thick layer of ash which preserved them to the present day.

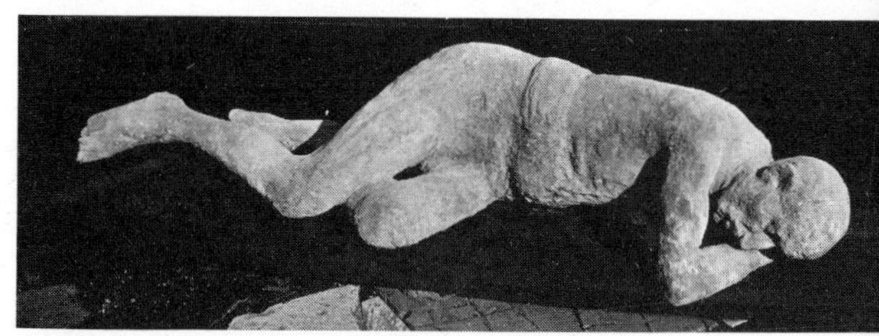

figure 4.4 Volcanic ash buried and preserved the body of this inhabitant of Pompeii

Application 4.1

USEFUL VOLCANOES

Although volcanoes can be very dangerous, we benefit from them in several ways.

- Volcanic ash is a very fertile soil. People farm the land around the base of some volcanoes even under the threat of an eruption.
- Volcanoes heat up underground water to produce steam. The steam can be brought to the surface and used to generate electricity. Water can also be pumped several kilometres below the Earth's surface where the rock is hot enough to boil it. This source of energy is known as **geothermal energy**.
- Useful minerals such as diamonds are formed by the intense heat produced by magma.

figure 4.5 Geothermal power station in California

TYPES OF ROCK

We can classify rocks according to the way they were formed.

Sedimentary rocks

All the material carried by rivers into lakes and oceans eventually settles on the bottom. Over millions of years this material builds up and is compressed into layers of rock. These are **sedimentary rocks**. Chalk (limestone) and coal are common examples.

Igneous rocks

Igneous rocks are formed when the molten magma beneath the Earth's crust cools down. This happens if the magma comes up to the surface of the earth. The speed of cooling produces different types of **igneous rock** (see investigation 4.1). Granite and basalt are common igneous rocks.

Metamorphic rock

When igneous or sedimentary rock is heated to a very high temperature or squeezed under great pressure, its chemical structure changes and a new type of rock, **metamorphic rock**, is formed. Marble is a metamorphic rock formed from limestone and slate is formed from clay.

INVESTIGATION 4.1

HOW IS THE FORMATION OF IGNEOUS ROCK AFFECTED BY DIFFERENT COOLING RATES?

When molten magma cools to form solid igneous rock, crystals form. You can investigate this process watching napthalene solidify.

a Half-fill two test tubes with napthalene and place them in a half-full beaker of water.
b Heat the beaker of water gently. The napthalene will melt at 80°C.
c Turn off the heat and then remove one test tube from the beaker. Hold it under cold running water until the napthalene solidifies.
d Leave the other test tube in the beaker of hot water until the napthalene solidifies.
e Observe the formation of crystals in both samples of napthalene.

Questions
1 Which test tube cooled most rapidly?
2 Which sample of napthalene has the smallest crystals?
3 What is the link between the rate of cooling and the size of the crystals?
4 Figure 4.6 shows magnified photographs of two igneous rocks, basalt and granite. Which one has the smaller crystals?
5 Which of the two rocks was formed by rapid cooling of magma?
6 Explain how this investigation supports the theory that granite was formed by the slow cooling of magma deep beneath the Earth's crust.

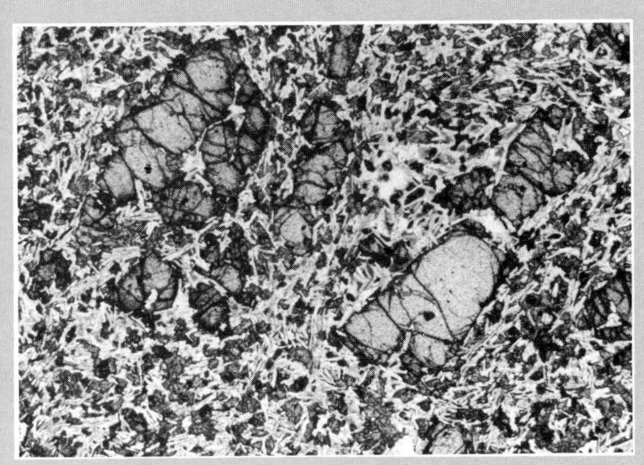

figure 4.6 a Basalt crystals magnified about 11 times

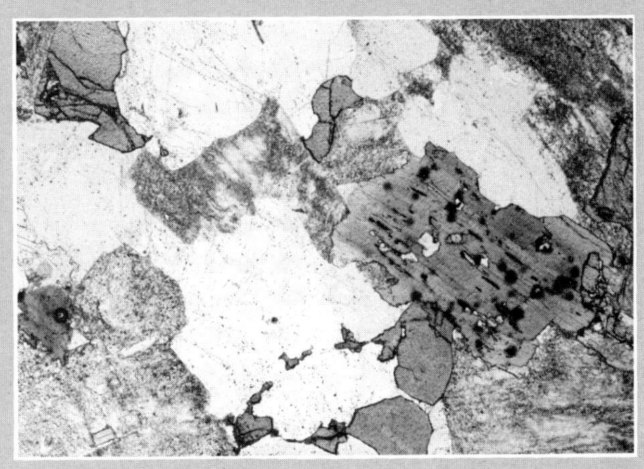

b Granite crystals magnified about 11 times

THE ROCK CYCLE

Weathering

Rock appears to be a very tough material. In fact, it is continually being broken down into smaller particles by forces in the environment. This process is known as weathering. Rocks battered by wind and rain are slowly worn away. Extremes of high and low environment temperature can cause rock to break up (see investigation 4.2). Certain alkaline or basic rocks such as limestone can be dissolved by weakly acidic rainwater (see page 340). Plant roots break up rocks by growing into cracks and enlarging them.

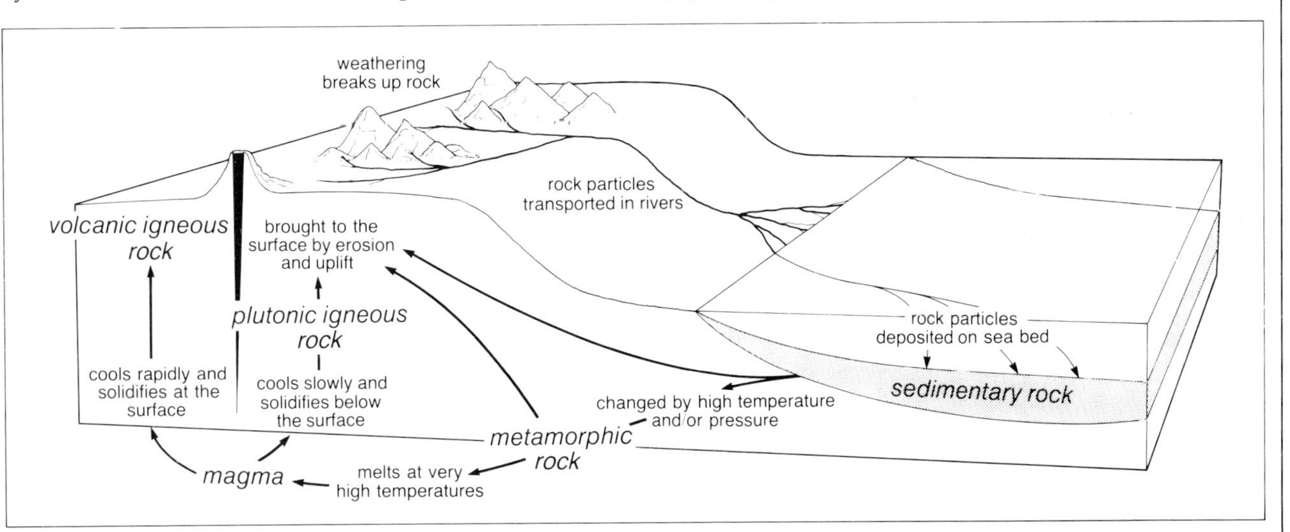

figure 4.7 The rock cycle

INVESTIGATION 4.2

HOW DO FREEZING TEMPERATURES AFFECT ROCKS?

When water freezes it expands (see page 8). In this investigation you can find out what effect this has on rocks.

a Take a sample of chalk and a sample of sandstone both about 2 cm^3 in size. Place them in a beaker of water and leave them to soak for 24 hours.
b Remove the rock samples from the beaker and place them in a freezer for 3–4 hours.
c Remove the rock samples from the freezer and allow them to thaw out.

Question

What changes have occurred in each sample? How do you explain these changes?

Erosion

The movement of rock fragments from one place to another is called erosion. Rivers play an important part in this, as a rapidly moving river can carry large rock fragments (see investigation 4.3). As the river slows down these large fragments settle on the river bed. Near the mouth of a river large quantities of fine particles can build up to form a delta. This has occurred where the River Nile flows into the Mediterranean Sea (see figure 4.8).

figure 4.8 The Nile Delta

INVESTIGATION 4.3

MEASURING THE SPEED OF WATER FLOW IN A STREAM

a Quarter-fill a plastic bottle with water and replace the bottle cap.
b Working in pairs, measure out a suitable distance (about 50 m) along the stream bank.
c Release the plastic bottle into the stream using a net and time how long it takes to travel the measured distance. Repeat 10 times.
(Do not forget to collect up your bottle using the net at the end of the investigation. Don't leave it to pollute the stream.)

Questions

1 How do you explain any difference in the times taken for the bottle to travel the same distance?
2 What was the average time taken for the bottle to travel the same distance?
3 Calculate the average speed of stream flow using this formula:

$$\text{average speed} = \frac{\text{measured distance (m)}}{\text{average time taken (sec)}}$$

4 How do you think the speed of stream flow will change over the course of a year?

WHAT IS SOIL?

Soil is formed when solid rock is broken down by weathering and mixed with the decayed remains of dead organisms. It has seven main constituents:

1 **Rock particles** – These vary in size. From smallest to largest they are known as clay, silt, sand and gravel.
2 **Soil water** – Soil particles are usually surrounded by a thin film of water. This is essential for all life living in or on the soil.
3 **Humus** – When animals and plants die in the soil their dead bodies gradually decay to form a sticky fibrous substance called **humus**.
4 **Living organisms** – Bacteria and fungi live in soil. They are responsible for breaking down dead animals and plants to form humus. They also carry out some of the chemical reactions which make up the **nitrogen cycle** (see page 384).
5 **Air** – In a good soil there are plenty of spaces between the soil particles. These spaces are filled with air which provides essential oxygen for plant roots and other living organisms in the soil.
6 **Mineral salts** – Dissolved in the soil water are mineral salts such as nitrates, phosphates and magnesium. These are essential for healthy plant growth.
7 **Lime** – Lime helps soil particles clump together. It also provides calcium. As it is an alkali, lime neutralizes soils that are too acid for plant growth (e.g. peat soils).

INVESTIGATION 4.4

SEPARATING THE CONSTITUENTS OF SOIL

a Quarter-fill a measuring cylinder with soil.
b Add water until the cylinder is three-quarters full.
c Put your hand over the open end and shake the cylinder well.
d Place the measuring cylinder on the laboratory bench and leave for 30 minutes.
e After 30 minutes, make a careful drawing to show the various layers that have settled.
f Measure the depth of each layer and calculate the percentage of each of the following components of your soil sample:

gravel sand silt clay

Questions

1 What is the soil constituent floating on the water's surface?
2 In what way is this substance different from the rock particles?

INVESTIGATION 4.5

HOW MUCH WATER IS THERE IN SOIL?

a Weigh an empty crucible.
b Half-fill the crucible with soil. Re-weigh the soil and crucible.
c Place the soil and crucible in an oven at about 100 °C for at least 24 hours.
d Cool the soil and crucible in a desiccator.
e Re-weigh the soil and crucible.

Questions

1 What happened to the soil water when the sample was placed in the oven?
2 What is the difference between the mass of soil before and after heating?
3 Use the method shown below to calculate the percentage of water in your own soil sample.

mass of wet soil sample and crucible (A) = 59.6 g
mass of empty crucible (B) = 54.0 g
mass of soil sample only (C) = (A – B) g
= (59.6 – 54.0) g
= 5.6 g
mass of dry soil sample and crucible (D) = 58.9 g
mass of water in the soil sample (E) = (A – D) g
= (59.6 – 58.9) g
= 0.7 g
percentage of water in soil sample $= \frac{E \times 100}{C}$
$= \frac{0.7 \times 100}{5.6}$
= 12.5%

INVESTIGATION 4.6

HOW MUCH HUMUS IS THERE IN SOIL?

a Use the soil and crucible from investigation 4.5.
b Place the crucible on a gauze and tripod.
c Heat the soil sample with a strong Bunsen flame.
d At intervals hold the base of the Bunsen burner and apply the flame directly to the surface of the soil. Continue to heat the soil for 15 minutes.

Questions

1 What is the difference in the mass of the dry soil before and after burning it?
2 What is the difference in the colour of the soil after heating strongly? What do you think has happened to the humus?
3 Use the method shown below to calculate the percentage of humus in your soil sample.

mass of soil and crucible after heating strongly (F) = 58.8g
mass of humus in the soil sample (G) = (58.9 – 58.8) g
= 0.1 g
percentage of humus in the soil sample $= \frac{G \times 100}{C}$
$= \frac{0.1 \times 100}{5.6}$
= 1.8%

4 Which component of the soil is left at the end of the investigation?

Different types of soil

Clay soil
Clay particles are very small. They stick tightly to each other and to water. The spaces between the particles also fill up with water. When it rains the water remains on the surface because it cannot drain away. The soil becomes **waterlogged**, and very difficult to drain.

Sandy soil
The larger particles in a sandy soil allow a lot of air into the soil. It is a light soil that is easy to dig but gives poor anchorage for building on. Water drains through a sandy soil very quickly. Useful minerals are washed out of the soil. This is called **leaching** and results in a soil with poor fertility.

Loamy soil
A loam is a fertile soil containing a good balance of sand and clay particles. It also contains a high proportion of humus. This acts as a sponge, holding in water and dissolved minerals. Plants can grow well in a loamy soil.

UNIT 4

Application 4.2

THE SOIL'S VANISHING TRICK

About a third of the world's topsoil is expected to disappear in the next 15 years. Where is it going? Soil is carried away by the natural forces of wind and water. Wind blows away dry soil near the surface to form 'dustbowls'. Rainfall washes soil away into streams, rivers and lakes.

Why is erosion getting worse?
Natural and man-made features of the landscape protect the soil from wind and rain. Woods and hedgerows provide shelter. Grass covers the soil and helps to hold it together.

Farming destroys these natural features when the land is used to grow crops. Hedgerows are removed to make large fields that can be ploughed more easily. Fields are left bare for up to six months before a crop grows to cover them. The unprotected soil surface is quickly eroded by wind and rain.

How can erosion be stopped?
Farmers can be encouraged to change their practices. They might be given grants to replace hedgerows that have been removed. A law requiring fields to be sown with grass every four years would also help soil conservation. If farmers were to plough along, rather than down, a sloping field the soil would not be washed away so easily.

figure 4.9 Terracing prevents soil erosion

WEATHER

A simple description of weather could be: the changes which take place in the atmosphere surrounding the Earth.

Changes in the weather take place when changes happen to the air or water vapour in the atmosphere. Energy causes these changes to take place. **Meteorology** is the study of weather.

Water and water vapour

There is an enormous amount of water spread around the surface of the Earth. Most of it is concentrated in the oceans and the arctic regions. What is not so easy to see is the water vapour contained in the atmosphere. A puddle, for example, will soon disappear into the atmosphere by **evaporation** (see page 7). This process speeds up if the temperature of the surroundings increases or the flow of air across the liquid increases.

Warm air can hold more more water vapour than col[d] air. When the atmosphere has absorbed as much as i[t] can at any particular temperature it is **saturated**.

Clouds, mist, fog and smog

When water vapour condenses in the sky it forms **clouds** which often condense to form **rain**. Near the ground, condensing water vapour produces **mist**. A thick mist is known as **fog** and when this contains **smoke** particles it is called **smog** (see page 113). Smog is no longer seen in the UK, may still be present in places such as Los Angeles in the USA.

Dew and frost

During the day the temperature may be warm enough to hold a considerable amount of invisible water vapour. If, during the night, the temperature falls then some of the water vapour may condense and be deposited as **dew**. If the temperature falls far enough the condensing water vapour may freeze. This is **frost** (**hoar frost**).

figure 4.10 a Fog

b Hoar frost

CLOUD CLASSIFICATION

Clouds can be named very easily using a set of simple guidelines.

cumulo or cumulus –	fluffy lumpy
strato or stratus –	layered
nimbo or nimbus –	rain
alto –	medium height (between 2000 and 5500 m)
cirrus –	thread (only found high up – made of ice crystals)
high clouds –	above 5500 m
medium clouds –	between 2000 and 5500 m
low clouds –	below 2000 m

figure 4.11 *Cloud types* *a Cirrus*

name	symbol	height (km)	description
cirrus	Ci	high, 5.5	delicate parallel threads or wisps, often turning upwards at the ends; often called 'mares' tails'.
cirrocumulus	Cc	high, 5.5	small lumpy clouds arranged in ripples or lines
cirrostratus	Cs	high, 5.5	thin layer or whitish veil of cloud which often makes a halo round the sun or moon
altocumulus	Ac	medium, 2–5.5	lumpy clouds usually in clouds or rows
altostratus	As	medium, 2–5.5	layer or veil of cloud at a high level
stratocumulus	Sc	low, 2–5.5	large lumpy clouds often with patches of blue sky
stratus	St	low, 2	continuous sheet of low cloud
nimbostratus	Ns	low, 2	stratus from which rain is falling
cumulus	Cu	low, 2	heaped cloud, often isolated, with rounded top and flat base
cumulonimbus	Cb	low, 2	towering, fully developed cumulus; the upper levels spread into anvil shape, often with heavy rain or hail and thunder

b Altocumulus

c Stratocumulus

d Cumulus

e Cumulonimbus

AIR CURRENTS

Air is constantly on the move. The movement of clouds across the sky is a good indication of this, though different layers of air may be moving in different directions at different rates. These movements of air are due to **convection currents**. Warm air is less dense than cool air, so it rises. As it rises in one part of the atmosphere so cool air will move in underneath it to take its place. These convection currents may take place at any level in the atmosphere and over a small or large area.

Lows and highs

Convection currents in the atmosphere create areas of **low pressure** and areas of **high pressure**. The atmospheric pressure in any one place varies with time. Pressure can be measured in units such as newtons per – centimetre squared (N/cm^2, the force, in newtons, acting on an area of $1\ cm^2$ – see page 000). For atmospheric pressures a different unit, the **millibar (mbar)** is used.

$$10\ N/cm^2 = 1\ bar$$
$$1\ bar = 1000\ millibars\ (mbar)$$

The average value for atmospheric pressure in the UK is 1013 mbar. This is measured at sea level. Atmospheric pressure usually ranges between 975 mbar (low pressure) and 1030 mbar (high pressure).

Barometers (bar-o-meters) are used to measure atmospheric pressure. Special barometers used in aircraft and by free-fall parachutists are called **altimeters**.

Isobars

These are lines drawn over a map of the Earth's surface which connect together areas with the same atmospheric pressure. They look rather like the contour lines on a map (see figure 4.12).

Wind direction always follows closely the line of isobars. In the Northern Hemisphere (the part of the Earth that is north of the equator)

a winds blow in an anticlockwise direction around a low pressure (**cyclone**).

b winds blow in a clockwise direction around a high pressure (**anticyclone**).

Warm and cold fronts

When cold air moves in underneath an area of warm air it is known as a **cold front**. These are often associated with widespread showers and some heavy rain. A **warm front** occurs when a belt of warm air moves in on cold air and rises above it. A warm front often brings rain (or snow). If a cold front 'catches up' with a warm front the result is called an **occluded front** (see figure 4.12).

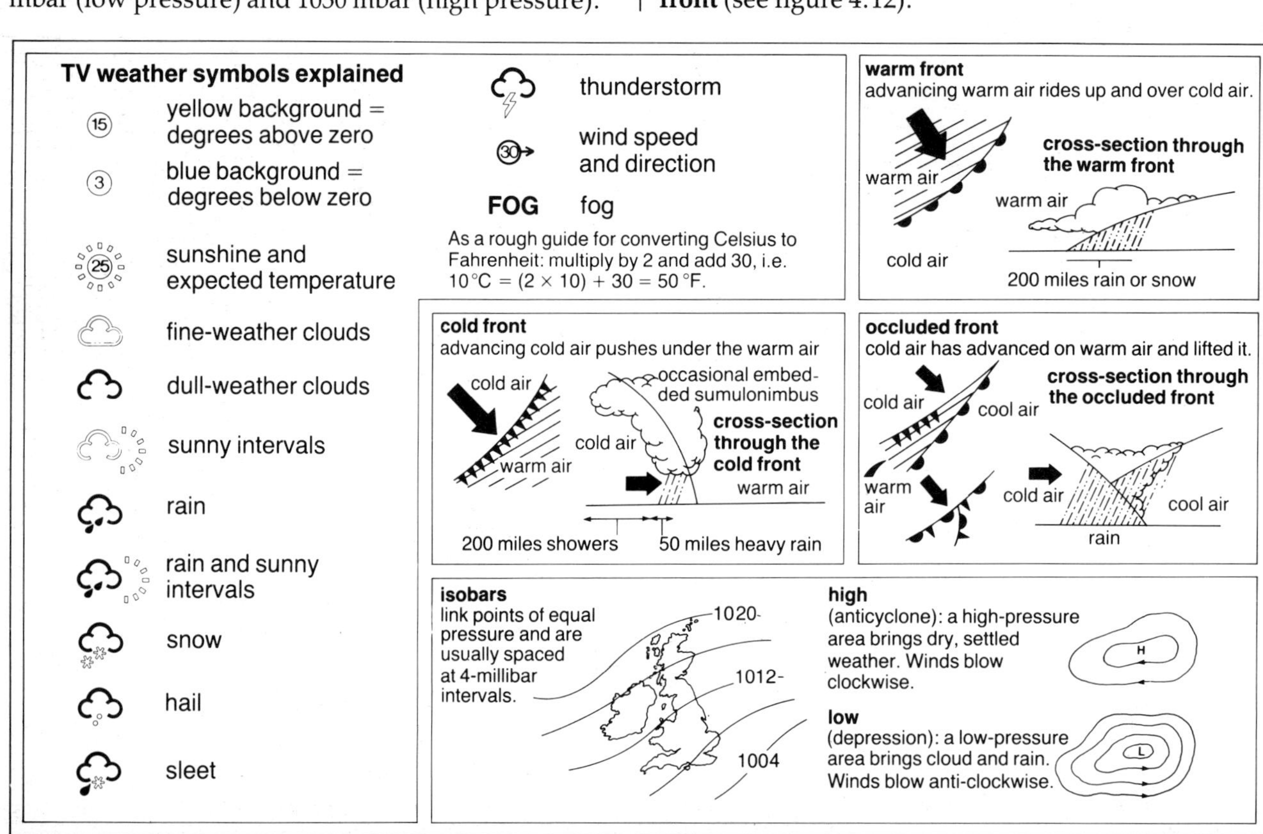

figure 4.12 Here is the weather

figure 4.13 Met. Office services – Weathercall and shipping forecasts

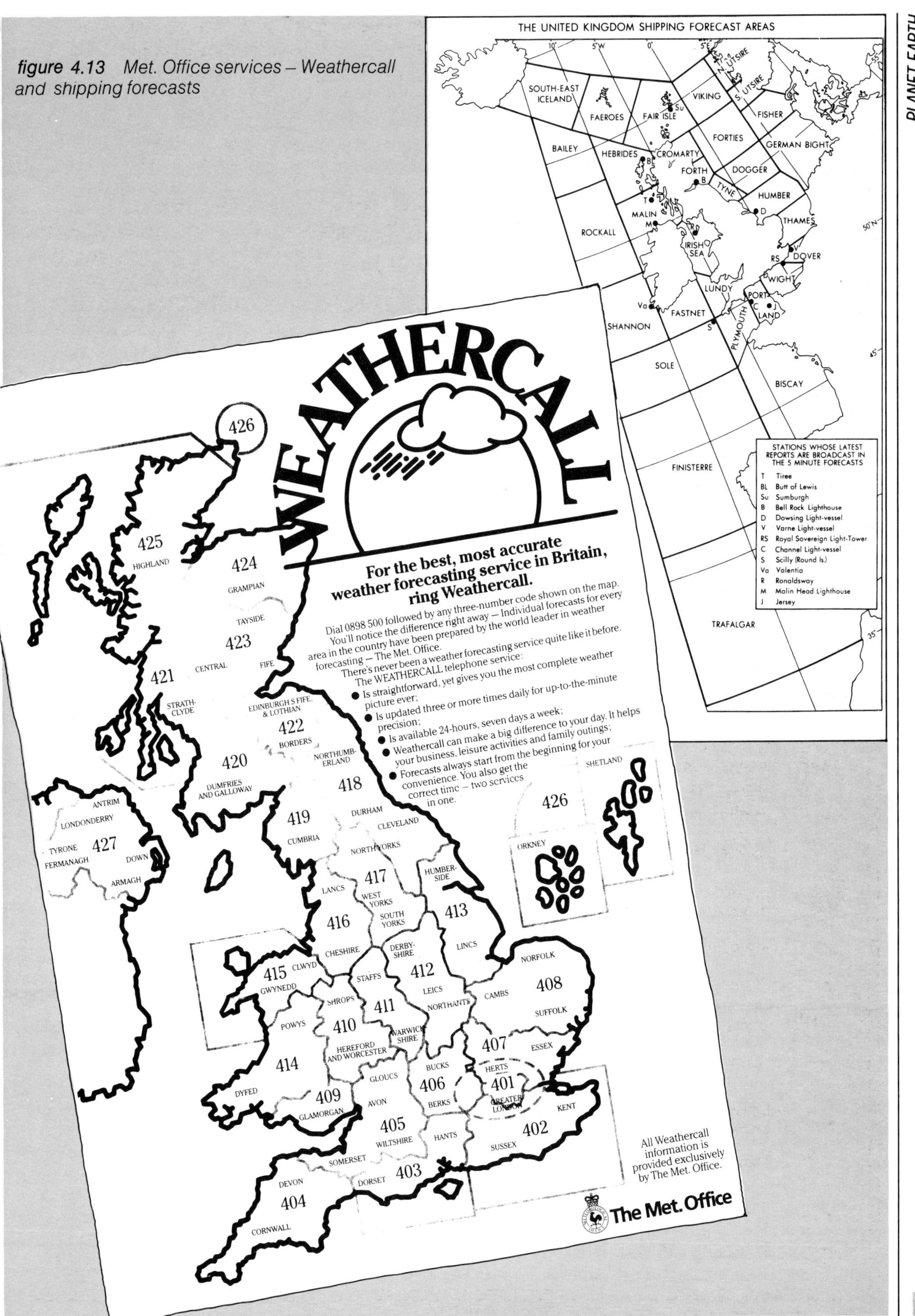
THE UNITED KINGDOM SHIPPING FORECAST AREAS
SOUTH-EAST ICELAND
FAEROES
FAIR ISLE
VIKING
N. UTSIRE
S. UTSIRE
FISHER
FORTIES
GERMAN BIGHT
BAILEY
HEBRIDES
CROMARTY
FORTH
DOGGER
TYNE
HUMBER
ROCKALL
MALIN
IRISH SEA
THAMES
DOVER
WIGHT
LUNDY
PORTLAND
PLYMOUTH
SHANNON
FASTNET
SOLE
BISCAY
FINISTERRE
TRAFALGAR
STATIONS WHOSE LATEST REPORTS ARE BROADCAST IN THE 5 MINUTE FORECASTS
T Tiree
BL Butt of Lewis
Su Sumburgh
B Bell Rock Lighthouse
D Dowsing Light-vessel
V Varne Light-vessel
RS Royal Sovereign Light-Tower
C Channel Light-vessel
S Scilly (Round Is.)
Va Valentia
R Ronaldsway
M Malin Head Lighthouse
J Jersey
WEATHERCALL
For the best, most accurate weather forecasting service in Britain, ring Weathercall.
Dial 0898 500 followed by any three-number code shown on the map. You'll notice the difference right away – Individual forecasts for every area in the country have been prepared by the world leader in weather forecasting – The Met. Office.
There's never been a weather forecasting service quite like it before. The WEATHERCALL telephone service:
● Is straightforward, yet gives you the most complete weather picture ever;
● Is updated three or more times daily for up-to-the-minute precision;
● Is available 24-hours, seven days a week;
● Weathercall can make a big difference to your day. It helps your business, leisure activities and family outings;
● Forecasts always start from the beginning for your convenience. You also get the correct time – two services in one.
426 SHETLAND ORKNEY
425 HIGHLAND
424 GRAMPIAN
TAYSIDE
423 CENTRAL FIFE
421 STRATH-CLYDE
EDINBURGH S FIFE. & LOTHIAN
422 BORDERS
420 DUMFRIES AND GALLOWAY
NORTHUMB-ERLAND
418 DURHAM CLEVELAND
419 CUMBRIA
427 ANTRIM LONDONDERRY TYRONE FERMANAGH DOWN ARMAGH
NORTH YORKS
417 WEST YORKS
HUMBER-SIDE
LANCS
416 CHESHIRE
SOUTH YORKS
413 LINCS
DERBY-SHIRE
412 LEICS
415 CLWYD GWYNEDD
STAFFS
411 NORTHANTS
SHROPS
410 HEREFORD AND WORCESTER
WARWICK-SHIRE
CAMBS
408 NORFOLK SUFFOLK
POWYS
414 DYFED
407 ESSEX
BUCKS
HERTS
401 GREATER LONDON
GLOUCS
406 BERKS
409 GLAMORGAN
AVON
405 WILTSHIRE
HANTS
402 KENT SUSSEX
SOMERSET
403 DORSET
404 DEVON CORNWALL
All Weathercall information is provided exclusively by The Met. Office.
The Met. Office

INVESTIGATION 4.7

WEATHER STATION

Only a few, simple-to-operate instruments are needed to observe and measure weather details (see figure 4.14). As weather is constantly changing it is best to observe weather patterns continuously over several days (or weeks). Instruments from a typical weather station might include

- i an anemometer (wind guage) – measures wind speed
- ii a weather vane – measures wind direction
- iii a Stephenson screen – containing thermometers to measure temperature
- iv a rain guage – measures amount of rainfall
- v an aneroid barometer – measures atmospheric pressure
- vi a hygrometer – measures amount of water vapour in atmosphere
- vii a sunshine recorder – measures hours of sunshine.

In addition to these instruments, weather recording and predicting rely on careful observations e.g. how much cloud and what type, previous weather, time of day.

- a Make a list of the observations and measurements you can make. This will make a weather data set.
- b Decide how many weather data sets you will record and when you will make them. Record with each of them the date and time at which they were made.

Questions

1. When you have a number of sets compare them to see whether patterns occur. (For example, does more cloud cover also mean higher pressure, etc.?)
2. Compare your data set with the weather forecasts from the media.
3. What do you think the weather will be like tomorrow?
4. Can you contact students in another school some distance away and compare your weather details with theirs?

figure 4.14 a anenometer

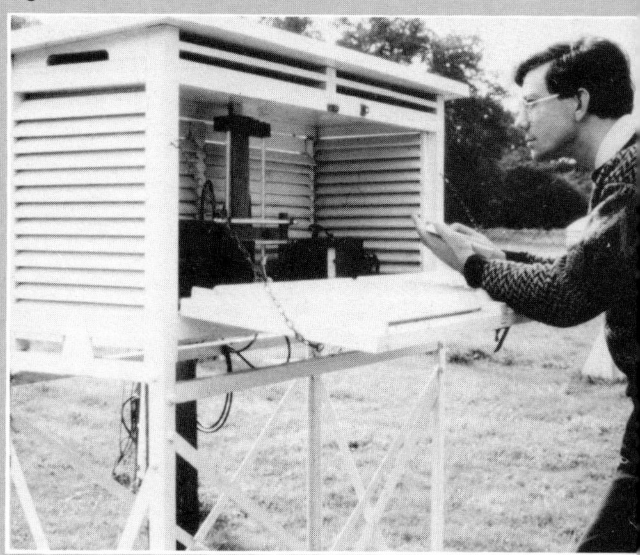

b Stephenson screen

HERE IS THE WEATHER

When all the weather symbols for isobars, depressions, fronts etc. have been added to a geographical map it makes a **weather chart** or **synoptic chart**. However, the charts which are usually shown on television are not true synoptic charts because many of the symbols have been simplified. It does not mean that they are any less accurate.

A weather chart can only be constructed from a large amount of meteorological data. These measurements include not only temperature and pressure but also a great many other observations and measurements such as cloud cover, cloud height and type, visibility and wind velocity (the speed and direction of the wind). These data are recorded by some 100 meteorological outstations of which about 65 make regular observations. Only when these data sets have all been collected can a weather map be constructed.

The Met. Office

The main meteorological office is at Bracknell in Berkshire. Data from weather stations, including satellites, are processed there to provide up-to-the-minute weather information. The Met Office provides a number of specialist services in addition to generating weather forecasts for the general public.

Although at first sight the following list may seem to contain a selection of unconnected ideas, they are all affected by the weather.

the construction industry insurance sea travel
the offshore industry farming road conditions
transport energy management in buildings pollution
flooding

Rain gauge

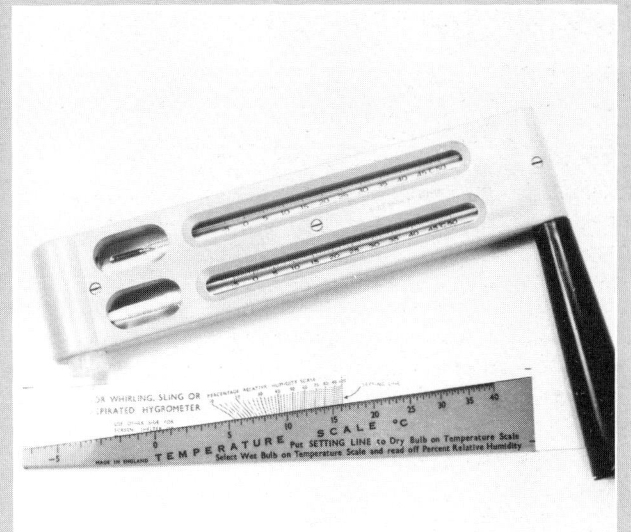

e *Whirling hygrometer*

Barometer

f *Sunshine recorder*

Global weather

As a weather pattern builds up over a small area of the Earth's surface, so this forms part of a much larger global weather pattern.

A very simple model to use would show a convection current rising as warm air in the tropics and then sinking as cold air in polar regions. Unfortunately, this simple idea does not work in practice because of the distribution and spacing of the land masses and the oceans, and the rotation and axis tilt of the Earth. However, large scale wind circulations persist and are predictable.

ALL YOUR OWN WORK

1 Sudden movements of the Earth's crust are more likely to occur at the boundaries of two ______. This may cause an ______.

2 Igneous rocks are formed when ______ cools. Sedimentary rocks form layers on the ______ bed. When igneous or sedimentary rock is changed by extremes of temperature or pressure it forms ______ rock.

3 Explain how modern agriculture can make soil erosion worse.

4 What are the main differences between a clay soil and a sandy soil?

5 Clouds form from ______ ______ in the atmosphere.

6 What type of clouds produce rain?

7 Wind always blows ______ around a high pressure and ______ around a low pressure in the Northern ______.

8 For what reasons (apart from making a profit) might the weather cause building firms to buy derelict houses to 'do up'?

UNIT 5

ORGANISMS AND THEIR ENVIRONMENTS

COMMUNITIES, HABITATS AND ECOSYSTEMS

An organism cannot survive on its own. It lives with many other species in a **community**. A community of organisms will be found living in the same area or **habitat**. A pond or a wood is a habitat that you will find easy to study. A habitat and the community of organisms that it contains is called an **ecosystem**. All the ecosystems on the Earth's surface form a delicately balanced system called the **biosphere**. A small disturbance in one part of the biosphere can have widespread effects.

The carbon cycle and how humans disrupt it

Figure 5.1 shows how carbon circulates between living and dead organisms and the atmosphere. You will notice that three important processes increase the amount of carbon dioxide in the atmosphere.

a respiration in living plants and animals
b decay of dead plants and animals by microbes called **decomposers**.
c burning of fossil fuels such as coal, gas or oil to provide the energy needed for modern societies

Carbon dioxide production from burning fossil fuels has greatly increased in recent years because of the demand for more energy. The build-up of carbon dioxide in the atmosphere has also been affected by the cutting down of natural forests. With fewer trees

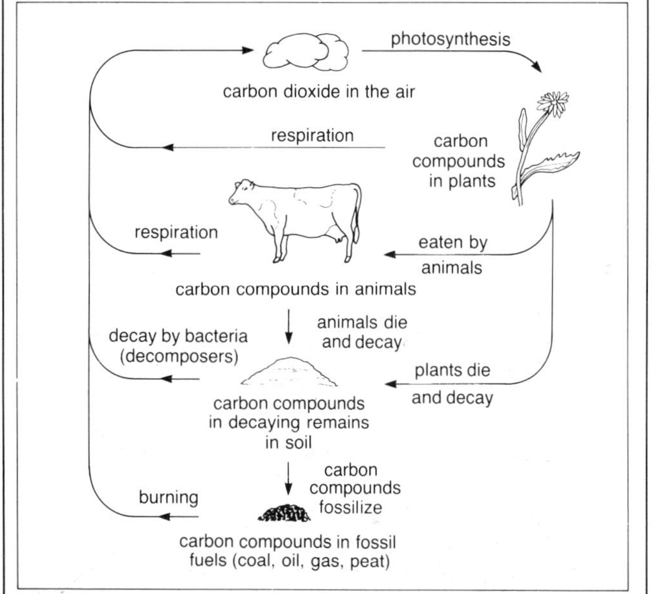

figure 5.1 The carbon cycle

the removal of carbon dioxide by photosynthesis is reduced. There is at present great concern about the effect that the destruction of the Amazonian rain forest will have on the rest of the biosphere.

Green plants: capturing energy

Animals can only obtain food by eating plants and other animals. Unlike animals, plants are capable of manufacturing their own food. Humans and all other animals are entirely dependent on the ability of plants to make food, and also to produce oxygen.

The green leaves of plants operate like tiny food factories (see figure 5.2). Raw materials (water and carbon dioxide) are supplied to the leaf. Water is carried up from the roots in tubes called **xylem vessels**. A ready supply of carbon dioxide is present in the atmosphere. Small pores (**stomata**) on the underside of the leaf allow gases to pass in and out. Like any factory process, the manufacture of food in the leaf requires an energy supply. Sunlight energy is captured by the green pigment **chlorophyll** contained in special structures called **chloroplasts** (see page 11). This energy is then used to drive the following chemical reaction.

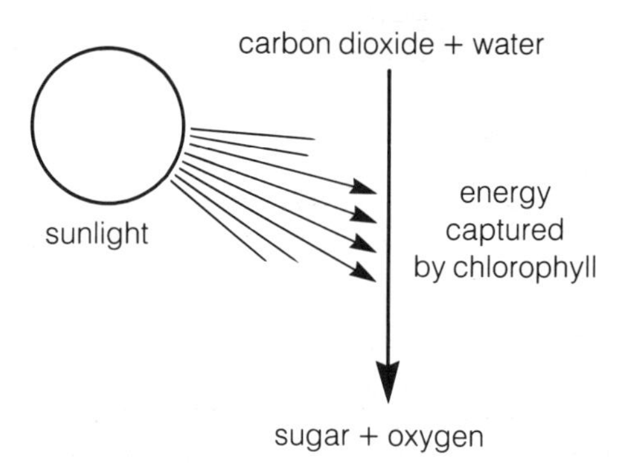

This reaction is known as **photosynthesis**. You will notice that photosynthesis has two products: sugar and oxygen. The oxygen is either used up in respiration or passed out of the leaf through the stomata. The sugar is converted to starch. Sugar molecules are soluble. At high concentration they might cause uptake of too much water by osmosis, which would damage the plant. Starch is insoluble so it can be easily stored in the form of tiny grains in the cytoplasm (see figure 1.18, page 11). Some of the sugar is carried along tubes (**phloem vessels**) to other parts of the plant.

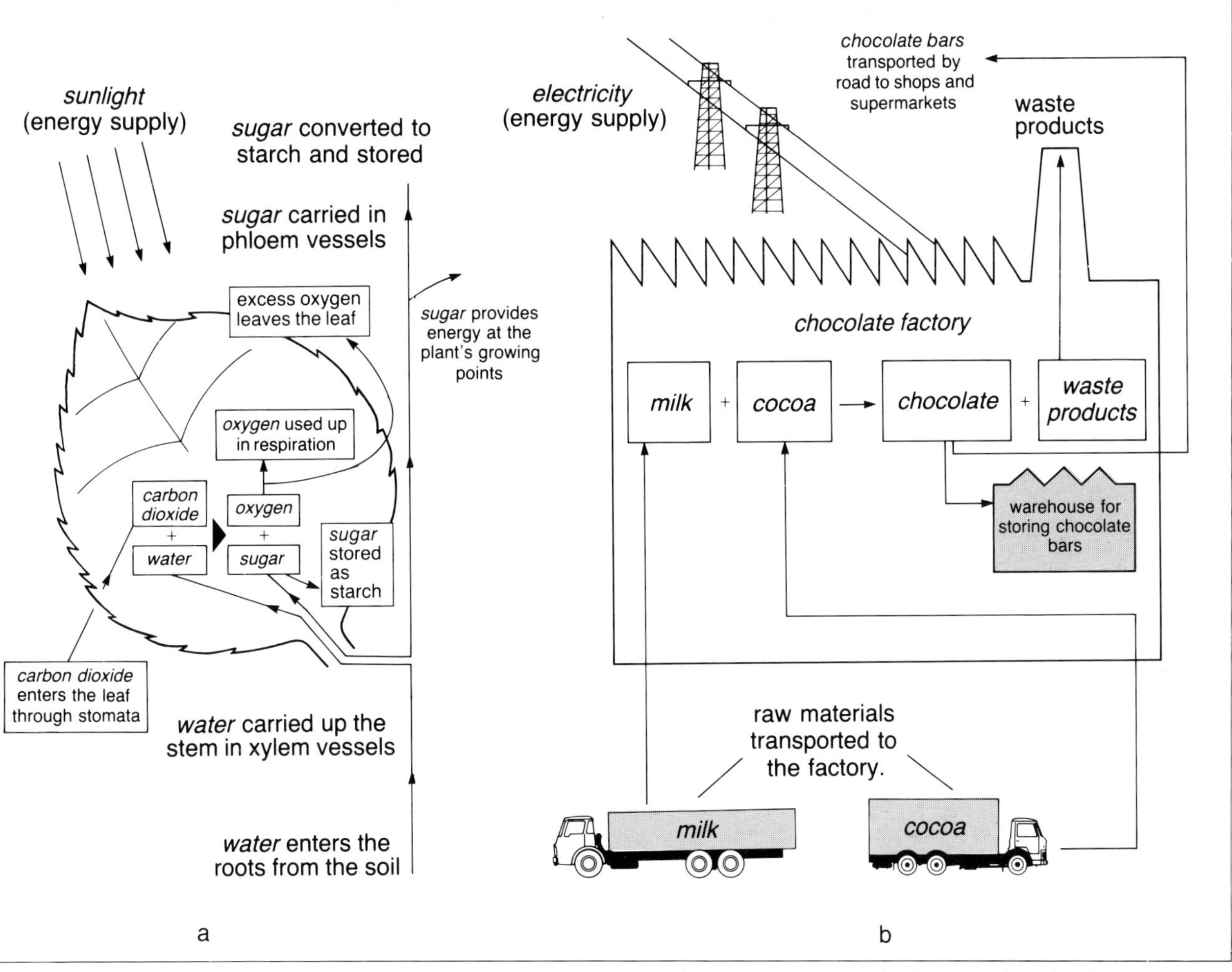

figure 5.2 Like a factory that manufactures food, a leaf needs an energy supply, transport routes to carry away the food, a means of storing it and a way of removing waste products

INVESTIGATION 5.1

IS SUNLIGHT NEEDED FOR PHOTOSYNTHESIS?

a Take two leaves: one from a plant that has been exposed to sunlight for three days and one from a plant that has been covered in dark paper or left in a dark cupboard during this time.

b Use tweezers to immerse both leaves in half a beaker of boiling water for about a minute. This kills the leaf cell membrane so that the iodine used later can enter more easily.

Now turn the Bunsen burner off

c Place the two leaves in separate boiling tubes. To each boiling tube add just enough ethanol to cover the leaves.

d Place the two boiling tubes in a beaker of (just) boiling water. The ethanol will heat up and remove the chlorophyll from the leaves. This makes it easier to see any colour change that occurs when the iodine is added.

e After 10 minutes carefully remove each leaf with tweezers and dip it in the boiled water to soften it.

f Place the leaves in separate dishes. Add two pipettefuls of iodine to the surface of each leaf. Note the colour of each leaf (see page 43).

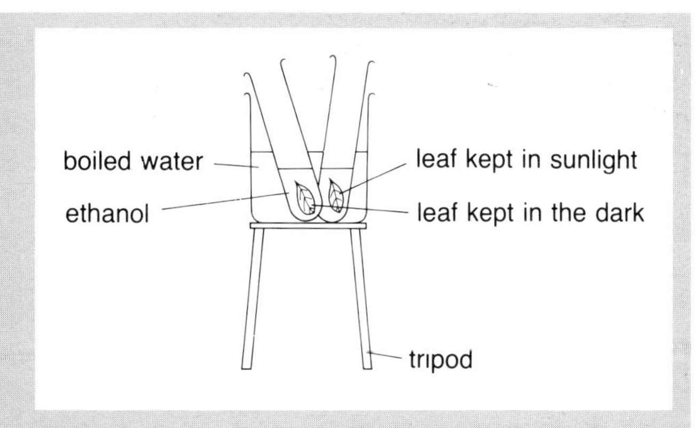

figure 5.3

Questions

1 Explain why the presence of starch is good evidence that photosynthesis has taken place.
2 What colour is the leaf that was kept in the dark?
3 What colour is the leaf that was exposed to sunlight?
4 How do you explain the difference between the two leaves?
5 Design an experiment to show that carbon dioxide is needed for photosynthesis. You may use two clear plastic bags and a small sample of potassium hydroxide which removes carbon dioxide from air.

FOOD CHAINS

If you spend some time in a wood during summer you will notice many examples of animals feeding. You may observe a caterpillar eating a leaf or a small bird eating a caterpillar. We can show these feeding relationships as in figure 5.4.

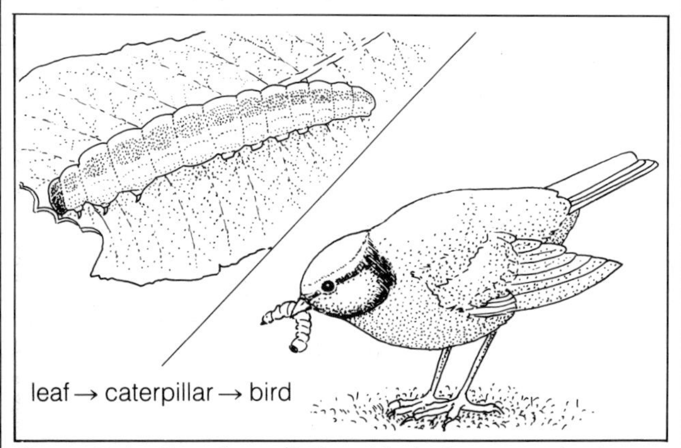

figure 5.4 *Feeding relationship between a plant, a caterpillar and a bird*

This sequence is known as a food chain. A green plant is always the first organism in the food chain. It is capable of using the energy from sunlight to produce food. Green plants are known as the **producers** in the food chain. In our food chain the caterpillar directly consumes green plants. It is called a **first consumer** or **herbivore**. The bird depends on the first consumers for its food supply and is known as a **second consumer** or **carnivore**. Large carnivorous birds may prey on smaller birds. They are the **third consumers**. The arrows linking the organisms in a food chain represent the flow of energy. Energy from sunlight can only enter the food chain through green plants. During photosynthesis this energy is stored in the plant's food molecules. Between five and 20 per cent of this energy store is passed on when the plant is eaten. The rest is lost as heat produced by respiration. Energy is lost at each level of the food chain, in the same way as money is lost in bank charges (see page 3). Only a small proportion of the energy that enters the food chain reaches the top of the chain. This explains why food chains rarely consist of more than five organisms.

There is also gradual reduction in the mass of living organisms (**biomass**) at each level in the food chain. A very large mass of green plants is needed to sustain a much smaller mass of animals at the top of the food chain (see application 5.3). This pattern can be represented in the form of a **pyramid of biomass** (see figure 5.5). The number of organisms at each level in the food chain also forms a pyramid – a **pyramid of numbers**. An understanding of food chains can help us to explain how the concentration of pesticides has built up to dangerous levels in certain animals (see application 5.1).

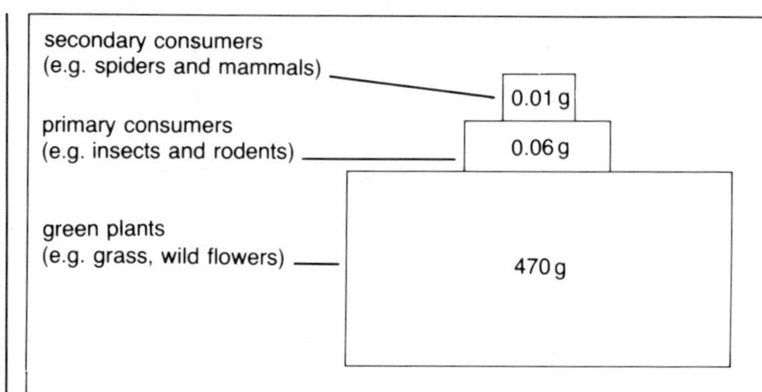

figure 5.5 *A pyramid of biomass (not to scale) for organisms in an uncultivated field*

INVESTIGATION 5.2

FOOD WEBS

In nature, food chains are linked together in a food web. This happens because the diet of a particular animal usually involves several different species. A given plant or animal may also be food for a number of other species. The food web in figure 5.6 could be found in a wood. Study it carefully and answer the questions that follow.

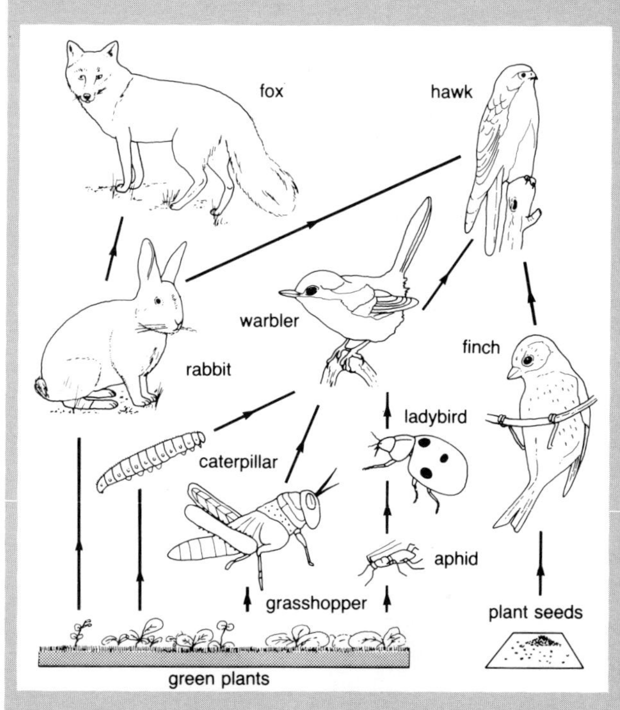

figure 5.6 *A food web in a wood*

Questions

1. Give the names of two primary consumers.
2. Give the names of two secondary consumers.
3. Write out two different food chains that include insect-eating birds.
4. What would happen to the number of ladybirds if all the aphids were wiped out by disease?
5. Explain how the removal of all the foxes from the wood could lead to an increase in the number of rabbits.
6. Describe how a disease that completely destroys the grasshopper population could affect the other organisms in the food web.

Application 5.1

FOOD CHAINS, DDT AND PEST CONTROL

Insects can be a threat to humans. Some insect species, such as locusts, devour crops. Others, such as the mosquito, carry disease (see pages 66–7). Scientists have been keen to develop chemicals to kill these insects. A chemical insecticide called DDT was first introduced in the 1940s. It was produced in vast quantities. At one time about 100 000 tonnes per year was being was being produced throughout the world. Scientists slowly began to realize that DDT and other insecticides could have very harmful effects too.

Rainwater can wash DDT from the soil where it has been sprayed. It is carried into lakes and rivers. At first, scientists were not worried about this. However, it soon became obvious that DDT did not break down very quickly. Gradually the concentration of DDT in lakes built up. The DDT was absorbed by microscopic plankton in the lake water. Fish feeding on the plankton were themselves eaten by carnivorous birds such as grebes. The DDT could not be excreted by the animals in the food chain. Instead it was stored in their fat reserves. As it passed up the food chain the DDT became more concentrated. At the top of the food chain the concentration was high enough to harm the grebes. Their eggs had very fragile shells and many chicks died. In some cases the adult grebes were killed by the high levels of DDT in their bodies. The dangers of DDT are now recognized and this has led to legislation being passed in many countries banning its use.

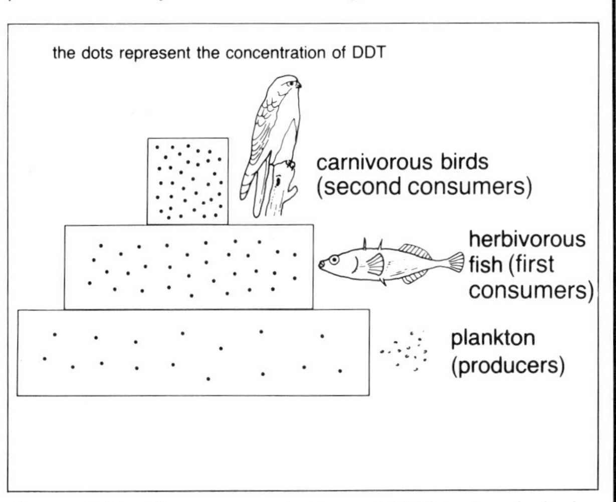

figure 5.7 The concentration of DDT in food chains

Application 5.3

VEGETARIANS VERSUS MEAT EATERS: WHO MAKES THE BEST USE OF THE AVAILABLE ENERGY?

Think about your daily diet. You will find that you are the final link in many different food chains. Some of these will be very short, involving yourself and a producer (a green plant). If you are a vegetarian all the food chains in your diet will be like this. If you are not a vegetarian you will also be involved in longer food chains. You may eat the flesh of an animal that itself feeds on plants.

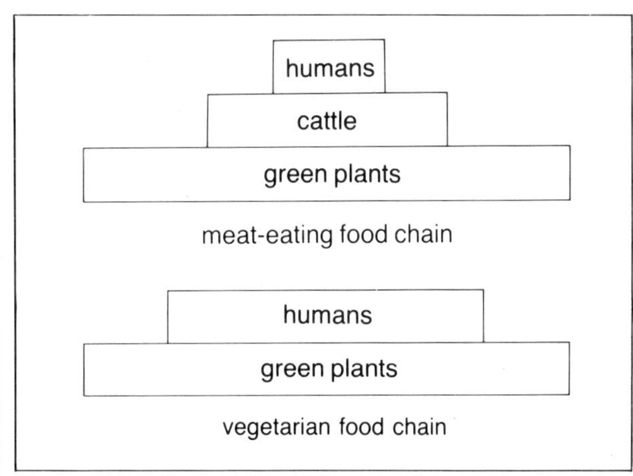

figure 5.8 Pyramids of biomass for meat eaters versus vegetarians

Figure 5.8 shows the pyramids of biomass for these two different types of food chain. Notice the difference in the human biomass that can be kept alive by the two different food chains. The vegetarian food chain can feed more people for the same mass of green plants. This explains why the diet of the world's poorer countries consist mainly of plant foods rather than of meat (see investigation 5.3). Imagine a plot of land (18 000 m^2) that is used to grow a crop such as potatoes. It provides energy to feed 20 people for a year. If the same plot of land is used as pasture to feed bullocks it could only provide enough meat to feed one person for a year.

The energy intake is not the only important feature we should consider. The amount of protein in the diet should also be taken into account. Meat has a higher concentration of protein than plant food. Vegetarians must therefore eat a larger quantity of food than meat eaters to achieve the same level of protein in their diet (see figure 3.1, page 41).

Application 5.2

CONTROLLING PESTS – ALTERNATIVES TO INSECTICIDES

Since DDT and other insecticides have such harmful effects, scientists have tried to find better ways of controlling pests. In Africa the tsetse fly carries a disease that causes sleeping sickness in humans, and can also infect cattle. It has been discovered that tsetse flies can be attracted from far distances by small quantities of a particular chemical. This chemical smells very similar to the smell given off by cattle. It is used to lure the tsetse flies into traps consisting of poisoned nets.

Using other living organisms to kill pests **biological control**. This method of pest control cannot completely eradicate the pest but it can keep the population small enough so that it does, not cause a problem.

INVESTIGATION 5.3

PLANT PROTEIN VERSUS ANIMAL PROTEIN

Table 5.1 shows the average daily protein intake of the population in different regions of the world. The number of grams of vegetable protein and animal protein are shown separately.

	East Africa	Asia	UK
vegetable protein (g)	32	43	31
animal protein (g)	6	10	53

table 5.1

Questions

1. Plot these data on a histogram.
2. Calculate the *total* protein intake for each region.
3. Which country has the lowest intake of vegetable protein?
4. In which region is kwashiorkor likely to be a major problem (see page 41)? Explain your answer.
5. Describe two ways in which the UK's protein intake differs from that of the other two regions.
6. What could be done to improve the East African protein intake?

INVESTIGATING ECOSYSTEMS

Scientists who investigate ecosystems are called **ecologists**. They have developed techniques to help them answer the following questions about a particular habitat.

How many individuals of each species are living in this habitat?
It is not possible to count each individual. Imagine counting every buttercup on the school playing field! Instead, ecologists select small areas using a square metal frame called a **quadrat**. This must be thrown at random onto the ground so that the areas studied represent a true picture of the species living in the habitat. This is an example of a **random sampling** technique (see investigation 5.4) and it is useful when studying the plant species in a habitat. Animals pose a more difficult problem, as they can move. Investigation 5.5 shows how you can estimate the size of a moving animal population.

How are organisms distributed in the habitat?
Organisms are not spread evenly throughout the habitat. Particular species will be common in some areas but not in others. We may suspect that a particular feature of the habitat is affecting the distribution. This can be checked by taking samples along a line that crosses this feature (see investigation 5.6).

What features of the habitat are affecting the distribution of organisms?
Ecologists take measurements of the physical environment at different places in the habitat, e.g. temperature, wind speed, light intensity. These measurements are then compared with the occurrence of different species. Sometimes there is a link between a physical factor and the types of species that are found. For example, in investigation 5.6 you will see that the types of plant species that can grow are affected by the amount of trampling of the ground.

INVESTIGATION 5.4

COMPARING THE VEGETATION

a. Choose an area where there is a variety of plants growing.
b. Make sure you can identify all the plant species in your area of study. Use a key if necessary.
c. List the names of the plant species in a chart like table 5.2.

name of species	quadrat number										total/10
	1	2	3	4	5	6	7	8	9	10	
buttercup		✓				✓			✓		0.3
daisy			✓	✓			✓	✓		✓	0.5

table 5.2

d. Throw a 0.5 m^2 quadrat at random. Do not move it from where it has landed. Study the plants within the quadrat.
e. When you find a particular species of plant in your quadrat place a tick against its name in the 1st quadrat column.
f. When you have identified all the plant species, repeat the procedure until 10 quadrats have been randomly thrown.
g. For each plant, add up the number of ticks to find out how many times it occurs. Convert this into a decimal between 0 and 1.

Questions

1. Which plant occurs most frequently?
2. Which plant is the least common?
3. Suggest possible reasons why some plants occur more frequently than others.
4. What is the point of using 10 quadrat samples rather than just one?
5. Would there be an advantage in increasing the number of quadrat samples, to say 20?
6. Repeat this investigation in a different area. Do you find the same plants?

INVESTIGATION 5.5

ESTIMATING THE SIZE OF A POPULATION

a Set up several pitfall traps in a sheltered area, for example, beneath a hedge or in a wood.
b Return not later than one day later and identify the insect species that have fallen into the pitfall trap.
c Choose one species from among the sample you have collected. A species well represented in your sample is most suitable.
d Mark all the individuals of this species with a small dot of light-coloured nail varnish on the underside of the body. Note the number of individuals marked.
e Release all the captured organisms. The next day return the pitfall traps to their original location.
f Return one day later and count how many of your chosen species have now been captured in the pitfall traps. Of these, count how many are marked individuals that were captured previously.

Question

Estimate the number of individuals in the population using this formula.

$$N = \frac{a \times b}{c}$$

Where N = estimated population size
a = number marked on first occasion
b = total number caught on second occasion
c = number of marked individuals captured on second occasion

COLONIZATION, SUCCESSION AND CLIMAX COMMUNITIES

Imagine that a patch of ground has been entirely stripped of vegetation and the soil treated so that all the seeds left behind are killed. Within a few weeks a few species of grass and weeds will begin to grow. They may be very different from the species originally growing there. These species are the pioneers (first to grow) and their growth is described as **colonization**.

figure 5.9 *Succession on bare ground will eventually lead to dense woodland*

INVESTIGATION 5.6

HOW DOES TRAMPLING AFFECT VEGETATION?

a Find an area where a path crosses some vegetation.
b Take a piece of string marked at 10-centimetre intervals. Use two pegs to stretch the string tightly across the path and into the vegetation on either side.
c Place a point frame at right angles to the string by one of the pegs. Try to identify the plants touched by each of the spikes. Make sure each spike of the point frame is resting on the ground.
d Using a table like table 5.3, record the number of times a particular plant is touched by the spikes of the point frame.
e Move the point frame 10 cm along the string. Again record the plants touched by the point frame. If a previously unrecorded plant species is found, use a new column of the table.
f Repeat the procedure until you reach the end of the string.

Questions

1 What species grow on the path?
2 What species do not grow on the path?
3 What features of the species found on the path help them to resist trampling by users of the footpath?
4 Are there any species that are only found at the edges of the path? Are there any advantages for the plants growing in this location?

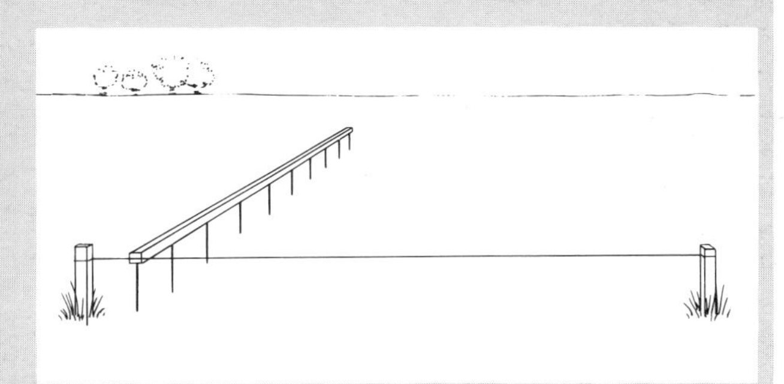

figure 5.10 *Using a point frame*

distance along transect	name of species	
	buttercup	bramble
0	0	10
10	6	3
20	5	0
30	0	0
40	0	0

table 5.3

These early species must be capable of surviving without the benefit of shelter from larger plants. As time passes, the first few species may die out to be replaced by a wider range of new species. This gradual change is known as **succession**. If the patch of ground is left alone the succession will follow a predictable pattern. The range of species in the community will increase and the food webs within the community will become more complex. After several decades, the patch of ground will resemble dense woodland. This very stable community is resistant to change. It is known as a **climax community**(see figure 5.10, previous page). Other types of climax community include hedgerows and rain forests. The enormous variety of species found in climax communities is a valuable asset. This is why it is important to protect them from destruction by human activities.

WHAT IS A POPULATION?

A **population** is a group of individuals of the same species living within a set boundary. The pigeons in Trafalgar Square, the rabbits in the New Forest and the humans in the world are all examples of populations. Scientists are particularly interested in the sizes of populations. For example, a small population of locusts may not be a serious problem but when the locust population of Africa increases rapidly it can destroy many acres of crops and cause famine. If we can discover why populations change in size, we might be able to solve problems like this.

Growing populations

A population will grow if the number of individuals entering is greater than the number leaving. Individuals enter the population either by moving into it from the outside (**immigration**) or by being born into it. Individuals leave the population either by moving away to join another population (**emigration**) or by dying. We can summarize these changes in the following equation.

change in population size = number of births
+ number of immigrants
− number of deaths
− number of emigrants

Populations have remarkable potential for increasing in numbers. The bacterium *Escherichia coli (E. coli)* is found in the human intestine. It divides once every 20 minutes. At this rate it could cover the surface of the Earth with a continuous layer in only 36 hours. Among the vertebrates, a fish such as cod lays over a million eggs. A pair of birds laying only five or six eggs a year could produce 10 million individuals in 15 years. In fact, these dramatic increases are never seen in natural populations. Populations are limited in size by the following factors.

- **Food supply**
The individuals of a population will compete for the limited amount of food available. The size of population is kept down by the deaths of individuals who fail to obtain food.
- **Disease**
Wild populations are vulnerable to fatal diseases that can drastically reduce their numbers.
- **Overcrowding**
When animal populations become overcrowded there is an increase in the death rate. Laboratory rats in these conditions suffer from stress-related illnesses such as heart disease.
- **Predators**
The population size of a predator species is closely linked to the population size of its prey. An increase in the predator population results in a fall in the prey population. This reduces the food supply of the predator population which also suffers. In such cases a regular cycle develops (see figure 5.11).

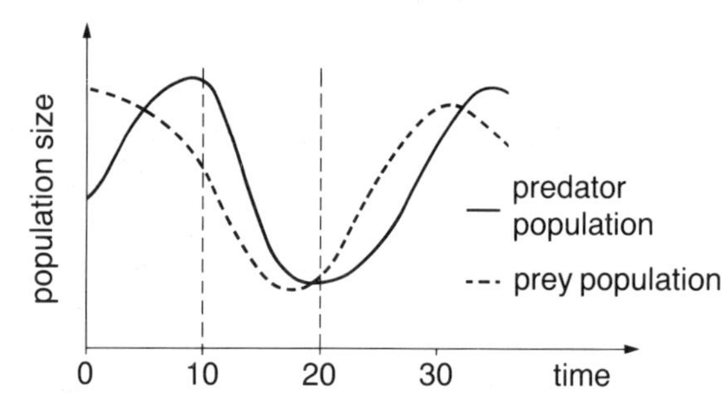

Between time 0 and 10 the predator population is increasing. As a result, the numbers of prey decrease sharply.

Between time 10 and 20 the prey population is so small that the predator population suffers a sharp fall because of the food shortage.

After time 20 the predator population is so small that it does not threaten the lives of the individuals of the prey population. This allows the prey population to increase.

figure 5.11

Human population growth

As with other animals populations, the size of the world's population of human beings is affected by the balance of deaths and births. Until we can send colonies to live on other planets, immigration and emigration will have no effect on the world's population. For much of human history a high death rate was normal. Famine and disease killed many people before they reached adulthood. This was balanced by a high birth rate. Women commonly gave birth to 10 children. As a result the world's population size was fairly stable.

INVESTIGATION 5.7

THE PROBLEMS OF AGING POPULATIONS

The age structure of a particular population can be shown as a series of rectangles. The length of each rectangle represents the number of people in a specific age group. Study the age structures of the two populations in figure 5.12.

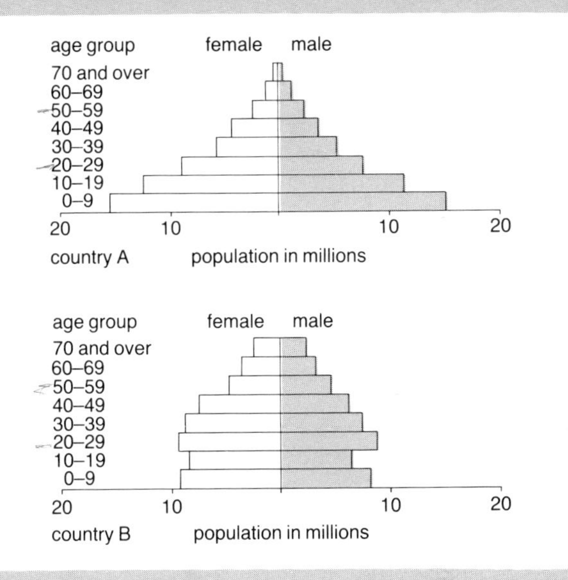

figure 5.12 Structures of two populations

Questions

1 Which of the two countries, A or B, has
   i the higher number of children in the 0–9-year-old age group
   ii the higher number of old people in the 60–69-year-old age group?
2 Which of the two countries has
   i the higher birth rate
   ii the lower death rate?
3 By measuring the size of the appropriate age groups calculate the number of adults (20–59-year-olds) in both countries.
4 Carry out a similar exercise to find out the number of elderly people (over 60 years) in both countries.
5 For each country calculate the ratio

$$\frac{\text{number of people aged 20–59}}{\text{number of people over 60}}$$

This ratio shows the number of adults of working age for each elderly person.

6 Which country has fewer working adults for each elderly person? What does this tell you about the problems of providing support to the elderly population in this country?
7 i Which of the two countries is most like Britain?
   ii In Britain, what is likely to happen to the ratio of adult workers to elderly dependents as people live longer lives?
   iii What effect is this likely to have on your own life?

over the last 200 years this situation has changed. Deaths from disease have been reduced, thanks to medical advances and improved public health measures. However, the birth rate has remained high. This has resulted in a rapid increase in the world's population. It is thought that if this increase continues it will double in the next 30 years.

It would be very difficult to feed the world's population if it grew to this size. In many countries people are encouraged to have fewer children.

ALL YOUR OWN WORK

In food chains ________ energy is absorbed by the ________ and used to make food. The first ________ or herbivores consume green plants. The second consumers or ________ depend on the herbivores for their source of food.
A population will grow in size if the number of individuals who leave it is ________ than the number of individuals who enter it.

3 Explain what is meant by a random sampling technique.
4 What are the raw materials needed for photosynthesis and how are they carried to the leaf?
5 Figure 5.13 below illustrates a food chain.
   a Where does the energy in the system come from?
   b Which is the producer?
   c Give two reasons why only a small proportion of the original energy reaches the end of the chain.

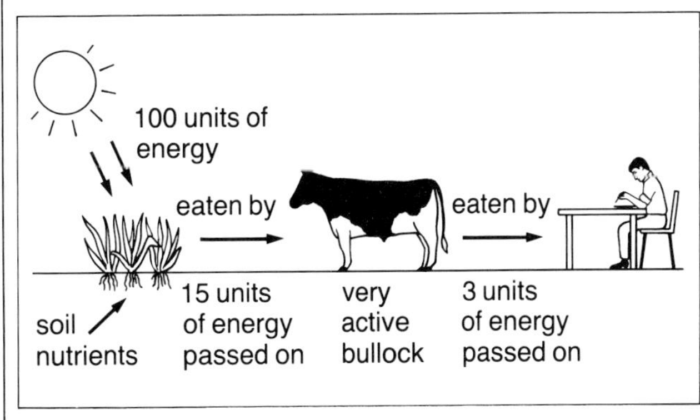

figure 5.13

ON THE MOVE

GOING OFF THE RAILS

Wherever you see movement taking place, a force (push, pull or turn) will be at work. Any movement you can think of is the result of a force. If forces did not act then movement would be impossible. Forces are responsible for

a starting something happening and

b stopping something happening.

In practice, all movements eventually cease because of the force of **friction** – which always opposes motion. The only objects which seem to defy this idea are those in space. Even they slow down eventually, because of friction with the minute amount of material in space, although this takes a long time.

Application 6.1

FRICTION AND ANIMAL MOVEMENT

As you walk or run, you depend on the friction between the soles of your feet and the ground. If there is too little friction your feet lose their grip and you will slip. Many animals have special features that help them to increase their grip, for example, claws.

Animals such as dolphins and fish have evolved a very smooth body surface and streamlined body shape. This keeps to a minimum the friction between their bodies and the surrounding water and enables them to move more swiftly through it.

We take movement for granted in our everyday lives. A bus journey to school, a day out by train to a large city or a flight on a holiday are thought of as quite normal – nothing special.

The methods of getting about shown in figure 6.3 are less than usual to most people.

From place to place

Travelling distances, moving from place to place takes time. Some journeys take longer than others. **Speed** or **velocity** is used to describe how quickly distance is covered. The difference between them is that velocity also shows in which direction the object is travelling. Most commonly, the units miles per hour (m.p.h. or m/h) or kilometres per hour (km/h) are used. These would be used for most practical travelling. The units commonly used in laboratories are metres per second (m/s) or centimetres per second (cm/s).

The distance – time graph in figure 6.4 has been constructed from a railway timetable that operated in 1986 – 7. (The distances shown are only approximate.) At various times of the day it is possible to cover the distance between Paddington (London) and Newbury (Berkshire) in different time periods. Train C takes the shortest time while train A takes the longest time and requires a stopover at Reading.

An **average speed** (or velocity) is one which remains the same and does not change throughout a journey. An average speed is a constant speed.

Application 6.2

REDUCING FRICTION IN JOINTS

When we move, the bones in our joints slide over each other. Many joints in our body, called synovial joints, have special features to reduce the friction between the bones. These joints are enclosed in a capsule that contains synovial fluid – this acts as a lubricant. The ends of the two bones in the joint have a covering of cartilage. This provides a more slippery surface so that they slide smoothly against each other.

Many elderly people suffer from **osteoarthritis**. They find it difficult and painful to move their joints. This is because the cartilage has become rough after years of wear. In severe cases a painful hip joint can be removed and replaced by an artificial one. The top of the thigh bone is replaced by a stainless steel ball. This moves smoothly inside a plastic 'cup' which replaces the hip socket.

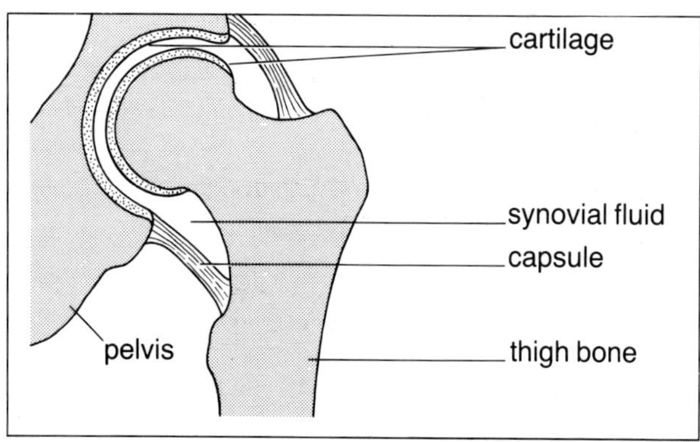

figure 6.1 *The structure of the hip joint – where the top of the thigh bone is connected to the socket of the pelvis*

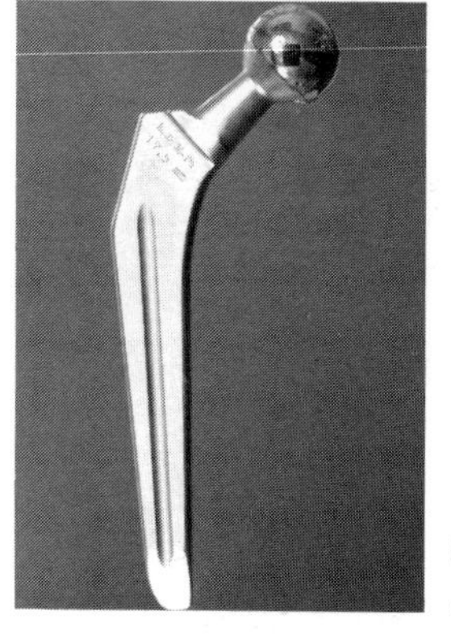

figure 6.2
An artificial hip joint

igure 6.3 *Three ways of getting about*

'o travel at constant speed is unrealistic. In ractice, all journeys have periods when the speed ncreases or decreases as well as periods when it is onstant. Figure 6.5 shows a pair of velocity – time raphs. Graph **a** shows a constant speed journey nd graph **b** shows a journey where speeds hange, which is probably more realistic. Graph **b** as been constructed for train C between addington and Reading, and assumes speeding p and slowing down times of 2 minutes each.

rain A travels the slowest and has a slow average peed (39 m.p.h.) throughout the journey. Train C as a higher average speed and is the fastest (64 n.p.h.). The average speed (or velocity) of any noving object is found from the following quation:

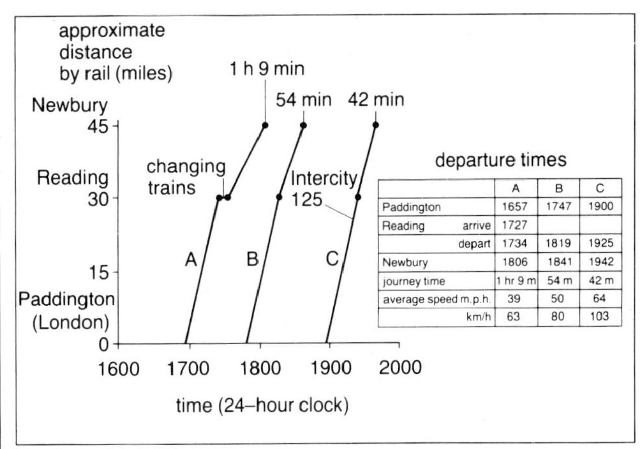

		A	B	C
Paddington		1657	1747	1900
Reading	arrive	1727		
	depart	1734	1819	1925
Newbury		1806	1841	1942
journey time		1 hr 9 m	54 m	42 m
average speed m.p.h.		39	50	64
	km/h	63	80	103

figure 6.4 *Let the train take the strain*

$$\text{average speed (unit m/s)} = \frac{\text{total distance covered (units m)}}{\text{total time taken (units s)}}$$

$$\text{average speed} = \frac{d}{t}$$

The equation can be rearranged:

$$d = \text{average speed} \times t \text{ or}$$

$$t = \frac{d}{\text{average speed}}$$

Accelerations and decelerations

Acceleration means speeding up and **deceleration** means slowing down. (**Forces** must be at work for either to happen.) A change in speed can usually be measured over a time period. The most commonly used unit is m/s^2 (m/s *per second*).

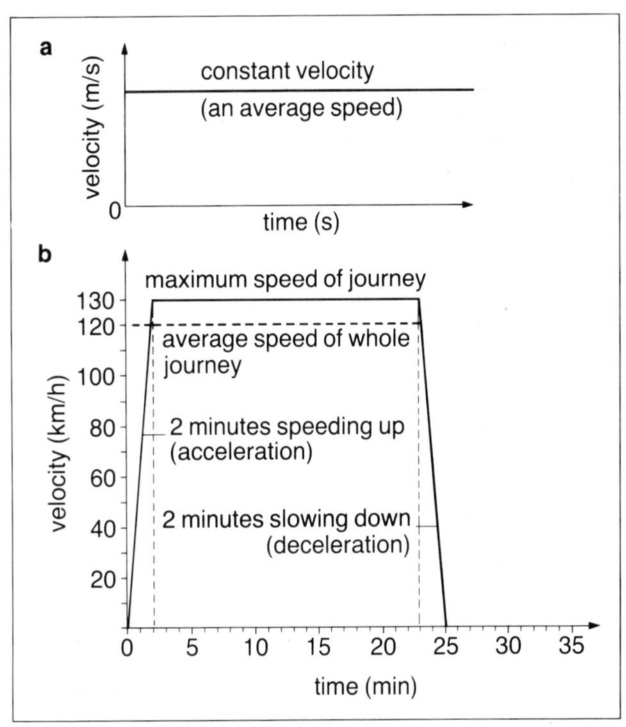

figure 6.5 *Velocity – time graphs*

$$\text{acceleration (units m/s}^2\text{)} = \frac{\text{change of speed (units m/s)}}{\text{time taken for change to happen (units s)}}$$

Speeding up is a positive (+) acceleration and slowing down (deceleration) is a negative (−) acceleration.

In the example, train C accelerates from being stationary to a speed of 36 m/s in 120 seconds.

$$\text{acceleration} = +\frac{36 - 0}{120}\ \text{m/s}^2 = +0.3\ \text{m/s}^2$$

(The deceleration at the end of this particular journey is −0.3 m/s^2.)

MEASURING MOTION

Ticker timers are easy instruments to use for measuring speeds and changes in speed. When a piece of tape is pulled through the machine it has 50 dots printed on it each second. They provide a 'picture' of motion, as table 6.1 shows.

dots	motion	speed
evenly spaced, close together	constant	slow
evenly spaced, far apart	constant	fast
getting closer together	decelerating	slowing down
getting further apart	accelerating	speeding up

table 6.1 Ticker tape 'pictures' of motion

The speed of the tape through the machine, in centimetres per second (cm/s), is found simply by measuring the dot spacing (i.e. the distance between a pair of dots):

speed of tape (cm/s) = dot spacing (in centimetres) × 50

Changes in speed can be easily measured by finding the speed of the tape twice! The first speed is measured at the beginning of the tape and the second 50 spaces further on (i.e. 1 second later). The acceleration equation mentioned earlier can then be used to find the acceleration (or deceleration). An example is shown in figure 6.6.

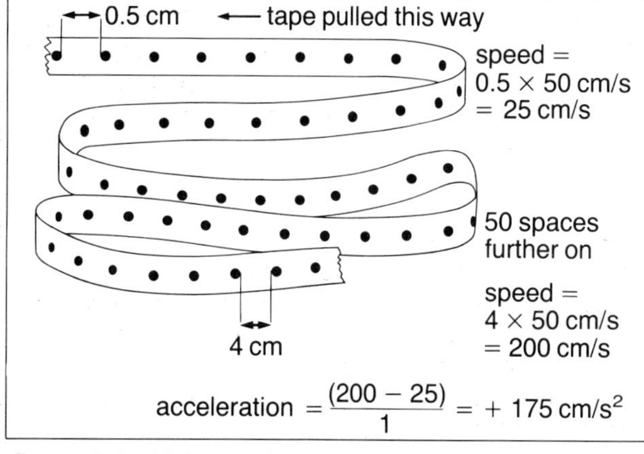

figure 6.6 Using a ticker tape to measure acceleration

INVESTIGATION 6.1

USING TICKER TIMERS

Before you start this investigation, satisfy yourself as to what constant speeds, accelerations and decelerations look like on tapes and practice measuring speed on them. You will probably need to work in groups for these experiments.

(Note: Useful (approximate) conversation factors: 1 cm/s = 0.036 km/h or 0.022 m.p.h.)

Design experiments to show and measure the following.

- a a person's average walking speed
- b the increase in speed of a cyclist starting off from rest
- c the maximum speed of a cyclist
- d the effect of gravity on an object

Question

What design problems did you have to overcome and how did you solve them?

Thinking, braking and skidding

No matter what form of transport is used, it has to be controlled. This may be by mechanical or electronic means or it may be by human effort and judgement. When a machine is under human control then human error may be involved. The thinking time needed to apply the brakes on a car is expressed as a distance. At 70 m.p.h. (113 km/h) you would travel 70 ft (21.3 m) in the time taken to decide to put the brakes on. See figure 6.7.

The stopping distances shown are only approximate and will depend on a number of things including

- **a** the state of the tyres
- **b** the material the road surface is made of
- **c** the condition of the road surface
- **d** the weather.

Successful braking depends on friction (see page 90). Unfortunately, it is all too easy to skid, especially in wet or icy weather conditions or when the road surface is loose. Antilock braking systems (ABS) are

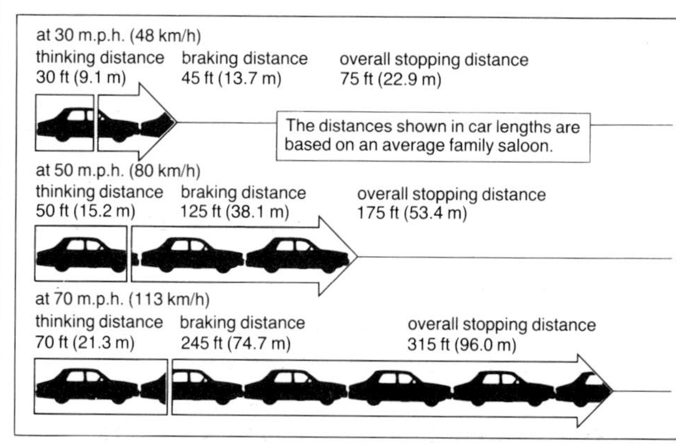

figure 6.7 Braking and thinking distances

ot the complete solution! They may stop the wheels ocking when the brakes are applied, but do not always prevent a skid.

gure 6.8 *Learning to control a skid*

MOVING ENERGY

Any object that moves has energy. The energy of movement is **kinetic energy** (KE – see pages 1 and 3). Increasing the mass (measured in kg) and/or the speed or velocity (measured in m/s) of an object increases its KE. Like all energy forms, kinetic energy is measured in joules (see page 3). The KE of a moving object is found from the equation

$$KE = \tfrac{1}{2} \times m \times v^2$$
(units J) (kg) (m^2/s^2)

$KE = \tfrac{1}{2}mv^2$ where m is the mass of the object and v is its velocity

For example, a typical mass for an HST (high speed train) locomotive is 70 000 kg. How much KE does it have travelling at 40 m/s (145 km/h)?

$$KE = \tfrac{1}{2}mv^2$$
$$KE = \tfrac{1}{2} \times 70\,000 \times 40 \times 40\text{ J} = 56\,000\,000\text{ J}$$

Because this is such a large number, we use another unit, the megajoule (MJ). Mega in front of a unit means a million, so 1 MJ = 1 000 000 J. So the KE is 56 MJ.

Always convert speed or velocity units to m/s and mass units to kg before calculating KE. This ensures that the answer is in joules.

INERTIA AND MOMENTUM

Objects which are stationary require an effort or **force** to get them moving. Similarly, objects which are moving require an effort or **force** to slow them down or speed them up. An object's reluctance to move, be moved or alter its motion is called **inertia**. A stationary truck would have inertia, as would a tanker coming into dock. Both examples would require a force to act in order to change their state of motion. When a piece of card with a coin on it is placed on top of a glass tumbler it is quite easy to flick the card and watch the coin drop into the glass. The coin's inertia makes it stay where it is while the card is suddenly moved. Magicians use the same trick when they pull a tablecloth from a table that is laid with crockery (see figure 6.9).

Passengers in a car accident provide an unfortunate example of inertia. When the car stops because of the crash, the occupants continue to move in the same direction as before because they have inertia. This is why it is important to wear seat belts – they oppose the inertia and stop the occupants when the car stops.

figure 6.9 *Inertia at work*

Application 6.3

ABSORBING ENERGY

Crumple zones

A typical family saloon car will have a KE of about 0.5 MJ when travelling at the maximum speed permissible in the UK. If this type of car has a crash at this speed the KE has to be absorbed somehow – it cannot just disappear. The front and rear sections of modern cars are made to act like 'crumple zones'. The KE is used to deform these sections of the car rather than the stronger 'passenger cage'.

Airbags

In a crash the occupants will appear to be flung forwards. They will have KE, but unfortunately human beings do not have 'crumple zones'. Airbags now in production inflate immediately in front of the passengers when the car is in collision. They act as shock absorbers and absorb the KE.

Impact-absorbing bumpers

These relatively new bumpers consructed from man-made materials are designed to absorb energy on impact if the impact speed is small. Cars fitted with them may now have 'little' crashes and suffer no damage! They are of particular use if the accident involves pedestrians – they absorb the energy so the pedestrian feels less impact.

Once an object is moving it can cause other objects to move. Snooker is one such practical example where movement is passed on. When one snooker ball hits another, the first ball stops and the second carries on. This shows the passing of **momentum**. Any object that is moving has momentum. Increase the mass of the object and/or its speed and its momentum increases. Boxers are very good at passing on momentum!

momentum = mass × velocity
(units kgm/s) (kg) (m/s)

When a bullet is fired from a gun, it emerges from the barrel with a high velocity. It has mass and because it is moving it has momentum. At the same time the gun recoils backwards (in the opposite direction.) It too has mass and velocity and so it has momentum. The total momentum 'created' in either direction is the same and so cancels out. No net momentum in one direction has been created.

FORCES MAKE CHANGES HAPPEN

braking bending twisting
accelerating decelerating attracting
repelling turning deforming shaping

Forces are responsible for all these changes. At its simplest a force is a push or a pull. Pushing down on the pedals of a bicycle causes the wheels to turn and the bicycle and rider to move. Friction is one of the more common forces at work all around you. To slide a book across a table top requires a force to be applied – to overcome the force of friction. (Force and energy are not the same. Energy is used up only if a force moves!) A not so obvious example of a force is the force required to make an electric current flow in circuit. An electric force is called **potential difference** and is measured in volts.

FORCE FIELDS

It is well known that some materials behave in a strange way when placed near a magnet. They are affected by magnetism. No one can really tell you exactly what magnetism is. At best we say it is an area 'where things happen', an area of influence where some materials have an effect on one another. The name given to such an area of influence is a **field**. A simple magnet produces such a field, known as a **magnetic field**. Static electricity produces a static field (see page 117) and currect electricity produces an electromagnetic field (see page 126).

Magnetic materials

The most common materials used to make magnets are steel, cobalt, nickel and iron. A material that keeps its magnetism (e.g. steel) is said to be **magnetically hard**. These materials are used to make **permanent** magnets. Iron, on the other hand, is **magnetically soft** and loses its magnetism easily. These materials are used to make **temporary** magnets. An electromagnet (see page 126) has a soft iron core which loses its magnetism as soon as the electric current is switched off.

Application 6.4

ROCKETS AND THE OCTOPUS

In space, a rocket has nothing to push against, so how does it move? It expels a small mass of hot, fast-moving gas from the engine nozzle – this gas has momentum in one direction. As it is not possible to create momentum in only one direction something must move in the opposite direction. The rocket moves forward! Nature uses this idea too. The octopus shoots jets of water in one direction and it moves off in the opposite direction.

figure 6.10 A rocket and an octopus have a lot in common

The field around a source of magnetism

he magnetic field around a source of magnetism can e shown easily with the aid of some iron filings or a lotting compass. A typical **magnetic field pattern** for simple bar magnet is shown in figure 6.11a. The nds of the bar, where the magnetic field pattern ppears to start and stop, are known as **magnetic oles**. A magnetic field pattern is given a direction – om the **north magnetic pole** to the **south magnetic ole.**

ttraction and repulsion

he simple rule linking the effect of magnetic poles n one another is :

like poles repel
unlike poles attract

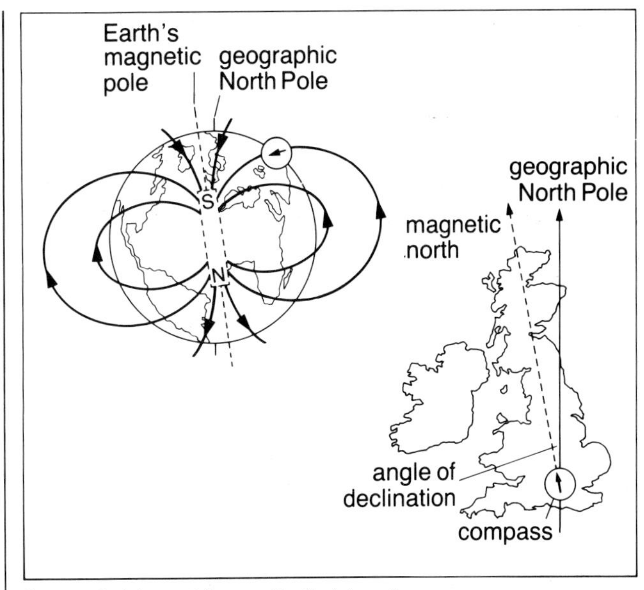

figure 6.11 a *Magnetic field patterns*

The three poles

he Earth is thought to act as if it had a huge bar agnet inside it. A **compass** is a small magnet that is ee to move. Each has a magnetic field. The two elds affect one another, but only the compass needle oves (because it is so small). The magnetic north ole of the compass points in a northerly direction. It a north-seeking pole. However the direction dicated by a compass is not true north and does not oint to the geographical North Pole.

here are three sets of poles:

a magnetic north poles and magnetic south poles
b north-seeking poles and south-seeking poles
c geographic North Pole and geographic South Pole

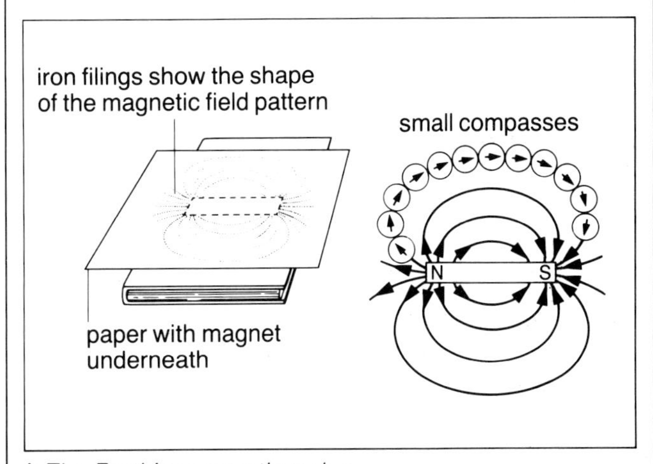

b *The Earth's magnetic poles*

INVESTIGATION 6.2

FORCES CAUSE ACCELERATION

(You should be familiar with the use of ticker timers and be able to measure accelerations from ticker tape before starting this investigation.)

Support a trolley runway at a gently sloping angle so that a trolley will just run down it without noticeably accelerating or decelerating. Fix the runway at this angle and do not change it.

- a At the top of the runway, with the ticker timer switched off, run a piece of tape through the timer and attach it to the rear of the trolley.
- b Hold the trolley firmly at the rear and attach one elastic cord to it.
- c Stretch the cord (say 20 cm) in front of the trolley.
- d Switch the ticker timer on and then release the trolley. It will begin to accelerate.
- e When the trolley has completed its run, save the tape.
- f Repeat the experiment with two elastic cords, keeping the stretched distance the same.
- g Repeat the experiment with three cords to provide a third tape.

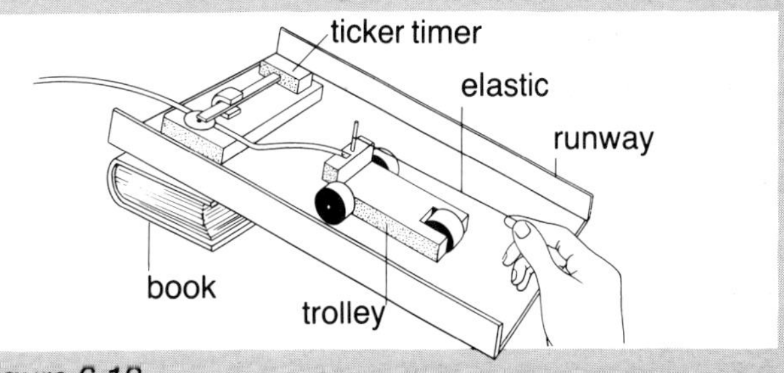

figure 6.12

Questions

1. How do the dot spacings change from the tape made with the smallest force (one cord) to the one made with the largest force (three cords)?
2. Do you have any evidence to support this idea: a larger force applied to a trolley causes a greater acceleration than a smaller force applied to the same trolley?
3. Calculate the acceleration created by each of the three forces.
4. Design an experiment to test the effect of 1 elastic cord on different masses, i.e. trolleys of mass 1 kg, 2 kg, 3 kg etc.

UNIT 6

The magnetic north pole of a compass points in a northerly direction. It can only be attracted by a magnetic south pole. The Earth's magnetic pole buried somewhere under the Arctic must be a magnetic south pole (see figure 6.11b, previous page).

Newton's Second Law

Whenever a force (measured in newtons) is applied to an object it will accelerate – so says Newton's Second Law. The larger the force the greater will be the acceleration.

$$\text{force} = \text{mass} \times \text{acceleration}$$
$$F = m \times a$$
(units N) (kg) (m/s^2)

e.g. A force of 6N will cause a mass of 3 kg to accelerate at 2 m/s^2. The acceleration will continue until the force is removed. (When this happens the velocity will remain constant.)

The force of gravity

Leaving an object stranded in mid-air on Earth is an impossible task. No matter what is done to the object it will always fall to the floor. As nothing moves unless a force is at work, then a force must be at work on the object to make it fall. We call this force **gravity**.

When something is released it begins to accelerate. The acceleration is caused by the force of gravity. This is a special example of Newton's Second Law at work.

force due to gravity (called **weight**) = mass of object × acceleration due to gravity (called *g*, usually taken as 10m/s²)

Mass is measured in kilograms; weight (a force) is measured in newtons. If a mass is placed where there is no gravity, e.g. in space, then the mass will be weightless.

For example, what is the weight of someone whose mass is 75 kg (approximately 10 st 7 lb)? Assume acceleration due to gravity on the Earth's surface is 10 m/s^2. What would be the weight of the same person in space?

weight (on Earth) = 75 × 10 = 750 N
weight (in space) = 75 × 0 = 0 N

Forces in balance

When two or more forces are in balance then nothing moves. An example of this would be a group tug-of-war. Only when one side pulls harder than the other do things start to move!

A laboratory spring balance works in this way. Hanging a weight on the bottom provides a downwards force (due to gravity). This causes the spring to stretch. When the tension in the spring (trying to pull it back to its original size) exactly equals the object's weight, no more movement will occur. A heavier weight will stretch a spring more than a lighter weight. The extension of the spring gives an idea of how heavy the weight is. **Hooke's Law** says that

The extension of a spring is directly proportional to the load force applied as long as the elastic limit is not exceeded.

The **elastic limit** is the point at which the spring will no longer return to its original length when the extending force is removed. (When the elastic limit has been exceeded the spring will never return to its original size.)

A spring obeying Hooke's Law would, for example, be expected to double its extension if the load force were doubled. See figure 6.13.

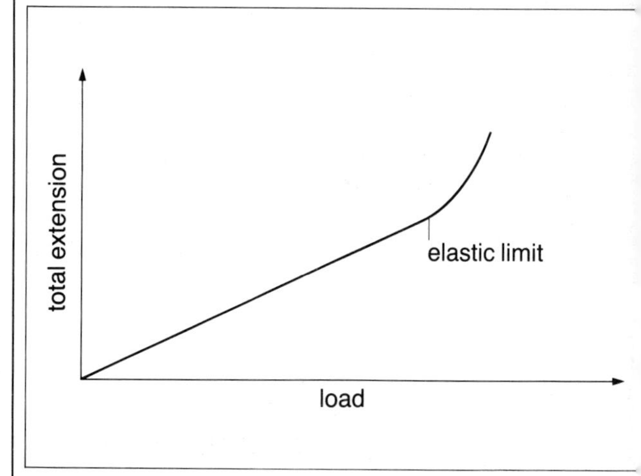

figure 6.13 *Hooke's law for a spring*

UNDER PRESSURE

When you stand up all your weight presses down on the floor. However, it is not all concentrated on one point (unless you stand on the heel of a stiletto) but is spread out over the area of your shoe that is in contact with the ground. The **spreading out of a force** over an area is known as **pressure**.

$$\text{pressure} = \frac{\text{force}}{\text{area}}$$

$$P \text{ (units N/m}^2\text{)} = \frac{F \text{ (N)}}{A \text{ (m}^2\text{)}}$$

The unit is the newton per metre squared (N/m^2). This is also called the pascal (Pa).

$$1 \text{ N/m}^2 = 1 \text{ Pa}$$

It is often more convenient to use N/cm^2.

INVESTIGATION 6.3

HOOKE'S LAW

Table 6.2 shows some useful approximate conversions.

mass (kg)	weight (N)
1	10
0.1 (100 g)	1

table 6.2

a Hang a spring from a stand. Attach a pointer to the bottom of the spring.
b Line up the zero mark on a ruler with the pointer (see figure 6.14).
c Hang a 1 N weight on the bottom of the spring.
d Record the extension of the spring.
e Add a second 1 N weight to the bottom of the spring and record the new total extension.
f Continue adding weights (maximum 10) and recording the total extension.
g Plot a graph of extension against load.
h Repeat the experiment using in place of the spring
  i a rubber band
  ii a circular loop cut from a plastic bag.

Questions
1 Do the graphs show the materials to obey Hooke's Law – even if only for a small extension?
2 What are the elastic limits for each material?
3 Could you now make predictions for the effect of using two springs or two rubber bands
  i in series ii in parallel?

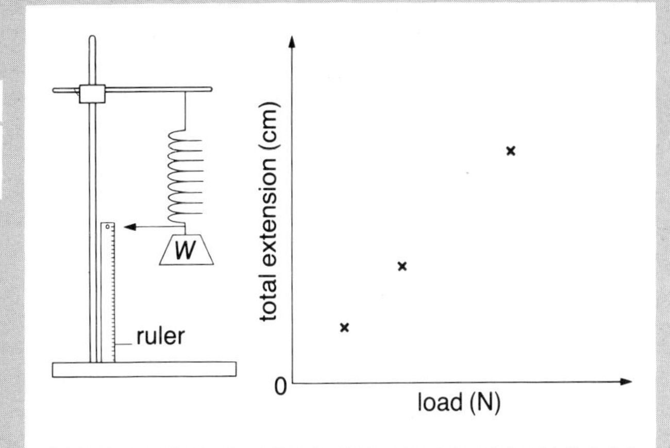

material 1		material 2	
force (N)	extension (cm)	force (N)	extension (cm)
1		1	
2		2	
3		3	
4		4	
5		5	
6		6	
7		7	
8		8	
9		9	
10		10	

figure 6.14 *Investigating Hooke's law*

Application 6.5

BOUNCING HEDGEHOGS – NATURALLY?

Has anyone seen a hedgehog bounce?

PRICKLY mammals have a neat line of self-defence. Their hair is modified to form protective armour – the hedgehog has its spines and the porcupine its quills. An investigation into the properties of these spiky objects, however, has shown that the design and probable function of spines and quills is quite different. J. F. V. Vincent and P. Owers of University of Reading suggest that porcupine quills are designed as defensive spears, but the hedgehog's spines probably act more as shock absorbers than as weapons.

Vincent and Owers determined the mechanical properties of these structures by compressing them end-on and measuring how much they bent or buckled, if at all. They found that the natures of the quill and spine were virtually opposite to each other: the quills were as long as possible but fairly stiff, whereas the spines were as short as possible yet still able to bend (*Journal of Zoology, vol 210, p 55*).

Quills must be long enough to fend off an enemy but not so slender as to bend too easily. Interestingly, the elasticity of the quill varied along its length; under pressure, the top quarter was likely to break off completely rather than buckle. So, the end of the quill can penetrate, break off, and remain in the tissues of an assailant. Inside, the quills are filled with a foam-like material, which the researchers suggest might support the keratin wall of the quill and so delay any damaging and irreversible buckling.

Hedgehog spines are short and relatively fat. They bend under pressure, but only because they are naturally curved. If they were straight, they would suddenly give way when compressed and buckle irreversibly. Their internal structure, comparable to a roll of corrugated cardboard, with supporting discs at intervals, maintains the circular cross-section of the spine and prevents localised collapse. The hedgehog spine, the researchers say, is designed to absorb energy by bending elastically.

Possession of all-round shock absorbers is presumably of great benefit to an animal which reportedly climbs trees and takes the quick way down by rolling into a ball and simply dropping to the ground. There is a slight problem here, however, as Vincent and Owers point out: the energy absorbed by the spines should be released ... has anyone seen a hedgehog bounce? □

The long and the short of it ... quills and spines are designed for different purposes

figure 6.15

Application 6.6

WINDOW PANES

High winds exert high pressures on window panes. The larger the window the more likely it is to shatter in a gale. When buildings are being designed it is necessary to use glass of suitable strength. The table shown in figure 6.16 lists probable maximum pressures exerted by winds of different speeds. The graph is used to indicate the thickness of glass needed to withstand a particular wind pressure for a particular window size. (There is a number of different graphs according to the type of glass required.)

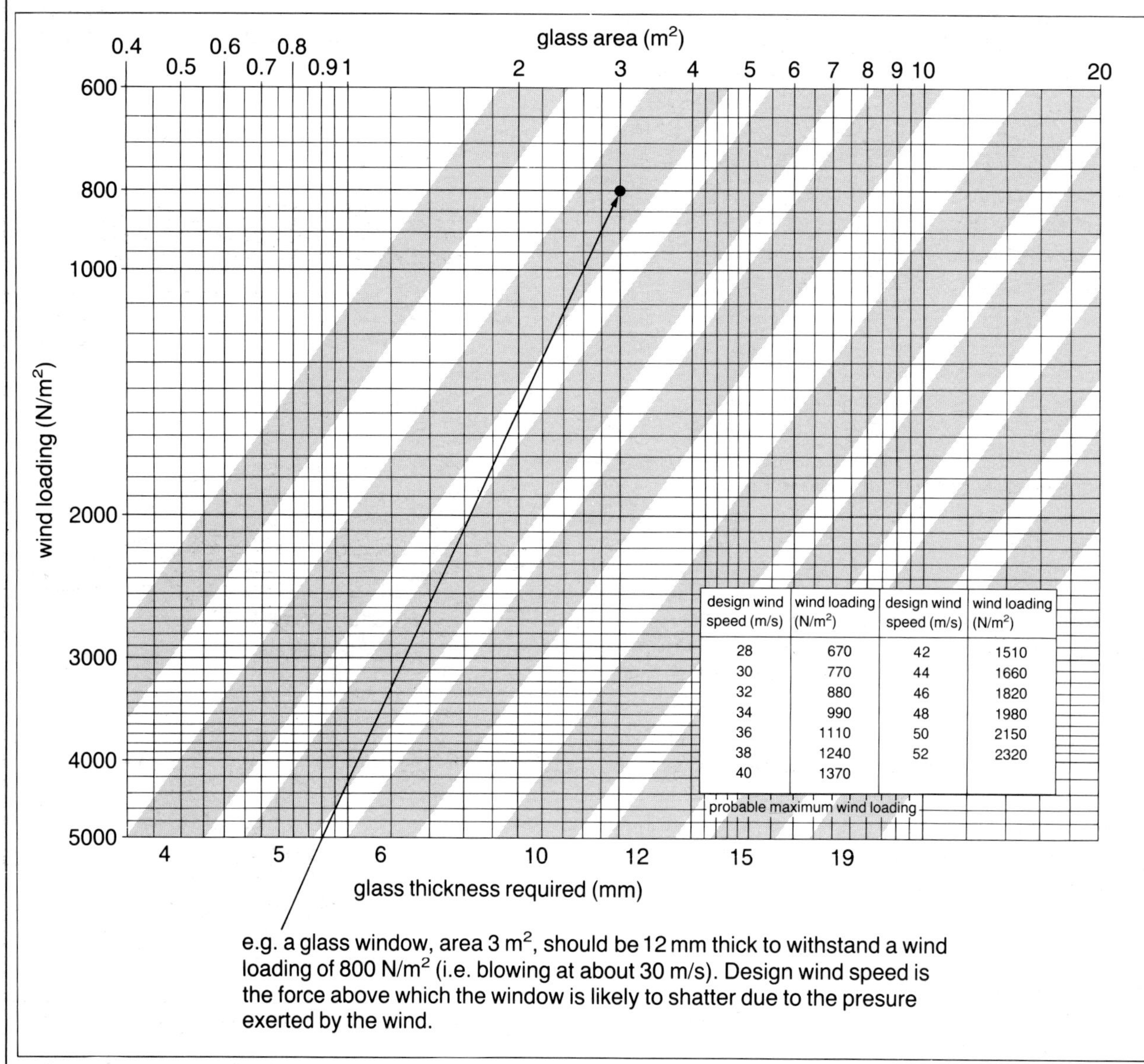

design wind speed (m/s)	wind loading (N/m^2)	design wind speed (m/s)	wind loading (N/m^2)
28	670	42	1510
30	770	44	1660
32	880	46	1820
34	990	48	1980
36	1110	50	2150
38	1240	52	2320
40	1370		

figure 6.16 *Choosing the right glass*

Measuring pressure

Pressure can be measured quite easily in a laboratory. An outline is drawn round the shoe of somebody standing on cm^2 graph paper. An estimate is made of the area of the shoe in contact with the ground. The instep area should not be included. The estimate of the area is doubled to allow for both feet. The person's weight in newtons may be known, or an approximation can be obtained from the following conversion chart.

1 kg	=	10 N
1 stone	=	70 N
1 lb	=	5 N

Using the equation $P = F/A$ the under-foot pressure can easily be calculated (in N/cm^2). Typical values would probably be pressures of around 3–5 N/cm^2). Someone standing on the heel of a stiletto could generate a pressure (e.g. 250 N/cm^2) far exceeding that under the foot of an elephant (e.g. 12 N/cm^2) though, of course, the total weight (force) of an elephant would be considerably larger!

Pressure in solids and fluids

The **transmission** of pressure is different in solids and fluids. In solids it acts only in one direction, the direction in which the force is acting. A drawing pin shows pressure transfer in solids. A force applied at the flat end of the pin (at low pressure) is transferred to the sharp end of the pin (at much higher pressure). Fluids, on the other hand, are able to move and flow in any direction. The force applied to water by a garden tap is transferred to the end of a hose pipe, no matter how bent or twisted the hosepipe is. In **fluids**, pressure acts in all directions. In fluids pressure also increases with depth. This is because the greater the depth of fluid in a container the greater will be the force at the bottom of the fluid, because a tall column of fluid has a greater weight than a short one.

Flight

Flight is the result of lift which is **created by a difference in pressure** between the two sides of a (wing) surface.

In figure 6.17a the pressure on the underside (pushing up) is greater than the pressure on the top surface (pushing down). The barrier or surface will tend to move upwards. The same effect can be produced by streams of fluid passing either side of the barrier. The stream which moves faster creates a lower pressure on the barrier. The barrier will tend to move in the direction from high to low pressure i.e. upwards (see figure 6.17b). This idea is known as **Bernouilli's Theorem**.

The two fluid streams can be created from a single stream by shaping the barrier (see figure 6.17c). A wing section splits the fluid stream into two parts. The part which travels over the top surface has further to go so it creates a lower pressure than the slower moving fluid travelling under the wing. The difference in pressure provides lift and the wing will rise. This lifting effect can be used just as easily with a liquid such as water as it can with air. A boat or yacht sail works on this principle to create forward motion (see figure 6.17d).

Lifting forces can easily be demonstrated by holding the edge of a sheet of paper close to your bottom lip and blowing steadily forwards over the top of the paper. The fast moving air will create a low pressure over the top surface of the paper. This will be less than the atmospheric pressure under the paper and so the sheet will begin to rise!

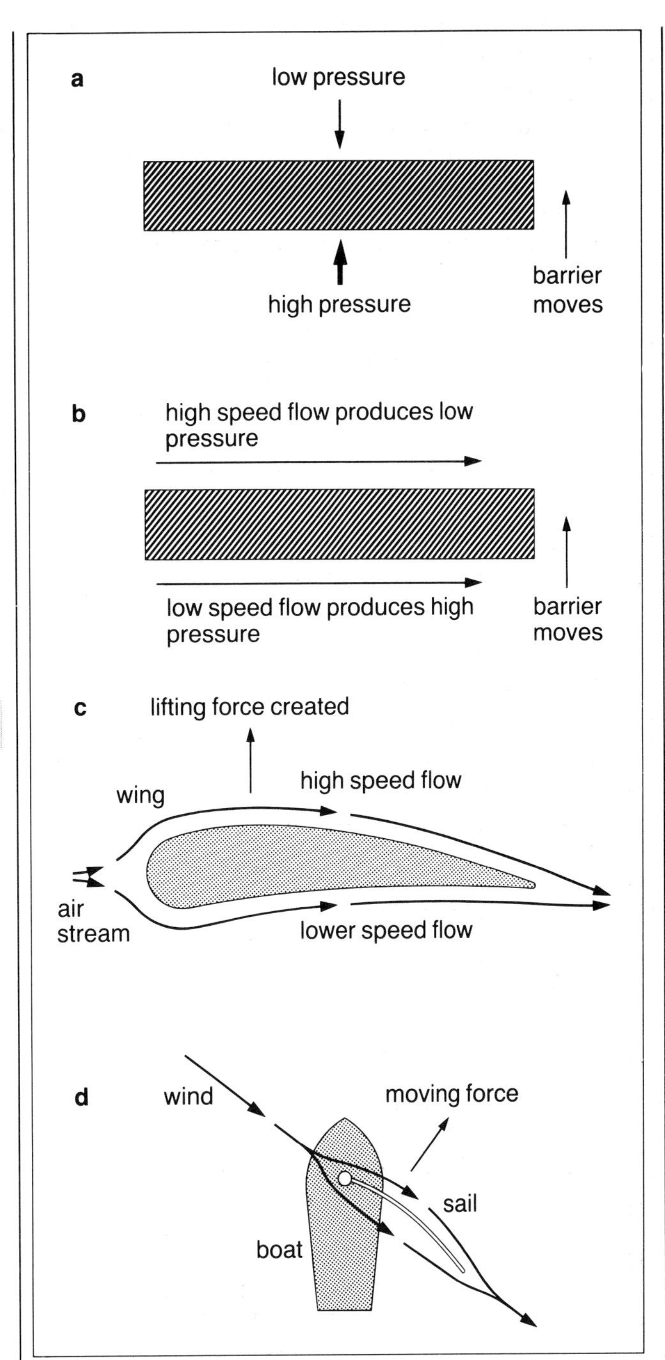

figure 6.17 Pressures involved in flight

figure 6.18 Voyager on her round-the-world non-stop flight

Application 6.7

LEVERS – MOVING LOADS AND MOVING JOINTS

A machine is something which makes a job easier. A lever is a very simple type of machine. It is a machine that helps us to move weight more easily. An **effort** is applied at one point on the lever. The lever **pivots** at the **fulcrum**. The effort force moves the load force at another point on the lever. Levers are valuable to us because we can use them to magnify the force we put in. They act as **force magnifiers**. Levers like this have the fulcrum much closer to the load force than to the effort force. Using a crowbar to lift a heavy stone is an example of a lever being used in this way.

Levers can be used in another way if the fulcrum is much closer to the effort force than to the load. The small distance moved by the effort force can be converted into a much larger distance moved by the load. A lever used like this is described as a **distance magnifier**. Someone using a fishing rod can draw in the line for long distances using relatively small movements of the arm.

Many joints in the body work as distance magnifiers. The joint itself acts as the fulcrum. A muscle is often attached to the bone close to the joint. The muscle contracts a small distance and pulls on the bone. The far end of the bone which is carrying the weight is moved through a much larger distance.

Levers can also be classified as one of three types according to the relative position of the fulcrum, load and effort. Examples of these three types are shown in figure 6.19.

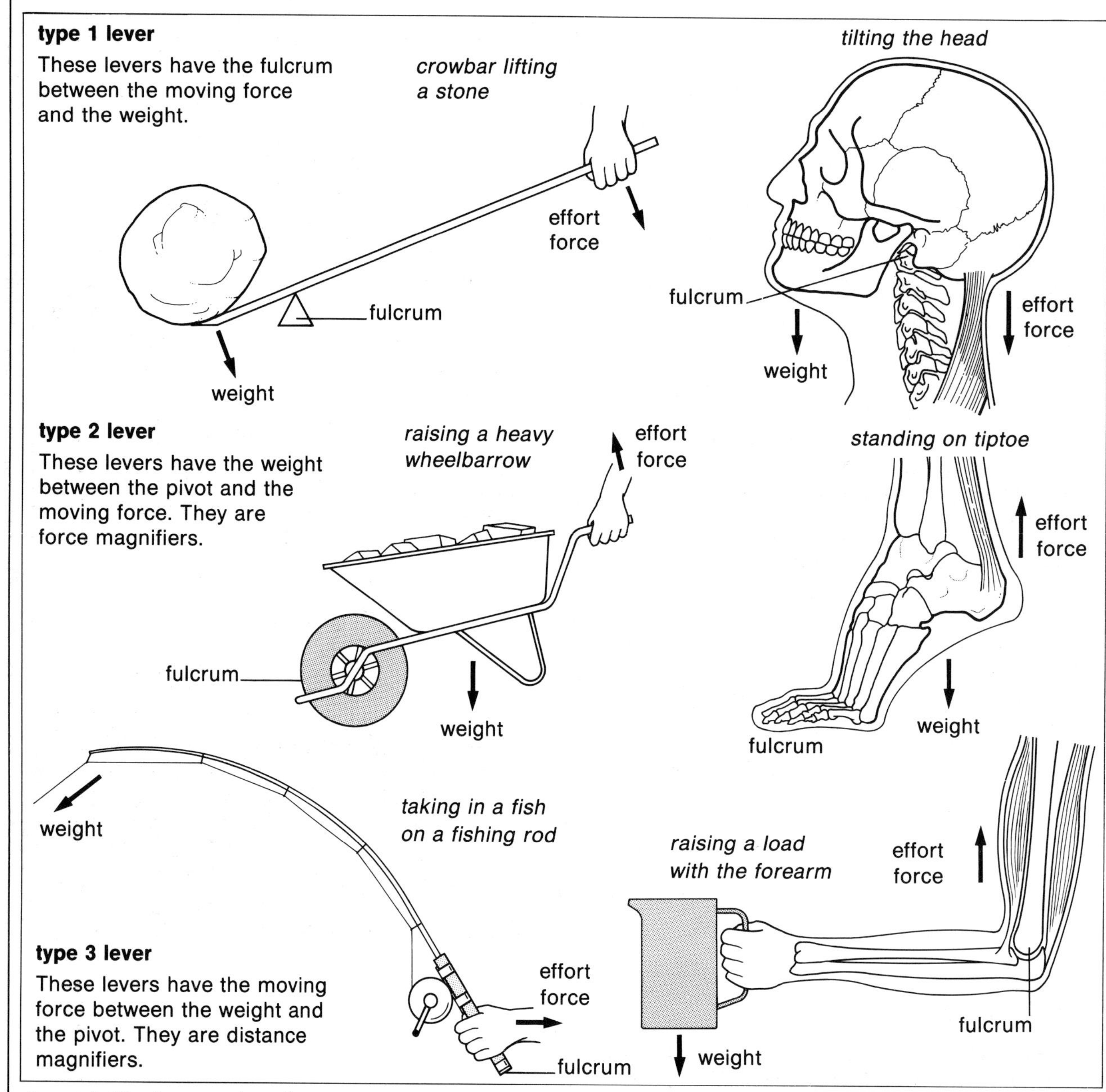

figure 6.19 Levers at work

INVESTIGATION 6.4

FORCES AT WORK

Another example of a machine is a slope – an object which is too heavy to lift can be raised from the ground by pushing it up the slope. The force required is then much less. (The Egyptian pyramids are thought to have been built in this way.) The diagrams in figure 6.20 show a number of simple machines.

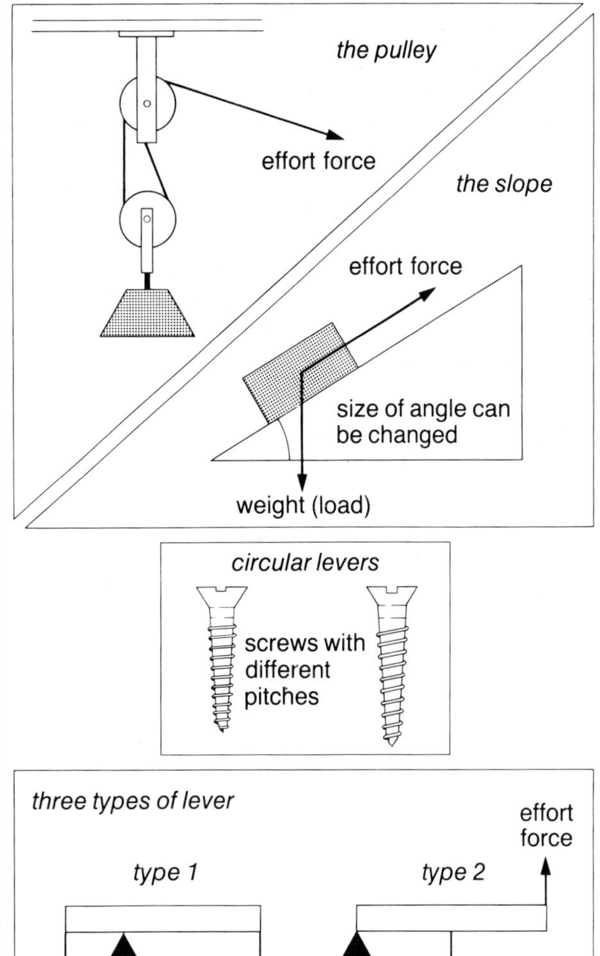

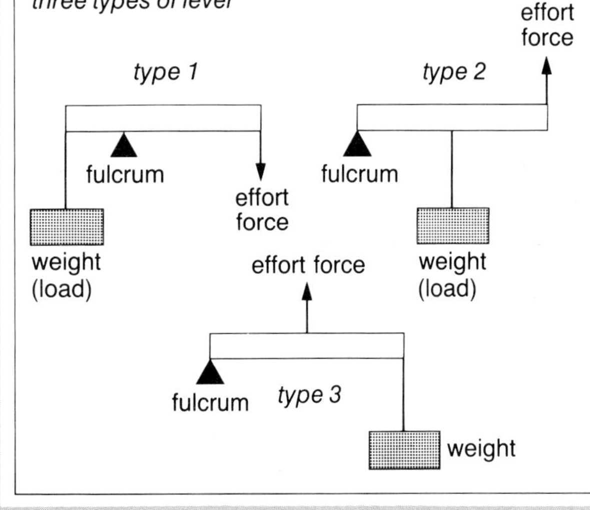

figure 6.20 Simple machines

a Design a series of experiments to test each of these machines in turn.

b By making suitable measurements compare the weight of the object moved with the effort force required to move it.

Questions

1 Can you make predictions about the effect of
   i increasing the angle of the slope
   ii increasing the number of pulley wheels in a pulley system
   iii moving the pivot to the right or left in *type 1* levers, *type 2* levers and *type 3* levers?

2 In effect each of these machines has
   i an advantage
   ii a disadvantage.
   What are they?

3 How many practical situations can you find which use these types of levers?

Time for thought

They say that Concord 'stretches' by quite an amount when it flies at supersonic speeds. Why do you think this might be? (Hint – friction causes heating.)

MOVEMENT IN ANIMALS

The skeleton

The human skeleton consists of 206 bones. Figure 6.21 shows the skeleton and the names of the important bones. As well as providing a rigid framework that gives shape to our bodies, the skeleton has several other functions.

a Delicate organs are protected by the skeleton. The brain is entirely enclosed within the **cranium** which is 1 cm thick. The heart and lungs are situated inside the **rib cage**.

b Red and white blood cells are manufactured in bone marrow located in the centre of the **ribs**, **sternum** and **femur**. A bone marrow transplant can sometimes save the lives of people with certain blood diseases (see application 3.8, page 60).

c The bones of the skeleton are connected at joints. Muscles attached to bones pull on them and allow us to move our joints in many different ways.

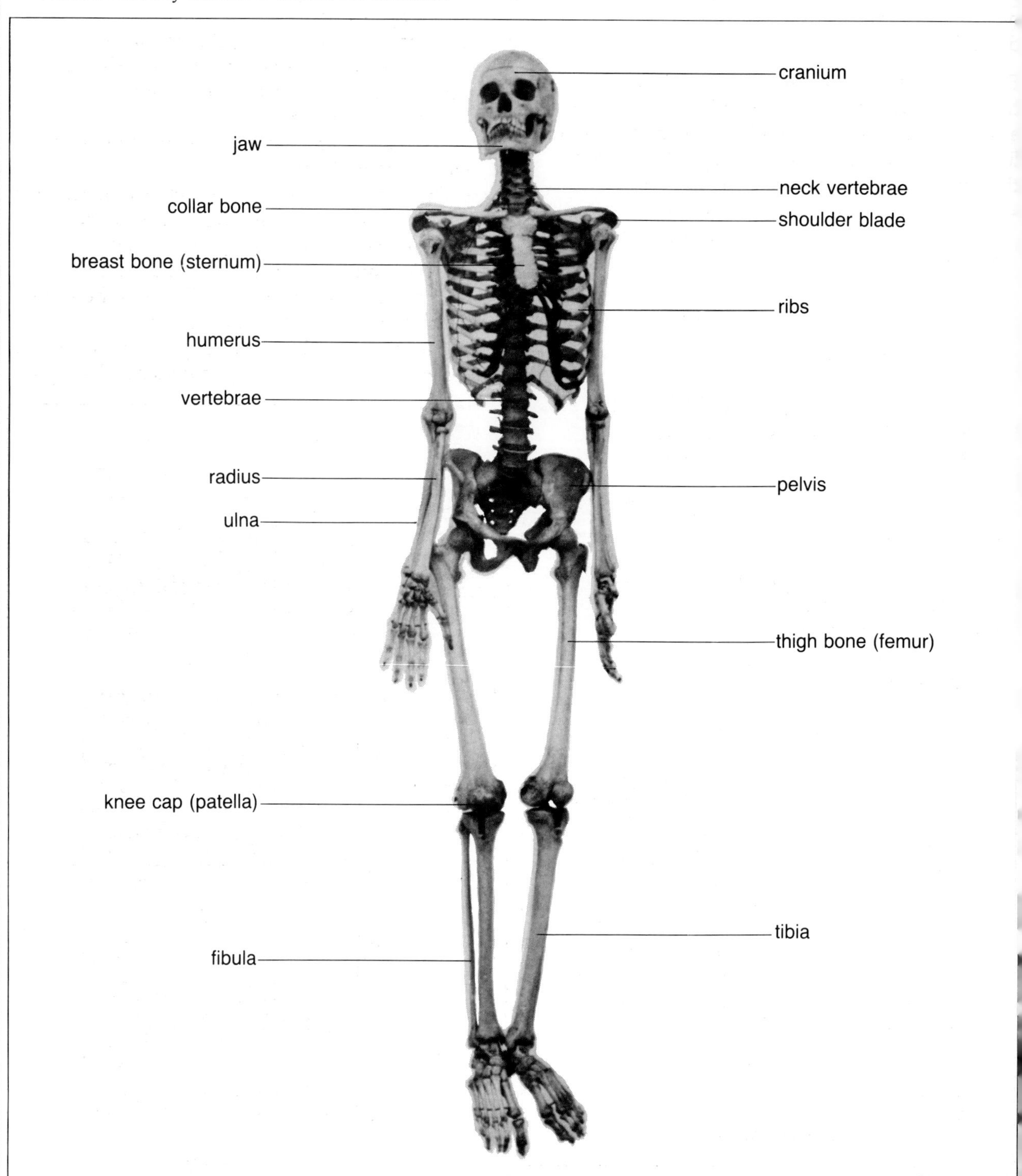

figure 6.21 *The human skeleton*

Muscles and movement

Each end of a muscle is connected to a bone by means of a tough **tendon** that will not stretch. One end of the muscle is fixed to a bone that remains still. This is known as the **origin** of the muscle. The opposite end which is fixed to the bone that moves is known as the **insertion**. The muscle pulls on the bone by contracting. This means that it shortens and gets fatter. When the muscle relaxes it lengthens but cannot exert a pulling force. Another muscle is needed to contract and pull the bone in the opposite direction.

Figure 6.22 shows how two muscles, the triceps and the biceps, work together to bend and straighten the arm. Muscles all over the body are arranged in pairs like this. Because they work in opposite directions they are known as **antagonistic** muscle pairs.

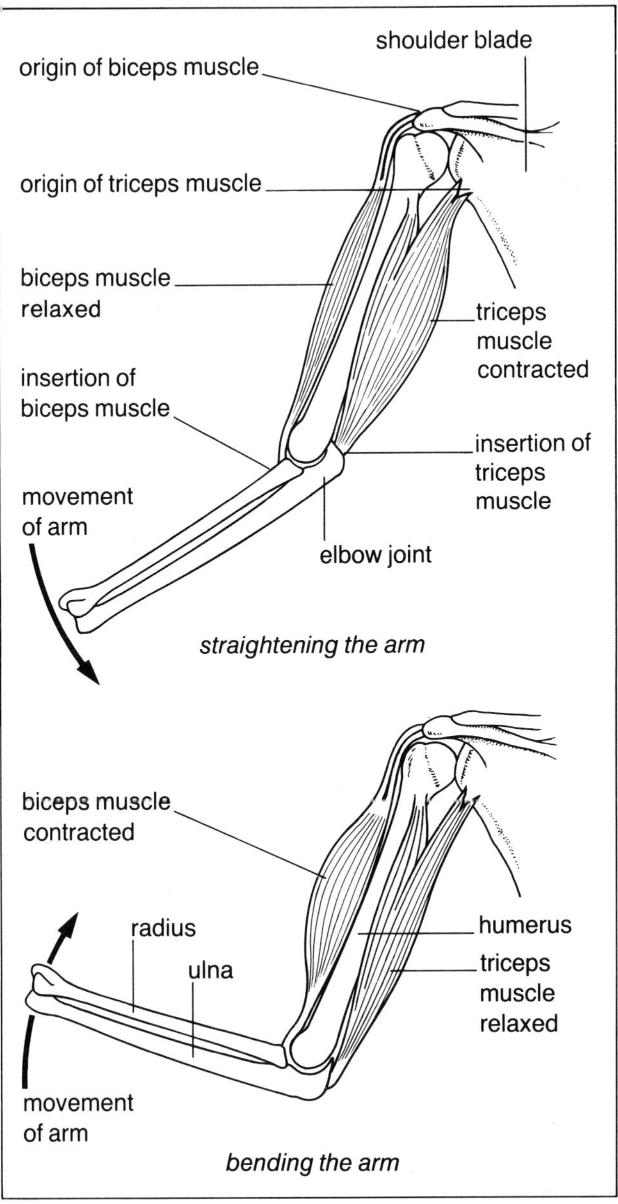

figure 6.22 Antagonistic muscles working together to bend and straighten the arm

ALL YOUR OWN WORK

1. A ________ or ________ speed is one that does not change.
2. An ________ is an increase in speed and is measured in ________ .
3. Why is the idea 'average speed' a false one but is useful in practice?
4. How can the KE of an object be changed?
5. How would you explain inertia to someone?
6. Why should a weighing machine not be called a weighing machine?
7. The driver of a car moving at 20 m/s along a straight level road applies the brakes. The car decelerates at a steady rate of $-5\ m/s^2$.
   a How long does the car take to stop?
   b What kind of force slows the car down, and where is it applied?
   c The mass of the car is 500 kg. How much kinetic energy has to be 'lost' and where does it go?
8. Another method for finding the distance travelled during a journey is to find the area under a velocity – time graph. This has the same numerical value as the distance. Find the distances travelled in the three journeys shown in figure 6.23 by finding the area under each graph.

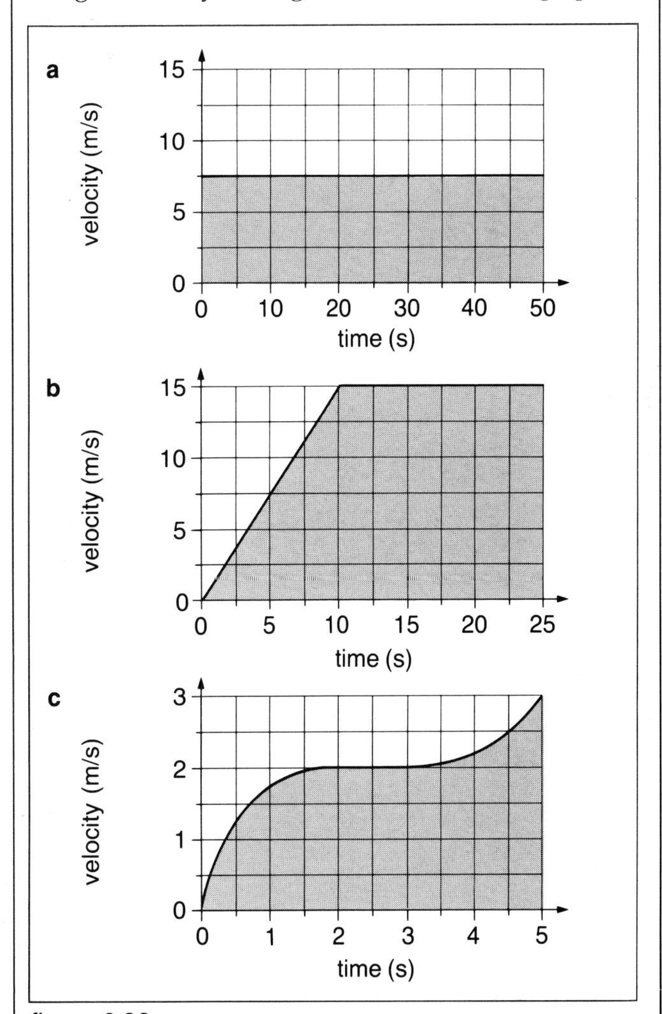

figure 6.23

UNIT 7

OUT OF THIS WORLD

GOING ROUND IN CIRCLES

Nature has a way of keeping things in order, even if chaos appears to rule occasionally. Nature is marvellous at creating ordered systems – look at the regular shape of a snow crystal, for example. How many natural things can you think of that are based on the circle or the ellipse?

figure 7.1 Nature keeps things in order

If the 'big bang' idea about the creation of the universe is correct, then all matter in the universe was thrown off in all directions at the beginning of time. Vast swirling clouds of material were sent off into space. After a while some of these clouds began to come together, rather like water going down a plughole, the material being pulled together by gravitational attraction. This was the starting point for the formation of the star system.

Stargazing

On a clear evening it is possible to see a vast number of stars. With the aid of only a small telescope or pair of binoculars it is possible to see them in much finer detail. What appears to the naked eye as a misty patch becomes a cluster of stars. These clusters are **galaxies**, and there are countless numbers of these too. The galaxy to which we belong is the Milky Way. It is sometimes referred to as **The Galaxy**. Our star system looks rather like a bicycle wheel laid on its side (see figure 7.2). It is enormous. One estimate suggests that there are as many as 100 000 million stars in the Milky Way alone, other estimates double this number. The sheer size of our galaxy is staggering. It is not really possible to picture just how

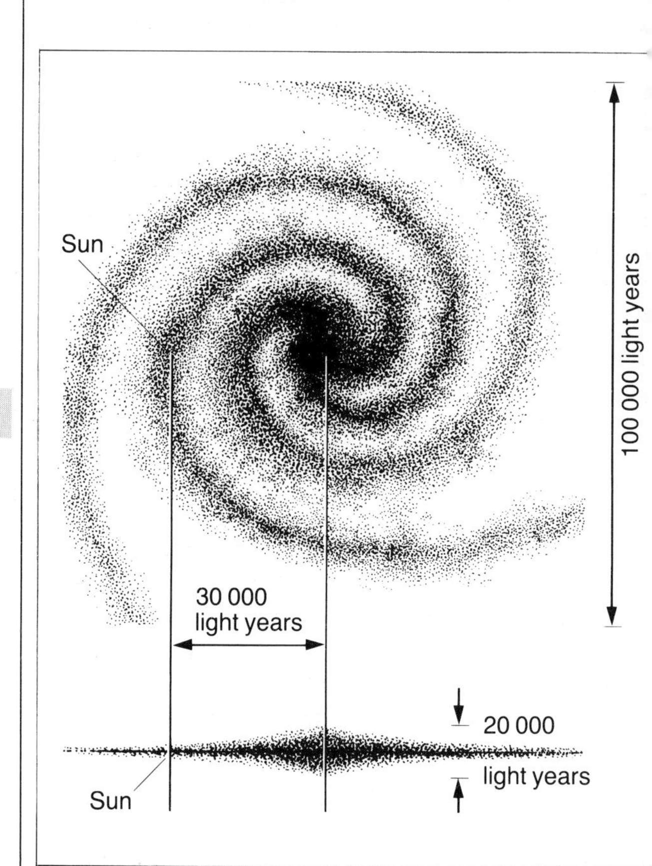

figure 7.2 The Milky Way galaxy

large it is, and yet the universe contains a vast number of galaxies like this. Current thought suggests the universe is expanding too!

On its passage through space the Milky Way revolves slowly. It will take an estimated 220 million years to revolve once completely. As anyone who watches the stars will tell you, the stars seem to move across the sky. Their motion looks relatively slow. This is because they are such vast distances away. As our senses don't register the Earth's rotation, we are fooled into thinking that it is we who are still and that the stars move.

A similar effect can be observed in a motor car. Suppose you are travelling in a car at 60 m.p.h. (approx 100 km/h) along the motorway. If a car were to pass at 70 m.p.h. (approx 120 km/h) it would only go past you slowly though both of you were moving at speed. (If you were in the faster car the slower car would appear to be going backwards!) Standing in the Northern Hemisphere (the half of the Earth north of the equator) the stars appear to rotate in an anticlockwise direction.

How large is large?

How long is a piece of string? To answer this rather trivial question is easy if you have a ruler. When thinking about astronomical distances the question 'How far?' becomes very difficult to answer. There are no real astronomical rulers. To begin to get an idea of astronomical size we need a measurement that has meaning but that is large enough to be used in the vast expanse of space. The distance between the Earth and the Sun is one such distance that can be used as a 'space ruler'. Unfortunately this is really only of any use in and around our solar system. This distance is approximately 150 000 000 km (93 000 000 miles). This is called 1 **astronomical unit** or 1 AU. (It certainly saves space on a page.)

When thinking of large distances outside our solar system a much longer 'space ruler' is needed. The speed of light has been known for a long time and it can be measured accurately in a school laboratory. Light travels at approximately 30 000 000 km/s (186 000 miles/s). In one tropical year (365.242 days), light travels 9 460 528 405 000 km (5 878 499 814 000 miles). This enormous distance is known as a **light year**. Inter-stellar (between-star) distances are described in light years. The nearest star to our solar system is Proxima Centauri at a distance of 4.28 light years.

The Sun

The star at the centre of our solar system is the Sun. On the surface it seems to be just a ball of fire. Inside it is a huge nuclear reactor (rather like a continually exploding set of hydrogen bombs, see page 186). It consumes several million tons of fuel per second, but even so it will be many thousands of years before its character changes noticeably. The Sun, like any star, has a life cycle, but unlike our own its lifespan is measured in thousands of millions of years. The Sun is thought to be approximately 5000 million years old and about half way through its life as a yellow dwarf. At the end of this stage it will begin to grow, slowly changing into a red giant. As it expands and changes colour it will engulf the inner planets and possibly in time the Earth too.

The Sun's appearance does change, even from day to day. Often, dark patches called **sunspots** appear on the surface. Most disappear fairly quickly but other larger ones may last for several months. This is very convenient as it allows us to observe and measure the Sun's rotation. It spins on its axis once every 25.4 days.

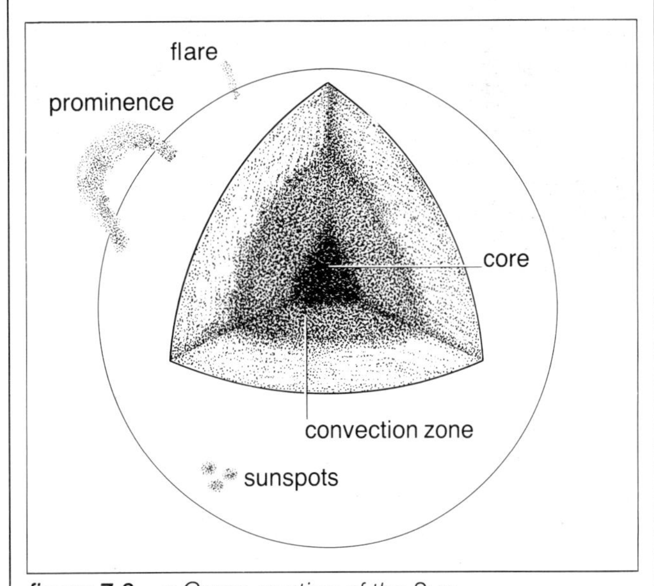

figure 7.3 a *Cross-section of the Sun*

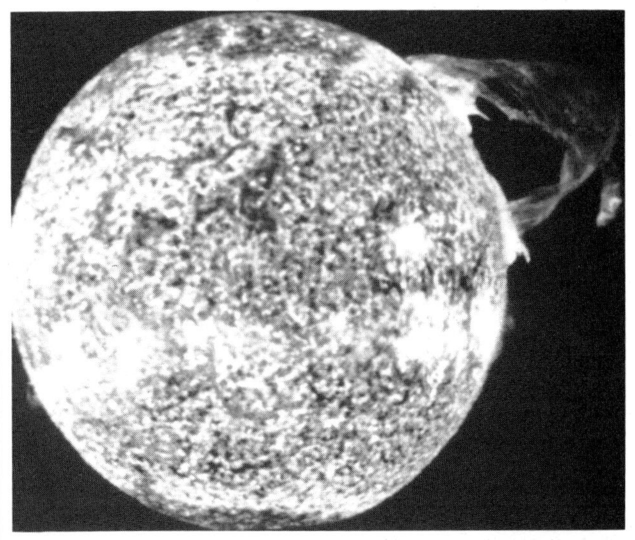

figure 7.3 b *Solar prominence taken from Skylab, December 1973*

Sunspots are not the only change we can see from Earth. Occasionally the surface will erupt and a huge gas cloud will rise from the surface (see figure 7.3b). These **solar flares** or **prominences** usually fall back to the surface, but sometimes they are emitted with such force that they escape into space. Apart from the solar flares and the sunspots, which seem to run in cycles of about 11 years, the Sun changes very little.

The solar system

In 1543 Nicolaus Copernicus proposed a major change in astronomical thought. He dared to suggest that a system of planets orbited a central body – the Sun. At the time this was thought outrageous – to suggest that the Earth was not the centre of the universe! It was certainly against the religious teachings of the time.

The solar system is the series of planets, planetoids and asteroids that orbit the Sun. It now also includes the hundreds of artificial satellites launched by man. There are nine known planets in the system in 1987. The hunt is on for the tenth planet which is thought to exist but has yet to be observed. Pluto, the outermost planet, was only discovered in 1930. For convenience the planets are divided into two sets.

the inner planets – Mercury, Venus and Earth

the outer planets – Mars, Jupiter, Saturn, Uranus, Neptune and Pluto

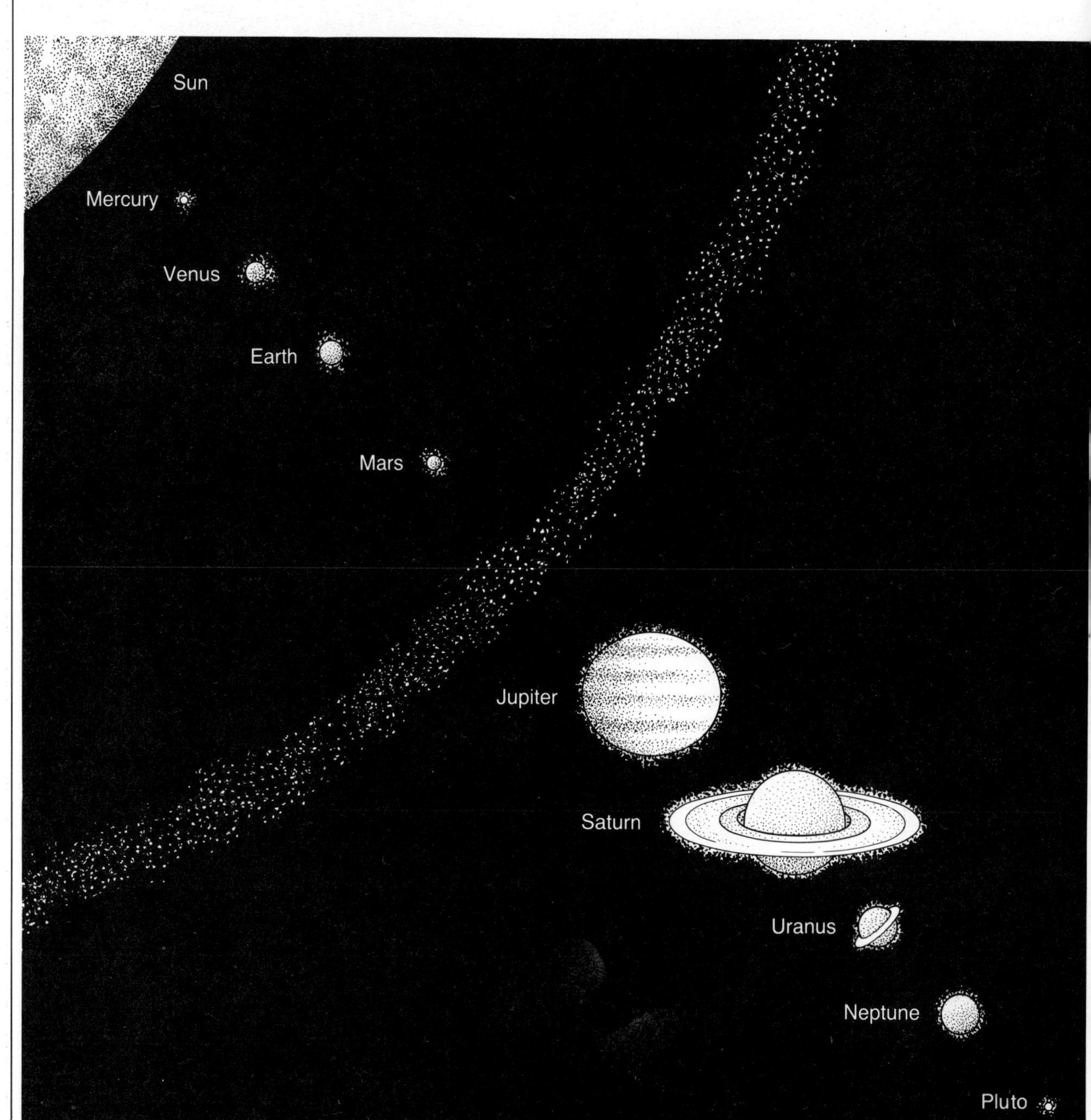

figure 7.4 The solar system

	length of day (measured in Earth days)	length of year (measured in Earth days (d) and years (y))	number of moons	axis tilt (°)	ringed planet	distance from Sun (AU)	diameter (measured in Earth diameters)	mass (measured in Earth masses)	density (times the density of water)	surface gravity (times the surface gravity of Earth)	atmospheric pressure (millibars)	surface temperature (°C) min	max	velocity needed to escape from planet's gravity miles/s	km/s	particular details
Sun	25.4	-	-	7		-	109	3.33 ×10⁶	1.4	28	-	+5660		383	617	
Mercury	176	88d	0	0		0.39	0.39	0.06	5.4	0.38	almost zero	−180	+420	2.6	4.2	
Venus	120 R	225d	0	2		0.72	0.95	0.82	5.2	0.89	100000	+470		6.4	10.4	
Earth	1	365d	1	23		1.00	1.00	1.00	5.5	1.00	1000	−88	+58	6.9	11.2	
Moon	27		-	-		-	0.273	0.012	3.2	0.16	-	−162	+117	1.5	2.4	
Mars	1.03	1.9y	2	25		1.52	0.53	0.11	3.9	0.38	<10	−120	+30	3.1	5.0	
Jupiter	0.41	11.9y	16	3	✓	5.20	11.27	318	1.3	2.54	-	−120		37.4	60.2	
Saturn	0.43	29.5y	20 min	26	✓	9.54	9.44	95	0.7	1.06	-	−110		27.5	36.3	
Uranus	0.65 R	84y	7 min	97	✓	19.2	4.10	15	1.31	1.07	-	−160		13.4	21.5	
Neptune	0.66	165y	2/3	28	✓ •	30.1	3.88	17	1.75	1.2	-	−160		14.8	23.8	• probable part ring
Pluto *	6.4 R	248y	1•	0		39.4	0.2	0.02	3.0	0.42	-	−220		2.2	3.6	• thought to be a double planet with its moon Charon
number 10	?	?	?	?	?	?	?	?	?	?	?	?			?	unofficial name 'Humphrey'

R = retrograde motion (spinning an axis in opposite direction to the direction of Earth's spin) * = data unreliable

table 7.1 The planets in the solar system

INVESTIGATION 7.1

THE PLANETS

From table 7.1 plot two bar charts on a set of axes similar to the ones in figure 7.5.

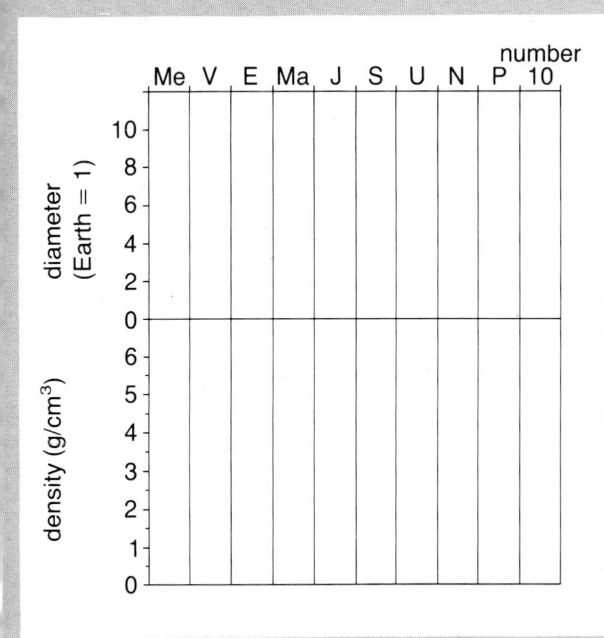

figure 7.5 Size and density in the solar system

Questions

1 Can you identify any patterns or odd-ones-out from your charts?

2 If planet 10 exists how might it appear on your charts? Could you make any suggestions to an astronomer as to what kind of planet to look for?

Mercury

In common with our moon, Mercury is small and covered with craters. Because it is so small (the Earth is approximately 20 times larger) its gravitational attraction is also small. Mercury is unable to hold on to an atmosphere. It does, however, experience an enormous gravitational attraction towards the Sun. Because of this gravitational attraction it has to whizz around the Sun at an incredible speed so as not to be pulled in. It travels at more than 110 000 m.p.h. (170 000 km/h) in its elliptical orbit. Because it is moving at such high speed and it is so close to the Sun, its years are quite short. A year on Mercury would last only about 88 Earth days. Strangely, it is spinning slowly on its own axis which gives Mercury a day that lasts 120 Earth days. (This gives Mercury a strange calendar with only two days in every three years.) It is a planet of extremes. At night the temperature falls to −180 °C and yet in the heat of the day it rises to +420 °C. With a temperature change of 600 °C and no atmosphere, Mercury is a very inhospitable world. The space probe Mariner 10 encountered the planet in 1973, 1974, and again in 1975. See figure 7.6a.

Venus

Venus can easily be seen with the naked eye, usually quite low on the horizon. With the exception of the Sun and the Moon it is the brightest object in the sky. The planet has fascinated people throughout the world for centuries. In Greek mythology Venus is the Goddess of beauty.

figure 7.6 *a Mercury from Mariner 10 showing craters*

figure 7.6 *b Venus taken from Venera 13*

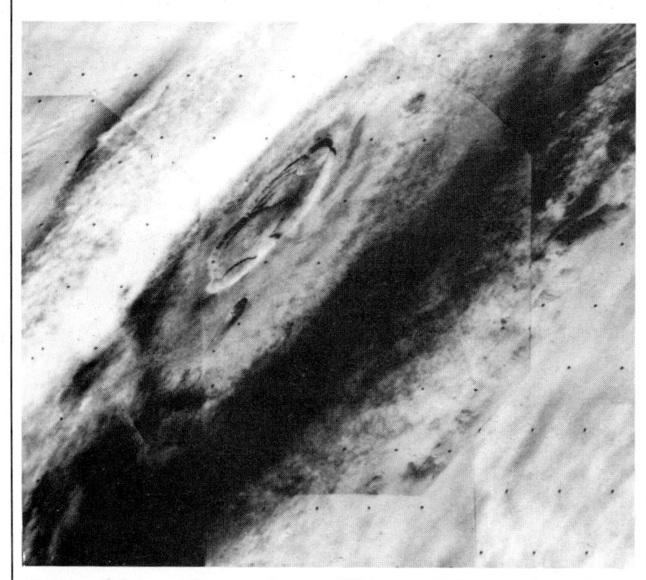

figure 7.6 *c Mars volcano Olympus Mons taken from Viking*

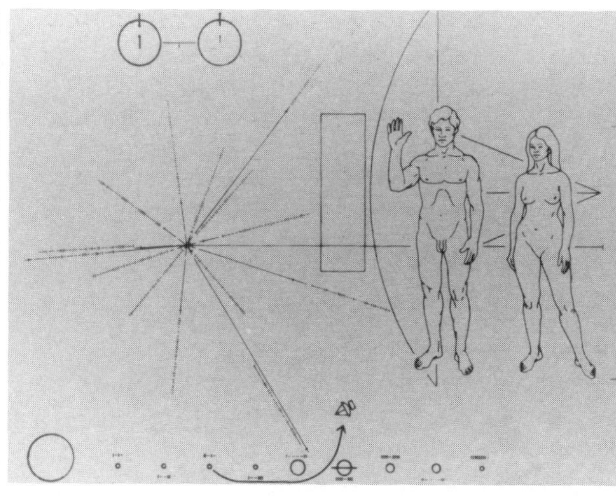

figure 7.6 *d Plaque from Pioneer 10*

figure 7.6 *e Saturn taken from Voyager 1 in November*

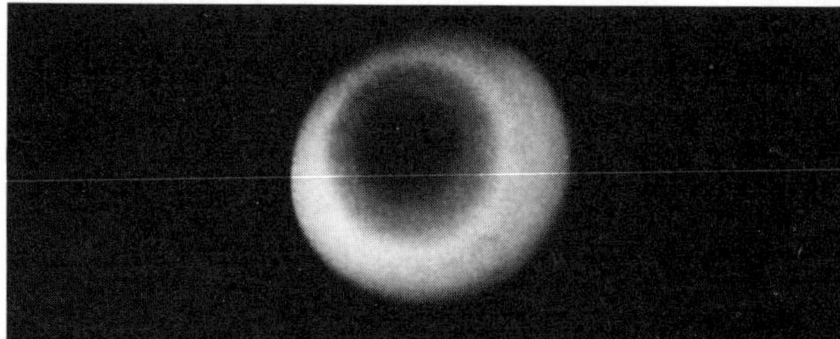

figure 7.6 *f Heat image of Uranus showing its rings*

figure 7.6 *g Moon Daedalus crater taken from Apollo 11*

Unlike the stars, Venus appears as a small dazzling white disc, and if viewed through a pair of binoculars or a telescope then its changes in shape (phases) also become apparent. It was the phases of Venus, first noticed by Galileo, that convinced him that it orbited the Sun and not the Earth. As our nearest neighbour it might seem the ideal planet to visit first, however, the Russian probes Venera 4, 5, and 6 were crushed by an atmosphere 100 times greater than that here on Earth. In addition, the atmosphere is mainly carbon dioxide with swirling clouds of sulphuric acid. The surface temperature rises to a sweltering +470 °C and drops to a cool +250 °C. If you tried to step onto the surface of this beautiful, charming planet you would be instantly crushed, fried and poisoned in one go! Apart from the natural hostility it would seem very strange to live on a planet where the 'Venusian day' (120 Earth days) lasts for almost half the 'Venusian year' (243 Earth days). If you could see through the clouds the Sun would rise in the west and set in the east and there would be no moon. Venus and Uranus have retrograde motion. They rotate on their axes in the opposite way to the other planets. The planet has been visited by a number of unmanned Russian (see figure 7.6b) and American probes. The American spaceprobe Pioneer Orbiter has been studying conditions on the planet since 1978.

Mars

On 30 October 1938 Orson Welles and the 'Mercury Theatre on the Air' broadcast a script by Howard Koch. It was based on H.G. Wells's *The War of the Worlds* – a science fiction story of an invasion from Mars. The broadcast was so lifelike that it caused a considerable amount of panic in New York.

Mars, at one time thought capable of supporting life, is too inhospitable to maintain any form of life, let alone potential invaders, Martian days are similar in length to ours and the axis tilt provides Earth-like seasons. However, there is only very little atmosphere which consists mainly of carbon dioxide. The two moons are Phobos and Deimos.

Known as the Red Planet, Mars has one particular feature that has puzzled astronomers for years. Its surface is covered with valleys that look as if they were caused by running water. See figure 7.6c.

Jupiter

Many astronomers believe there should be another planet between Mars and Jupiter. There is a stream of large rocks and boulders between the two planets known as the **asteroid belt**, but it does not contain anything resembling a planet. Jupiter is by far the largest planet in the solar system. It is largest in size and mass. The dense layers of cloud in the upper atmosphere make it impossible to see any details on the planet's surface. The layers of gas, mainly hydrogen and helium, and a layer of ammonia crystals hide a small dense core. On the surface of this huge puffball is the famous **red spot** which is visible from Earth. As it is part of the upper atmosphere, the red spot moves around the planet. It is thought to have been present for three centuries. Jupiter is a ringed planet with a number of moons.

Io, one of the more interesting satellites, shows evidence of volcanic activity. Europa, another moon, is covered with a smooth layer of ice making it like a galactic snooker ball. It is possible there is water underneath the ice layers. If this is so then Europa is the only other known planet with a supply of water. Pioneers 10 and 11, which were the first man-made objects to break free of the Sun's gravity, flew past in 1973. See figure 7.6d.

Saturn

In 1610 Galileo made the following announcement, "I have observed the furthest planet to be a triple." However, it was not a triple planet, but a single planet with a bulge on either side. These bulges were only a result of the poor quality of early telescopes. With modern equipment the 'rings' of Saturn can be studied easily. In November 1980, Voyager 1 took a series of spectacular close-up photographs, separating the rings into over 1000 ringlets. See figure 7.6e.

The surface of the planet is covered with a very dense cloud formation. Those at the equator race around at over 1100 m.p.h. (1800 km/h). Under that protective blanket are thought to be oceans of liquid hydrogen and helium. Saturn has an enormously strong magnetic field (1000 times that of the Earth's magnetic field). Unusually, the magnetic poles are at the geographic poles. (The Earth's north magnetic pole is actually in the Arctic, but some way from the North Pole. See pages 94 – 5.)

Titan is one of Saturn's more interesting moons. It has a nitrogen-rich atmosphere, and in this way resembles the Earth. In addition though, there is a considerable amount of methane (our 'natural gas') present. It has been suggested that the temperature of −180 °C could turn this onto 'methane rain or snow' which could fall onto 'methane oceans' containing 'methanebergs'.

Uranus and Neptune

Little is known about the seventh planet which was discovered by William Herschel, on the evening of

Tuesday 13 March 1781, at 19 New Street, Bath. Uranus's main claim to fame is that it lies on its side. Its poles are where you would normally expect the equator to be. Launched in 1977, Voyager 2 intercepted Uranus on 24 January 1986. Photographs taken while flying past show the planet to have 10 rings and seven moons, two of which are extremely small. Travelling at the speed of light, the satellite signals took 45 minutes to reach Earth. See figure 7.6f.

The eighth planet was predicted by John Couch Adams in England in 1845. Unfortunately, his ideas were ignored. Neptune was discovered by the Berlin Observatory on 23 September 1846. Little is known about it except that it might have an atmosphere of methane, and it has part of a ring system. More will be known after Voyager 2 reaches it in 1989.

Pluto and planet number 10

Just as the Sun and the Moon have a gravitational effect on the Earth (the tidal system, see page 113), 'wobbles' in Uranus's orbit led to the suspicion of additional planets further out in the solar system. The discovery of Neptune did not provide enough evidence to explain the erratic behaviour of Uranus, and in January 1930 Clyde Tombaugh announced the presence of the ninth planet, Pluto. Very little is known about it except that it has quite a large satellite (Charon – discovered in 1978) and may have retrograde motion. Because its moon is so large it is probably more correct to call Pluto and Charon a double planet. Even so, sufficient evidence does not exist to explain fully the 'wobbles', and so the hunt goes on for planet number 10, which already has a nickname, 'Humphrey'.

The dynamic duo

The Earth and Moon (see figure 7.6g) system is unique. Nowhere else in the solar system is there anything to resemble it. All the planets have some characteristics in common with one another and they all follow the accepted laws governing motion, etc. Only Earth has the particular set of circumstances that allow the formation of various life forms. Water is one of the key substances needed for the existence of life. If our planet were a little nearer the Sun, water would exist only as steam. If we were just a little further away, water would exist only as ice. The Earth's position is very finely balanced. Like a number of other planets, Earth has an atmosphere containing carbon dioxide. This is poisonous to animals in large quantities, but fortunately plants use it as a 'food' with which to fuel themselves and grow (see page 82). In turn, we use their waste product oxygen! It is only because the Earth is the size it is that it has enough gravity to hold on to the atmosphere. (This is why there is no atmosphere round the moon.) Not only does the atmosphere support the countless number of lifeforms, but it also acts as a blanket, protecting the Earth's surface from the harmful effects of the Sun's rays. Without this blanket, life could not exist.

Earthly cycles

A number of familiar cycles or patterns involve the Earth, the Moon and the Sun. The first and most obvious of these is what we call a 'day'. Because of the Earth's rotation, the Sun appears to rise and set in a regular manner. Breaking this pattern down into 24 hours is an artificial, but convenient arrangement. The second natural cycle is the time taken for the Earth to travel in orbit once around the Sun. This time period, of 365.242 days, is a solar (or sidereal) year. While all this is going on, the Moon is whizzing round the Earth. It takes approximately 28 days to complete one revolution of the Earth. Unfortunately, this is the same as the time needed for it to rotate once on its own axis. The result is that we only ever see one side of the moon (approximately 59% of its total surface). The far side (the remaining 41%) is never visible from anywhere on Earth.

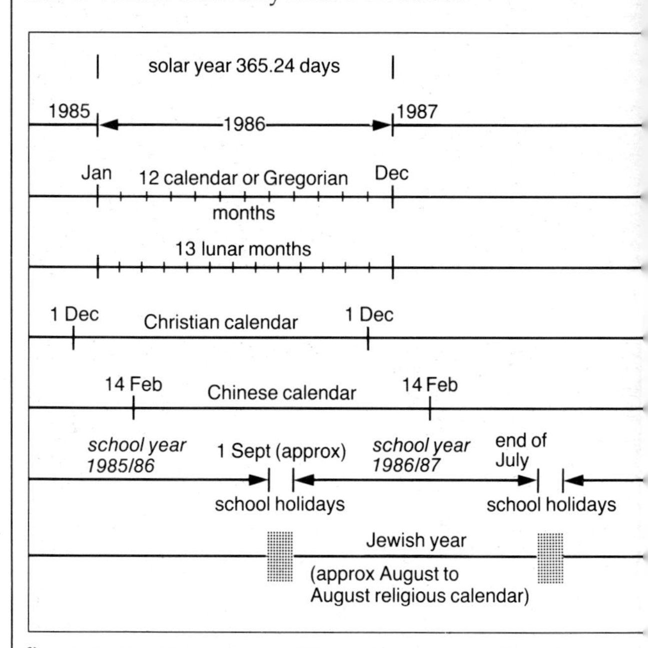

figure 7.7 How many different years are there?

Eclipses

The Sun is the source of light in the solar system, and everything else that glows is just a reflector of that light. As the Earth and the Moon continually whizz around the Sun they sometimes get in each other's way and block out the sunlight. When the Moon

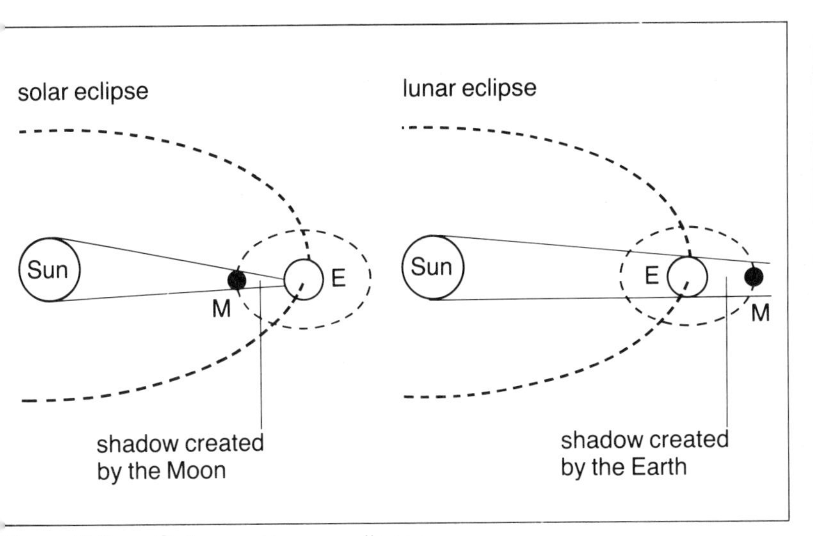

figure 7.8 **a** *Solar and lunar eclipses*

figure 7.8 **b** *Solar eclipse taken from Wallops Island, Virginia, USA, on 7 March 1970*

figure 7.8 **c** *Phases of the moon*

i six days,

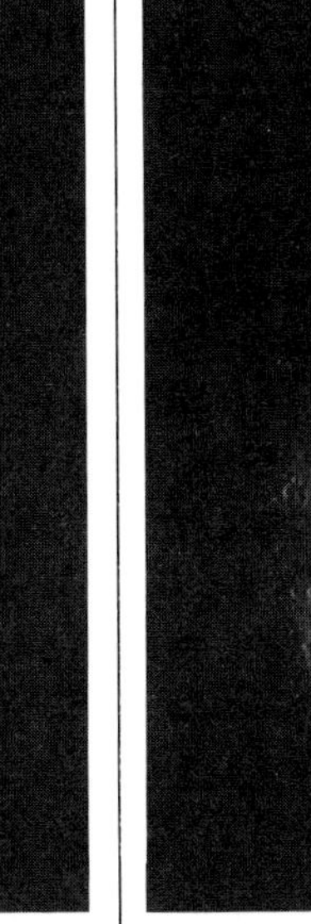

ii 10 days,

iii 19 days,

iv 24 days

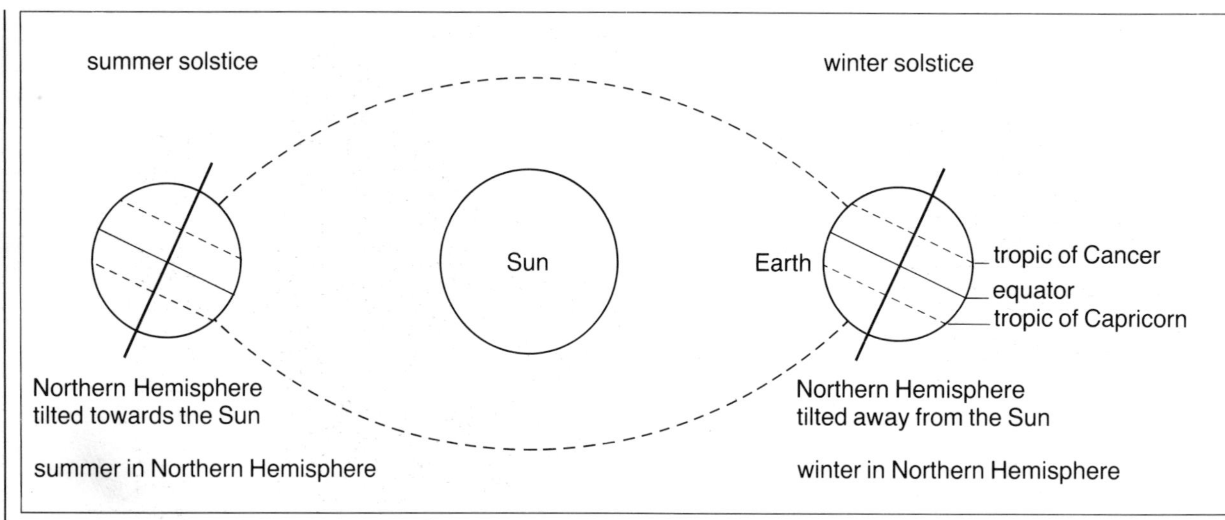

figure 7.9 Earthly seasons

moves directly between the Sun and the Earth a **solar eclipse** occurs. Anyone standing on a particular part of the Earth's surface (in line with the Sun and the Moon), will see the Sun 'go out'. When the Moon moves into the Earth's shadow, sunlight will no longer be reflected from it and the Moon will 'go out'. On leaving the shadow, the Moon reflects light once more. In the normal course of events the various phases of the Moon can be seen during a revolution of the Earth by the Moon (the 28-day lunar cycle). See figure 7.8.

INVESTIGATION 7.2

PRACTICAL ECLIPSES

Figure 7.8a shows the positions of the Sun, Earth and Moon during solar and lunar eclipses. Design an experiment to investigate the shadows formed when eclipses occur. Some possible ideas to consider might include

a objects to represent the Earth, Moon and Sun
b type of light source
c is distance important?
d method of recording the shadows
e photography
f list of apparatus
g laboratory surroundings.

figure 7.10 a The midnight sun – taken at midnight at Nordcap, Norway. This would not happen if the Earth were not tilted on its axis

figure 7.10 b Icebergs on the beach – Svalbard (Spitzbergen)

figure 7.10 c The Arctic Icecap

Season's greetings

If the Earth were not tilted on its axis the Sun would be directly overhead at the equator all year round. There would be no real change in the weather pattern throughout the year. The **seasons** would not exist.

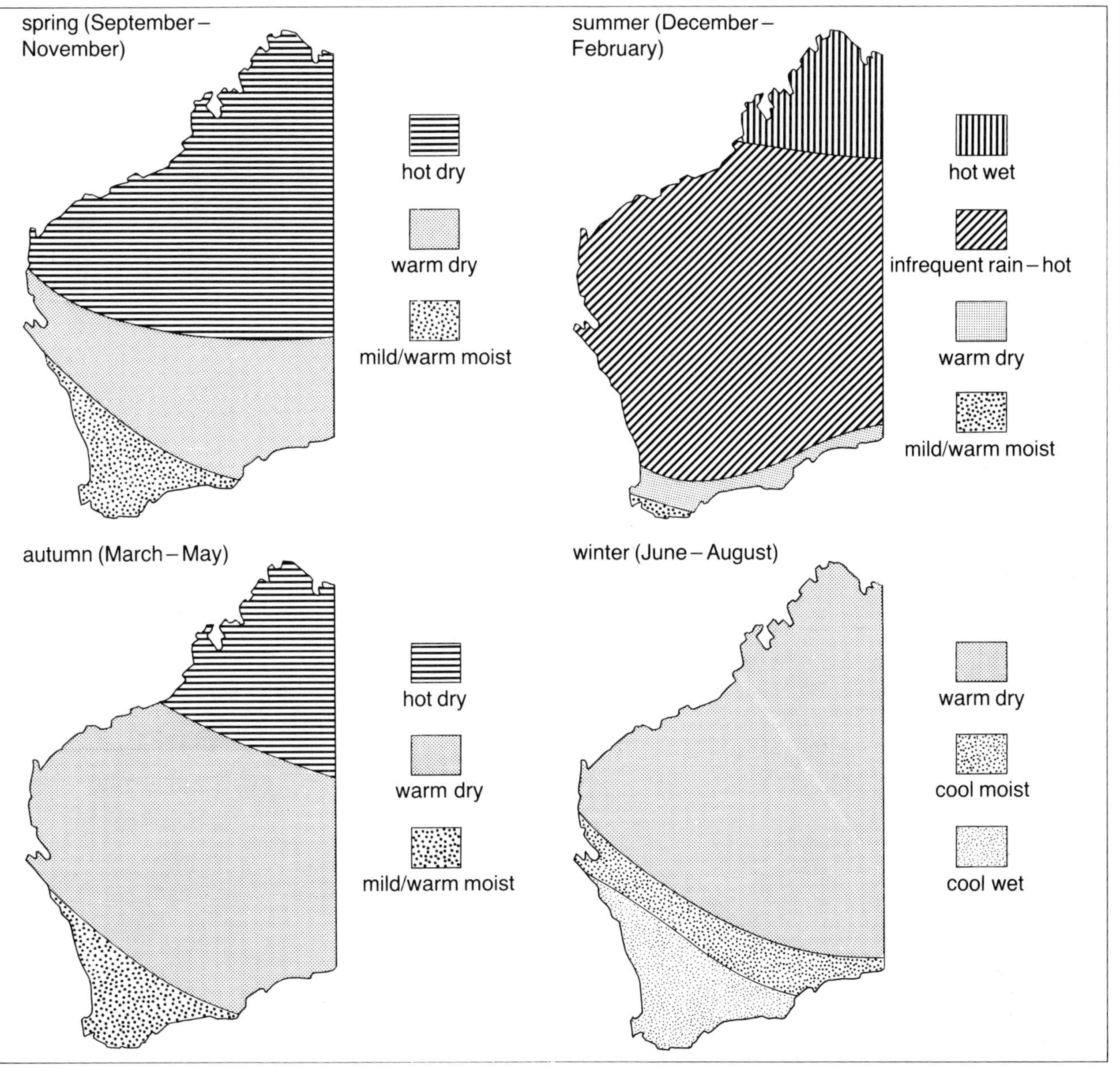

figure 7.11 *Seasons in Australia. Spring and autumn are fairly similar to those of the UK. Australia's winter is the UK's summer!*

The seasons are created by the **23° tilt of the Earth's axis**. As the Earth rotates at this angle to its path around the Sun, different areas of the planet's surface are directed towards the Sun at different times of the year. The two imaginary lines drawn parallel to the equator (the tropic of Cancer and the tropic of Capricorn) are the limits at which the Sun appears vertically overhead. See figure 7.9.

When there is summer in the Northern Hemisphere there is winter in the Southern Hemisphere. Spring and autumn occur when the sun is directly overhead at the equator. See figures 7.10 and 7.11.

On 21 and 22 June (the **summer solstice**) the effect of the Earth's inclination is that the Sun will appear to be directly overhead at the tropic of Cancer. This is the time when the Northern Hemisphere will be most tilted towards the sun so it receives the maximum amount of sunlight and the days are longest. The reverse happens on 22 and 23 December (the **winter solstice**). The seasons have little to do with the distance the Earth is from the Sun. The closest approach to the Sun occurs in January when the Northern Hemisphere has its winter!

Out with the tide

Eclipses occur irregularly. The to-and-fro action of the Earth's tides are a good example of a much more frequent interaction between the Sun and the Moon. They are a result of the **gravitational attraction** between the Sun and Moon and the Earth. Water is a fluid and will flow easily. If the Sun and Moon are effectively pulling together they will cause the oceans to move a considerable amount. The range of tides (difference between high tide and low tide) will be great. Tides like this are known as **spring tides**. Small tidal ranges or **neap tides** occur when the gravitational attraction of the Sun and Moon are cancelling one another out.

INVESTIGATION 7.3

SEASONS

Most of the planets spin in an anticlockwise direction on their own axes and travel in an anticlockwise direction around the Sun. Use a lamp, and, for example, a ping pong ball, to demonstrate the motion of a planet around the Sun. Can you explain what kind of seasons the poles of this planet would have? Now reverse the planet's spin direction. What effect does this have on the planet's day? Investigate the effect of a small tilt in the planet's axis. Test the ideas given in figure 7.9 (page 112).

Question

What happens to the seasons if a planet's tilt is changed to **a** small, **b** large, **c** as large as that of Uranus?

FLY ME TO THE MOON

Up until the beginning of 1987 12 men had walked on the surface of the Moon, 16 astronauts and cosmonauts had died for space exploration and a further 10 died in plane crashes while gaining experience of flying in preparation for space travel. Space is an extremely hostile environment. Any journey through space involves considerable danger, (see tables 7.2 and 7.3).

problem	result
no atmosphere	death by suffocation
no atmospheric pressure	death as the body fluids would evaporate (you'd explode)
low temperature extreme	death by freezing
solar radiation	death by radiation sickness

table 7.3 *Problems for humans in space*

date	mission	crew	details
AD 1042			war rockets described by Tseung King Liang in China
Summer 1687			Sir Isaac Newton published proposed laws governing the flight of satellites
8 October 1806			'six pound rocket' used by Royal Navy against French at Boulogne
16 March 1926			first launching of a liquid fuelled rocket
1928			Fritz von Opel built the first rocket-powered car
4 October 1957	Sputnik 1		first artificial satellite went into orbit
24 September 1959	Luna 2		12.02 a.m. (Moscow time) first artificial satellite lands on the moon
7 October 1959	Luna 3		first pictures returned of the far side of the moon
12 April 1961	Vostok 1	Yuri Gagarin	first space flight, duration 1 hr 45 min.
18 March 1965	Vostok 2	Belyayev Leonov	Leonov makes first space walk
27 January 1967	Apollo 1	Grissom, White, Chaffee	crew killed on launch pad by oxygen fire
21 December 1968	Apollo 8	Borman, Lovell, Anders	first manned space flight around the Moon
16 July 1969	Apollo 11	Armstrong, Aldrin, Collins	first moon landing (21 July – Neil Armstrong)
11 April 1970	Apollo 13	Lovell, Haise, Swigert	Service module explodes 55 hours into the mission – Emergency return to Earth
6 June 1971	Soyuz 11	Dobrovolsky Yeliseyev Patsayev	Crew killed as craft decompressed on re-entry of Earth's atmosphere
2 March 1972	Pioneer 10	———	First artificial satellite to leave the Earth's orbit (at 32 114 m.p.h., 51 682 km/h) on journey out of solar system
15 July 1975	Soyuz 19	Leonov, Kubasov	Joint docking manoeuvre
	Apollo 18	Stafford, Brand, Slayton	Joint docking manoeuvre
12 April 1981	STS-1 Colombia	Young, Crippen	First space shuttle flight
8 November 1984	STS-51A Discovery	Hauck, Walker, Allen Gardner, Fisher	Lost satellites recaptured by space walks
28 January 1986	STS-51L	Scobee, Smith, Resnik, McNair, Onizuka McAuliffe Jarvis	shuttle craft exloded at 47000 ft (14.5 km) during take-off killing all the crew – American shuttle programme suspended

table 7.2 *Space flight log*

INVESTIGATION 7.4

PLANET AND STAR SPOTTING

The **ecliptic** is an imaginary flat layer or surface in which the solar system travels among the stars. Most of the planets also orbit the Sun in a narrow band close to the ecliptic (9° either side). The stars and constellations (arrangements of stars) which appear to be in this band are known as the **signs of the zodiac**, e.g. Aquarius, Scorpio.

Only about two-thirds of the star map shown in figure 7.12 can be seen at any one time. The following guidelines will help you identify some of the stars, constellations and planets.

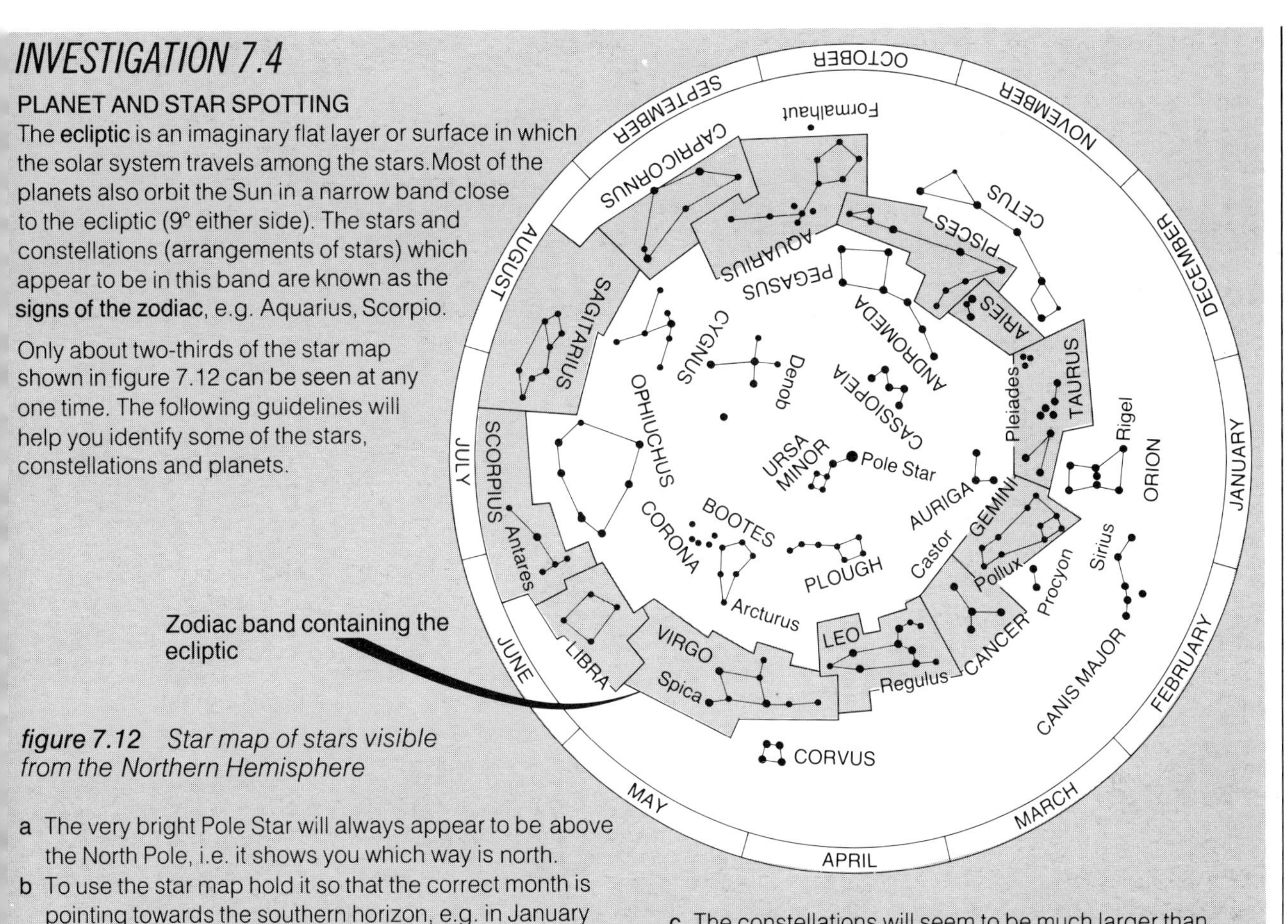

figure 7.12 Star map of stars visible from the Northern Hemisphere

a The very bright Pole Star will always appear to be above the North Pole, i.e. it shows you which way is north.

b To use the star map hold it so that the correct month is pointing towards the southern horizon, e.g. in January the constellation Orion will be seen above the southern horizon.

c The constellations will seem to be much larger than figure 7.12 suggests.

d To get your bearings look for an easy-to-recognize constellation such as Cassiopeia; it is shaped like a large 'W' and is visible all year round.

e Use one constellation to guide you to another.

f Most planets travel around the Sun in the ecliptic so they will be seen in the zodiac band.

g Stars twinkle, planets don't.

h Some daily papers have information about current planet movements.

Artificial gravity

Weightlessness can cause bone tissue to lose calcium and become brittle. Some people believe this change to be permanent. Using artificial gravity would make long-term space exploration easier. In the same way as a conker will pull on a string when you whirl it round, so a large, slowly rotating space station will provide artificial gravity. Arthur C. Clarke imagined a space station the shape of a bicycle wheel. At the centre of the wheel there would be no gravity and you would be weightless. However, if you were to move towards the rim of the wheel the effect of the artificial gravity would increase. The rim would act as a 'floor' that would go all the way round. (It would make an interesting swimming pool!)

INVESTIGATION 7.5

LIVING IN A VACUUM

A very simple demonstration can be performed to show the effects of a vacuum on, say, a partially inflated balloon or a ripe tomato with the skin completely intact. The balloon or tomato is placed inside a sealed jar and the air removed. Guess what happens. Try to explain your ideas using the kinetic theory (see page 8) to help you.

ALL YOUR OWN WORK

1 Look at the photographs in figure 7.6 (page 108) containing cratered landscapes. What information do the shadows give you about crater depths? How can you tell which were formed first and which last?

2 What clues are hidden in the photograph of Mercury (figure 7.6a) about its
  a age
  b history
  c atmosphere?

3 What other calendars can you think of apart from those in figure 7.7 (page 110)?

4 One suggestion made in this unit is that the phases of the moon play a part in the Earth's tidal system. If you had access to a library how would you check this idea? (No dictionaries or encyclopaedias allowed!)

UNIT 8

ELECTRICITY

ELECTRICKERY

Shocking experiences

How many times have you walked into a room, usually across a carpet, touched a chair or table and suddenly received an electric shock? You are probably very familiar with **static electricity**! Static means still or stationary. Static electricity is electricity which does not move – until something provides it with an escape route, which can often be you. If an object collects static it is said to be **charged**. There are many simple everyday examples of static charges. For example, a television screen will build up a static charge on itself. You can feel the effect of the charge by placing your hand nearby. A pullover, especially if it is made of a synthetic fibre like acrylic, will often crackle with static when you take it off.

If the static charge is very large there may be sparks. Very large charges can be very dangerous. Lightning is nature's way of releasing huge charges that have been built up in the sky by clouds rubbing past one another.

Positive and negative charges

Static is usually produced by the process of **friction**. Just as the clouds can create static when they rub together, so too can other materials. If two surfaces rub together then a static charge may appear between them when they are separated. A **negative charge** will appear on one surface and a **positive charge** will appear on the other. Static charges can only appear on **insulators**. Insulators will not allow the charge to flow away. (Charge will flow through a **conductor**.) One of the most successful ways of creating static is to rub a plastic rod through clean hair.

Static charges follow these simple rules for attraction and repulsion.

like charges repel one another
unlike charges attract one another
both charges attract an insulator that has no charge

Figure 8.1 shows a number of simple experiments that illustrate static at work. Each one provides evidence that a force is at work (the forces are created by the separation of charges). This means that energy is about. Static electricity is an energy form because it can make things happen.

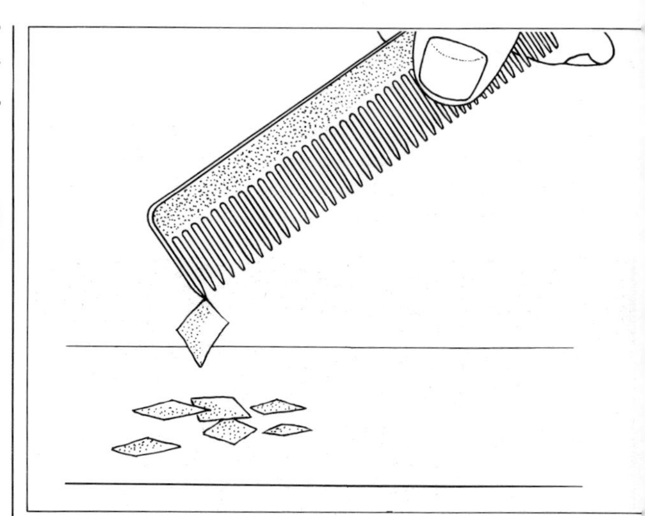

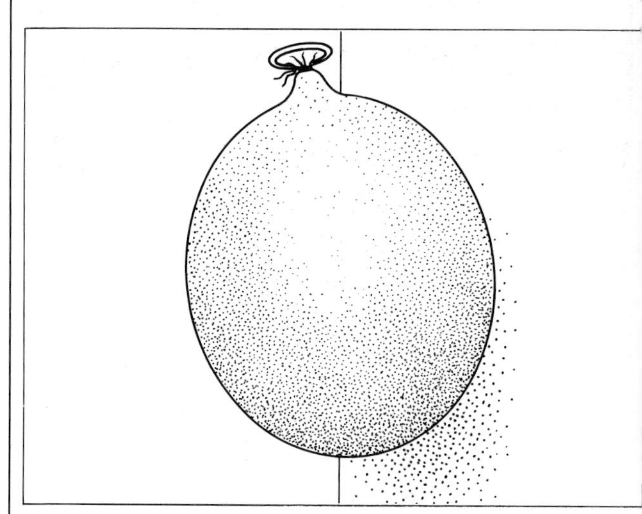

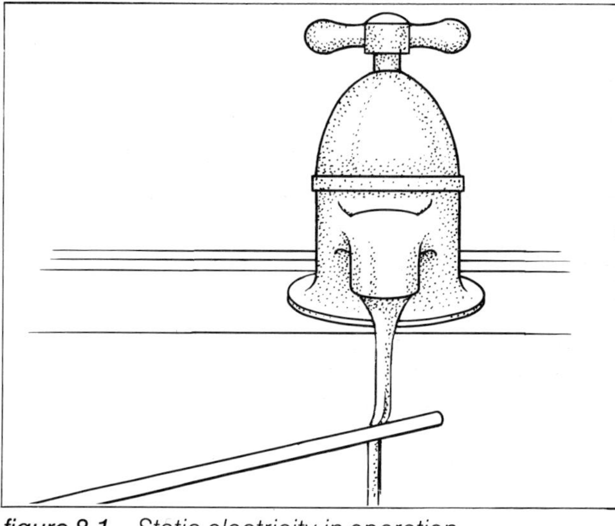

figure 8.1 Static electricity in operation

INVESTIGATION 8.1

STATIC CHARGES

'Repulsion can only happen between two charged objects if both charges are the same.' Using any or all of the apparatus shown in figure 8.1 devise a series of experiments to test this idea. (Try the experiments from figure 8.1 first – then invent some more.)

Application 8.1

STATIC

1 Car bodies are charged before they are dipped into a vat of paint. The added attraction created helps to bond the paint to the body surface, creating a more durable finish (see figure 8.2a).

figure 8.2 *a Electrostatic painting of a car body*

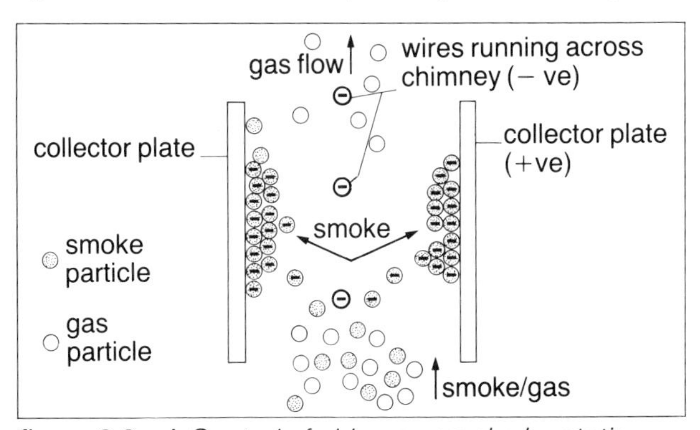

figure 8.2 *b Control of chimney smoke by static*

2 Figure 8.2b shows the internal arrangement of plates and wires inside a chimney used for removing smoke particles from flue gases. As the smoke particles rise they are given a negative charge by the wires placed across the chimney. The negatively charged particles are naturally attracted to the plates on the inside of the chimney. Periodic collection of the dust from the plates keeps them in operation.

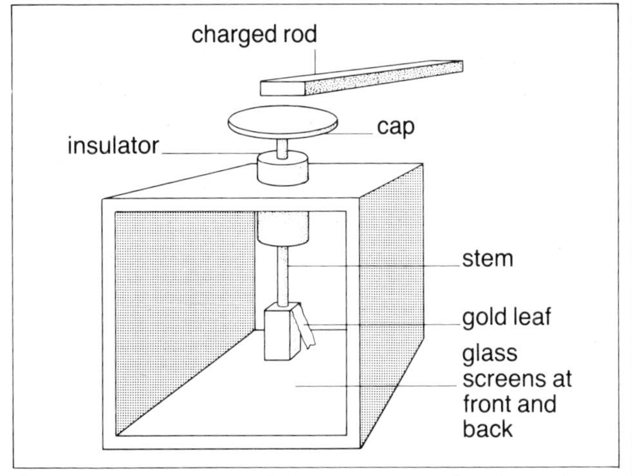

figure 8.3 *A gold leaf electroscope – for detecting static*

DETECTING STATIC AND STATIC FIELDS

A **gold leaf electroscope** is often used to detect a static charge (see figure 8.3). Whenever a charged object, such as a polythene rod or rule, is brought close to the cap of the electroscope the gold leaf will move away from the stem – it is repelled. It is not necessary to touch the cap with the charged object for this to happen. The effect is noticeable even though the charge is some way from the electroscope. The charge is effective at a distance. (Magnetism will also act at a distance.) The effect of the charge seems to travel through the air. The space around a charged object is known as a **static field**.

INVESTIGATION 8.2

STATIC FIELDS

A **field** is the area of influence around an object in which things happen. Using an electroscope and any or all of the objects from figure 8.1, investigate the static field around a charged polythene rod. (Start by making sure you can create and keep a field around a polythene rod.)

Questions

1 Does the field surround the rod completely? Where is it – in one place, along one side, on one edge, or somewhere else?
2 Does the strength of the field change at a particular distance away from the rod, or does it die away gradually?
3 If left alone, is static permanent?
4 If you were now given a strong bar magnet, how would you test the idea that a magnetic field has an effect on a static field?

ON THE MOVE

The following sentence may seem rather silly: 'When charges move from place to place they are no longer static'. The word **current** can be used to mean **flow,** and a current of electricity is a flow of electricity. While a set of charges are stationary they are known as static charges, but when they are on the move they are known as a current of electricity. More usually you would call them an **electric current**. An electric current is just a flow of charges from place to place.

Electric current

On the opposite page, two kinds of charge were mentioned – positive and negative. Is it possible to have two kinds of electric current? It is – in practice it is the negative charges that move. (The movement of positive charges will be dealt with in unit 13. A single negative charge is called an **electron**. An electric current is the flow of electrons from place to place.

Just as length has a unit of measurement, so too does flow of current. The symbol used to indicate a current flow is *I*. The flow of current (*I*) is measured in **amperes**. This is often shortened to amps and sometimes it is written as A. For example, a current flow of 2 amperes may appear as 2 amps or 2 A. Electric currents are measured with an **ammeter**.

Using ammeters correctly

Like most measuring instruments, ammeters need to be used correctly. The job they have to do is to measure the flow of electricity in a wire. They must be part of that wire to do their job. Electricity must flow into an ammeter and then out again. Most important of all is that all the current that flows into an ammeter flows out again. Ammeters are always connected **in series**. A series circuit is one in which the various parts, or **components**, are all connected one after the other. (A TV series shows one episode after another.) As most ammeters are one-way devices it is vital they are connected the right way round. (Most ammeters will break if you try to run them backwards!)

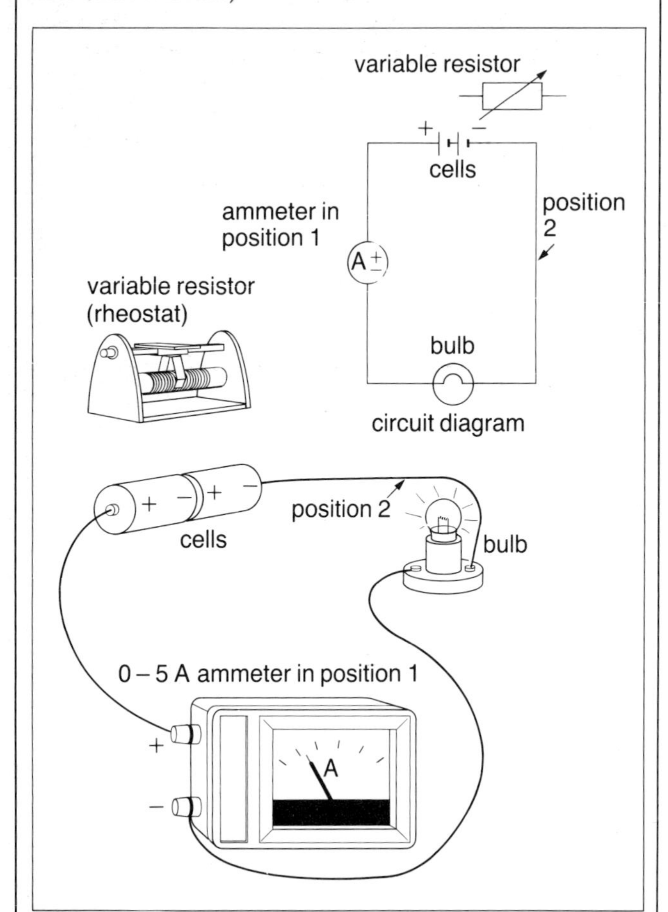

figure 8.4 A simple series circuit with an additional variable resistor (rheostat)

INVESTIGATION 8.3

USING AMMETERS

Figure 8.4 shows a very basic series circuit as it might look in the laboratory and as it would be drawn as a circuit diagram. Assemble the circuit and by reading the scale on the ammeter find the current value in amperes. Re-assemble the circuit with the ammeter in position 2. Look at the new value on the scale.

Questions

1. Are both ammeter values approximately the same, or are they very different from one another? Explain your observations simply.
2. Re-assemble your series circuit to include the variable resistor (rheostat). What effect does this have on the circuit? Repeat the exercise of changing the position of the ammeter.
3. Draw your new circuit diagram and briefly explain what each part of the circuit does.

The in-out law

No matter where you place an ammeter in a simple series circuit the current value will always be the same. Whatever electricity flows into one end of a wire must flow out again at the other end. (A piece of string behaves in a similar way when you pull one end of it. Could you imagine it tight at one end and loose at the other?) The current flowing away from a component must have the same value as the current entering the component. This is **Kirchhoff's Law**. For example, if 5 A flow into a component then 5 A must flow away again even if it has been shared out between a number of wires. After all, the idea of losing electricity doesn't really make any sense!

The electric storage jar

The flow of electricity in a circuit can be large or small – it can have any value. A larger current simply means that more electricity flows around the circuit. A smaller current means less electricity flows around the circuit. If you could put a bucket on the end of the wire you might imagine that it would fill with electricty. A larger current would fill the bucket faster than a smaller current. A small current would need to be switched on for longer to fill the same sized 'electric bucket'. A single quantity or amount (one bucketful!) of electricity is known as a **coulomb**. One coulomb (symbol C) is the amount of electricity that passes round a circuit if a current of one ampere (1 A) is left switched on for one second (1 s).

$$\underset{\text{(units C)}}{\text{charge}} = \underset{\text{(units A)}}{\text{current } I} \times \underset{\text{(units s)}}{\text{time } t}$$

or example, if a current of 6 A flows for 10 s then a uantity of 60 coulombs of electricity will have passed round the circuit.

here is no such thing as an 'electric bucket', but here is a practical alternative: the **capacitor**. As its ame suggests it has a capacity – a capacity for storing lectricity. (You might suggest that a battery is a store f electricity but this is not really so. A battery is only store of chemicals which release electricity when hey react with one another. This can be quite a slow rocess.)

> ## *Application 8.2*
> **CAPACITORS**
>
> A camera flashgun contains a battery and a capacitor. When it is switched on it takes time to charge up. You can usually hear it humming for a little while before it is ready to use. Batteries slowly supply electricity to the capacitor. Once stored, it can be released as quickly or as slowly as its needed – as quick as a flash!

Capacitors

capacitor contains two or more metal plates. lectricity flows on to one of these plates (or sets of lates) until the plate is 'full'. The capacitor is then aid to be fully charged. A large current will charge a apacitor quickly. A smaller current will take longer o fully charge the capacitor. Both currents will result in the same total charge on the capacitor (see also driving forces, overleaf).

Larger currents

From your own experiences you will already know that electricity can create heat, an electric fire for example. A reasonable suggestion to make would be that the greater the current flowing through something the hotter it becomes.

INVESTIGATION 8.4

LARGER CURRENTS

The circuit shown in figure 8.5 is a simple series circuit. By moving the slider on the rheostat (or by turning the control knob on the circular varieties) the amount of electricity flowing in the circuit can be controlled.

The component being heated is a '10 ohm resistor' (whatever one of those is!) It doesn't matter what a resistor is for the moment – think of it as any electrical device which allows electricity to flow through it. (The results of this experiment are the same for any device if too much current flows through it.)

Questions

1 Observe the effect of slowly increasing the current on the resistor (by moving the rheostat slider). A scientific observation is not necessarily something you see. What observations can you make?

2 Re-assemble the circuit without the rheostat for this next investigation. In figure 8.5 is an additional piece of equipment – a 0.5 m length of Nichrome wire. How would you use this wire to investigate the effect of a change in length or a change in thickness on the ability of the wire to conduct electricity? Are there any simple relationships, for example, if you double the length of the wire does the current flow double, halve or what? Table 8.1 might help.

current (A)	length (m)	current (A)	thickness

table 8.1

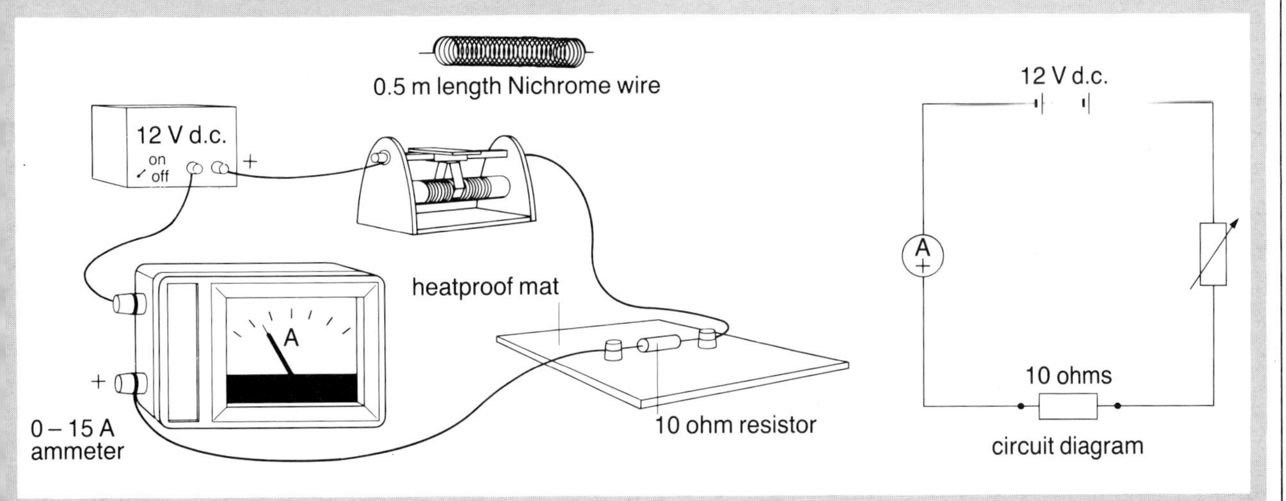

figure 8.5 *Investigating the heating effect of a current*

UNIT 8

THE NEED FOR FUSES

In the previous investigation you met the **heating effect** of a current. The greater the current, the greater the heating effect. Some devices are particularly sensitive to heat. If their temperature rises too much they will burn out. This is also true of connecting wires. If wires are placed beneath carpets and burn out they may start a fire. The replacement of burnt-out cables in walls is an expensive and messy business.

The prevention of these possible dangers is simple. A **fuse** is designed to **blow** or burn out when a certain maximum current flows through it. For this reason **fuses are labelled in amperes**. By including a suitable fuse in a circuit you can make sure the fuse blows first, thereby breaking the circuit before any of the other components in the circuit, overheat when the current gets too big. (The main function of a fuse is to protect the electricity supply wiring, but obviously it also helps to protect the components or devices using the electricity.) Common values for fuses include 1 A, 2 A, 3 A, 5 A, 10 A, 13 A and 15 A.

In practice, fuses are designed to allow a slightly higher current to flow through them than their stated rating.

DRIVING FORCES

It is very easy to show that a current flows in a circuit. A simple bulb is evidence enough of that. A current does not flow on its own. Something must make it flow – something must drive it or force it around the circuit. The current seems to be pushed out of the battery (or supply) from one terminal, only to be pulled in at the other terminal. A flow of electricity always has a **driving force** behind it to make it move. The driving force is known as **potential difference (p.d.)**. It is measured in **volts** (V) using a **voltmeter**. A source with a large potential difference (a high **voltage**) will be able to drive a larger current around a circuit than a source that has only a small potential difference (a low voltage). The reason why you do not receive an electric shock from a small battery is because the driving force is too small. If you touched mains electricity the driving force would be much larger and quite capable of driving a large current through you. A driving force (potential difference) of only 50 volts is more than enough to drive a current through people and kill them. As a general rule you should always disconnect the electrical source from a circuit before touching any part of the circuit. Switching off is not enough.

If a current is to pass through a device it will need a driving force behind it. If there are several devices in a line then a driving force will be needed for each one. You could say that the driving force is 'used up' as a current flows around a circuit. The potential difference on one side of a circuit will be maximum and on the other side of the circuit it will be minimum (nil, because there will be nothing left to drive the current through). The series circuit diagram in figure 8.6 has three identical components in it. The p.d. between the two ends of the circuit is 12 V. The red positive terminal (+ or +ve) is labelled +12 V. The black negative (− or −ve) terminal is labelled 0 V.

Put simply, the whole circuit has a p.d. of 12 V to 'use up', and there are three identical bulbs, so each bulb uses 4 V.

		measured between
bulb 1	p.d. = +4 V	A and B
bulb 2	p.d. = +4 V	C and D
bulb 3	p.d. = +4 V	E and F
total	p.d. = +12 V	A and F

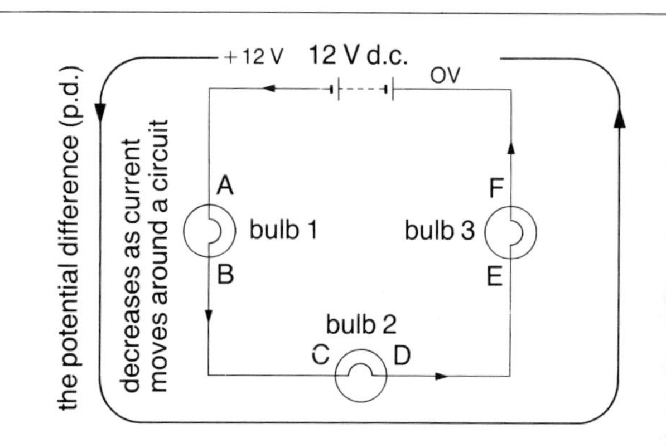

figure 8.6 *a Potential difference decreases as current moves around a circuit*

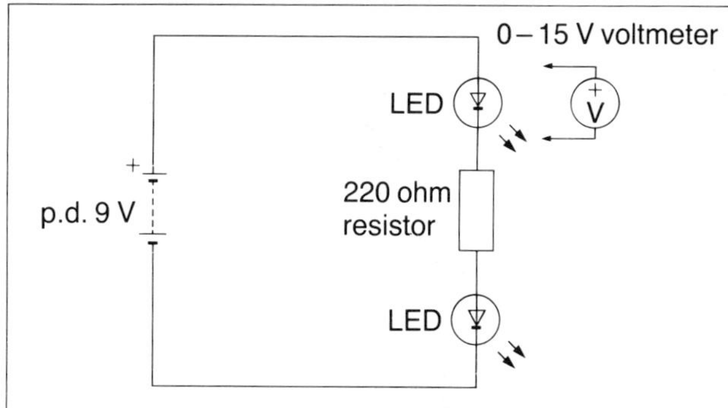

figure 8.6 *b Voltmeters ready for connection to the light emitting diode (LED)*

ulb 1
he driving force at position A is +12 V.
he driving force at position B is +8 V.
he p.d. for bulb 1 = +4 V.

ulb 2
he driving force at position C is +8 V.
he driving force at position D is +4 V.
he p.d. for bulb 2 = 8 − 4 = +4 V.

ulb 3
he driving force at position E is +4 V.
he driving force at position F is 0 V.
he p.d. for bulb 3 = 4 − 0 = +4 V.

f the three bulbs (or any three devices) were all ifferent they would each take a different share of the).d. available. (Remember, the same current is driven hrough each bulb and the current is the same all ound the circuit!).

Fixed voltage values

/lost electronic and electrical devices are designed to vork correctly when connected to a particular otential difference (voltage). This is the maximum safe limit for the device and should not be exceeded. For example, light bulbs are marked with voltage values. If you connect them to a lower value they will glow dimly. If you connect them to a higher than normal value they will glow extremely brightly. They will not last as long and may blow straight away. Always connect a device to the correct supply voltage to avoid damage.

Using voltmeters correctly

A voltmeter measures a difference in driving force, the difference between two positions in a circuit. In figure 8.6a the difference between A and B is 4 V. 4 volts will be displayed on the voltmeter. Voltmeters are usually one way devices and must be connected correctly. A voltmeter is only added to a circuit when the circuit has been completely assembled. The two connections from the voltmeter are placed one on either end of the component under investigation. Figure 8.6b shows the two connections ready for attachment. Note that the circuit built on a Spring-board Electronics board, is working, the light emitting diodes (LEDs – electronic lights) are lit without the voltmeter being connected.

INVESTIGATION 8.5

MEASURING VOLTAGES

Series circuits

Assemble any simple series circuit, such as one of the examples shown in figures 8.4, 8.5 or 8.6, and check that it is working. Start the investigation by becoming familiar with the use of a voltmeter. By 'dipping' the two connections either side of a component you should obtain a reading. If the voltmeter needle tries to swing the wrong way, reverse your connections and try again. Sketch a circuit diagram and label it in a similar way to figure 8.6a.

Questions

1. What are the p.d. values of the components in the circuit? Do they add up to the p.d. of the battery?
2. Does the position of each component in the circuit affect its p.d. value?

PARALLEL CIRCUITS

Use three components to build a circuit like the one shown in figure 8.7. The top part of the circuit has two LEDs arranged in parallel. This pair is then connected in series with a 220 ohm resistor.

Questions

1. What are the p.d. values for each circuit component? Do they add up to the p.d. of the battery?
2. Are p.d. values affected if the parallel LEDs change place with the resistor?
3. What is the connection between the p.d. values of the two parallel components?

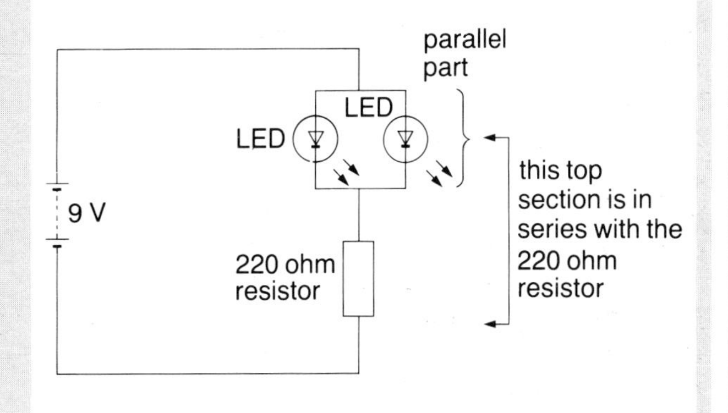

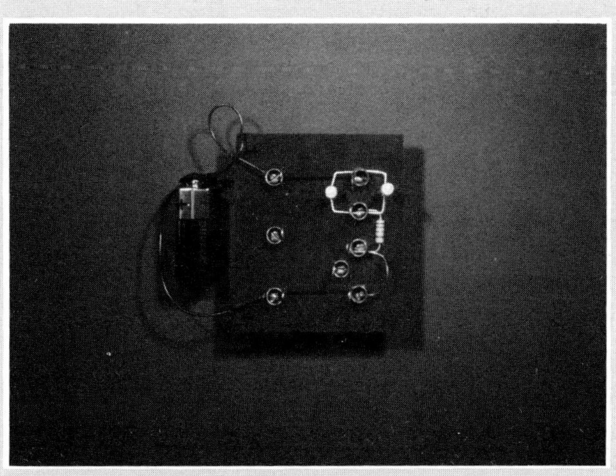

figure 8.7 The p.d. of parallel components

UNIT 8

STOPPING CURRENTS

Anything that gets in the way of electricity will tend to stop it. Every component in a circuit will have an effect on the flow of current. (Even the connecting wires have an effect, but this is so small that it is not worth mentioning further.) Anything that tries to stop or reduce the current flowing in a circuit is called a **resistance**. Resistances are measured in **ohms**. The symbol for ohms is Ω (the last letter in the Greek alphabet). A bulb is a resistance, so is a bell or buzzer. Some devices are better resistances than others. There are high, medium and low value resistances. If a circuit contains a very high value resistance then very little current will flow. A resistance with a small value will allow more electricity to flow through it. A **resistor** is a component designed to provide resistance.

This idea can be tested very easily using the apparatus shown in figure 8.8. (The LED should never be placed in the circuit without a resistor.)

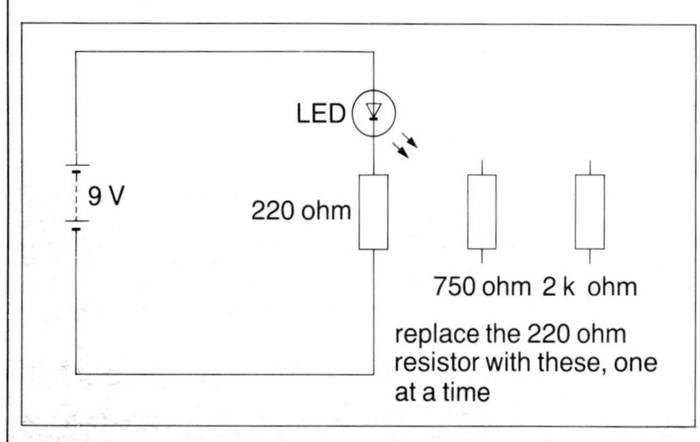

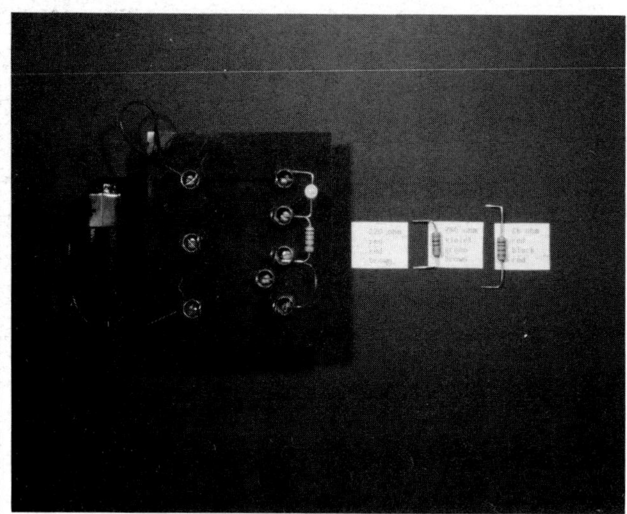

figure 8.8 Resistors limit the current flow

With the 220 Ω resistor in the circuit, the brightness of the LED will be maximum. By swapping this resistor for the 750 Ω resistor and then the 2 Ω (2000 ohm) resistor you can make the LED become quite dim. The resistors are current limiters. They control how much current flows in a circuit.

INVESTIGATION 8.6

RESISTANCE

Series

Assemble the circuit shown in figure 8.8 with the resistance values suggested (or similar ones). Satisfy yourself that by using the larger resistances you can limit the current flow and make the LED become dimmer. Set your observations out as in table 8.2.

resistance (Ω)	LED brightness
220	
750	
1000 (1k)	

table 8.2

Questions

1 By using only these resistances it is possible to pair them up to make other resistances. What other *series* resistances can be made and what effect do they have on the LED?
2 Can you describe the effect of placing two resistances in series? What is the connection between the total resistance of the pair and the individual value of each resistance?
3 Repeat the experiment and questions 1 and 2 with the LED connected the other way round. What does this tell you about the resistance of the LED?

Parallel

The other possible arrangement for a pair of resistances is a parallel arrangement. If you have completed the first part of this investigation you will know how each single resistance affects the LED. Replace the single resistance in figure 8.8 with a parallel pair. Suggested pairs are shown in table 8.3.

parallel pair (Ω/Ω)	LED brightness
220/750	
220/1000	
750/1000	

table 8.3

Questions

1 What happens to the brightness of the LED when a second resistance is added in parallel?

Adding resistors

Series resistors

These add together very simply. The resistors in figure 8.8 are 220 Ω, 750 Ω and 2 k Ω. To find the total resistance (R_{total}) of these in series just add them together, for example

220 Ω and 750 Ω

R_{total} = 220 Ω + 750 Ω
= 970 Ω

220 Ω and 2 k Ω

R_{total} = 220 Ω + 2000 Ω
= 2220 Ω

Parallel resistors

Resistors in parallel are not quite so easy to add together. The equation used is

$$\frac{1}{R_{total}} = \frac{1}{R_1} + \frac{1}{R_2}$$

A more convenient rearrangement of this equation is

$$R_{total} = \frac{R_1 \times R_2}{R_1 + R_2}$$

For example, with two resistances 220 Ω and 750 Ω in parallel

$$R_{total} = \frac{220 \times 750}{220 + 750}\ \Omega$$

$$R_{total} = \frac{165\,000}{970}\ \Omega \quad = 170\ \Omega$$

For 750 Ω and 2 k Ω resistances in parallel

$$R_{total} = \frac{750 \times 2000}{750 + 2000}\ \Omega$$

$$= \frac{1500\,000}{2750}\ \Omega \quad = 545\ \Omega$$

In both examples the final resistance value is less than either of the two original resistances! This is rather surprising, but nevertheless true. When resistors are in parallel their total resistance decreases.

OHM'S LAW

Once a resistor is put in a circuit it immediately fixes the current value in that circuit. The same resistor connected to a battery (or source) with a higher p.d. will allow a larger fixed current to be driven through it. When the resistor is connected to a lower p.d. the smaller driving force will cause only a small current to flow.

Combining resistance, p.d. and current produces a very useful equation, known as Ohm's law.

$$\text{p.d.} = \text{current} \times \text{resistance}$$
$$V = I \times R$$

One of the more useful arrangements of this law is

$$\text{resistance} = \frac{\text{p.d.}}{\text{current}}$$

$$R = \frac{V}{I}$$

INVESTIGATION 8.7

OHM'S LAW

Assemble the circuit shown in figure 8.9. You should find that movement of the slider control on the rheostat produces a noticeable change on the ammeter and on the voltmeter. By making suitable adjustments to the rheostat, take a selection of readings from the meters to allow you to fill out a table and plot a graph similiar to those shown in figure 8.9.

Repeat the experiment twice – once with two bulbs in series and once with two bulbs in parallel. Complete two further tables of results and plot these on your original graph.

Questions

1 Place these in order of resistance (high, medium, low).
   two bulbs in parallel
   two bulbs in series
   one bulb on its own
2 In effect, you have used three different resistance values. What effect did each resistance value have on the shape of graph plotted from its current and voltage values?
3 Having completed this experiment, design another to
   a test the true current rating of these standard fuses: 1 A, 2 A, 3A
   b show by experiment which of the three fuses has the highest resistance value.

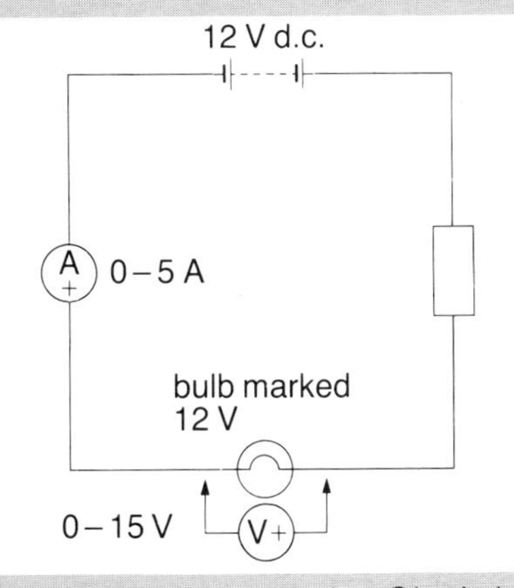

for a single bulb		
reading	p.d. (V)	current (A)
1	3	?
2	4	?
3	5	?
4	6	?
5	7	?

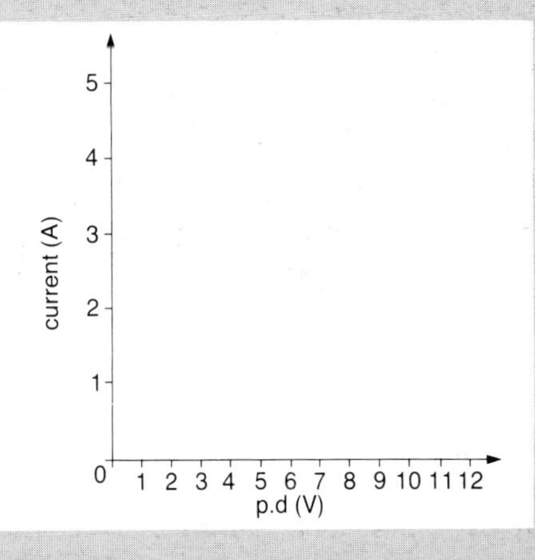

figure 8.9 Investigating Ohm's law

resistance R	potential difference p.d. or V (driving force)	current flow (driven through the resistance by the driving force)
↑ fixed at one value ↓	high	high
	medium	medium
	low	low

table 8.4 Ohm's law

Application 8.3

ELECTRIC FIRES

A one-bar electric fire run from the mains requires a p.d. of 240 V. Suppose this drives a current of approximately 4 A through the heating coil. When a replacement coil is required the resistance of the new one will have to match the old one if it is to perform as well. Ohm's law is used to calculate how much resistance the original coil had.

$$R = \frac{V}{I}$$

$$R = \frac{240}{4}\,\Omega = 60\,\Omega$$

The new coil will also need a resistance of 60 Ω. Of course, these calculations are not done in local electricity shops but in the factories where the fire bars are made by the hundred.

Current – voltage graphs

The graph in figure 8.10 shows three typical examples of the relationship between current and voltage for three different fixed resistors. They could easily have been plotted from the results of an experiment similar to investigation 8.7. Each of the three graph lines demonstrates the idea of **direct proportion**. Proportion means increasing in even steps producing a straight line graph. Direct proportion means the straight line passes through the origin.

The slope (gradient) of a current – voltage graph supplies information about the resistance of the devices in the circuit. The steeper the graph is, the lower the resistance.

circuit A has a smaller resistance
circuit B has a medium resistance
circuit C has a larger resistance

This is easily checked by looking at the three points (1, 2 and 3) marked on the graphs. The voltage at each of these points is the same. At position 3 on graph A more current flows than at either position 2 or position 1. The resistance of the device used in circuit A must have a smaller value than the others.

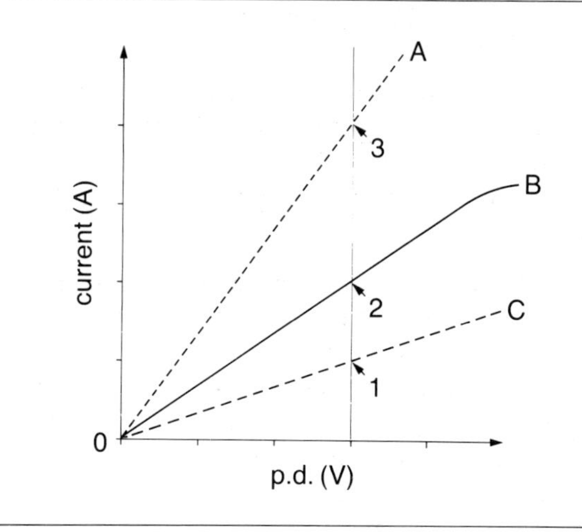

figure 8.10 Typical current – voltage graphs

Graph B changes near the end. The resistance seems to increase. In practice, the **resistance** of something like a light bulb **increases with temperature**, so the graph will begin to change shape as the current/voltage values increase (see page 120).

ALTERNATING SUPPLY

All the work about electricity so far has involved **direct current (d.c.)**. As the name suggests, the current moves directly from one place to another. The supply terminals are labelled positive and negative and for convenience, we think of direct current moving on one direction only, from the positive terminal to the negative terminal. (Batteries, for example, only supply d.c.)

Many devices are made that will work only one way round, that is with a direct electricity supply. In some circumstances it is useful to use electricity in a slightly different way. An **alternating supply** does exactly that: it alternates. If you imagine swapping over the terminals of a battery the electricity would be changed round, or alternated. The current would move around the circuit in the opposite direction. If you continued to swap the terminals, the current would also continue to change direction. An alternating supply does this automatically.

Mains electricity is alternating. It changes direction, and then changes back again, 50 times every second. Each complete swing backwards and forwards is known as an **oscillation**. The frequency, or number of complete oscillations that take place in one second, is measured in **hertz (Hz)**. The frequency of the mains electricity supply is 50 hertz (50 Hz). The following symbol is usually printed on devices that require alternating current from the mains.

50 Hz a.c.
or
50 Hz~

Measuring a.c.

'he meters used so far have all been d.c. meters. It is nportant they are connected the right way round in circuit to avoid damage. A new type of meter must e used if it is to be able to swing backwards and orwards.

Jnfortunately, as a.c. supplies, such as the mains, lternate so quickly (50 times every second), you vould have difficulty watching the meters. (They vould be moving so fast they would probably break nyway!) A new type of meter is required. There are neters with pointers, similar to d.c. meters, that can e used, but a much more useful tool is the **cathode ay oscilloscope (CRO)**.

The CRO

'igure 8.11 shows a CRO in its correct working osition with the important controls labelled.

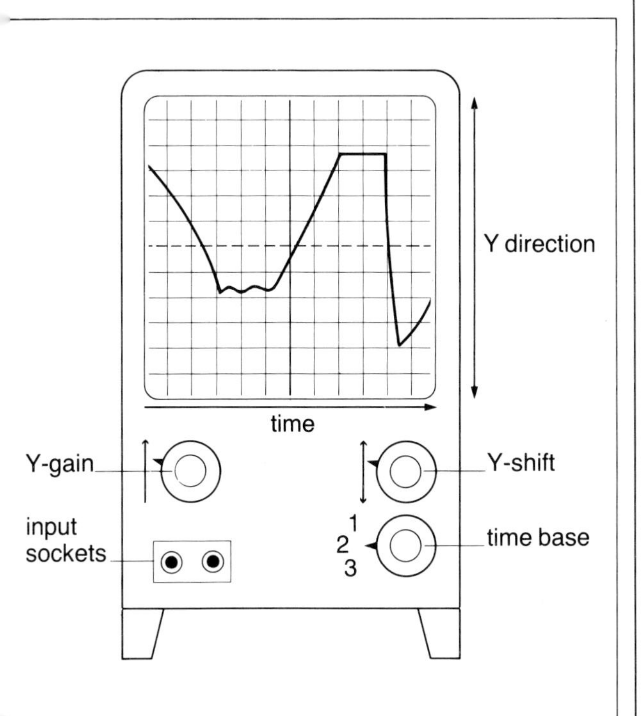

igure 8.11 CRO controls

Reading a trace

The path followed by the beam or spot is called a **trace**. In figure 8.11 the trace represents a voltage picture. The dotted centre line is the zero mark. Everything above this line indicates a current flowing in one direction around the circuit. Everything below the line indicates that a current is travelling the opposite way around the circuit. This particular trace shows a very uneven selection of voltage changes. The trace takes time to move across the screen so it is showing you changes which took place a short time ago, when the trace started. A CRO is a very fast acting voltmeter and can work with rapidly changing voltages.

Using a CRO as a voltmeter

In any of the circuits mentioned earlier in this unit a CRO could have been used in place of any of the voltmeters and would have functioned equally as well.

CRO controls

(See figure 8.11.)

Time base
This is used to change the time taken for the trace to sweep across the screen. The faster the trace the shorter the time taken – it can freeze changes that happen more quickly. A slow time base value is useful for changes that take a long time to happen.

Y-shift
This moves the entire trace up or down the screen. If you think the CRO is not working it may just be that the trace has been moved too far up or down!
Y-gain
If a trace is too small or too large in the Y direction this control will magnify or shrink it to any size you like. In this way you can make the trace fill the screen to best advantage.

ELECTRICAL ENERGY

Electricity is an **energy** form because it can make something happen (see page 1). Table 8.5 gives a number of examples of energy changes in which electricity plays a part.

device	major energy change(s)
loudspeaker	electricity → sound
battery	chemical energy * → electricity
lamp	electricity → light and heat
torch	chemical energy * → electricity → light
heater	electricity → heat
motor	electricity → kinetic energy
dynamo	kinetic energy → electricity
transformer	electricity → magnetism → electricity

table 8.5 Energy changes involving electricity

*Chemicals store energy and so they are another variety of potential energy – see page 2.

To make a bulb glow, a p.d. has to drive a current through the filament coil. The amount of light given off by the bulb depends on the p.d. and the current.

INVESTIGATION 8.8

CATHODE RAY OSCILLOSCOPE

Whilst it is very difficult to do much harm to a CRO (electrically!) it should be treated with respect.

a Take a pair of connections from a 12 V a.c. power pack and plug them directly into the input sockets of a CRO.

b Experiment with the CRO controls until a waveform, similar to the one in figure 8.12a, appears on the screen.

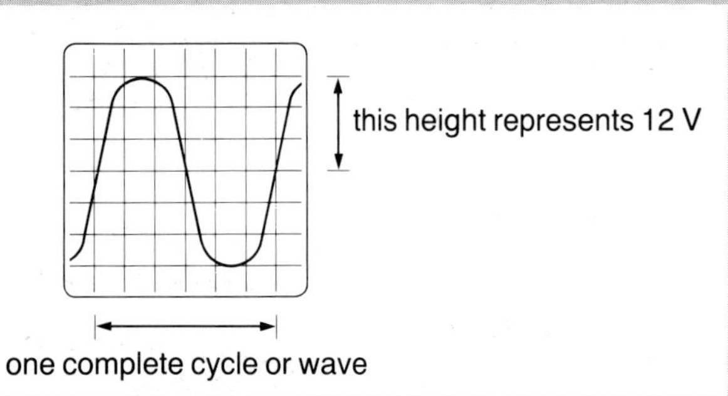

figure 8.12 a A simple a.c. trace

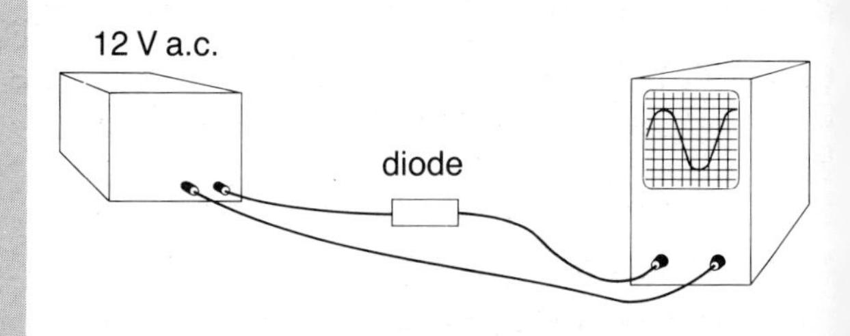

figure 8.12 b A diode in series with the power pack

Questions

1 What differences are there when the CRO is connected to lower a.c. voltages? What do d.c. voltages look like?

2 Assume that the 12 V a.c. output from your power pack is correct. How would you go about checking the other voltage values available on the same power pack?

3 Connect a diode in series with the power pack (see figure 8.12b). What changes does this device cause on the screen?

4 What function does a diode serve?

If you leave the bulb on for a long time more energy will be consumed. Electrical energy involves

p.d. (V)
current (I) — electrical energy depends on these three
time (t)

Calculating electrical energy should not be confused with Ohm's law which is used to calculate voltages, currents and resistances. Like all other forms of energy, electrical energy is measured in joules.

energy = current × p.d. × time
(units J) (units A) (units V) (units s)

$$E = IVt$$

Power

Power is the rate at which energy (electrical or otherwise) is used or transformed (see page 138). It is the amount of energy used per second.

$$\text{Power} = \frac{\text{energy}}{\text{time}}$$

The units of power are **watts (W)** (joules per second).

$$\text{electrical power} = \frac{\text{electrical energy}}{\text{time}}$$

$$\text{electrical power} = \frac{IVt}{t}$$

$$\text{electrical power} = IV \text{ watts}$$

One watt of electrical power is a very small quantity. A thousand watts, or 1 kW, is often a more convenient amount to consider. Most household electrical appliances have their power rating clearly marked.

Energy transfer

A simple 40 W light bulb demonstrates energy transfer. 40 joules of energy are consumed by the bulb every second. That energy is not lost every second but transferred or changed into 40 joules of light and heat. An energy transfer has taken place. When designing electrical devices (kettles, lamps, computer chips etc.) it is important to consider the energy changes that will take place. For example, it would be rather silly to design a lamp that heated up rather than glowed! The filament material is carefully chosen and constructed (it is long, thin and coiled) to give out energy in the form of light rather than heat.

ELECTROMAGNETISM

When an electric current passes along a wire it creates a **magnetic field** (see page 94.). The magnetic field surrounds the wire completely. The arrangement of the magnetic field pattern is rather like a set of tubes, one inside the other (see figure 8.14).

This magnetic field is an area of influence where things happen (see page 117). For example, if a compass is brought near the wire the compass needle will move. The direction of the compass needle follows the direction of the magnetic field around the wire. This is also shown in figure 8.14. The magnetism will be there as long as current flows. It will disappear when the current is switched off. If the direction of the current is reversed the direction of the magnetic field will be reversed too. Because the

INVESTIGATION 8.9

ELECTROMAGNETIC FIELDS

Electromagnetism is produced wherever an electric current flows through a conductor.

a Connect approximately 1 m of insulated copper wire to the 1 V terminals of a laboratory power pack. (Batteries can be used but they will not last long.)
b Hold part of the wire vertical.
c Use the compass to follow the magnetic field around the wire (see figure 8.13a). Check out the suggestion that an electromagnetic field appears around the wire.

Questions

1 How does the electromagnetic field change when the wire is coiled up (see figure 8.13b)? (Make the coil large enough for the compass to go through the centre.) Make a drawing of the field pattern.
2 An **electromagnet** is a coil of wire with a metal centre. Design an experiment to test the magnetic strength of an electromagnet with
i an iron rod in the centre
ii a steel rod in the centre.
What advantages and disadvantages do iron and steel have over each other?
3 What happens when both metal cores are in the coil and the power is switched on?
4 The experiments that involve a compass will only work with d.c. yet the others will work with both a.c. and d.c. Why is this?

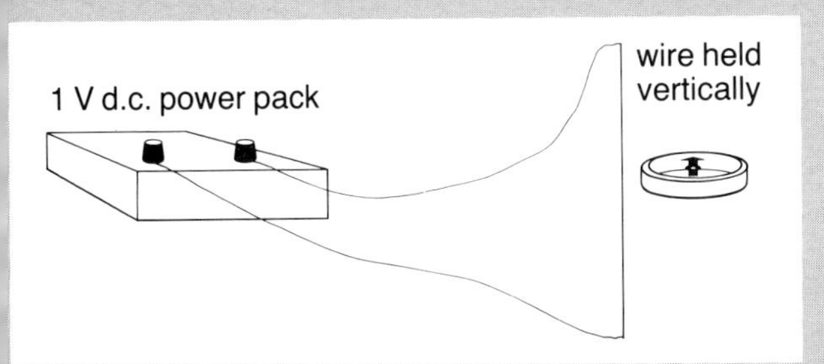

figure 8.13 a Following the magnetic field

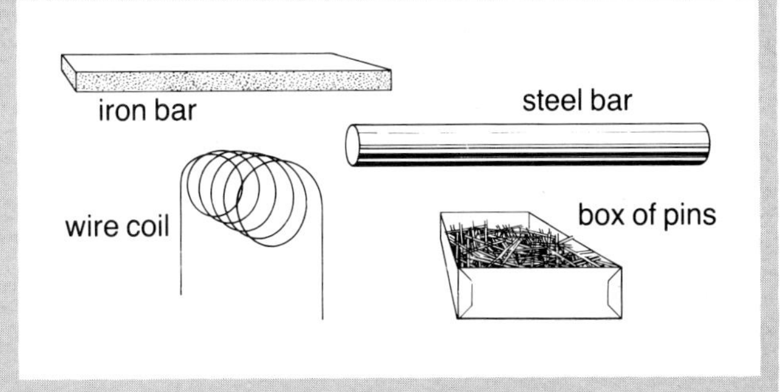

figure 8.13 b Making an electromagnet

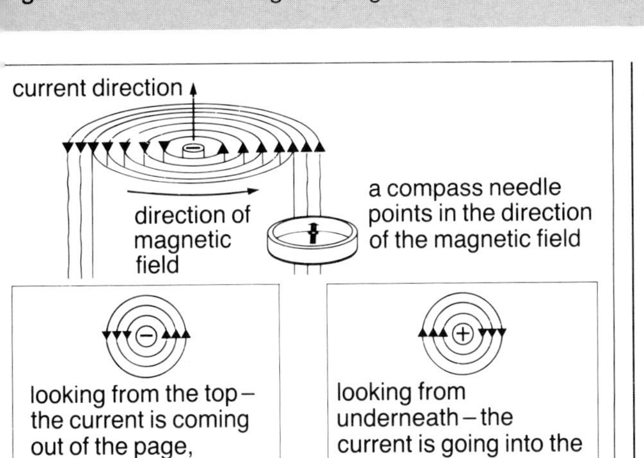

figure 8.14 Some of the many 'tubes' of magnetism around a wire carrying current

magnetic field has been created from electricity it is known as an **electromagnetic** field. Electromagnetism is temporary magnetism.

Electric motion

If two strong magnets were brought together, one of two things would happen.

like poles would repel
unlike poles would attract

Attraction happens with electromagnets too. It also happens with a mixture of a permanent magnet and an electromagnet.

When a magnetic field and electromagnetc field are brought together either attraction or repulsion will result. If the fields are strong enough the result will be movement. Figure 8.15 shows a simple laboratory arrangement that demonstrates movement from combining a magnetic field and an electromagnetic field.

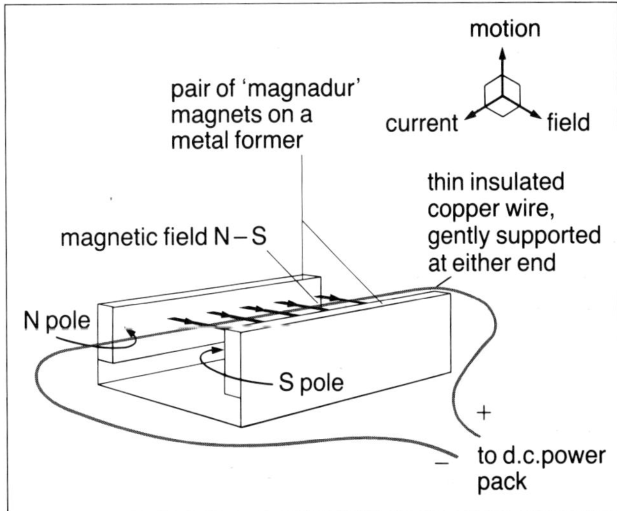

figure 8.15 Movement from magnetism

When the current is switched on, an electromagnetic field is created around the wire. This reacts with the magnetic field between the two magnets and the result is movement of the wire. In this particular arrangement the movement will be upwards. If the direction of the current or the direction of the magnetic field is changed the wire will move in the opposite direction: downwards. The relative directions of motion, current and magnetic field are given by the 'corner of a box' section of figure 8.15.

If alternating current is used instead of direct current, the wire will vibrate up and down with a frequency of 50 Hz (see page 124). This movement can be seen with the aid of a stroboscope.

Electric motors

The electric motion we have looked at so far has not been continuous. In order to use the motion created by an electromagnet and a permanent magnet it must be possible to create continuous circular motion, which can be used practically.

The apparatus is shown in figure 8.16. The wire which produces the electromagnetic field is wrapped around a core which can rotate freely. The wire on the right side of the core carries current from the front to the back of the motor. The wire on the left side of the core carries current from the back to the front. So if the left side of the core moves upwards, the right side moves downwards, and the motion continues on in a circle.

This particular type of motor will not work on a.c. for the reasons mentioned earlier in the chapter. The arrangement of bared wires at the front of the motor is known as a **commutator**.

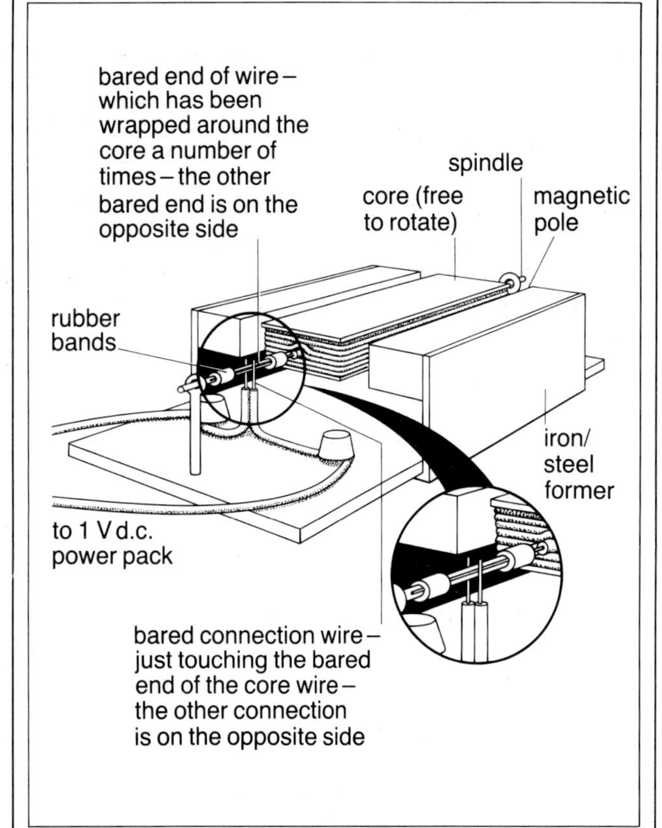

figure 8.16 An electric motor

INVESTIGATION 8.10

THE ELECTRIC MOTOR

Use the kit of parts to make up the motor described in figure 8.16. (Connecting the commutator arrangement can be awkward.)

It is possible to vary the speed at which the motor turns. There are three possibilities

a change the strength of the magnetism
b change the current value
c change the number of coils on the core.

Questions

1 Work through these changes to find how they are related to the speed of the motor. What design problems will you have to solve?
2 You are told that a machine called a dynamo is an electric motor which works in reverse (i.e. the energy changes happen in reverse, the motor does not move backwards!). How would you go about investigating this idea?

Transformers

If electricity produces magnetism, is the reverse true? Can magnetism create electricity? A **transformer** is a simple device which does both. At one side of the transformer a coil (the **primary coil**) creates magnetism. Inside the coil is a soft iron core which acts as a guide for the magnetism – it channels it to a second coil (the **secondary coil**). Making the soft iron core out of laminations (thin strips) assists the passage of magnetism. As the magnetism flows past this second coil it is transformed (changed back) into electricity. The energy changes that take place in a transformer are *electricity into magnetism and then back into electricity* (see figure 8.17).

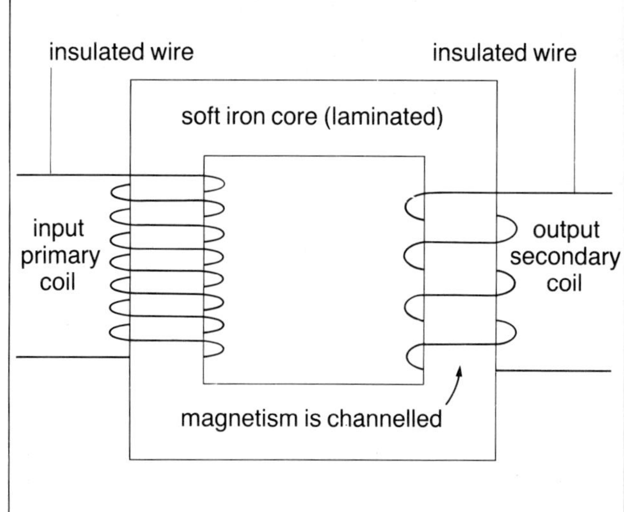

figure 8.17 A transformer works with alternating current

.t first thought there would seem to be little point in ıaking a device which seems to get nowhere. After ll you feed a transformer with electricity and it ͽturns electricity to you. The benefits of ·ansformers are quite simple. They allow you to ·ansform (change) voltage and current values. For ·xample, it would be quite possible to make a ·ansformer to carry out the changes in table 8.6.

primary coil voltage/current (V)	(A)	changes to	secondary coil voltage/current (V)	(A)
240	1	→	24	10
240	1	→	10	24
240	1	→	12	20
240	1	→	6	40
240	10	→	2400	1

ɪble 8.6

'hese changes are very simply made by altering the .umber of turns on the secondary coil. There are two arieties of transformer: **step-up** and **step-down**. The .ame step-up is given to a transformer which ıcreases the voltage value at the secondary coil. The .nal example in table 8.6 is the only example of a tep-up transformer. The other examples are step-.own transformers.

'ransformers will only work while the magnetic .elds are changing. For this reason **transformers will nly work on alternating current**.

INVESTIGATION 8.11

DOES A TRANSFORMER WORK?

a Use a CRO to look at the potential difference (p.d.) between the 1 V a.c. terminals on a laboratory power pack.
b Using a simple transformer like the one in figure 8.17, disconnect the CRO and attach the primary coil to these terminals.
c Reconnect the CRO (without changing any of its controls) to the secondary coil.

Questions

1 Is the CRO trace from the secondary coil larger or smaller than the trace from the primary coil?
2 By experimenting find which words from the following list will most correctly complete this sentence. (There are two correct answers.)
 _______ turns on the _______ coil will _______ the potential difference at the _______ .
 Choose words from this list: secondary, more, decrease, output, input, less, primary, increase.
3 What happens to the secondary trace if d.c. is used at the input?
4 What happens to the trace if a coil of wire is connected directly to the input of a CRO and a magnet slowly or quickly moved in or out of the coil?

Application 8.4

PLASTIC MONEY

The magnetic strip on the back of a credit card creates a tiny electric current when it is moved past a sensitive 'read head' (like the record/playback head in a tape recorder). This electrical signal is converted electronically into a numbered code. This type of code is used in other ways too, for example for telephone units for British Telecom card phones, bank accounts and credit limits.

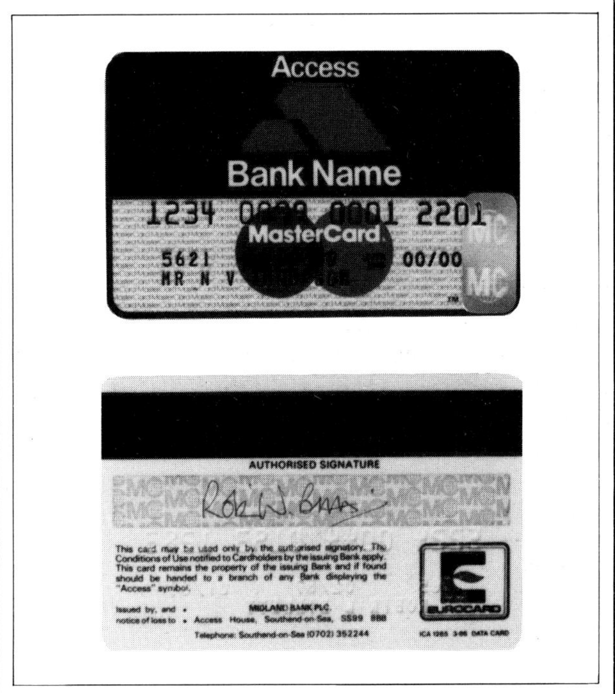

figure 8.18 Plastic money

TRANSMISSION IN PRACTICE

The idea of transmitting electrical energy by wire from a power station to people who use it is very simple, but in practice there is a number of practical problems to solve. Here is a roundabout – where do you get off?

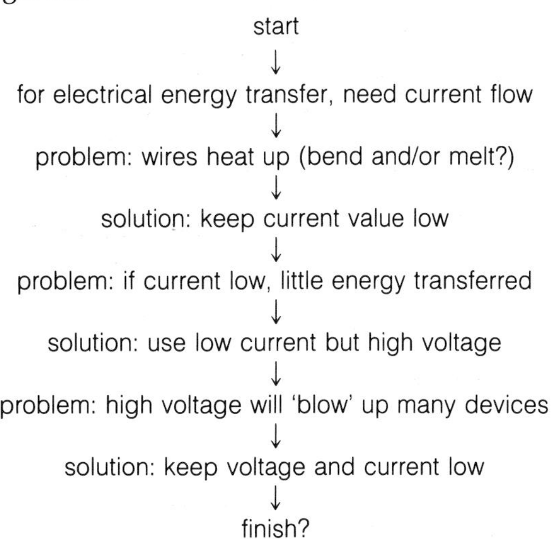

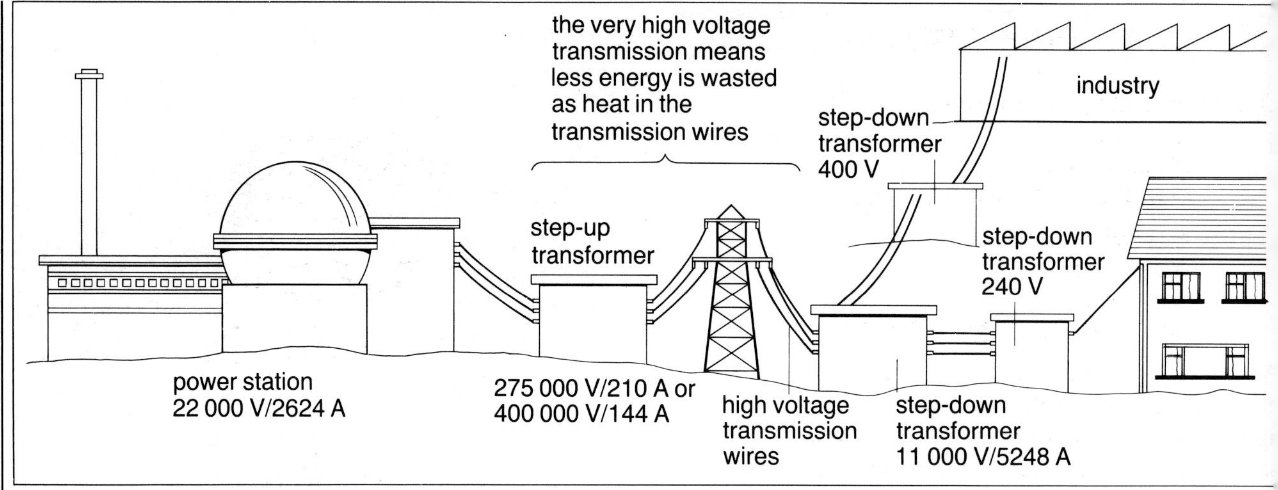

figure 8.19 *Using transformers*

By trying to transfer electrical energy we have met more problems than we have answers for! Efficient electrical energy transfer only became effective and practical with the use of transformers. The transmission system in the UK is arranged as shown in figure 8.19.

The use of a transformer allows the voltage from the power station to be increased considerably. At the same time the current value decreases. With a lowered current value the overhead transmission wires will not heat up so much and less energy will be wasted. The voltage value is reduced in a number of stages before it reaches the consumers. If the consumers are heavy industry they may require 11 000 V. Light industry only requires 400 V and our homes only 240 V.

At home

A household **distribution box** supplies electricity at a voltage of 240 V. The total current may be as much as 60 A. There is a single large household fuse (or trip switch) which will blow (or trip), cutting off the supply if this maximum current rating is exceeded. To do this you would need the equivalent of 15 one-bar electric fires or five washing machines on at once! Trip switches can be reset, fuses must be replaced.

Power is distributed throughout the house by a number of individual circuits called **ring mains**. They get their name because the wiring for these circuits 'rings' the house (see figure 8.20). Each ring main has its own separate fuse.

A typical fuse box might contain the circuits shown in table 8.7.

ring main circuit	fuse (in fuse box)
upstairs lights	5 A
upstairs power sockets	13 A
downstairs lights	5 A
downstairs power sockets	13 A
separate electric cooker circuit	15 A

table 8.7

By separating certain functions (lights from power sockets etc.) then if something should go wrong, only that ring main will cease to work. The others will be unaffected.

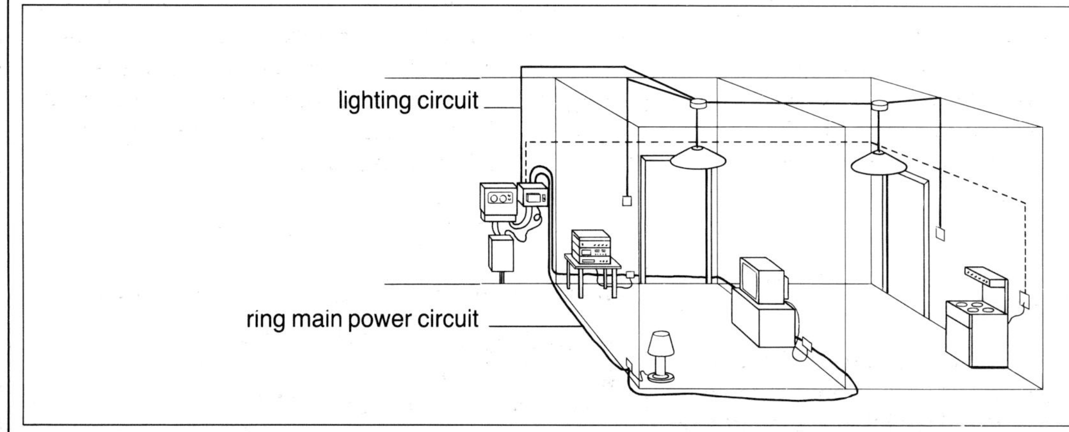

figure 8.20 *The household distribution network – ring mains and fuse box system*

Three wires for safety

The colour coding for household circuits is

red – live
black – neutral
bare copper – earth *safety wire*

The colour coding for appliances is

brown – live
blue – neutral
green and yellow – earth *safety wire*

As electricity needs only two wires, one to carry electricity to the device and one to carry it away again, why bother with a third? The third wire is the safety wire. It is known as the **earth connection**.

The earth connection is particularly valuable on items with metal casings. If a fault develops inside an appliance, the casing may become connected to the live wire from inside. This would be very dangerous if you came along and touched the casing. The earth wire is connected to the inside of the casing at one end. The other end is connected directly to the earth which is a conductor, so it automatically provides a pathway to the earth for the electricity if the casing becomes live.

Why live and neutral?

A source of direct current has easily identified terminals: positive and negative. It is quite straightforward to show which is which. Why bother to change the labels when using alternating current? Why give the two terminals new names – **live** and **neutral**? The word live suggests danger, the word neutral might even suggest safety. As alternating current travels one way around a circuit and then the other, you might expect the terminals to alternate and become live, neutral, live, neutral, live, neutral, etc. Perhaps they should both be labelled 'dangerous'. The answer to the problem can be found with the help of a drinking straw. It's possible to blow and suck liquid with a straw. The live terminal acts in a similar way. First it 'blows' out electric current and pushes it through the circuit into the neutral terminal, then it draws electric current from the neutral terminal through the circuit. The live connection is the active one – but you should not touch either! Fuses and switches are always connected to the 'live wire'.

Plugging safety

A correctly wired plug is shown in figure 8.21. It is shown in black and white to illustrate one possible view of it as someone with colour blindness might see it. The earth wire is clearly visible because it is striped. The live and neutral wires are different shades of grey.

figure 8.21 **a** *A correctly wired plug*

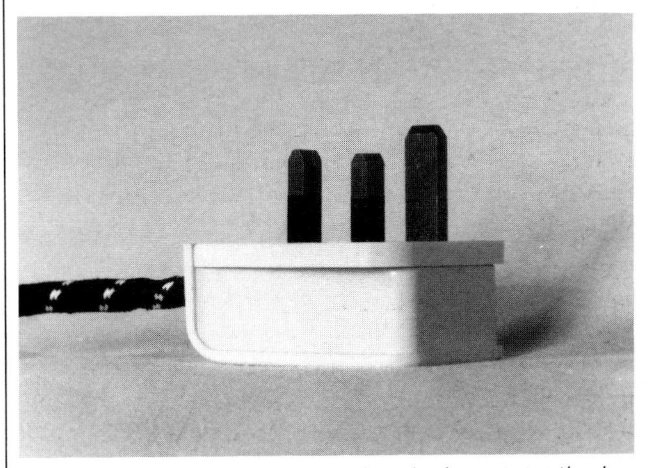

figure 8.21 **b** *Side view showing the longer earth pin*

The side view shows clearly that the earth pin is longer than the others. On a standard three-pin socket the earth pin has to be inserted first. As the earth pin enters the socket it lifts two flaps, protecting the live and neutral connections, and allows the live and neutral plug pins to enter. When the plug is removed the two safety flaps automatically close again.

Whenever a new appliance has a plug fitted a fuse of the correct rating should be used. It is quite normal for plugs to be sold with 13 A fuses in them, even though this fuse value may be wrong for a particular application. The correct fuse rating value may be found from this simple equation:

$$\frac{\text{wattage}}{\text{voltage}} = \text{current (fuse rating)}$$

Find the correct fuse from the list that follows for

a a washing machine (3 kW or 3000 W)
b a 100 W lamp
c a one-bar fire (1 kW or 1000 W)

Fuses available are 1 A, 2 A, 3 A, 5A, 7 A, 10 A, 13 A.

Example **a** washing machine

$$\frac{\text{wattage}}{\text{voltage}} = \text{current}$$

$$\frac{3000}{240} = 12.5\text{ A}$$

Nearest higher fuse value = 13 A

Note: In practice, most appliances such as washing machines and televisions require a 13 A fuse. A surge of current may pass through these appliances as they are switched on and smaller but correctly sized fuses would blow. Always check the manufacturer's label first.

Example **b** lamp

$$\frac{100}{240} = 0.4\text{ A}$$

Nearest higher fuse value = 1 A

Example **c** one-bar fire

$$\frac{1000}{240} = 4.2\text{ A}$$

Nearest higher fuse value = 5 A

Time for thought

Do you think that current electricity and friction produce anything in common?

ALL YOUR OWN WORK

1 The two types of static ______ are ______ and ______.

2 Electricity flows through a ______ but not through an ______.

3 A flow of electricity is called a ______ and is measured in ______. A ______ measured in ______ is needed to make electricity flow from place to place.

4 The flow of electricity through a material can have one or more of the following effects: ______, ______, ______.

5 Why
  a are fuses rated in amperes
  b are Christmas lights wired in parallel
  c are switches placed in the live side of a circuit?

6 What effect would be noticed on a CRO if
  a the input voltage decreased
  b the frequency of the signal increased?

7 A washing machine marked 'made in Transylvania: 2.5 kW, 240 V' arrives from your Uncle Alucard. What size fuse would you use?

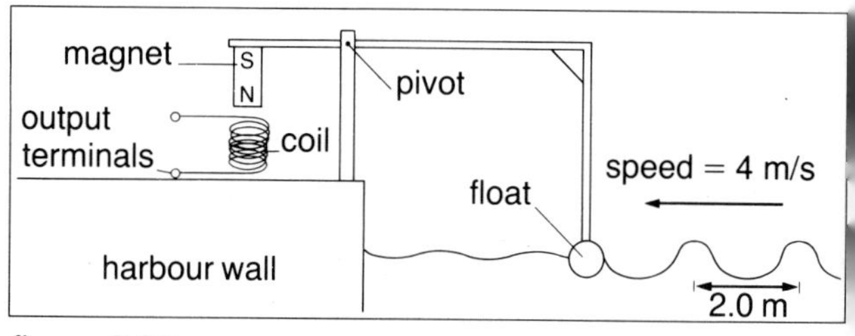

figure 8.23

8 Figure 8.23 shows a simple generator designed to convert the energy from waves into electricity.
  a Bearing in mind how the float moves, how does the generator produce an alternating voltage?
  b How might the output voltage be increased?
  c What is the frequency of the output and how would your check your answer experimentally?

Application 8.5

POWER PACKS

Power packs, such as the 'transformer units' that run calculators, model racing cars and trains, as well as the ones used in school laboratories, would not be possible without transformers, diodes and capacitors. When these three are combined it is possible to convert 240 V mains to any desired voltage, a.c. or d.c. Figure 8.22 shows a typical such arrangement.

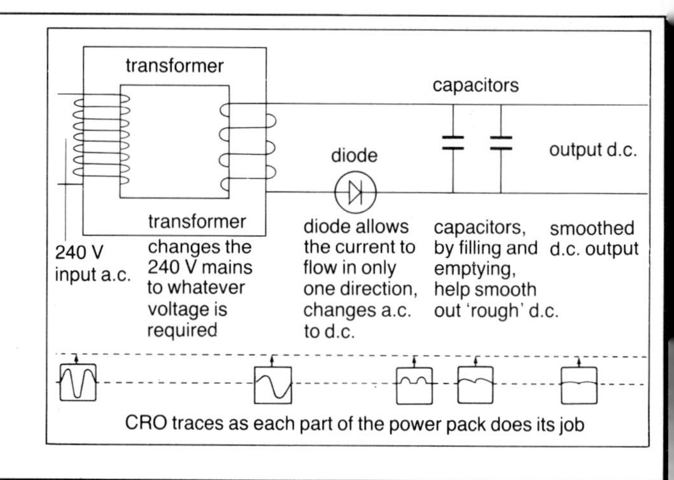

figure 8.22 The internal arrangement of a simple power pack

ENERGY

NATURAL ENERGY RESOURCES

Energy, the main energy forms and simple energy changes were mentioned in unit 1, Starting science. We shall now look at where energy comes from.

Global energy resources

There are two kinds of energy source: **non-renewable** and **renewable**. Examples of non-renewable energy sources include coal and gas. These are destroyed when they are used and take millions of years to be replaced. Renewable energy sources are not destroyed when they give up their energy. A hydroelectric power station takes the stored energy from water so it can be used elsewhere, but does not

figure 9.1 Earth's energy flows

destroy the water in the process. Table 9.1 lists a number of non-renewable and renewable energy sources.

non-renewable		renewable
coal	fossil fuels	solar energy
oil		energy from the tides
gas		energy from the wind
peat		geothermal energy
nuclear energy		

table 9.1

Fossil fuels

Most of the energy we use for industry and in our homes comes from the fossil fuels: coal, oil and natural gas. These are the remains of dead plant and animal material buried millions of years ago.

Coal

In prehistoric times much of the Earth's surface was covered with vegetation. The warm climate provided ideal growing conditions. When it died, much of this vegetation sank to the bottom of the rivers and lakes. There it formed layers at it was slowly decayed by bacteria. These layers, known as peat, are the beginnings of coal.

The layers of peat were covered over in time by mud and silt brought down by the rivers. The pressure of this covering squeezed out water and gas from the peat, and it was also compressed by changes in the Earth's surface. The peat first turned to **lignite** which is soft and brown. This changed to black **bituminous** coal when more water and gas had been squeezed out of it by the pressure of the earth above. Finally, if the seam was buried very deeply **anthracite** was formed. This is very hard and shiny. It is very different from the original peat.

Britain has coal reserves enough to last for 300 years if we continue to use it as we are now, and Australia has enough for 1000 years. It is mined either from underground mines or from opencast mines if the coal seams are close enough to the surface.

figure 9.2 *Coal reserves will not last forever*

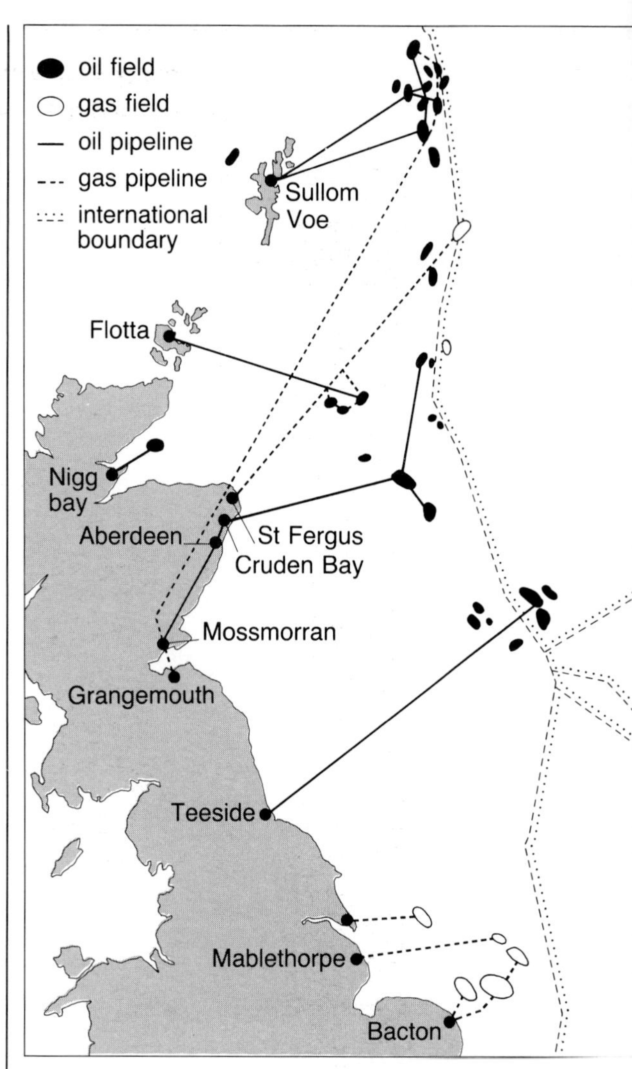

figure 9.3 *Oil reserves may not last your lifetime*

Oil and natural gas

The countries surrounding the North Sea are very fortunate to have a supply of oil and gas on their door steps. The reserves that have been discovered will last only into the next century. Oil and gas deposits are the remains of marine organisms which sank to the bottom of the oceans millions of years ago. Like the land vegetation which produced coal, the marine organisms were covered and compressed into the organic layers known collectively as petroleum, or crude oil. Often a pocket of natural gas is trapped above the petroleum layer. Liquefied petroleum gas (LPG) is widely used as a convenient, portable heat source. It is gas that has been turned to a liquid under high pressure.

Supplying the base load

The easiest form of energy to use is electricity. Unfortunately it is not practical to keep vast supplies of it for our everyday needs, since it cannot be stored in large quantities. It has to be generated

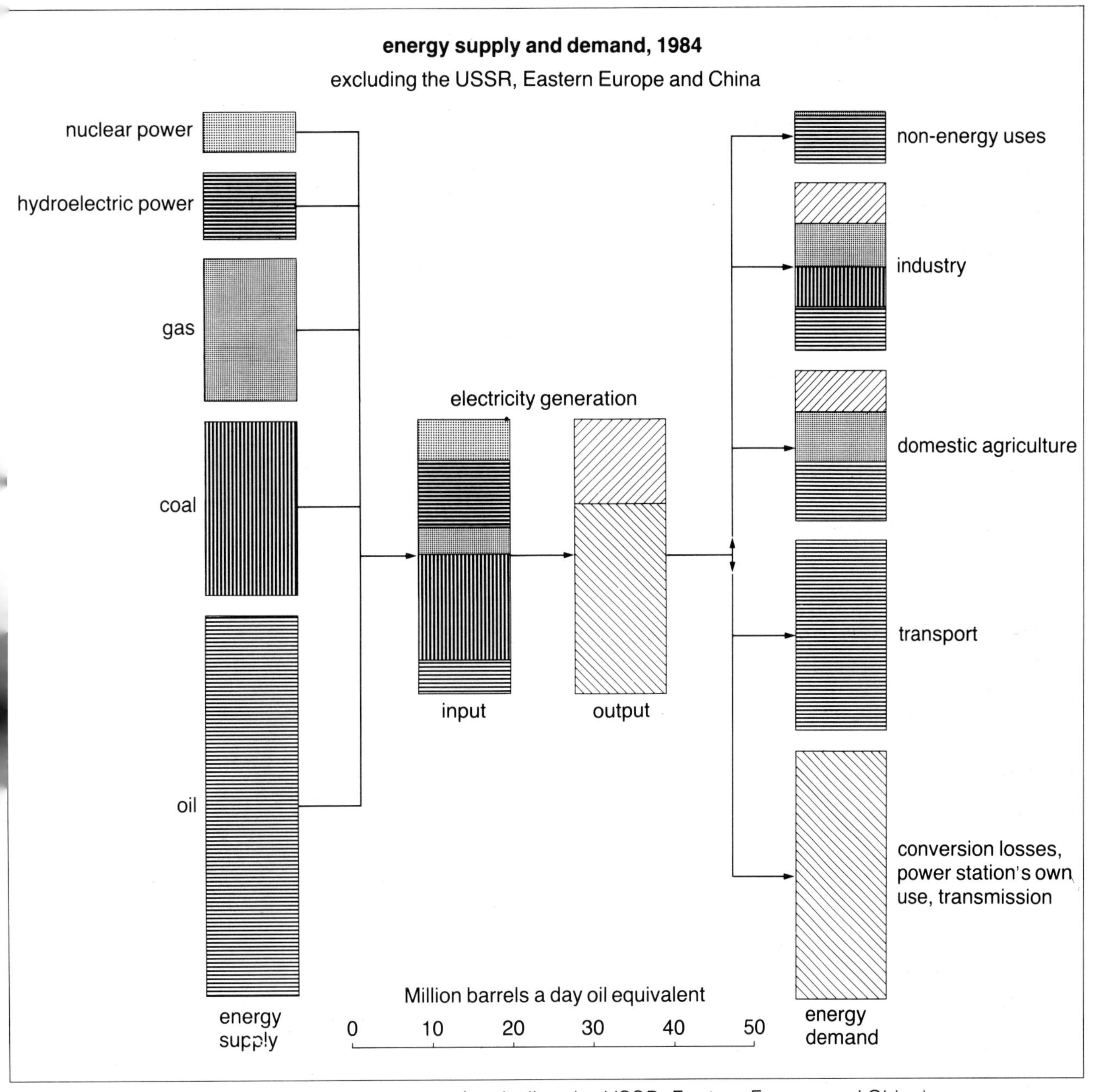

figure 9.4 *Energy supply and demand, 1984 (excluding the USSR, Eastern Europe and China)*

continuously. The energy we need to have available, night and day, to allow the country to function is known as the **base load**. This is supplied mainly by **coal-fired** and **nuclear power stations**. **Gas-fired** and **hydroelectric power stations** are more easily started and stopped which makes them useful for supplying additional energy at times of peak demand.

Supply and demand

Constructing new power stations around the country is not an easy task. There is a number of things to think about.

a Is the primary energy (e.g. coal, gas) priced cheaply enough to provide electricity economically?

b Is it an easy process to change the primary energy into electricity?

c How much of the primary energy will be wasted in the conversation process?

d Will there be problems transporting the primary energy to the station or storing it there?

e Are any health hazards to the public, the power station workers or the environment likely as a result of building a particular type of power station or using a particular type of fuel?

f Will there be any long-term effects on the local community or the environment?

g Will the whole system (station and fuel) be able to meet changing patterns of demand?

h What political considerations are involved?

INVESTIGATION 9.1

ALTERNATIVE ENERGY

There is a number of possible alternatives to the fossil fuels already used in power stations. Listed for each of the alternatives mentioned below are possible advantages and disadvantages. Think about each one in turn. As a starting point, ask yourself the questions raised earlier, concerning supply and demand, about each as a practical alternative energy source.

solar	free, renewable, unreliable, geographical problems, very visible.
tidal	free, renewable, only available at certain times, geographical problems, restricts marine traffic
geothermal	free, renewable, small scale
nuclear	non-renewable, enormous reserves, geographical problems, storage of waste products, possibility of radiation leakage
wind	free, renewable, unreliable, geographical problems, very visible, large number of generators needed
hydroelectric	free, renewable, geographical problems

figure 9.5 *a Solar furnace at d'Odeillo, France*

figure 9.5 *b California – where wind is more profitable than agriculture*

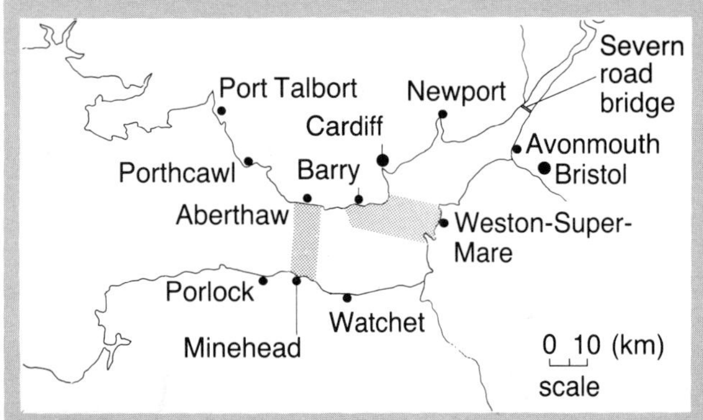

figure 9.5 *d Two of the areas where a Severn barrage might be built*

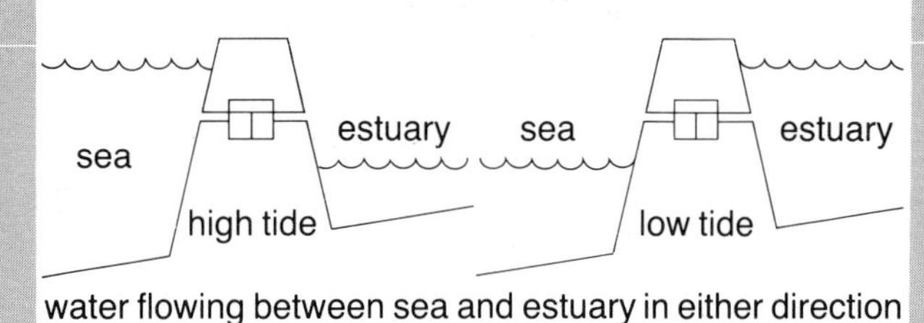

figure 9.5 *c The Rance tidal power station, France*

Questions

By discussion, how would you defeat or defend these statements?

1. 'The only practical alternative energy source for the future is nuclear power.'
2. 'Any Welsh valley large enough to contain a hydroelectric power station should be purchased by the government for such a scheme.'
3. 'Areas such as Dartmoor in Cornwall should be used exclusively for wind-generated electricity.'
4. 'Solar panels should be included in all new houses and added to all existing buildings.'

ENERGY MEASUREMENT

Four types of energy that can be measured easily are **kinetic energy, potential energy, electrical energy** and the **work done** by an energy form. Each is measured in joules. (A joule is a small energy unit and it is often more convenient to use kilojoules (kJ) or megajoules (MJ)).

$$1 \text{ kJ} = 1000 \text{ J}, \ 1 \text{ MJ} = 1\,000\,000 \text{ J}$$

Kinetic energy (KE)

This is the energy of movement (see pages 1 and 93). Increase either the mass or the velocity of an object and the total KE will increase. Common sense will tell you that a high speed train will have more energy than a bicyclist! As we saw on page 93,

kinetic energy (units J) = ½ × mass (units kg) × velocity^2 (units (m/s)^2)

$$KE = \tfrac{1}{2} m v^2$$

Example
How much energy does a typical family saloon car (mass 1000 kg) have travelling at 90 km/h?

mass = 1000 kg
velocity = 90 km/h (change km/h to m/s)
$= \frac{90\,000}{3600}$ m/s (there are 1000 m in 1 km) (there are 60 × 60 = 3600 s in 1 hr)
= 25 m/s
energy = 0.5 × 1000 × 25 J
= 312 500 J = 312.5 kJ

Potential energy (PE)

One variety of PE is the energy resulting from position. Walking upstairs increases your potential energy. The PE of an object increases with increasing mass or increasing height.

potential energy gained (units J) = mass (units kg) × acceleration due to gravity (units m/s^2) × height raised (units m)

$$PE = m g h$$

Example
How much PE does a passenger (mass 75 kg) gain travelling upstairs (3 m high) on the 'minivator' in figure 9.6? (The acceleration due to gravity $g = 10$ m/s^2, see page 96)

mass = 75 kg
height = 3 m
PE gain = $m g h$
= 75 × 10 × 3 J
= 2250 J
= 2.25 kJ

figure 9.6 Helping to gain PE?

All work and no play...

Work is done only when a force moves. Pushing a bicycle pedal down or pulling a crate along the floor are two examples of work. Energy is needed to do work.

work (units J) = force (units N) × distance (units m)

work – $F d$

Example
How much work is done by two household removal men when they both have to push against a heavy trolley with a force of 250 N for the 15 m from the house to the removal van?

force = 250 N × 2 (2 men)
= 500 N
distance = 15 m

work = $F d$
= 500 × 15 J
= 7500 J

Each man supplies 3750 J of energy.

INVESTIGATION 9.2

MEASURING WORK

A car with a flat battery may be 'bump' started. The car is first put into a low gear and the clutch pedal depressed. Willing volunteers push the car to get it moving. When the car is rolling steadily the clutch pedal is quickly released and the gear engages. This turns the engine, which may fire.

Questions

1 Design an experiment to measure the amount of work a volunteer (or volunteers) would need to do to bump start a small car such as a Mini. What problems need to be solved to
  a measure the force applied to the car to keep it rolling steadily
  b decide where to start measuring distance from and what distance to measure
  c make sure the car is not marked in any way?

2 The energy used to start the car would have originated as sunlight! How does it end up? Where does it go?

Electrical energy

The amount of electrical energy that flows in a circuit depends on three things: current, potential difference and time. Increase any of these and the total energy value will increase (see page 126).

electrical energy (units J) = current (units A) × potential difference (units V) × time (units s)

$$E = IVt$$

Example

Suppose the local council offices contain 20 fluorescent strip lamps connected to the mains and each of them conducts a current of 0.25 A. If they all left on how much energy would be used during

a one lunch hour

b 50 weeks of lunch hours?

potential difference = 240 V
current = 20 × 0.25 A = 5 A
time = 3600 s

a energy used in one hour

$E = IVt$
$= 5 \times 240 \times 3600$ J
$= 4\,320\,000$ J $= 4.32$ MJ

b energy wasted in 50 weeks

energy wasted in 1 week = 5 × 4.32 MJ
energy wasted in 50 weeks = 5 × 4.32 × 50 MJ
= 1080 MJ

ENERGY AND POWER

Energy and power are linked but they are not the same. A problem concerning energy may involve energy changes and energy transfers. For example, the main energy changes which take place in a light bulb are as follows.

electrical energy → transferred or changed to → light energy + heat energy

Power is how quickly the energy is transferred or changed. Power is the **rate of energy transfer**.

$$\text{power} = \frac{\text{energy}}{\text{time}}$$

The units of power are joules per second or watts.
1 joule/second = 1 watt (W)
The watt is a small unit. It is often more convenient to think of energy transfer rate in kilowatts.

1 kW = 1000 W

Example

Three different machines (A, B and C) each require different amounts of energy to complete the same task in different time periods.

machine A uses 24 kJ in 120 seconds
machine B uses 30 kJ in 180 seconds
machine C uses 20 kJ in 150 seconds

Which machine is the most expensive to run, and which is the most powerful?

		machine A	machine B	machine C
power output	=	$\frac{24\,000}{120}$ W	$\frac{30\,000}{180}$ W	$\frac{20\,000}{150}$ W
	=	200 W	166.7 W	133.3 W

The most powerful machine is machine A (its power output is 200 W).
Machine B uses most energy and will be the most expensive to run.

Typical household electricity bulbs are labelled with power ratings:

40 W (uses 40 joules per second)
60 W (uses 60 joules per second)
100 W (uses 100 joules per second)

KEEPING WARM

Figure 9.7 highlights an important human need – to keep warm. The enzymes in our bodies function only in a small temperature range. 'Normal' body temperature is 37 °C. Our bodies produce a certain amount of heat from the chemical reactions which take place, but this is lost to the surroundings if they are cold. We can take certain steps to reduce this heat loss (see page 260-61), but in extreme cold these are not enough and hypothermia may result, which may lead to unconsciousness and even death.

THE SUNDAY TIMES

COLD COMFORT

At No 6 Redan Road, Ada Hankin sits alone and tries to warm herself by the fire. The open coal-burning grate is one of the cheapest sources of heat, yet she can't keep the cold at bay.

When the fire has been burning for eight or nine hours the living room is still a long way from warm. As temperatures rose outside after last week's severe conditions, Ada still couldn't reach the level the experts say is safe. At 16 °C, according to doctors, respiratory disease may set in. But downstairs at No 6, it is only 11 °C.

If Ada lived at the end of her terrace, she would most likely be dead by now. But No 6 is somewhere near the middle and so she gains 1.5 °C of heat from the man who lives next door. His central heating is absorbed through their common wall and helps keep Ada alive.

"This lady may develop a chronic form of hypothermia living at this temperature," says a geriatrician consulted on Ada Hankin's case. "She is lucky that her house is protected by homes on either side. I certainly don't think she should be allowed to live there without at least one room being very much warmer than this."

Last week Ada's bedroom temperature did not rise above 8 °C. In the kitchen, where she strip washes every day, the temperature was 4.5 °C. In the lavatory, which in her unmodernised home is still outside in the yard, the temperature we measured was 3.5 °C.

TO FIND out just what measures would be needed to rescue Ada, our audit looked at how much money she would have needed to raise her home's temperature. As it was, with a coal fire downstairs and direct electric heater in the bedroom, she spent £14.38 last week to sustain a mere 11 °C in the living room and 4.5 °C elsewhere.

If she burned more coal and kept the heater on longer, these temperatures could be raised. But to get the house above the hypothermia line of 16 °C in the living room and 11 °C elswhere, it would cost her £21.27 a week. To raise No 6 to the levels enjoyed by comfortable families in centrally heated homes – 21 °C in the living room and 18 °C elsewhere – she would need to spend £44.36, an increase of £29.98 on last week's fuel bill.

In the search to find a solution for Ada, our audit next examined how insulation might help. If No 6 Redan Road had basic draught-proofing and the recommended 100 mm thickness of roof insulation, she could stick to her 11 °C living room and spend £1.41 less than she does now. Raising the temperature to the 16 °C hypothermia line would cost her £19.10 a week, £2.17 less than if she tried to do the same in her uninsulated home.

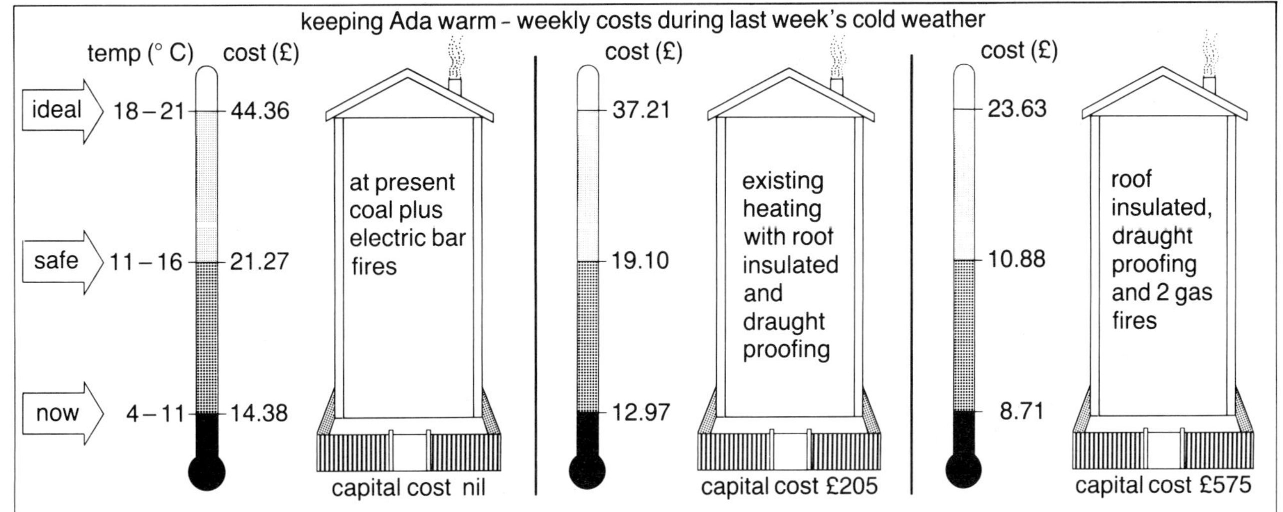

Ada's living room is 11 °C at present and the rest of the house is 4 °C above freezing. Heating cost her £14.38 last week. To get her living room up to a safe 16 °C and the rest of the house to 11 °C would have cost her £21.27. The ideal temperature of a centrally-heated house (18 – 20 °C) would have cost £44.36. Improving insulation and changing the heating method would cut costs dramatically. But can Ada afford the outlay?

calculations by Energy Conservation and Solar Centre

figure 9.7 Cold comfort

Heat transfer

The idea of heat transfer is mentioned a number of times in the article:

'His central heating is absorbed through their common wall', 'draught proofing', 'roof insulation'.

There are three methods of heat transfer: **conduction, convection** and **radiation**.

A **conductor** is a material which allows heat to travel through it easily (an **insulator** does not). (Compare this with electrical conductors and insulators, see page 116). Heat from Ada's neighbour was transferred by conduction. Glass fibre roofing insulation would help stop heat escaping by the same method. The many air pockets trapped between the glass fibres are effective insulators. Birds puff up their feathers to trap insulating air. A duvet works in the same way but if you use one in summer and don't want to overheat, then flatten it a little to remove the trapped air pockets.

Houses which have had draught-proofing fitted to the doors and windows benefit from reduced heat loss. Hot air rises because it is less dense and cold air comes in from underneath it to replace it. A **convection current** starts if cold air is allowed in.

The twin glass layers of double glazing prevent draughts, and are kept fairly close together to stop convection currents from starting between them. Convection currents can only take place in fluids (liquids and gases).

The third method of heat energy transfer is by **radiation** or **infra-red radiation**. Infra-red radiation is an energy wave and travels directly from the source to the receiver. An electric fire is a good example of an infra-red radiator.

Summary

Conduction - the material stays still, heat energy passes through. Bad conductors are known as insulators. Conduction happens mainly in solids. It is like a line of people standing still passing a ball from hand to hand.

Convection - the material moves and takes heat with it. Convections take place only in fluids. It is like a line of people each carrying the ball to the next person in the line.

Radiation - energy waves travel directly between the source and the receiver. Radiation needs no material to travel in and can travel through a vacuum. It is like a ball being thrown through the air.

INVESTIGATION 9.3

EXPENSIVE WASTE

The two bills shown in figure 9.8 were paid during the period in which this book was being written. Study them carefully.

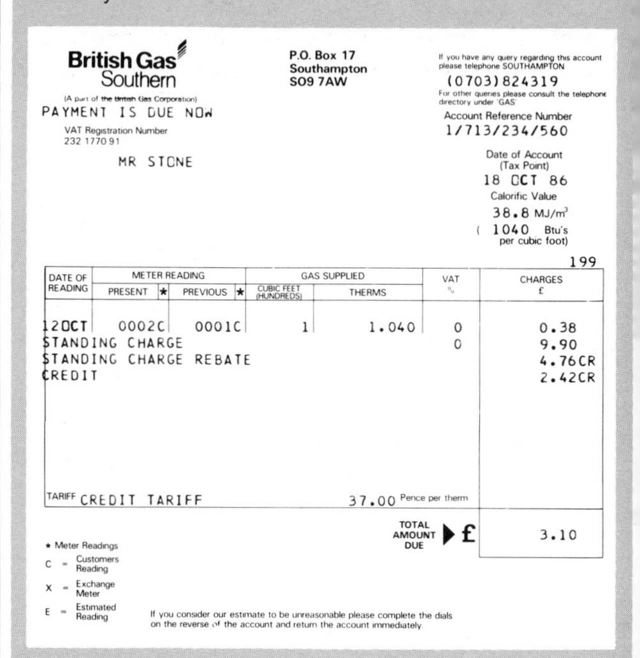

British Gas Southern
(A part of the British Gas Corporation)
PAYMENT IS DUE NOW
VAT Registration Number 232 1770 91

MR STONE

P.O. Box 17
Southampton
SO9 7AW

If you have any query regarding this account please telephone SOUTHAMPTON
(0703) 824319
For other queries please consult the telephone directory under 'GAS'
Account Reference Number 1/713/234/560
Date of Account (Tax Point) 18 OCT 86
Calorific Value 38.8 MJ/m³
(1040 Btu's per cubic foot)

199

DATE OF READING	METER READING PRESENT ★	PREVIOUS ★	GAS SUPPLIED CUBIC FEET (HUNDREDS)	THERMS	VAT %	CHARGES £
12OCT	0002C	0001C	1	1.040	0	0.38
STANDING CHARGE					0	9.90
STANDING CHARGE REBATE						4.76CR
CREDIT						2.42CR

TARIFF CREDIT TARIFF 37.00 Pence per therm

TOTAL AMOUNT DUE ▶ £ 3.10

★ Meter Readings
C = Customers Reading
X = Exchange Meter
E = Estimated Reading

If you consider our estimate to be unreasonable please complete the dials on the reverse of the account and return the account immediately.

figure 9.8 a Gas bill

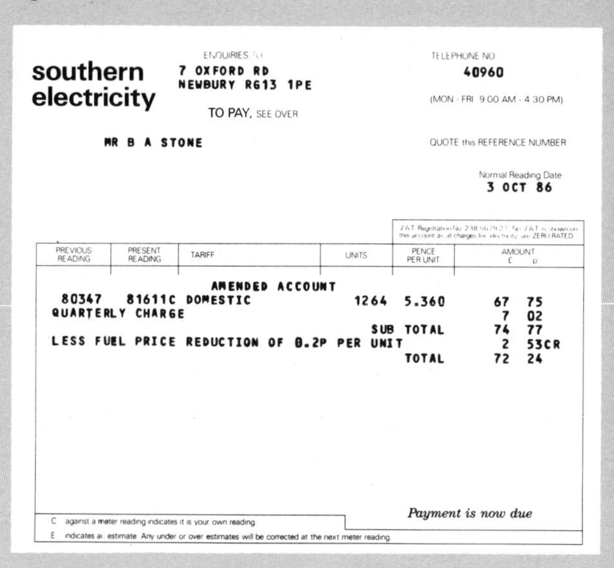

southern electricity

ENQUIRIES TO
7 OXFORD RD
NEWBURY RG13 1PE

TO PAY, SEE OVER

TELEPHONE NO
40960
(MON - FRI 9.00 AM - 4.30 PM)

MR B A STONE

QUOTE this REFERENCE NUMBER

Normal Reading Date
3 OCT 86

PREVIOUS READING	PRESENT READING	TARIFF	UNITS	PENCE PER UNIT	AMOUNT £ p
		AMENDED ACCOUNT			
80347	81611C	DOMESTIC	1264	5.360	67 75
QUARTERLY CHARGE					7 02
				SUB TOTAL	74 77
LESS FUEL PRICE REDUCTION OF 0.2P PER UNIT					2 53CR
				TOTAL	72 24

Payment is now due

C against a meter reading indicates it is your own reading
E indicates an estimate. Any under or over estimates will be corrected at the next meter reading

figure 9.8 b Electricity bill

Questions

1 What do you think the following terms mean?
   a calorific value
   b pence per unit
   c quarterly charge
   d standing charge rebate
2 Why do you think there is a standing charge rebate on the gas bill?
3 What are the current costs of gas and electricity? What would be the total cost of these bills today?
4 Research your old energy bills. Estimate what your next energy bills will cost.

Household insulation

There is a number of different types of household insulation and they perform different tasks. When installing energy-saving devices at home it is important to think about value for money. For example, draught-proofing your home will be cheap to do and it will pay for itself in a matter of days! It may take 10 years to recover the costs of double glazing. Table 9.2 shows what to consider when choosing insulation.

insulation	value for money (* good, ***** poor)
draught proofing	?
double glazing	?
loft insulation	?
cavity wall insulation	?
carpet underlay	?
What council grants are available? What insulation is already installed?	

table 9.2 Energy sense

1246 units (of electricity)

The electricity bill in figure 9.8b shows that 1246 units of electricity were used. What is a unit? It is an amount of electrical energy. The electricity board sells electrical energy.

1 unit = 1 kWh
1246 units = 1246 kWh

To use 1 kWh of electrical energy you could have:
25 × 40 W light bulbs on for 1 hour
OR 1 × 2 kW electric fire (2 bars) on for ½ hour
1 × 1 kW electric fire (1 bar) on for 1 hour
1 × 100 W light bulb on for 10 hours
1 × 20 W stereo on for 50 hours

Each of these examples would use 1 unit of electrical energy. How might the 1246 units have been used in the three months up to 3 October 1986?

Time for thought

'All senior citizens (of pensionable age) should have their homes insulated by the local council free of charge and have their winter heating bills paid for from government funds.' Do you agree?

ENERGY FOR LIFE

A dog, like a washing machine, requires an input of energy. Both convert their energy input into a number of energy outputs. The washing machine converts electrical energy into the movement of the steel drum. The dog also produces movement as an energy output. It converts the chemical energy in its food into the movement of its muscles. Both the dog and the washing machine produce sound energy and heat. You will notice however that the dog is able to use its energy input in ways that the washing machine cannot. The dog can build new cells and so grow and reproduce. It is this use of energy that distinguishes living organisms from machines (see page 10).

figure 9.9 Energy is used by dogs and washing machines

figure 9.10 Organisms use energy in many different ways

UNIT 9

RELEASING ENERGY FROM FOOD

How do living organisms release the chemical energy contained in food? They break down the large complex food molecules into simple substances. The energy previously used to hold the food molecule together is released. This energy is not released all at once – if it were it would produce a minor explosion! Food is broken down in a series of steps releasing a small amount of energy at a time. The breakdown of food and the release of energy by living cells is known as **respiration**. There are two types of respiration.

1 Aerobic respiration
This needs oxygen in order to release energy from food.

2 Anaerobic respiration
This can release the energy from food *without* oxygen.

Respiration using oxygen – aerobic respiration

Aerobic respiration is very similar to another chemical reaction – **combustion**. During combustion a fuel such as coal or petrol is oxidized. This breaks down the fuel into simpler substances, some of which are poisonous.

The fuel used in respiration is the sugar **glucose**. Glucose is broken down in cells to form water and carbon dioxide. Water can be used within the cell. However, carbon dioxide is poisonous and must be removed quickly. The following equation summarizes the changes that occur when glucose is used to produce energy in aerobic respiration.

glucose + oxygen → water + carbon dioxide + energy

Respiration without oxygen – anaerobic respiration

In a 100 m race lasting about 10 seconds, the world class athletes in figure 9.11 require 30 kJ of energy. Yet during this time their muscles will use up very little oxygen. Some may not even take a single breath during the race. When their muscles suddenly need more energy, their lungs and circulatory system cannot respond fast enough to provide the necessary oxygen. Fortunately, for short periods the muscles can obtain energy without using up oxygen.

Glucose is used up during the race but it is only partly broken down, without oxygen, into **lactic acid**. Lactic acid is a poison and so after the race it must quickly be broken down further into carbon dioxide and water. Oxygen is needed to do this. This explains the rapid breathing of sprinters after the race is over. The build-up of lactic acid is described as an **oxygen debt**.

figure 9.11

The oxygen debt can build up during a race but it must be paid off at a later stage.

If you exercise too vigorously you will experience the effect of too much lactic acid. After a while there is enough to make the muscles ache. Eventually it causes cramp and the muscles will no longer contract. You have to stop until the blood can bring fresh supplies of oxygen to break down the lactic acid.

So this is what happens when your oxygen debt catches up with you.

figure 9.12

INVESTIGATION 9.4

RESPIRATION AND DISTANCE RUNNING
Table 9.3 shows the proportion of energy obtained from anaerobic respiration in an athlete's muscles during races of varying distances.

race distance (m)	100	800	1500	10 000	marathon
percentage energy from anaerobic respiration	95-100	65	45	10	2

table 9.3

Questions

1 What percentage of energy comes from anaerobic respiration
   a in the muscles of an 800 m runner
   b in the muscles of a marathon runner?
2 Describe how the level of anaerobic respiration changes as the race distance gets longer.
3 How would you explain the pattern you can see in these data?

HOW MUCH ENERGY DO YOU NEED?

The amount of energy you need depends on a number of factors.

1 Energy for growth

Building new body cells needs energy. The faster you are growing the more energy you need. An active growing teenager needs to take in more energy in food than an adult who has stopped growing.

2 Energy for reproduction

A pregnant woman needs additional energy in her diet. Some of this energy will be used up by the rapidly growing foetus. A mother who is breast feeding also needs a high energy intake in order to produce breast milk.

3 Energy for activity

All the body's activities use up energy. Even while asleep we use energy. We move slightly, our muscles are slightly contracted (muscle tone) to keep our position, our brain is still active and we might need some energy to produce snoring sounds! During the day your need for energy will depend on how active you are. Table 9.4 illustrates the amount of energy used up by different activities.

activity	energy used (kJ)
1 hour sleeping	300
1 hour sitting down	350
1 hour standing up	430
1 hour walking slowly	750
1 hour walking quickly	1260
1 hour playing tennis	1600
1 hour playing football	2200

table 9.4 *Approximate amounts of energy used up in different activities by an average adult male*

4 Energy for your job

If a person's job or occupation involves very little activity, less energy will be needed in their diet. More active jobs need a higher energy intake. A bricklayer will need more energy than a VDU operator.

INVESTIGATION 9.5

DIFFERENT PEOPLE NEED DIFFERENT ENERGY INTAKES

Table 9.5 shows the different amounts of energy in kJ required by different individuals.

	male	female
eight-year-old	8 800	8 800
15-year-old	12 600	9 600
adult – light work	11 500	9 500
adult – manual work	12 100	10 500
pregnant woman		10 000
breast-feeding mother		11 300

table 9.5

Questions

1 How much energy should each of the following take in each day in their diet?
  a 15-year-old schoolgirl
  b female bricklayer
  c male nurse
  d female typist
2 Explain each of the following.
  a A 15-year-old schoolgirl needs more energy than her mother who works as a secretary.
  b A woman needs more energy during pregnancy than when doing light work.
  c A manual worker needs more energy than a sedentary worker.
3 What are the main differences in the energy needs of the average man and woman? Suggest reasons.
4 Think about your own leisure activities. How do they affect your energy requirements?

KEEPING THE ENERGY BALANCE

It is important that we balance the amount of energy in our diets with the amount of energy we use up. What happens if we do not balance these energy values? There are two possible outcomes.

1 Not enough energy in the diet

In this situation we can rely on the fat layer that is stored beneath the skin surface. Fat can be converted to glucose and broken down to provide energy. A person on a slimming diet eats food containing less energy than they need so that their fat layer is reduced.

If this shortfall in energy intake continues for long periods the fat stores will be completely used up. The only energy source left is the person's own body tissue. Protein from muscles in the arms and legs is broken down to supply the body with energy to stay alive. This results in the frail and emaciated bodies typical of famine victims (see investigation 9.6, page 145).

2 Too much energy in the diet

Fat and carbohydrate provide us with our main source of energy. In Britain the average diet contains a high proportion of these foods. A certain amount of carbohydrate can be stored as glycogen in the liver and muscles. However, most excess carbohydrate is converted to fat and builds up beneath the skin. Fat contains twice as much energy as carbohydrate and any surplus to our needs is readily stored. If the food we eat contains more energy as carbohydrates and fats than we need we will eventually put on weight. If this continues we may suffer from being grossly overweight. This is known as obesity (see application 9.2).

Application 9.1

ANOREXIA NERVOSA – DELIBERATE STARVATION

Anorexia nervosa, the 'slimmers' disease', occurs when someone thinks that he or she is very overweight. They deliberately starve themselves and will refuse to accept food even when they become dangerously thin. Even at this stage they may still think they are fat. The causes of anorexia are very complex but it seems likely that the slim body images presented in the media may contribute to many cases.

figure 9.13 The media portray thin figures as the ideal

Application 9.2

GETTING THE ENERGY BALANCE WRONG – OBESITY

Obesity is defined as when a person's weight is 20% above their ideal weight according to their height (see figure 9.14). The following would increase a person's chances of obesity: a job, such as office work, that involves a lot of sitting down; leisure time spent in ways involving little physical effort, such as watching TV, and a diet including large amounts of fatty and sugary foods.

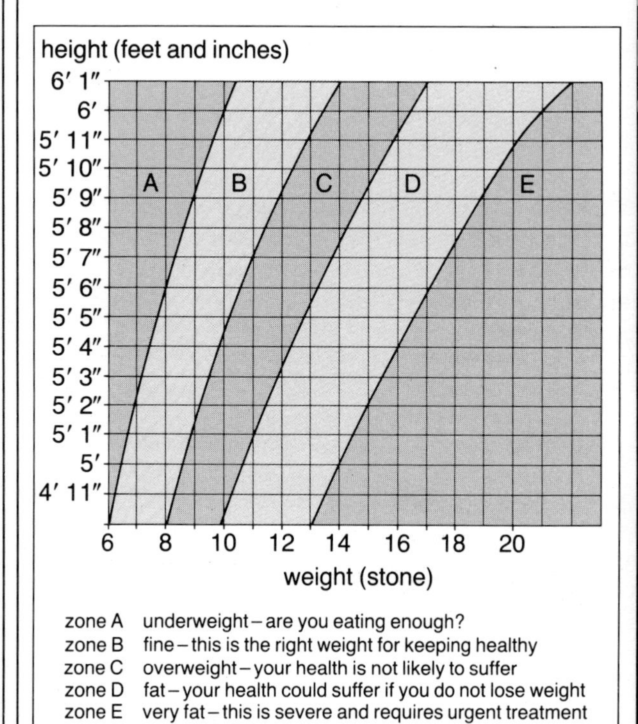

figure 9.14 Use this graph to check your weight. The range of ideal weights takes account of different builds

Obesity often starts when children develop the wrong eating habits. Unfortunately fattening foods such as sweets and chocolate are traditionally given as treats. Perhaps this is why these foods are popular ways of giving ourselves reassurance when we are worried. Anxiety, boredom and loneliness can all lead to overeating.

There are some people who can eat large quantities of food without putting on weight. This is because they have inherited a particularly fast metabolism that 'burns up' food quickly. If you have a slow metabolism, you will find that you have to control your energy intake carefully.

Doctors treat obesity as an illness because it can have many serious effects. An obese person is more likely to suffer from diabetes, heart disease, high blood pressure and joint problems. The treatment of obesity usually consists of a specially restricted diet and a programme of exercise. It is also important to help the person understand the reason for their overeating. Overeating can be an addiction. Sometimes a complete change of attitude is needed to achieve a permanent loss of weight.

INVESTIGATION 9.6

THE WORLD'S ENERGY GAP

Table 9.6 compares the average energy intake (kJ) in people's diet in various countries in 1961 and 1971.

	country	average daily intake (kJ) 1961	1971
developed countries	United Kingdom	13 400	13 200
	United States	13 100	13 700
	Spain	11 050	11 000
developing countries	Nigeria	10 250	9 600
	Mexico	10 500	10 750
	India	8 800	8 600
	China	8 400	9 950
	Pakistan	8 150	9 600

table 9.6

Questions

1 In 1971 which country had
   a the highest energy intake
   b the lowest energy intake?

2 What important difference between these two countries might explain the gap between them?

3 The minimum recommended daily intake of energy is 9200 kJ. A person whose energy intake is below this level can be said to be starving. Which countries had an average energy intake below this level in
   a 1961
   b 1971?

4 Calculate the change in energy intake (+ or −) between 1961 and 1971 in each country (e.g. United Kingdom −200 kJ).

5 In which countries has there been a significant increase in the energy intake between 1961 and 1971?

6 Which developing countries have *not* shown a significant increase in energy intake over the decade? What conditions in these countries might explain this trend?

7 In 1971 the average daily energy intake of the people of Pakistan improved to a level above the recommended minimum of 9200 kJ, yet some of the population were still starving. How would you explain this?

INVESTIGATION 9.7

HOW MUCH ENERGY IS THERE IN A PEANUT?

a Set up the apparatus shown in figure 9.15.

b Take one peanut, weigh it and record its mass in grams.

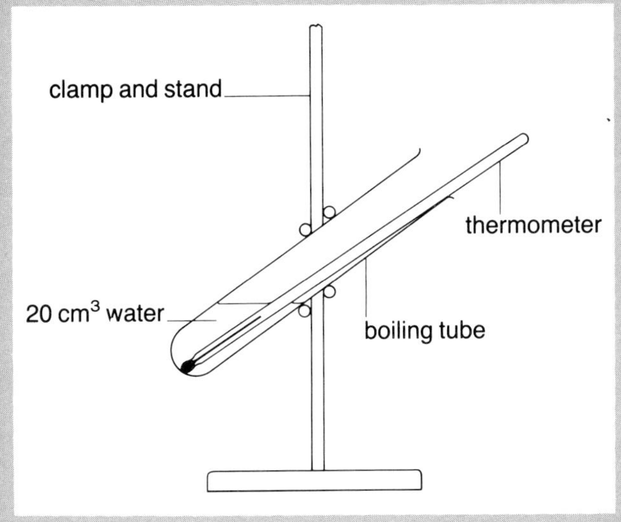

figure 9.15

c Take a reading of the water temperature and record it.

d Carefully fix the peanut on the end of a mounted needle.

e Set light to the peanut in a Bunsen flame.

f As soon as it ignites, hold the peanut under the boiling tube containing water.

g Keep the peanut under the boiling tube until it has completely burnt.

h Continue to observe the water temperature. Record the highest water temperature reached.

i Use the following formula to calculate the amount of energy contained in your peanut.

A = temperature of water at start of investigation (°C)
B = highest temperature of water after investigation (°C)

$$\text{energy} = \frac{20 \times 4.2 \times (B - A)\,\text{kJ}}{1000}$$

j Divide this result by the mass of your peanut. This gives the energy produced by one gram of peanut.

k Compare your result with others in your class. Can you explain why they differ?

Questions

1 When very accurate equipment is used the amount of energy contained in one gram of peanut is measured as 25.2 kJ. Why are your results different from this?

2 You will only get an accurate result if *all* the heat energy from the burning peanut is used to raise the water temperature. If you repeated this investigation, what would you do to improve its accuracy?

ALL YOUR OWN WORK

1 There are two types of energy source: ________ and ________. List the main varieties of each source.
2 Coal, oil and gas are forms of ________ fuel.
3 What are the arguments for and against alternative energy sources?
4 An electric water pump raises a bucket, containing 25 kg of water, from a well 15 metres deep in 10 seconds. How much PE is gained by the bucket? How much energy is supplied to the pump if it is only 75% efficient? What are the power input and output values of the pump? (Acceleration due to gravity = 10 m/s^2.)
5 Given a 1 kg mass and a stopwatch design an experiment to investigate human power output.
6 What questions would you include on a questionnaire which would allow you to survey the insulation measures taken by an 'average' household? What would be the most effective way of presenting the information collected?
7 In 1986 it was **monergy year**. This was a campaign set up by the Department of Energy (DEn) to encourage energy efficiency. If you were responsible for the next campaign what would you do?
8 What is your reaction to this statement 'Energy efficiency has nothing to do with me!'.
9 With no local electricity how might you supply pumped water in places such as Ethiopia?
10 Figure 9.16 shows the relationship between obesity and the risk of mortality.

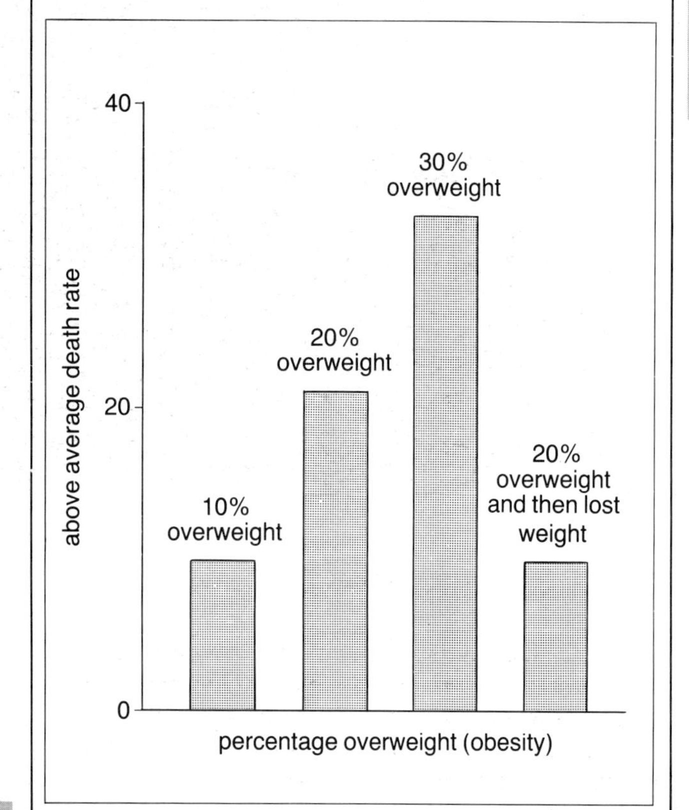

figure 9.16

a From the evidence provided by figure 9.16 give two relationships between being overweight and above average death-rate.
b Both exercise and diet can affect weight control. Discuss why it might be unwise for an obese person to rely on vigorous exercise as the main way to reduce weight.
c The recommended daily energy intake for an office worker is 11 300 kJ. An office worker is 20 kg overweight and has a daily food intake as follows.

BREAKFAST AND SNACKS

		energy in kJ
cereal	50 g	780
milk	100 g	270
sugar	10 g	160
eggs	(2)	660
bacon	50 g	900
white bread	2 slices	970
butter	25 g	800
marmalade	50 g	500
chocolate	150 g	3 300
	TOTAL	8 340

EVENING MEAL

		energy in kJ
soup	100 g	300
white bread	2 slices	990
fish fingers	100 g	600
chips	150 g	1 500
apple pie	100 g	1 200
custard	100 g	500
cheese	100 g	1 600
biscuits	50 g	250
butter	25 g	800
	TOTAL	7 740

Suggest remedial action for the office worker. In your answer consider the following points and refer to specific foods in the diet where appropriate:
i the effects of continuing this diet
ii how the diet may be improved to meet the recommended intake.

[SEG specimen question]

WAVES

UNIT 10

WAVES TRANSFER ENERGY

You can probably think of a number of different waves: water, radio, microwaves, sound, light and X-rays are just a few. Any wave belongs to one of two groups and although these two groups share some patterns of behaviour they do not share them all. All waves have energy and are able to **transfer energy from place to place**. Energy is being transferred in figure 10.1

figure 10.1 *Energy transfer at work*

In each of the three examples energy is transferred from place to place without material being transferred. The blooming flowers will only grow if their leaves receive energy in the form of sunlight. The energy transferred from the Sun allows photosynthesis to take place. The noise from the rock singer is transferred through the air via vibrating molecules and the energy of the water waves reaches the shore but the water itself only bobs up and down.

Oscillations

A second characteristic that all waves share is that of **oscillation** (see page 124). An oscillation is part of a regular repeating pattern. The swinging motion of a clock pendulum is a simple example of an oscillation. A wave is a series of repeating patterns or oscillations.

INVESTIGATION 10.1

OSCILLATIONS

Figure 10.2 shows a number of devices which can be used as oscillators. Try each one in turn.

Questions

1. Which ones could be used as the basis for a timing device?
2. Which ones are not regular oscillators?
3. How would you modify one of the experiments to produce a trace or picture of the oscillations it makes?

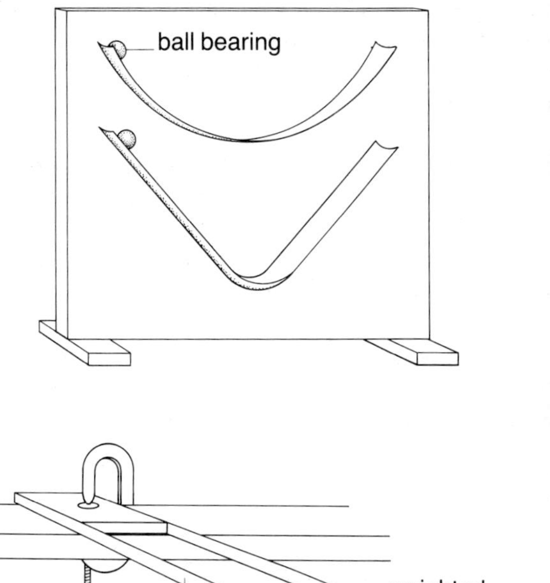

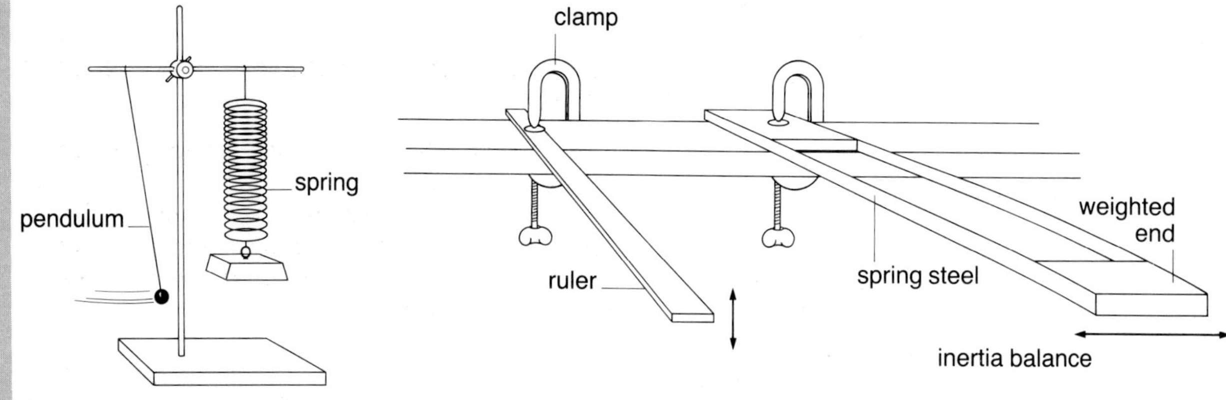

figure 10.2 Which of these could be used as the basis for a timing device or clock?

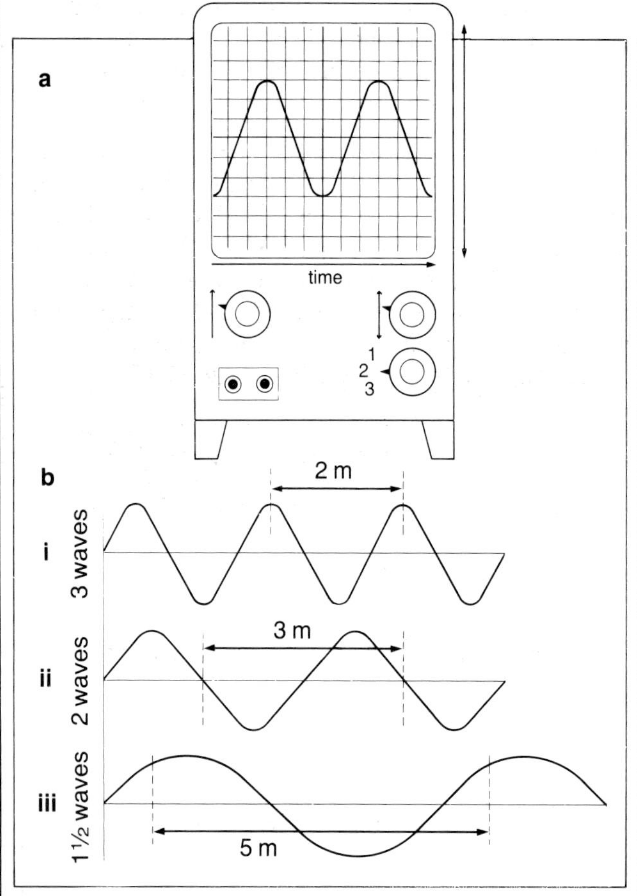

figure 10.3 a Waves on a CRO
b Different waves, different wavelengths

Oscillations on a CRO

A very convenient way to show oscillations or waves is to use a cathode ray oscilloscope (CRO, see pages 125 – 26). The trace on the oscilloscope in figure 10.3a is part of a repeating pattern. This instant picture shows two complete waves or oscillations. A CRO trace is a useful model to use when describing waves.

WAVELENGTH AND FREQUENCY

Each of the three wave traces in figure 10.3b shows a section of a wave. Each is different. The distance between repeating parts of the pattern in each trace is known as the wavelength, symbol λ (lambda) and is measured in metres. The examples shown have wavelengths of 2 m, 3 m and 5 m.

Something which happens often may be said to happen frequently. Frequency is a measure of how often something happens in one second. Frequency is measured in hertz (Hz, see page 124). If the top trace in figure 10.3b were to cross a CRO screen in one second it would contain three complete wavelengths and the wave would have a frequency of 3 Hz. If the middle trace were to cross the screen in one second the wave would have a frequency of 2 Hz. What would be the frequency of the third wave?

adio and TV stations broadcast their signals at very igh frequencies. (High frequency waves are usually ore penetrating than waves with a lower equency). However, their broadcasting avelengths are quite long.

radio station	broadcast frequency (kHz)	broadcast wavelength (m)
BBC Radio Clwyd	657	457
Viking Radio (Humberside)	1161	258
BBC Radio London	1458	206
Capital Radio (London)	1548	194
Radio Forth (Edinburgh)	1548	194
Radio City (Merseyside)	1548	194
1 kHz = 1000 Hz		

able 10.1 *How can three radio stations broadcast at he same frequency?*

THE SPEED OF WAVES

nformation about the wavelength and frequency of a vave will allow you to work out how fast it is ravelling. A wave with a frequency of 3 Hz and a vavelength of 4 m would travel (3 × 4) m = 12 m in ne second. The wave would have a speed of 12 m/s. his method can be used for any wave.

$$\underset{\text{(units m/s)}}{\text{speed}} = \underset{\text{(units Hz)}}{\text{frequency}} \times \underset{\text{(units m)}}{\text{wavelength}}$$

he speed of the waves in table 10.1 can also be alculated in the same way.

Capital Radio: frequency 1548 kHz, wavelength 194 m
speed = 1 548 000 × 194 m/s
= 300 312 000 m/s = 300 312 km/s
= 300 000 km/s (approximately)

Amplitude and loudness

The loudness of a musical note has nothing to do with the wavelength or frequency. The word loudness is used to describe how much energy is carried by a sound wave. It is the huge rising and falling motion of a water wave which causes damage (carries energy) when the waves crash on the shore. It is the same large rising and falling effect that causes loudness in a sound wave. **Amplitude** is used to describe the how much a wave rises and falls. A sound wave with a large amplitude will carry more energy and be louder than a sound wave with a small amplitude (see figure 10.4).

WAVES AND MATERIALS

Some waves involve materials (e.g. water), some do not (e.g. light). It is often easier to use a model to describe and explain what happens than to describe

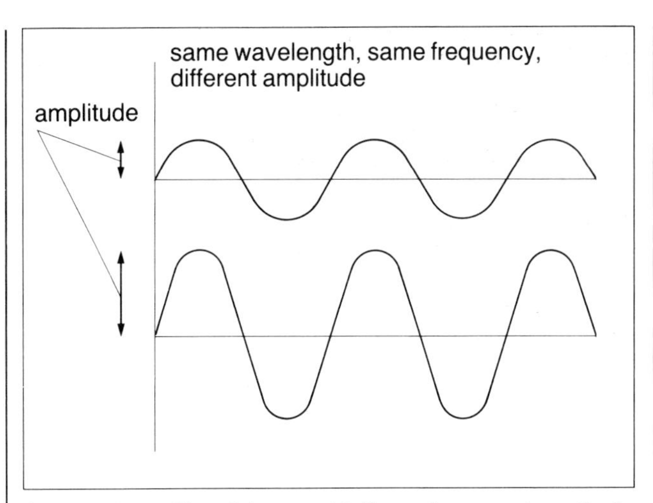

figure 10.4 *The rising and falling of a wave is called amplitude*

the real thing. Water waves make a very good model. It is possible to use them to describe a number of the different effects of waves and then to suggest that what happens in water happens in other kinds of waves as well. For other effects light waves make the most convenient model.

Reflections

You will know from looking in a mirror that light waves can be reflected. All waves can be reflected. However, not all surfaces will reflect all kinds of wave. All waves travel in straight lines unlesss something causes them to change direction.

For a shiny flat (regular) surface such as a mirror the **law of reflection** states that the **angle of incidence** (i) of a wave is equal to the **angle of reflection** (r). The incident ray is the wave that goes into the surface and the reflected ray is the one which bounces off. An imaginary line called the **normal** is just a construction line used to help with the measurement of the incident and reflected angles (see figure 10.5).

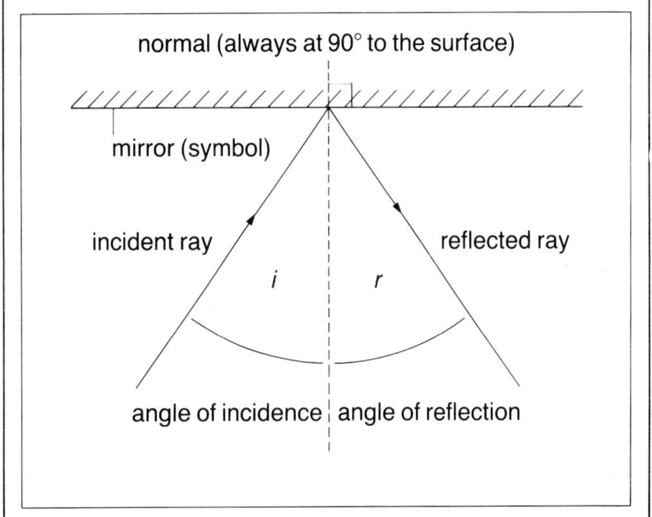

figure 10.5 *Angle of incidence = angle of reflection*

INVESTIGATION 10.2

THE LAW OF REFLECTION

The purpose of the investigation is to see whether or not the law of reflection works for all angles of incidence. Set up the apparatus as in figure 10.6.

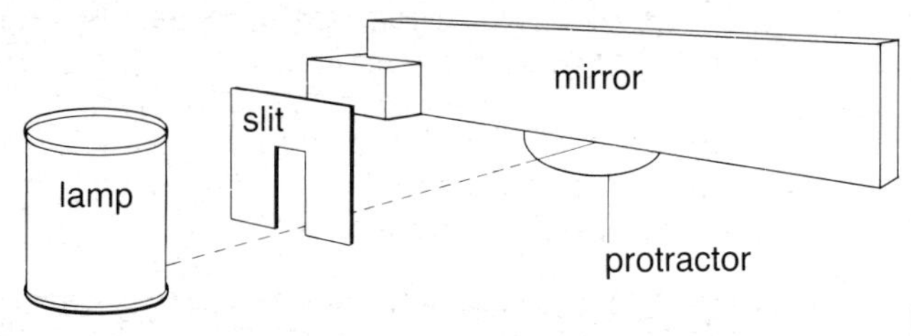

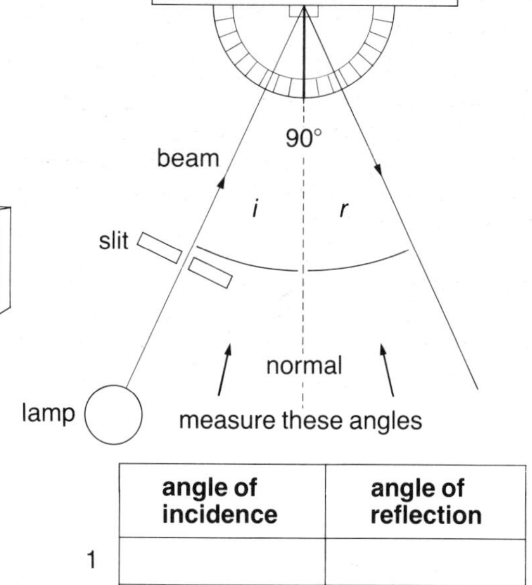

	angle of incidence	angle of reflection
1		
2		

figure 10.6 Investigating the law of reflection

a Arrange for a thin beam of light to strike the mirror at the centre of the protractor with a small angle of incidence (say 10°).
b Measure the angle of reflection and record both angle measurements.
c Repeat steps **a** and **b** a number of times with increasing values of *i*.

Questions

1 It is unlikely that the two sets of measurements will match exactly – why?
2 What effect does the thickness of the mirror have on the reflected ray?
3 Do you think you have enough evidence to support the law of reflection – can you say with confidence that it is true?
4 How many other situations can you think of where the law of reflection might be useful or important?

Application 10.1

LIGHT TRAVELS IN STRAIGHT LINES

The fact that light waves travel in straight lines unless they are made to change direction can be put to good use. The diagram in figure 10.7 shows a pair of telescopes being used to help in ship construction. If the propeller shafts and other engine parts were not properly lined up the support bearings and engine would soon wear.

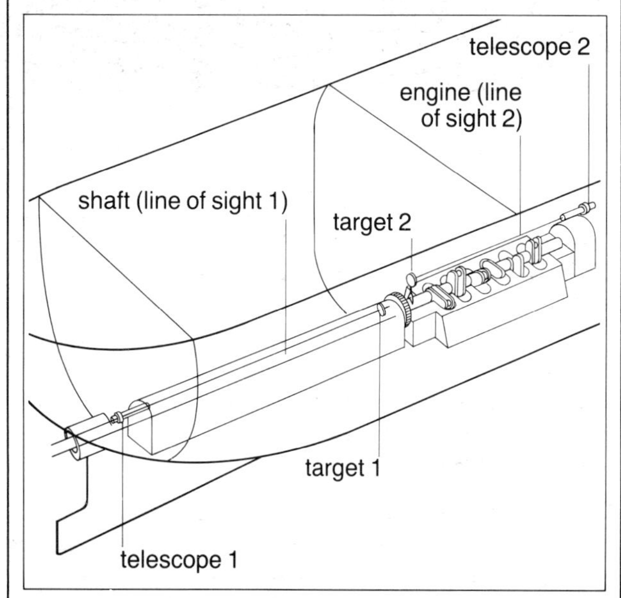

figure 10.7 Light travels in straight lines

Refraction

Refraction is the changing of a wave's direction as it enters or leaves a material. Refraction only happens **when a wave crosses a boundary**. For example, light waves change direction when they cross from water to air or from air to water. These changes are shown in figure 10.8. The imaginary line, the normal, has

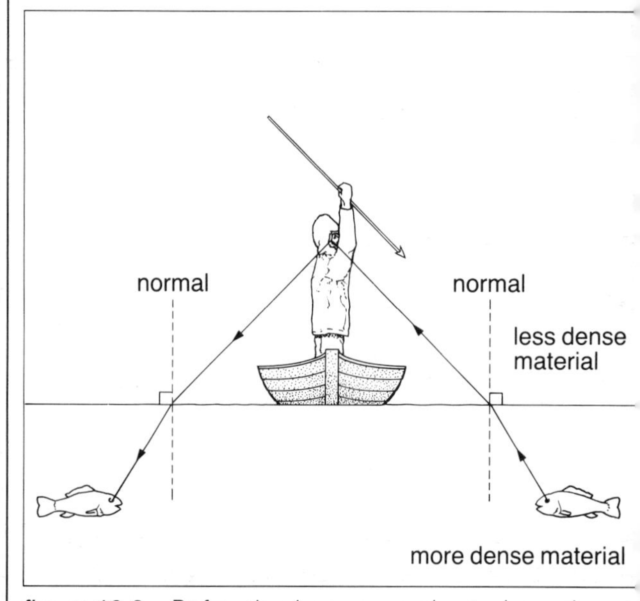

figure 10.8 Refraction happens only at a boundary

INVESTIGATION 10.3

REFRACTION

When light travels through a glass block it will be refracted twice: once as it enters the block and a second time when it leaves. It is quite easy to demonstrate this. The photograph in figure 10.9 shows how a thin beam of light might travel through a block of glass, viewed from above.

a Arrange the apparatus to show this effect.
b By changing the size of the incident angle observe and record carefully the effect on the refracted angle.

Questions

1 What is the link between the incident angle and the refracted angle as the size of the incident angle increases?
2 How would you record practically the direction of the light path through the block?
3 What practical problems will you have to overcome to measure a series of values for the incident angle and the refracted angle for the light ray?
4 Design an experiment to test this idea: 'The angle between the normal and the refracted ray *r* is always half the size of the angle between the normal and the incident ray *i*.' Is the idea correct?

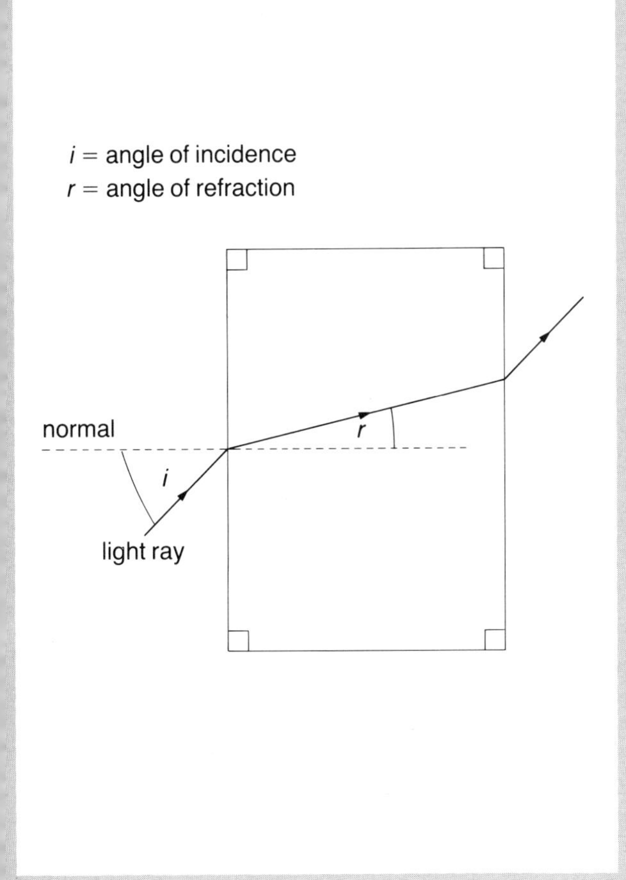

figure 10.9 Two boundaries, two refractions

been used again to help measure the changes in direction. Light going from a more dense material (e.g. water) to a less dense material (e.g. air) will always be refracted away from the normal.

The opposite is true too, as light waves follow the same path when travelling in the opposite direction.

Total internal reflection

If a ray of light is about to emerge from a glass block into air it must cross the glass/air boundary. When it does so it will be refracted (see figure 10.9), but only if the striking angle (the incident angle) is not too large (see figure 10.10a, overleaf). If the angle of incidence increases the angle of refraction increases (see figure 10.10b). If the striking angle increases too much it gets to the **critical angle** and refraction no longer happens. Instead the light ray is **reflected from the inside glass surface**, i.e. the surface now acts like a mirror (see figure 10.10c). The critical angle for glass is approximately 42°.

Experiments done to show total internal reflection and to measure the critical angle for a material usually use a semicircular block. There is no special reason for this, it is just more convenient (see figure 10.10d).

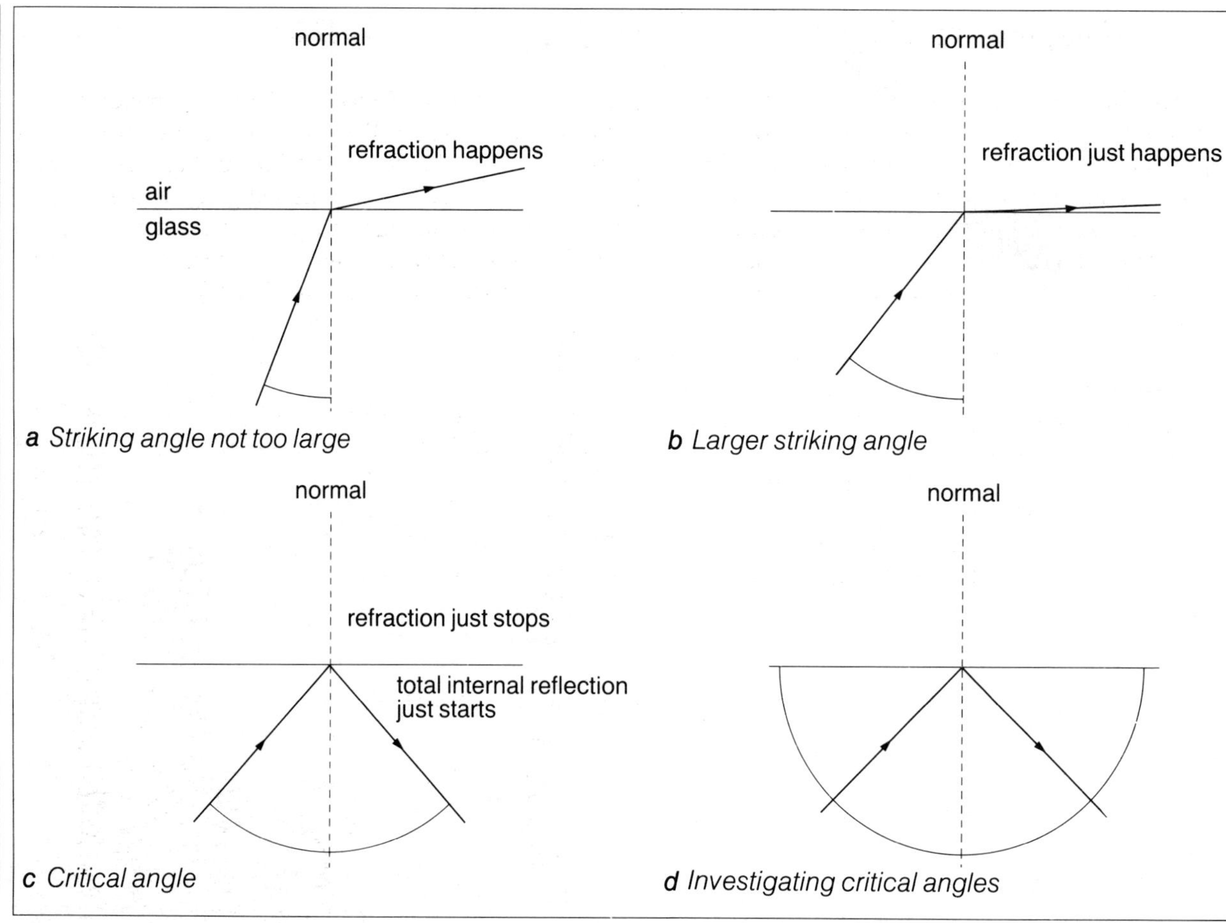

figure 10.10 *Refraction versus total internal reflection*

Application 10.2

TOTAL INTERNAL REFLECTION

Because the critical angle for a material such as glass is less than 45°, a number of useful devices can be made from it, see figure 10.11.

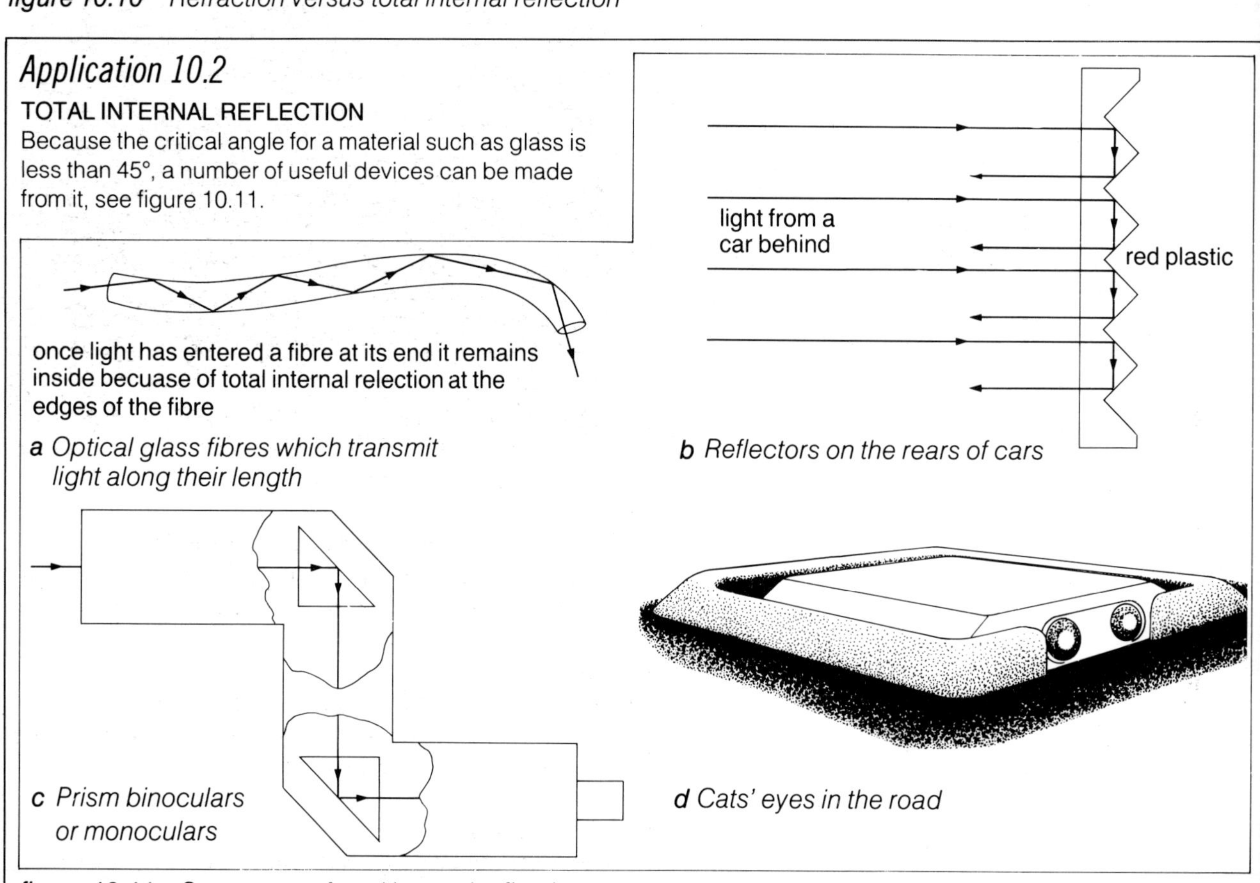

figure 10.11 *Some uses of total internal reflection*

Absorption and reflection

On a sunny day a dark coloured car will be hotter inside than the same sort of car in a lighter colour. Houses in hot countries are painted white to help keep them cool by reflecting the strong rays from the sun. Different materials and surfaces have different effects on striking waves. For example, ordinary glass is quite good at letting infra-red rays pass (be transmitted) through it. It is not so good at transmitting ultraviolet (uv) light. In fact glass is quite a good absorber of uv. It has been known for people to sunbathe privately in a greenhouse in the hope of getting an all-over sun tan. All they get is hot and bothered! There are some very simple guidelines for the absorption and reflection of light waves.

The best reflectors are bright and shiny.

The best absorbers are dark and dull and they are also the best emitters.

There is a number of simple experiments that can be done to test this idea. One experiment is shown in figure 10.12. Two flasks containing equal amounts of water are placed equal distances away from a heater. The temperature of each flask is recorded before the heater is switched on. At intervals of one minute the two temperature readings are repeated. When the two sets of temperature measurements are plotted on the same set of axes it is quite easy to compare the temperature changes of the flasks.

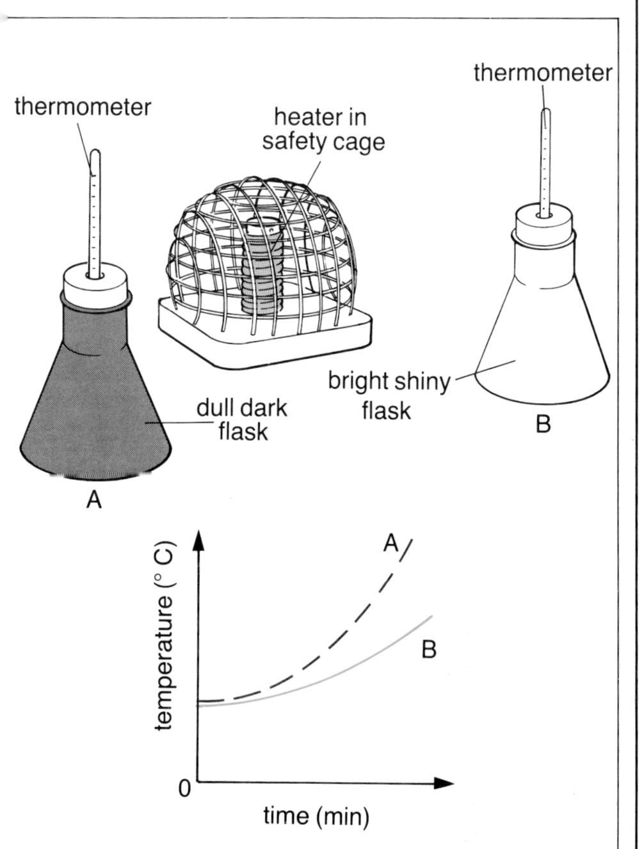

figure 10.12 Testing absorption and reflection

IMAGES

You will have experienced looking in a mirror or going to the cinema. What you see in the mirror or up on the screen is known as an **image**. Light is the most convenient wave to use to investigate images for the simple reason that light images can be seen. (You can produce images with other waves such as sound but they are much more difficult to find.)

figure 10.13 Distorted virtual images

Images can be produced by an **optical device** (a **mirror** or **lens**). They all belong to one of two groups: **real** or **virtual** images. They are easy to tell apart.

Real images

These can be projected onto a screen. They are always upside down when compared with the object e.g. the image on a cinema screen.

Virtual images

These cannot be projected onto a screen. They are always the right way up when compared with the object e.g. your image in a plane flat mirror.

No matter how hard you try you cannot project a virtual image onto a screen. Do not confuse images with shadows. The shadow created by an object placed in front of a light source is an area of darkness or semi-darkness behind that object. Shadows cannot be reflected or refracted to create an image.

Images from plane mirrors

The image formed by a plane mirror is virtual. It always appears to stay in the mirror, or rather behind it. Plane mirror images are **laterally inverted**, i.e. back to front. Look into a plane mirror and blink your left eye. The image will appear to blink its right eye back at you. Because a plane mirror image is virtual it is not possible to investigate the image directly. However, it is possible to find out where the image appears to be. This is described in investigation 10.4.

figure 10.14 *A fire engine has a laterally inverted sign so it will appear the right way round in a driver's mirror*

Images from curved mirrors and lenses (shoot the projectionist!)

How many times have you been to a film or slide show and the first images on the screen have been blurred and out of focus? Perhaps it even happens in

figure 10.15 *Out of focus?*

INVESTIGATION 10.4

LOCATING PLANE MIRROR IMAGES

As it is not possible to investigate a plane mirror image directly we have to find a way to get around the problem – in this case we try to get into the mirror! Figure 10.16 shows a suitable arrangement for locating a plane mirror image.

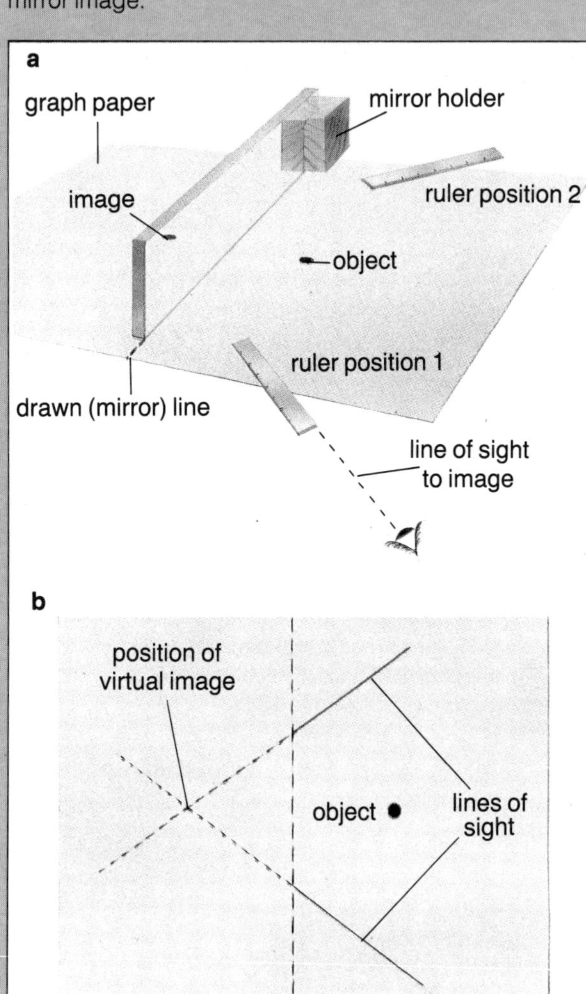

figure 10.16 *Locating a virtual image*

- a Draw a line across the graph paper and place the mirror on it.
- b Use a ruler in position 1 as a line of sight to look at the image.
- c Having found the image, and without moving the ruler, draw a line along the line of sight on to the paper.
- d Repeat steps **b** and **c** with a ruler in position 2.
- e Remove all apparatus from the paper.
- f By continuing the two lines of sight backwards they should cross the mirror line to meet behind it. Where they cross is the position of the virtual image.
- g There is a connection between the position of the image and the position of the object. They should each be the same distance from the mirror line. Check, for yourself, your practical skill by comparing the positions of the image and the object.

chool! Focusing is done with the help of curved nirrors and **lenses** (usually lenses). A lens is an ptical device, usually made of glass, which alters the path of light going through it.

Curved lenses and mirrors are either **concave** or **convex**. They get their names from their shape, not rom what they do! Concave lenses and mirrors cave nwards like the inside of a bowl. Convex lenses and nirrors bulge outwards like the outside surface of a alloon. Figure 10.17 shows the basic shapes and ymbols for curved lenses,and mirrors.

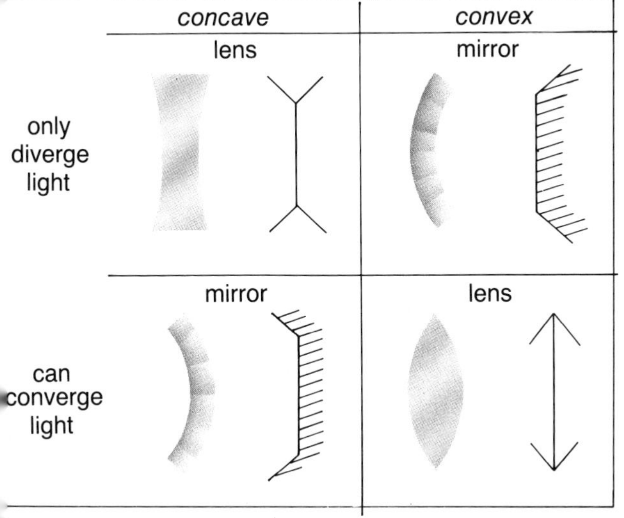

gure 10.17 Symbols and shapes for curved mirrors nd lenses

Converging mirrors and lenses

If a beam of light passes through a lens or is reflected from a mirror and **converges** (see figure 10.18) to form an image, the light is **focused at a point**, and the image is real, because it really exists. (It has been projected!) If the beam of light is originally a parallel beam, then it will be brought to a focus at the **focal point** of the lens or mirror. The distance between the focal point and the lens or mirror is known as the **focal length** (see figure 10.18a).

Converging mirrors and lenses form virtual images. The following are examples showing how converging mirrors and lenses are used.

A convex lens producing a real image: the convex lens of the eye produces an upside down real image on the retina. A camera lens produces an upside down image on the film.
A convex lens producing a virtual image: a magnifying glass creates an upright virtual image.
A concave mirror producing a real image: an astronomical reflecting telescope produces an upside down real image.
A concave mirror producing a virtual image: the images created by make-up and shaving mirrors are upright virtual images.

Concave mirrors are also used in torches, car headlamps, and as radar dishes.

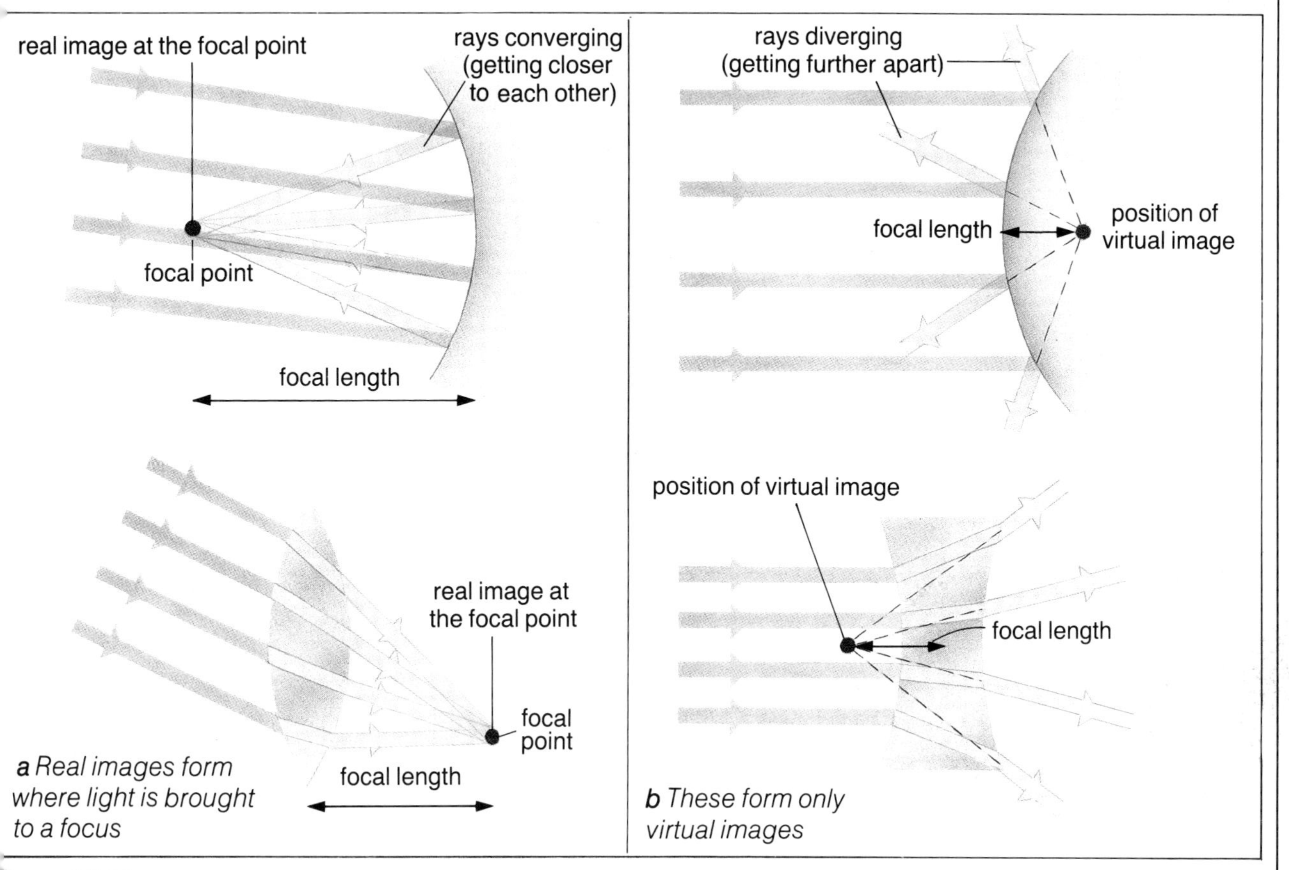

a Real images form where light is brought to a focus

b These form only virtual images

igure 10.18

Diverging mirrors and lenses

If a beam of light passes through a lens or is reflected from a mirror and it spreads out or **diverges** the light cannot meet at a point. Diverging mirrors and lenses are **unable to produce real images**. They produce only virtual images. Here are some examples showing when diverging mirrors and lenses are used.

A concave lens creating a virtual image: the small concave lens used as a security device (peep hole) on a door creates an upright virtual image.

A convex mirror creating a virtual image: the images produced by car driving mirrors and the large mirrors high up on the walls in shops and supermarkets are upright and virtual.

WHAT ARE WAVES?

So far we have seen that all waves carry energy from place to place and that they are all oscillations (repeating patterns). However, any particular wave belongs to one of two groups of waves, depending on whether its oscillations are transverse or longitudinal. These two kinds of oscillations share a number of things in common, but they also have some differences.

Transverse and **longitudinal** oscillations can easily be demonstrated with the use of a 'slinky'. Figure 10.19 shows both types of oscillation. In both cases the energy moves from one end of the 'slinky' to the other. The difference is in the actual movement of the 'slinky'. In demonstrating transverse waves the 'slinky' seems to move sideways.

The longitudinal wave is a set of compressions, because the 'slinky' has been compressed in some places.

Transverse waves

Waves in this group include the **electromagnetic spectrum** (EMS) and water waves. The EMS includes a number of waves you will have come across, e.g. radio waves, microwaves, light waves and X-rays. These waves differ in their wavelengths and penetrating powers. As all the waves in the EMS can pass through a vacuum they do not need a material to travel in. They are called by various names: waves, rays or radiations. See figures 10 .20 and 10.21.

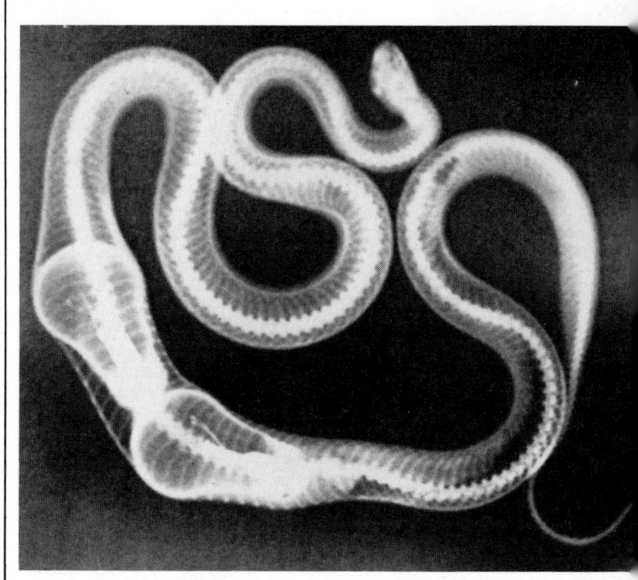

figure 10.20 *Some waves can travel through 'solid objects'*

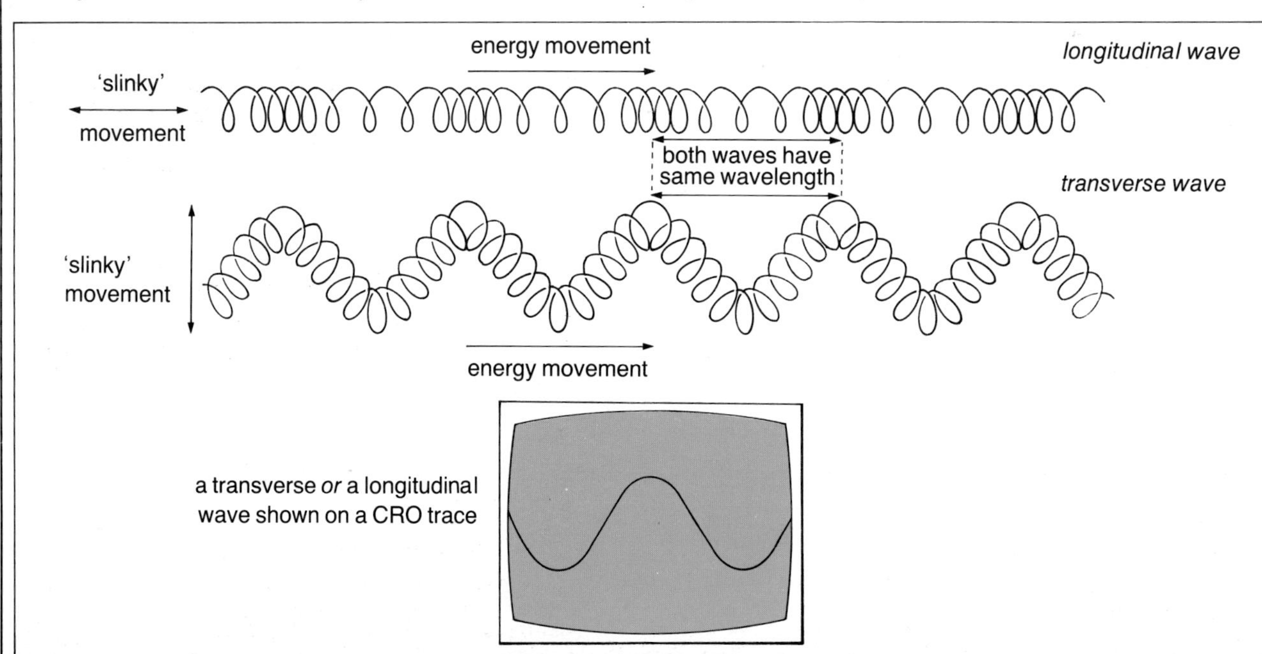

figure 10.19 *Longitudinal and transverse waves*

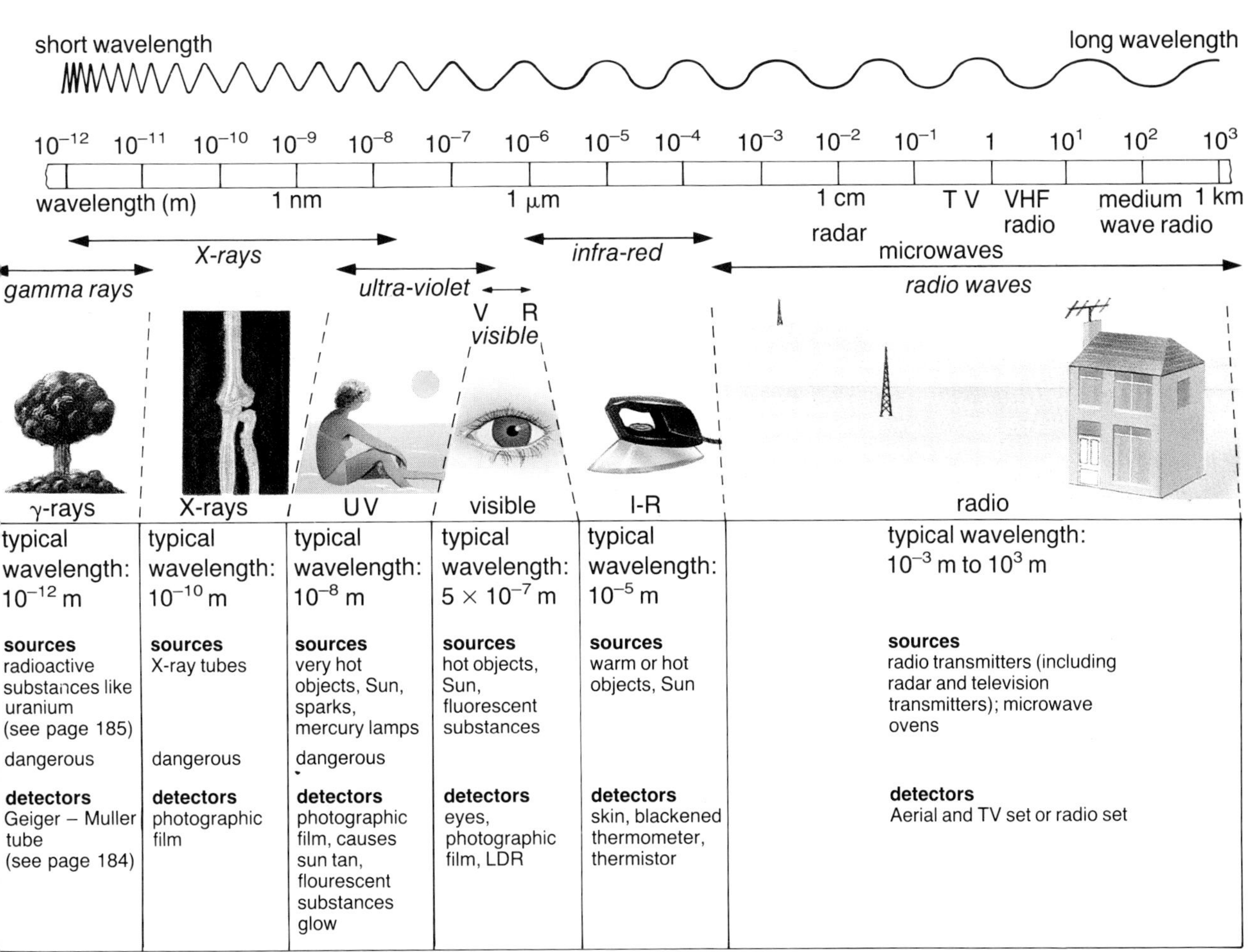

γ-rays	X-rays	UV	visible	I-R	radio
typical wavelength: 10^{-12} m	typical wavelength: 10^{-10} m	typical wavelength: 10^{-8} m	typical wavelength: 5×10^{-7} m	typical wavelength: 10^{-5} m	typical wavelength: 10^{-3} m to 10^{3} m
sources radioactive substances like uranium (see page 185)	**sources** X-ray tubes	**sources** very hot objects, Sun, sparks, mercury lamps	**sources** hot objects, Sun, fluorescent substances	**sources** warm or hot objects, Sun	**sources** radio transmitters (including radar and television transmitters); microwave ovens
dangerous	dangerous	dangerous			
detectors Geiger – Muller tube (see page 184)	**detectors** photographic film	**detectors** photographic film, causes sun tan, flourescent substances glow	**detectors** eyes, photographic film, LDR	**detectors** skin, blackened thermometer, thermistor	**detectors** Aerial and TV set or radio set

All travel at the same **speed**, in a vacuum, of 3×10^{8} m/s (300 million metres per second).

figure 10.21 *The electromagnetic spectrum*

INVESTIGATION 10.5

DISPERSION OF WHITE LIGHT

Rainbows are a natural example of **dispersion** – white light is broken up into the seven colours of the spectrum. The apparatus (viewed from above) shown in figure 10.22 shows a suitable practical arrangement for dispersing white light. The convex (converging) lens is used to help focus the spectrum on the screen.

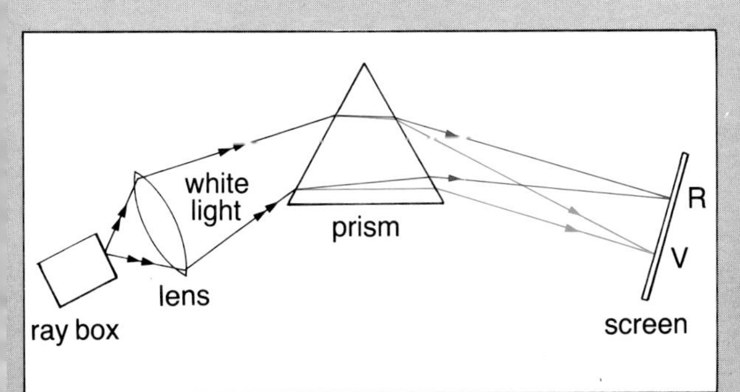

figure 10.22

Part 1

a Turn the prism to obtain the best spectrum – move the screen to catch it.

b Use the lens to focus the final spectrum.

c Check that the direction or path of the red end of the spectrum has been changed the least and the blue end has been changed the most. The order of the spectrum should be: red, orange, yellow, green, blue, indigo and violet.

Question

1 By using a second prism how would you show that
 i the spectrum could be dispersed or spread out more
 ii the spectrum of colours can produce white light?

Part 2

a Using a set of six 'standard' filters (red, green, blue, yellow, cyan and magenta) place them, in turn, between the prism and the screen.

b Record the colour(s) of the spectrum which each filter allows through.

Questions

Primary colour filters allow only one distinct colour group to pass, e.g. reds. **Secondary** colour filters allow two distinct colour groups to pass, e.g. reds *and* blues.

1 List the six filters as primary or secondary colours.
2 What pairs of colours do the secondary filters allow to pass? What colours are they made up from?
3 What colour groups do the filters stop?

Primary and secondary colours

The **three primary colours** are red,green and blue. (They are not the same as the primary colours used in painting.) From these three any other light colour can be made. Adding two primary colours together will produce a **secondary colour**. Adding all three together produces white light. Mixing colours in this way is known as **colour addition** (see figure 10.23).

Filters and pigments

Filters do exactly what their name suggests. They stop certain colours getting through – in rather the same way as a tea strainer works (see investigation 10.5, previous page). Dyes or pigments add colour to cloth.

A green filter stops all light except green. A cyan filter stops everything except green and blue. Yellow light is a mixture of green and red light so a yellow filter will stop all colours except green and red.

Pigments such as dyes and paints absorb the light and then give it out again rather than just letting it pass straight through. A green pigment or dye absorbs all light but only reflects green light, so it looks green. Red paper absorbs all light but only gives out red. By using filters (or dyes, paints or pigments) you are removing colours from the original light source. This is known as colour subtraction. If you place a blue filter and a yellow filter in front of each other they share no colours in common and so all the light will be filtered out and nothing will get through. The result is black (darkness). 'Blue' (strictly speaking cyan) paint reflects blue light and a little green. Yellow paint reflects green and red. So blue and yellow paint share green in common and when these paints are mixed the common reflected colour of green is seen. The blue light is absorbed by the yellow paint and the red light by the blue paint.

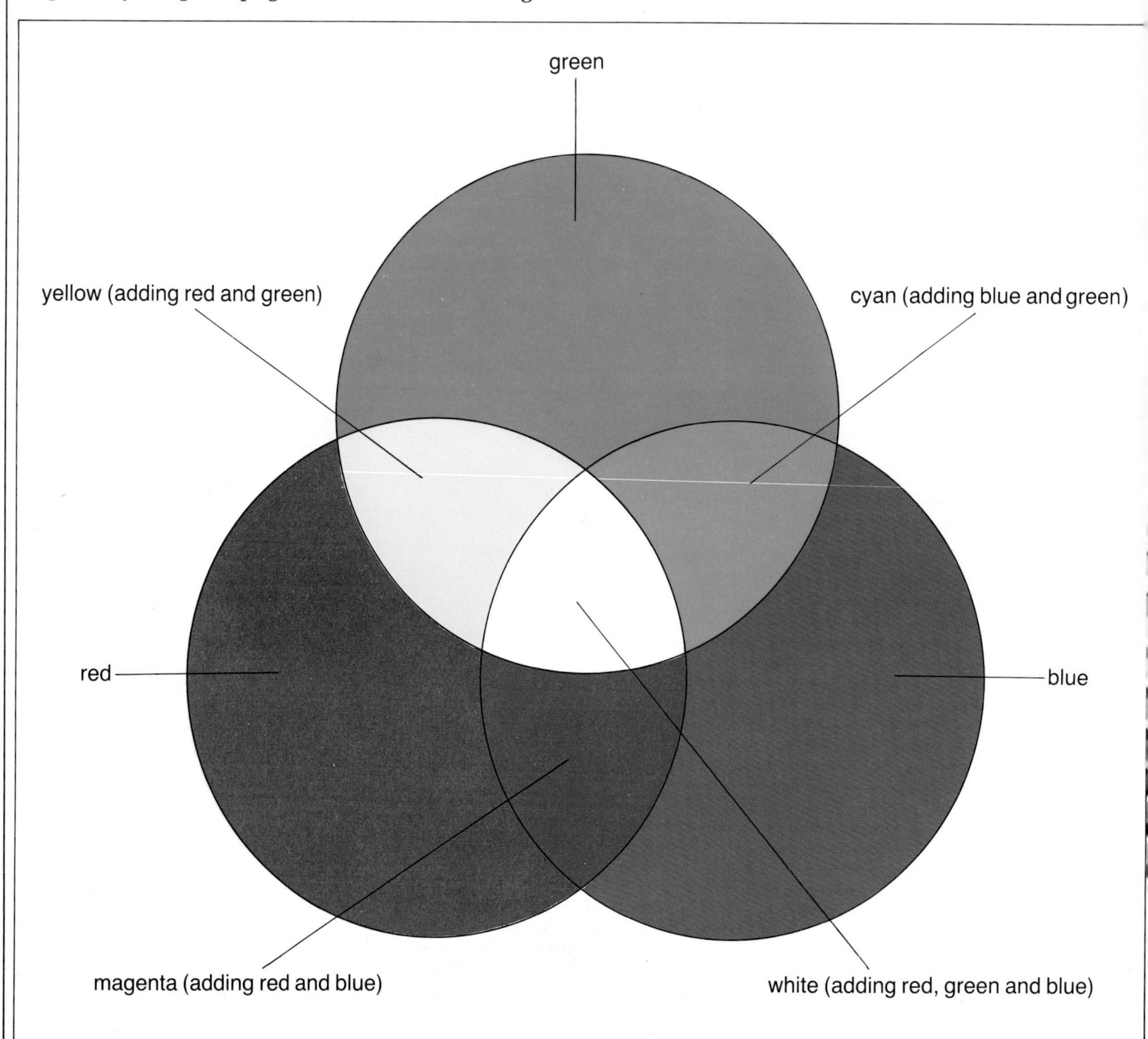

figure 10.23 *Adding primary and secondary coloured lights*

Photographic colour slides produce images by the method of colour **subtraction**. A slide is made from three coloured layers of gelatine. In the development process these coloured layers are removed one at a time to leave behind the correct colour filter or combination of filters. This can easily be shown by carefully scratching the layers of gelatine away and viewing the scrapings under a microscope.

Coloured pictures are printed on paper using pigments. Whatever pigment is used absorbs a particular colour or set of colours. When the picture is viewed in white light various colours are subtracted (or filtered out) by the pigments. What is left is reflected (see figure 10.24) and this us what you see. Different pigments can be mixed together to create new pigments which filter out other colours.

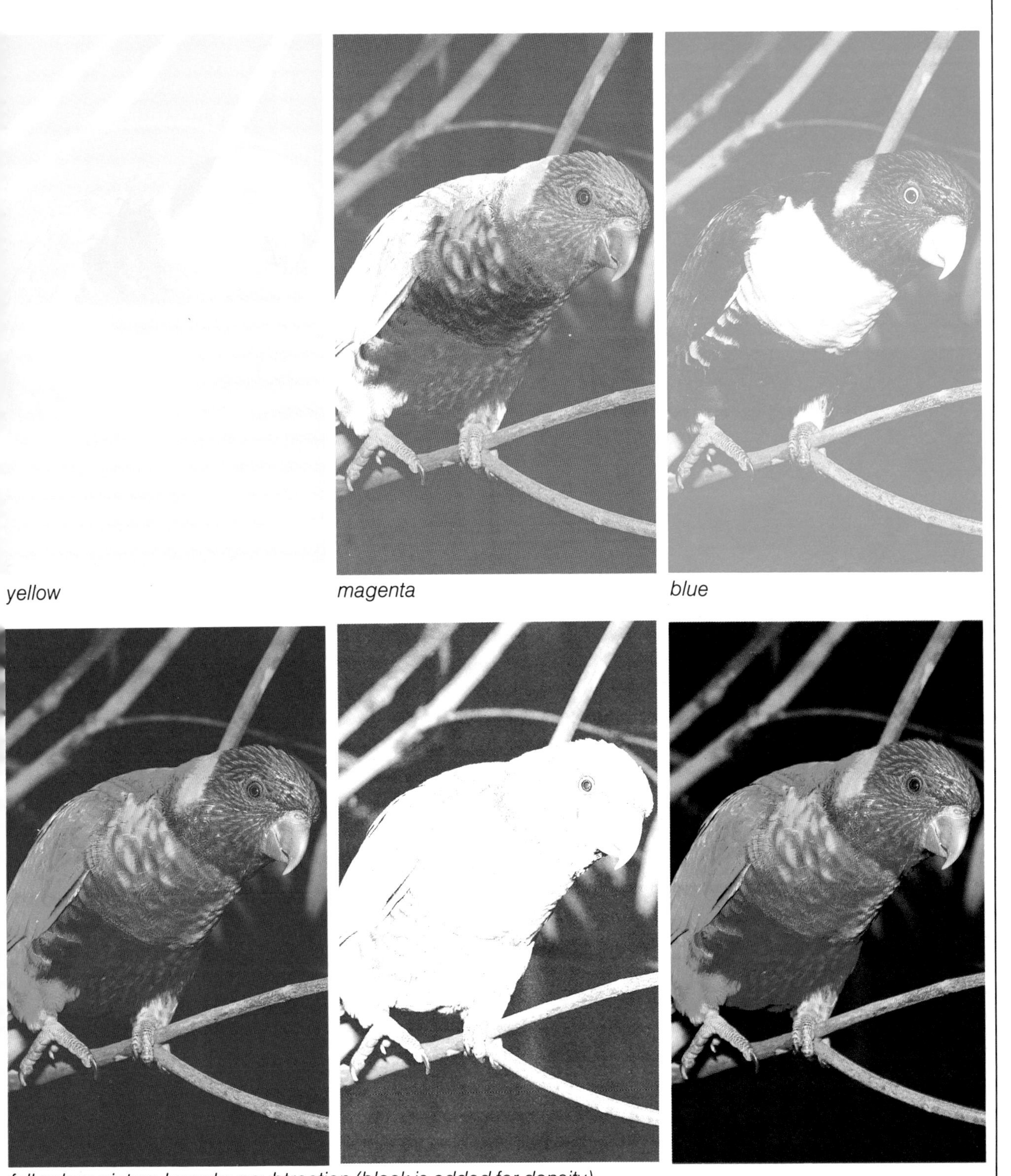

figure 10.24 Colour subtraction

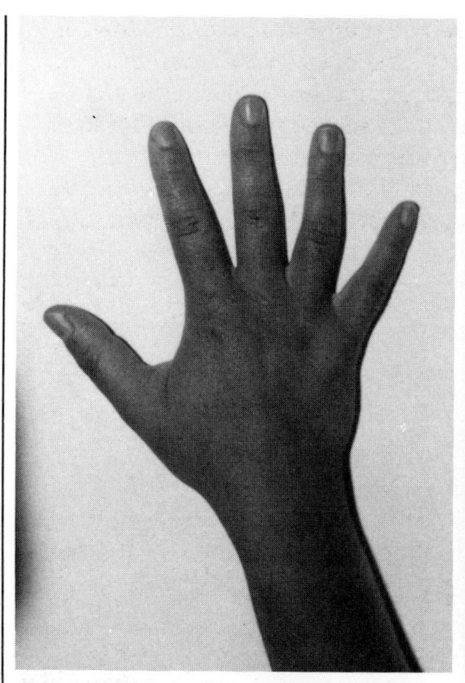

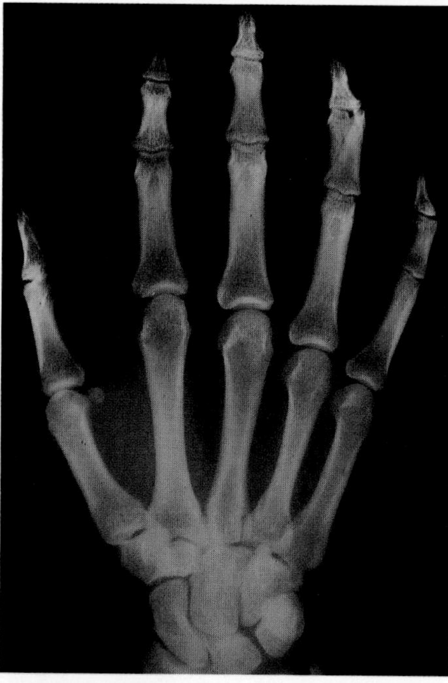

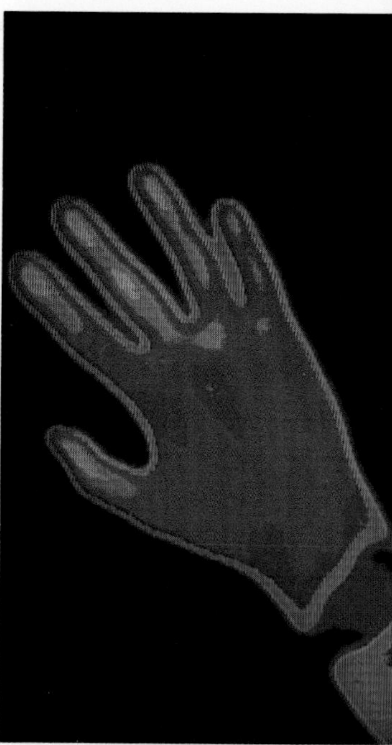

figure 10.25 **a** *Photography is very versatile – a hand shown by light, X-ray and infra-red photography*

figure 10.25 **b** *These two photographs were both taken at night. The one on the right used infra-red film*

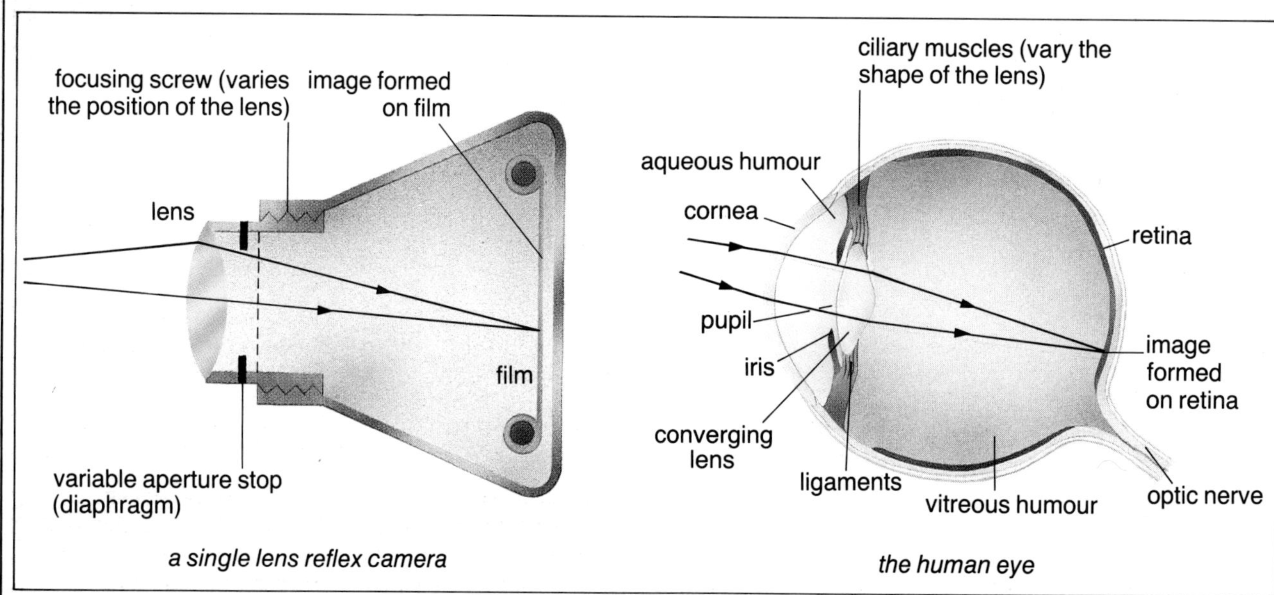

figure 10.25 **c** *Spot the difference*

DETECTION OF EMS WAVES

Different detection methods are needed for different parts of the EMS. For example, eye sight is suitable for detecting waves which belong to the visible light section of the EMS but it is no use at all for the detection of waves such as X-rays or infra-red rays. Photography is a very versatile detection method. Films can be manufactured to detect most varieties of wave in the EMS. The three photographs in figure 10.25a were taken using light, X-rays and infra-red (heat) radiation. Figure 10.25b shows how infra-red can be used at night when photography by visible light would be impossible.

LONGITUDINAL WAVES

Sound waves are longitudinal waves. A sound wave travelling in a material is a series of compressions moving through the material (just like the compressions in the 'slinky' in figure 10.19, page 156). The compressions are transferred through the material by the molecules. These compressions are more usually called vibrations. Sound waves can only be transferred from a source to a receiver if there is some sort of material in between. Vibrations cannot be passed on in a vacuum. Space is silent. Air is usually the material used to transfer sound from a vibrating source to a human ear.

Detection of sound

The human ear is a very sensitive detection device. Sound waves are guided by the shape of the ear towards the **ear drum** and cause it to **vibrate**. A series of small bones on the other side of the ear drum transfer the vibrations to fluid contained in the inner ear where they are converted to electrical signals. Nerves carry these signals to the brain. If the ear drum is perforated or split in any way a loss of hearing and perhaps even complete deafness will result, because the vibrations will not be transferred properly.

Human hearing changes with age. Young people can generally hear higher frequency vibrations than older people. A typical frequency range for a young person would be 20–20 000 Hz. A number of animals can detect vibrations with frequencies above this range – **ultrasonic** vibrations.

Application 10.3

SOUND RECORDING

There are four main ways to record sound: by record, tape, laser disc and optical film soundtrack. In each method sound has been converted into another form suitable for storage. When this is reconverted, amplified and the signal fed to loudspeakers the original sound is recreated.

In a record; sound is stored in the form of vibrations and 'bumps' in the record grooves.

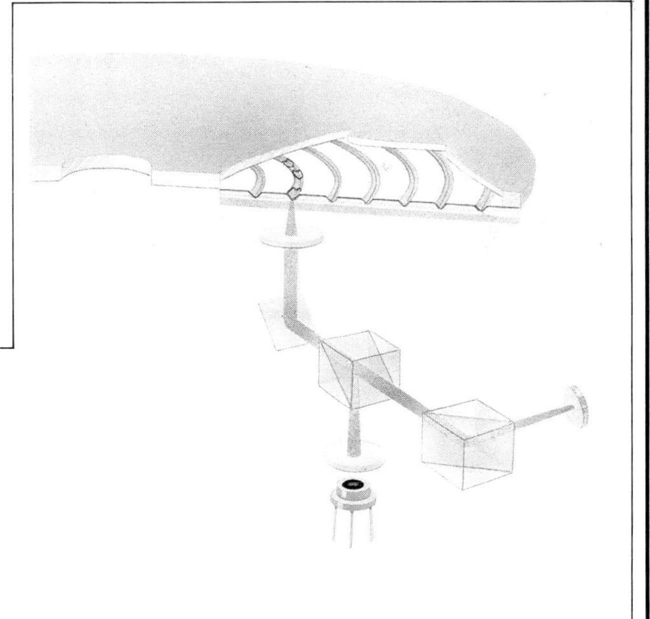

structure of a laser disc

optical soundtrack

In a tape; varying strengths of magnetism are stored on a very long piece of magnetic material.

In a laser disc; information is stored on the underside of a disc in the form of reflective pits which are read by a laser.

In an optical soundtrack; sound is stored as a varying transparent track along one edge of the film. The shape and size of the track determines how much light will pass through it and what type of sound is produced from it.

figure 10.26 *Sound recording*

Some animals use ultrasonics as their own inbuilt radar sets (sonar). Certain bats emit ultrasonic vibrations and by listening for the echo can locate lunch (see figure 10.27a). Ultrasound is used to examine the foetus in a mother's uterus (see figure 10.27b) – avoiding the potential dangers of using X-rays (see page 187).

Trawlers use sonar tracking devices to locate shoals of fish (see figure 10.27c).

figure 10.27 **a** *Gone to lunch*

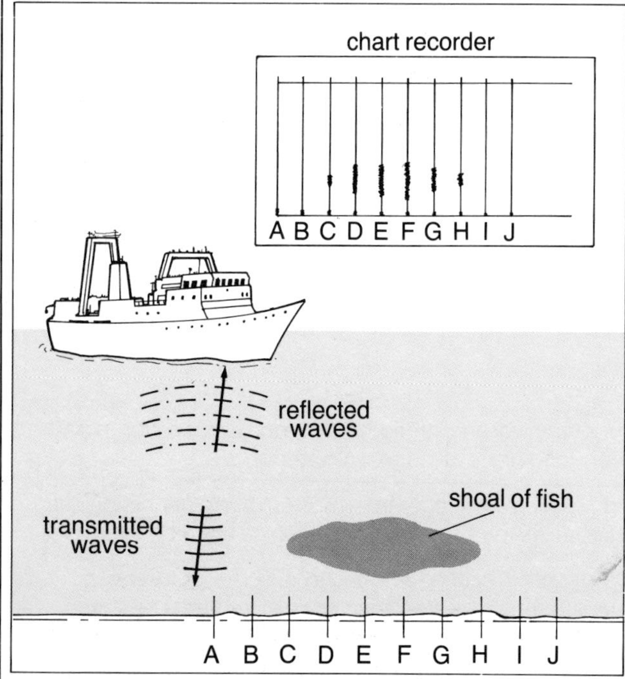

figure 10.27 **c** *Gone fishing*

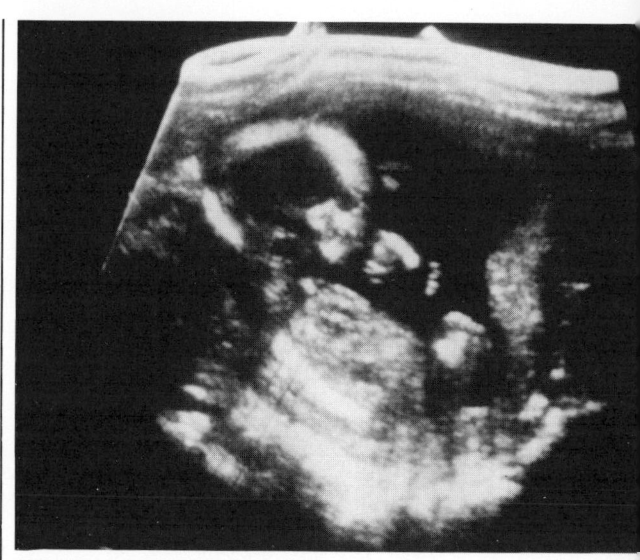

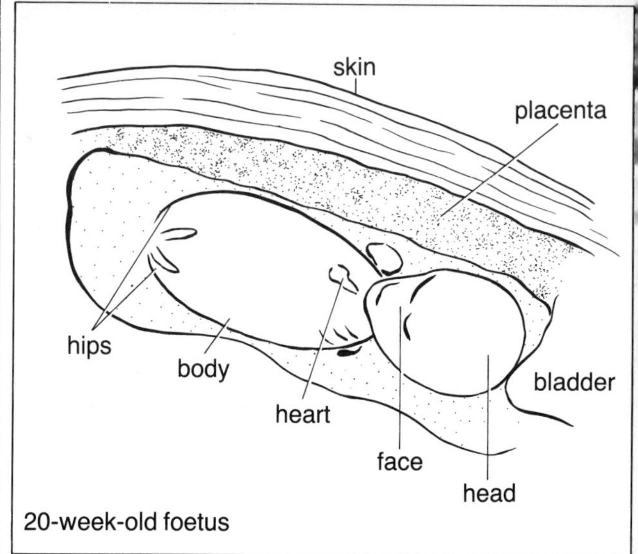

figure 10.27 **b** *Ultrasonic foetal scan*

RIPPLE TANKS

A ripple tank is a very convenient model to use to investigate waves. Although it is only a model, it does allow predictions to be made. The general arrangement of a ripple tank is shown in figure 10.29.

Straight barriers are used to imitate plane mirrors, curved barriers represent curved mirrors and an area of shallow water represents a difference in density like a glass block. The effects of reflection and refraction can be seen clearly as shadow patterns under the ripple tank (see figure 10.30).

Using water waves a new effect can also be seen – diffraction. When a series of straight waves strikes the edge of a reflector head on some part of the wave will pass by. It does not pass straight by as you might expect but appears to bend behind the reflector (see figure 10.31, page 164). This effect is known as **diffraction**. If a second barrier is added to create a

Application 10.4

HELPING THE HARD OF HEARING

Unfortunately not everyone has perfect hearing. Apart from the loss of information supplied by sound, partial or total deafness can create problems with speech, particularly if the hearing problem starts when very young. Speech is learned by repeating what is heard. If speech is not heard correctly it may not be repeated correctly. However, just because a person who is hard of hearing may have a speech problem it does not mean they are any less intelligent!

There is a number of devices now available to assist people who experience hearing difficulties. Those shown in figure 10.28 convert sound waves into other waves that can be detected by deaf people.

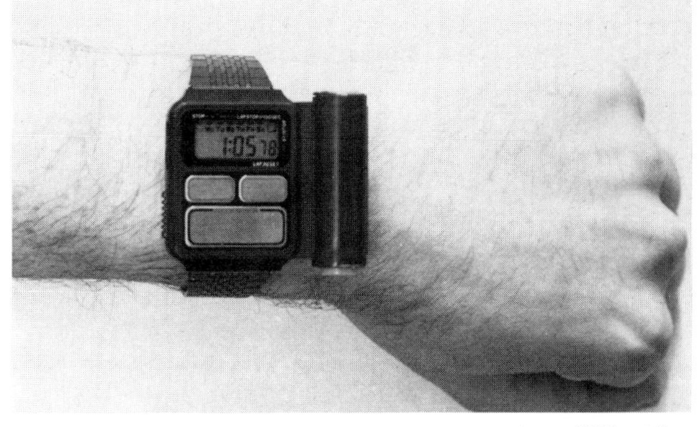

figure 10.28 Help for people with hearing difficulties **a** Vibrating alarm watch

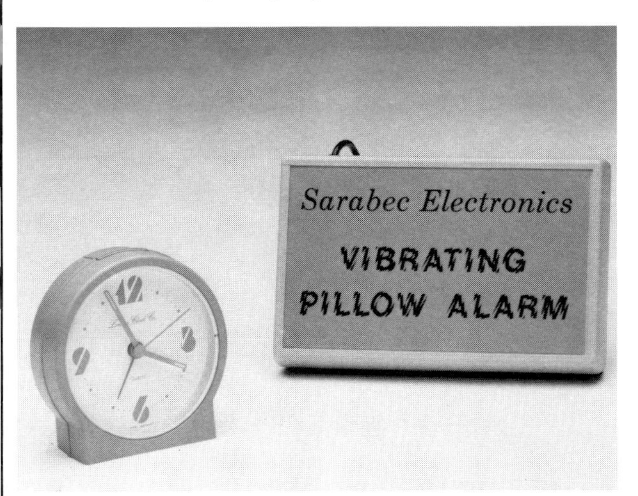

figure 10.28 **b** Vibrating pillow alarm

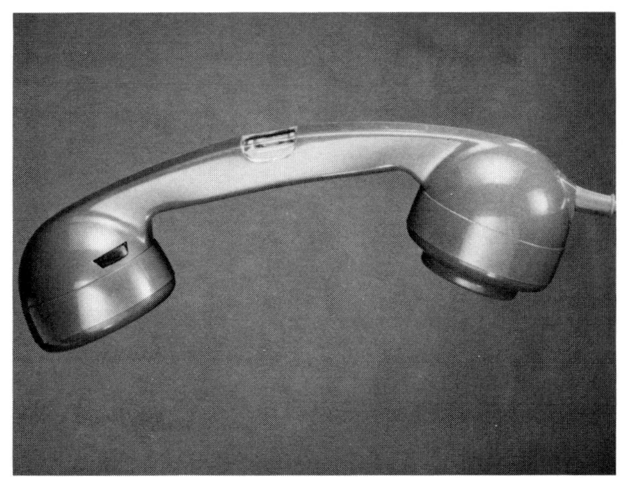

figure 10.28 **c** Telephone hearing aid – a light flashes when the phone rings

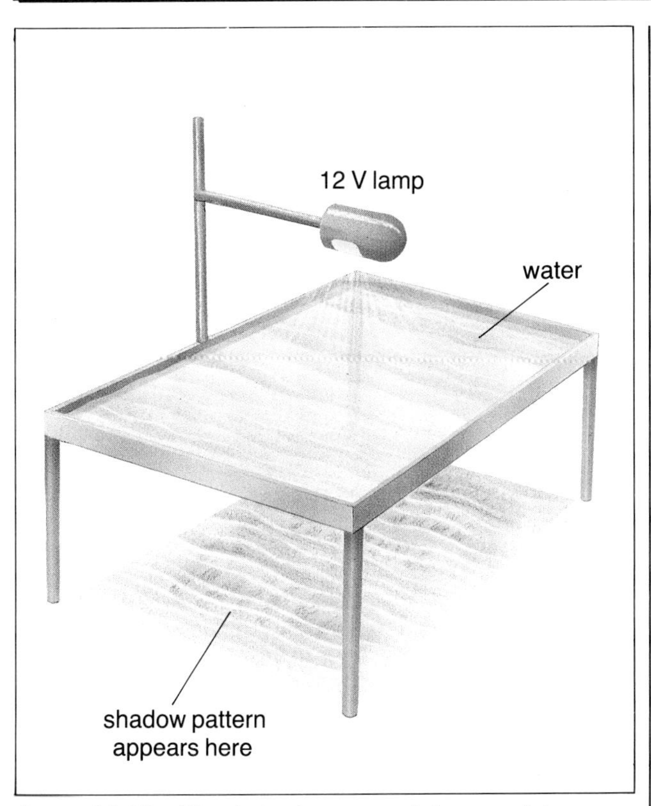

figure 10.29 Ripple tanks are useful as models

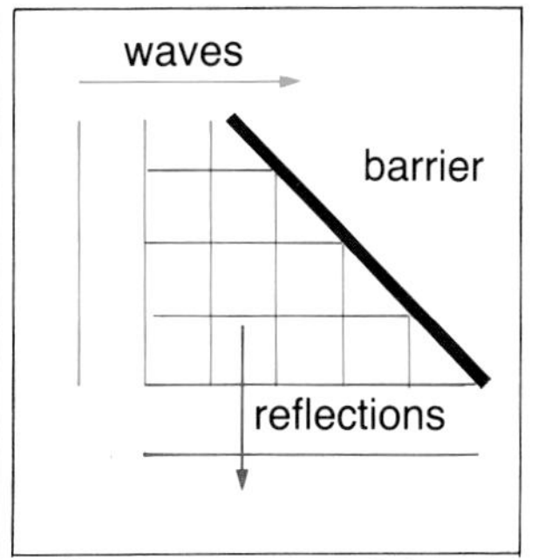

reflections at a straight barrier

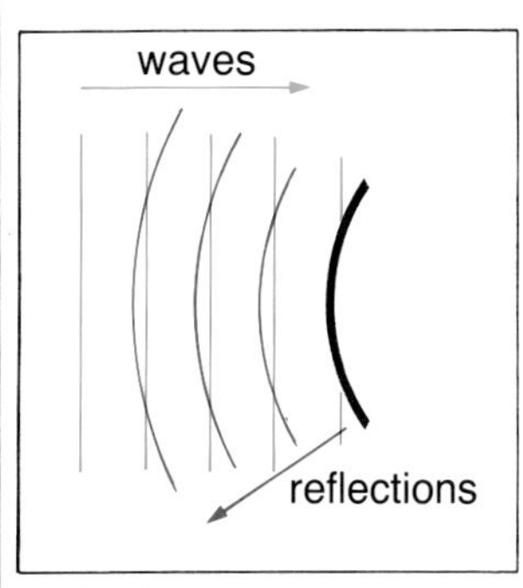

reflections at a convex barrier

reflections at a concave barrier

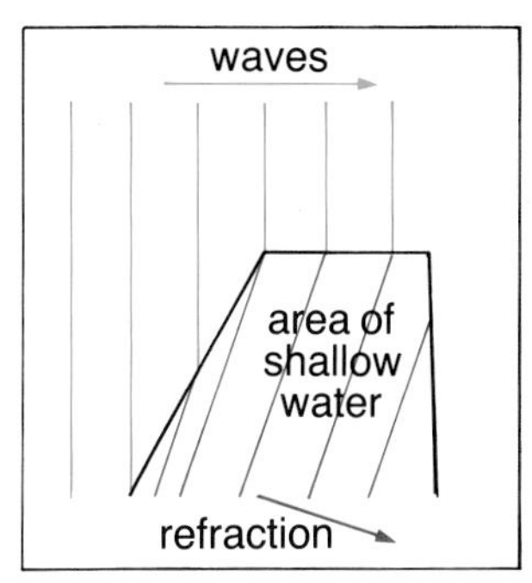

refraction at a boundary

figure 10.30 Reflection and refraction in the ripple tank

gap between the two barriers then waves passing through the gap spread out on either side (see figure 10.31c).

Diffraction is only really noticeable when the waves under investigation have relatively large wavelengths. The diffraction of waves such as light can be observed, but this is quite difficult. However, the diffraction of sound waves is quite common. Shouting down a corridor or along one edge of a building easily demonstrates this. The sound will bend around any corner where someone else will be able to hear you quite easily.

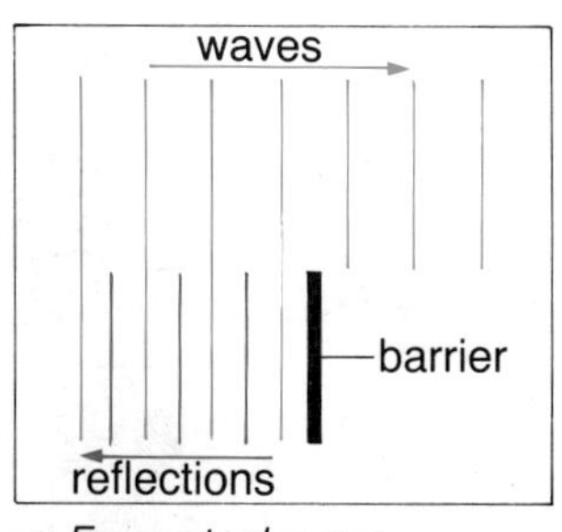

a Expected wave behaviour

waves

reflections

b Diffraction effect

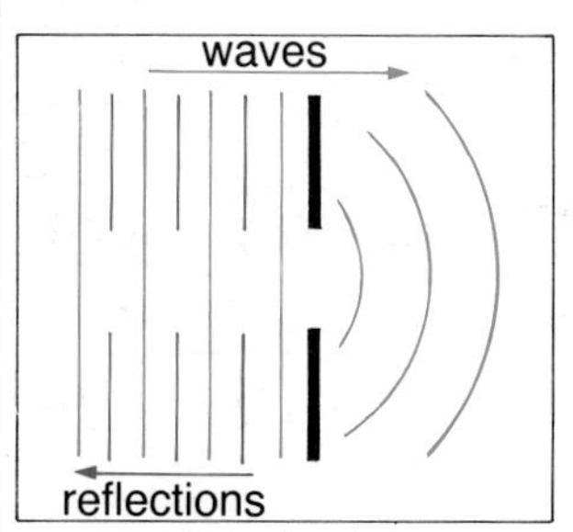

c Diffraction at a gap

figure 10.31 *Diffraction – not what you might expect*

OTHER OSCILLATIONS

There is a number of other interesting types of oscillation. Mechanical oscillations, for example, are in use every day. The suspension units on cars and motorcycles are oscillators. They are, however, designed to provide smaller amplitudes as the oscillations continue: they **dampen** down. A car will bounce up and down only a few times before coming to rest.

The results of undamped oscillations building up can be quite dangerous. In 1940 the bridge at Tacoma Narrows in America began to oscillate at its own natural frequency. It began to **resonate**. Unfortunately the oscillations did not dampen down and the bridge finally vibrated itself to destruction.

Strings and pipes have their own natural frequency too. Pluck a string or blow into a pipe and it will produce a musical note of some kind. Making a change to the length of the string or pipe will cause a change in the musical note. This is the basis of all stringed and wind instruments.

figure 10.32 *The Tacoma Narrows bridge collapse*

Time for thought

What do you think about the following statement?

Noise is just another form of pollution!

ALL YOUR OWN WORK

1 Waves transfer ________ from place to place.
2 Infra-red radiation is a wave which belongs to the ________ ________. All members of this group are ________ waves. Sound waves do not belong to this group. They are ________ waves.
3 The distance between the repeating parts of a set of oscillations is known as ________. The number of complete oscillations which take place in one second is known as the ________ and is measured in ________.
4 The law of reflection states that the angle of ________ equals the ________ of ________.
5 What are the main differences between real and virtual images?
6 Why might diffraction be important to civil engineers when designing harbours and beach defences?
7 How does the absorption of light create colours such as **a** white **b** black **c** red **d** yellow when seen, for example, in a colour photograph?
8 A fishing boat uses sonar to locate shoals of fish. It does this by emitting a short pulse of sound waves and recording the time taken for the echo to return. If sound travels at 1500 m/s through sea water and the echo time is 0.1 s,
   a how far does the pulse travel?
   b how far below the boat is the shoal of fish?
   In practice the echo lasts longer than the emitted pulse. Suggest a reason for this.

CHEMICAL CLASSIFICATION

BASIC BUILDING BRICKS

Although there is a great number of different substances around us they are all made up from basic building bricks called **elements**. Just as you can make different shapes from a Lego set (see figure 11.1) so elements can be joined together to form all the different substances we can see.

figure 11.1 *Models from Lego*

Elements

Elements are single pure substances that cannot be split into anything simpler by chemical means. There are just over 100 different elements of which about 30 are fairly common.

The periodic table is a list of all the known elements. Look at the periodic table (figure 11.26, page 177) and see how many of the elements you have heard of. As you look at a list of the elements you will notice that each element has been given a **symbol**.

Symbols

Chemists have found it very useful to give symbols to the different elements. In the early days of chemistry many strange shapes and letters were used to represent substances (see figure 11.2). Today scientists thoughout the world use the same symbols for the same elements. The symbol consists of one or two letters. The first letter is a capital letter and the second letter, if there is one, is always a small letter. There is a second letter if the first letter has already been used for a different element. It is often obvious that the letters used in the symbol come from the element's name. In other cases, the symbol seems to bear no resemblance to the name. In these cases the letters have come from another name for the element.

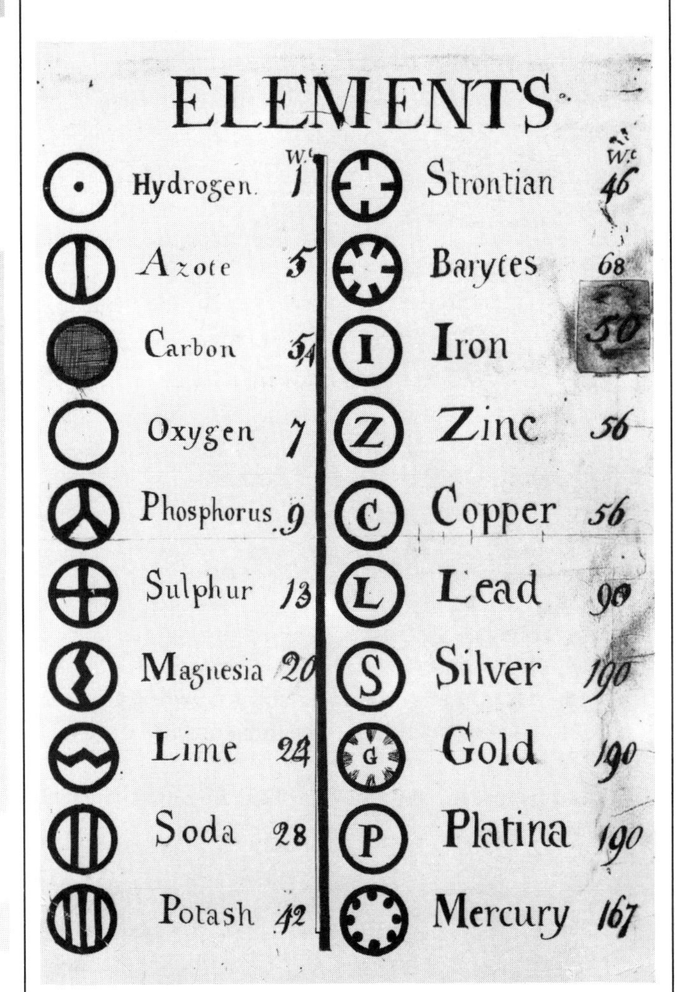

figure 11.2 *Early chemical symbols*

Some common elements and their symbols

hydrogen	H	silicon	Si	bromine	Br
helium	He	sulphur	S	krypton	Kr
carbon	C	chlorine	Cl	silver	Ag
nitrogen	N	potassium	K	tin	Sn
oxygen	O	calcium	Ca	iodine	I
fluorine	F	chromium	Cr	gold	Au
neon	Ne	iron	Fe	mercury	Hg
sodium	Na	nickel	Ni	lead	Pb
magnesium	Mg	copper	Cu	uranium	U
aluminium	Al	zinc	Zn	plutonium	Pu

Compounds

Most substances we come across are not elements but are **compounds**. A compound is formed when two or more elements join together. These new substances formed are often very different from the elements from which they are made.

UNIT 11

Compounds have the following characteristics.

a They have properties of their own and not the properties of the elements from which they are made.
b They contain fixed amounts of the elements in them.
c They are formed by a chemical reaction and so there is often an energy change as they are made.
d They are difficult to split up and cannot easily be made back into the elements from which they were formed.

For example, when sodium metal is heated in chlorine, sodium chloride (common salt) is formed. Sodium is a soft grey metal that has to be stored in oil because it is very reactive. Chlorine is a green poisonous gas. Sodium chloride is a white crystalline solid and contains one part sodium and one part chlorine.

Water is a compound of hydrogen and oxygen. What do you know about the properties of the elements and the compound? Are they different?

Formulae of compounds

The **formula** of a compound is made using the symbols of the elements in the compound. The formula tells you which elements the compound is made from and how much of each element there is.

Some common substances and their formulae

water	H_2O
carbon dioxide	CO_2
ammonia	NH_3
methane	CH_4
sulphuric acid	H_2SO_4

Work out from the formulae which elements each of the above compounds contain. (You will find out in unit 16 how to work out the formulae of compounds.)

Making compounds

One way of making compounds is by the direct combination of elements. This type of reaction is called a **synthesis** reaction.

The reaction between iron and sulphur is a good example (see investigation 11.1). It demonstrates the differences between mixtures and compounds.

A reaction which illustrates the energy change when a compound is formed is the reaction between aluminium powder and iodine (see investigaton 11.2). Your teacher may show you this as a demonstration. The reaction needs to be carried out *outside* because of the fumes produced.

INVESTIGATION 11.1

MAKING IRON SULPHIDE

a Place one spatula measure of iron filings and one measure of sulphur on separate pieces of paper. Describe the elements.
b Mix the two elements and describe the mixture.
c Design a simple experiment to separate the mixture into iron and sulphur.
d Make some more mixture and place about 1 cm depth of the mixture in a clean dry ignition tube. Heat the mixture until there is evidence of a reaction (see figure 11.3).
e When the ignition tube is cool, examine the contents.

iron + sulphur → iron sulphide
$Fe + S \rightarrow FeS$

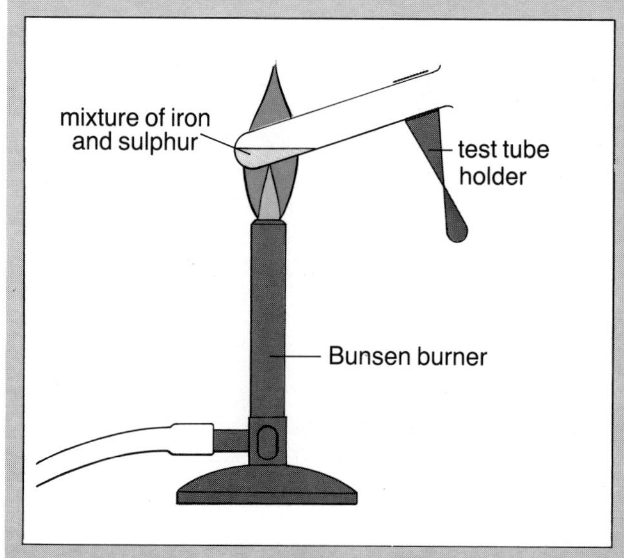

figure 11.3 Heating iron and sulphur

Questions

1 How did you separate the mixture of iron and sulphur?
2 What evidence is there that a reaction is taking place?
3 Both the mixture and the compound will react with dilute acid but they produce different gases. See if you can write word equations for each reaction and so identify the two gases.

Names of compounds

Some simple rules are used when deciding how to name a compound. The most important ones are listed here.

a If a compound contains a metal, the name of the metal comes first, e.g. *iron* sulphide.
b The name of a simple compound containing two elements ends in the letters *-ide*, e.g. aluminium iod*ide*.
c A compound of two elements and oxygen often ends in the letters *-ate*, and oxygen does *not* appear in the name, e.g. copper sulph*ate*.

Try these few examples – name the following compounds.

1 a compound of chlorine and lead
2 a compound of magnesium and sulphur
3 a compound of silver, oxygen and nitrogen
4 a compound of oxygen and copper
5 a compound of carbon, zinc and oxygen

Mixtures

Mixtures consist of any two or more substances just mixed together. No new substance has been formed and so the mixture looks and acts like the things in it. Also, because the substances in a mixture are not joined together, the mixture can be fairly easily separated into its constituents.

INVESTIGATION 11.2

MAKING ALUMINUIM IODIDE

Warning – this experiment must not be carried out in the laboratory and should only be performed by a teacher.

a About 5 spatula measures of iodine crystals are ground with a pestle and mortar.
b The iodine and about twice as much aluminium powder are mixed together on a gauze on top of a tripod.
c 1–2 drops of water are added to the centre of the mixture and then the reaction starts. On a cold day this can take a few minutes. The reaction can also be started by warming it with a match. (See figure 11.4.)

aluminium + iodine → aluminium iodide
$2Al$ $3I_2$ $2AlI_3$

mixture of aluminium and iodine

aluminium and iodine reacting

figure 11.4 *The reaction between aluminium and iodine*

Questions
1 What did you see during the reaction?
2 What does aluminium iodide look like?
3 How hot do you think this reaction became?

SEPARATING MIXTURES

In order to separate any mixture, you first of all need to find a difference in the properties of the substances in the mixture – you have to find something that one part of the mixture does that the other part does not.

INVESTIGATION 11.3

SEPARATING SAND AND SALT

This is a very easy exercise but it shows how a difference in property allows the separation to be carried out.

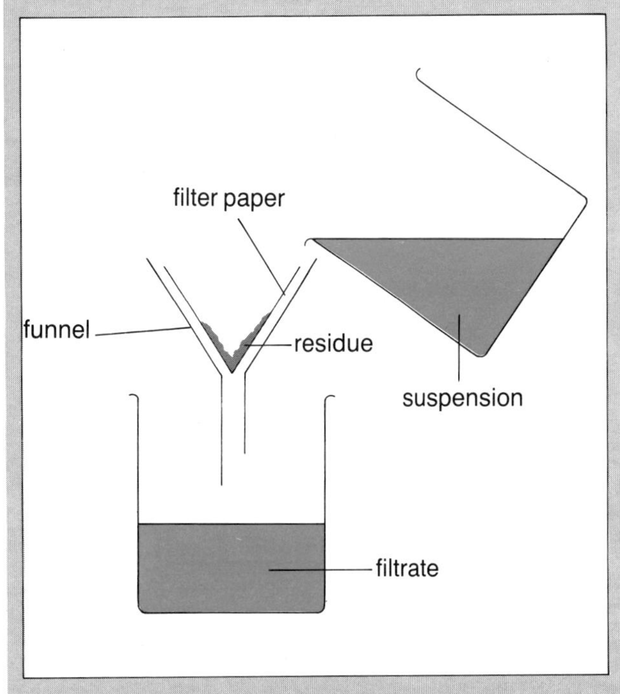

figure 11.5 *Filtering sand and salt solution*

a Place about 5 spatula measures of a mixture of sand and salt in a plastic beaker.
b Add about 25 cm^3 of water to the beaker and stir the mixture for a few minutes. (The salt will dissolve because it is **soluble** in water, but the sand is **insoluble**. The water is acting as a **solvent** for the salt.)
c Carefully pour the mixture into a filter paper in a filter funnel (see figure 11.5). Collect the salt solution in a clean evaporating basin. The liquid that has been filtered is called the **filtrate** and whatever is left in the filter paper is called the **residue**.
d Wash the residue with clean water to obtain a sample of pure sand.
e Evaporate the water over a medium Bunsen flame using a tripod and gauze. If you heat it too strongly it will spit. When all the water has evaporated you should be left with a sample of pure salt.

Questions
1 Why do you have to stir the mixture?
2 Why does the sand not pass through the filter paper? Why does the salt pass through?
3 Name a solvent other than water.
4 How can salt be obtained from sea water?

Distillation

Distillation is a way of obtaining the solvent from a solution. When a solution is evaporated, only the pure solvent evaporates leaving the dissolved substances behind. The solvent vapour has to be cooled down so that it can condense and be collected, rather than being lost in the air. Figure 11.6 shows a typical laboratory distillation apparatus.

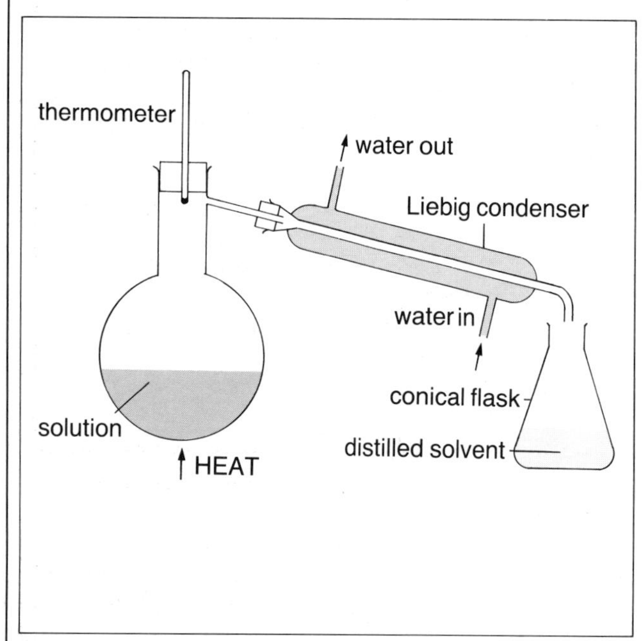

figure 11.6 *Distillation in the laboratory*

> **Application 11.1**
>
> Distilled water is pure water. It is used in car batteries, some steam irons and as a solvent for laboratory work.

Fractional distillation

This is a very important separation technique. It is a way of separating mixtures of **miscible** liquids (liquids that mix together). It makes use of the fact that the different liquids in the mixture have different boiling points.

Liquids do, however, evaporate at temperatures below their boiling points and an additional piece of apparatus called a **fractionating column** is used to improve the separation (see figure 11.7). As the vapour moves up the column it cools and will tend to turn back to a liquid again. So only the vapour with the lowest boiling point will get to the top.

> **Application 11.2**
>
> Fractional distillation is important in two major industries – the oil industry and the manufacture of spirit drinks e.g. gin, whisky and rum.

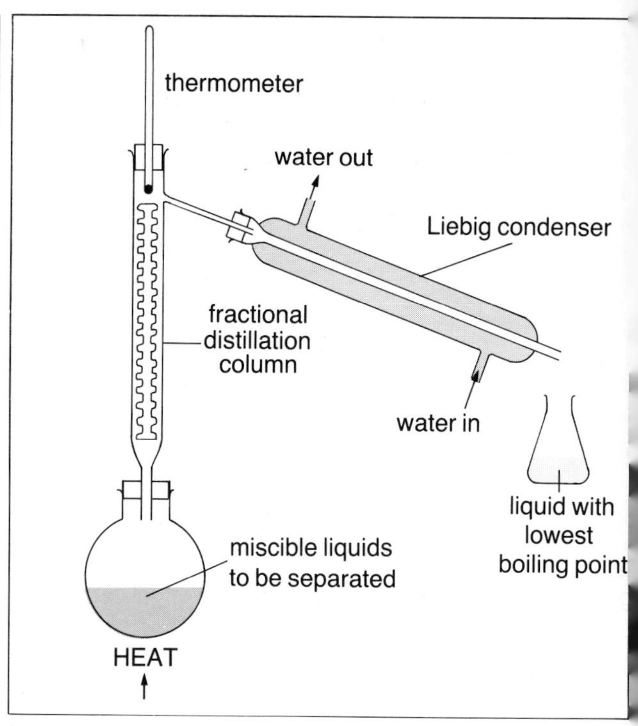

figure 11.7 *Fractional distillation*

Chromatography

Chromatography is a way of separation that was originally developed for separating coloured substances in dyes and pigments. Nowadays it is an important process in the separation of gases as well as dissolved substances, and is a valuable technique in **analysis** (finding out what compounds are present in a mixture).

The general idea in chromatography is to allow a solvent to wash over the mixture of substances on a filter paper or similar surface. The chemicals in the mixture cling to the surface and dissolve in the solvent with different strengths and so the solvent carries the chemicals different distances. Figure 11.8 shows an **ascending chromatography** experiment (the solvent moves *up* the paper). Notice how the original dye spots were above the surface of the solvent so they would not be washed away. Figure 11.9 shows the result of a chromatography experiment (a **chromatogram**). The components in the mixture can be identified by using spots of the pure dyes alongside the mixture.

If the chemicals are not coloured, the chromatogram has to be **developed** by spraying it with a chemical that will show where the spots are.

A more accurate analysis to match the spot can be done using the R_f value. This is the fraction of the distance that the substance travels up the paper (see figure 11.10). The R_f value is a constant for a given substance under stated conditions (the temperature and solvent used, for example).

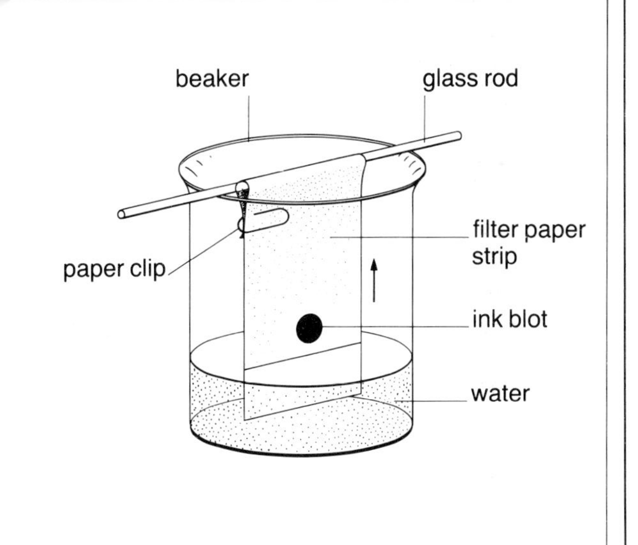

gure 11.8 *Ascending chromatography*

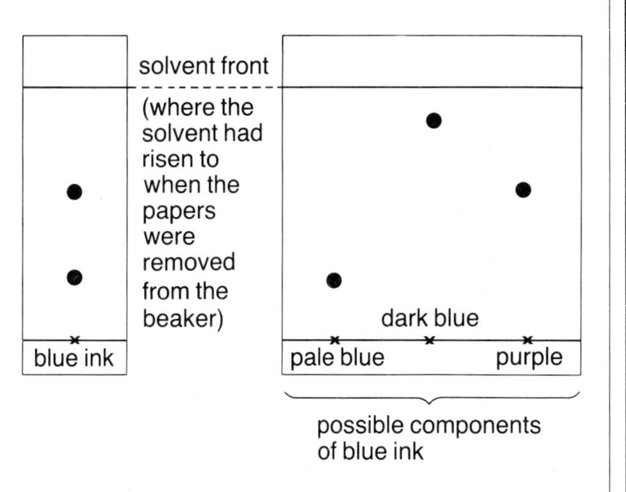

gure 11.9 *Chromatography results – a hromatogram*

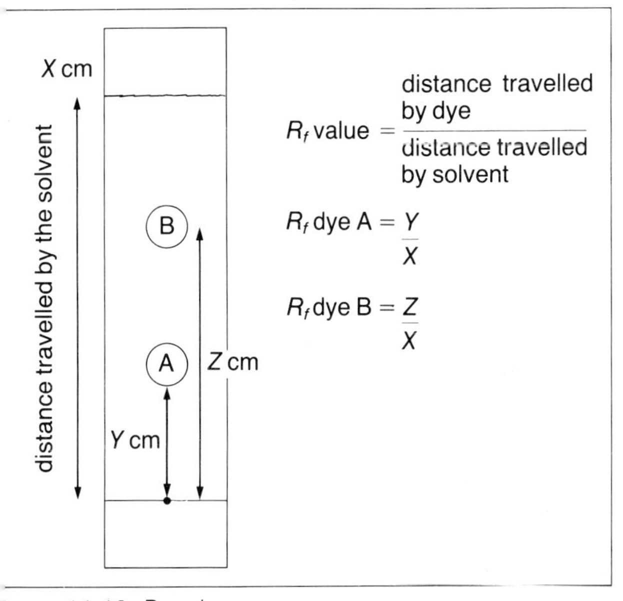

gure 11.10 *R_f values*

Application 11.3

INDUSTRIAL APPLICATIONS OF SEPARATION TECHNIQUES

Problem How do you extract salt from underground deposits? How do you extract and purify sugar?
Solution Dissolve and crystallize the salt and sugar.
Problem How do you remove solids from drinking water, industrial waste and sewage?
Solution Filter the mixtures – the solutions pass through and the solids are left behind.
Problem How do you make pure water for laboratory use and to go in car batteries? How do you make drinking water from sea water?
Solution Distil the water and the impurities are removed from it.
Problem How do you obtain petrol, paraffin etc. from crude oil? How do you obtain alcohol from fermentation mixtures in the manufacture of spirit drinks?
Solution Fractional distillation separates liquids with different boiling points.
Problem How do you analyse the dyes in an unknown sample?
Solution Chromatography separates the components and the R_f value identifies them.

Time for thought

Distillation is a way of obtaining pure water fit for drinking from sea water. Why do you think that it is not used more often?

Separating liquids that do not mix

Liquids that do not mix are said to be **immiscible**. Instead of mixing they form two different layers. The liquid with the highest density forms the bottom layer. The mixture can be separated by using a separating funnel (see figure 11.11). The bottom layer can be run off first and then the upper layer can be run off into a different container. Oil and water are examples of liquids that do not mix.

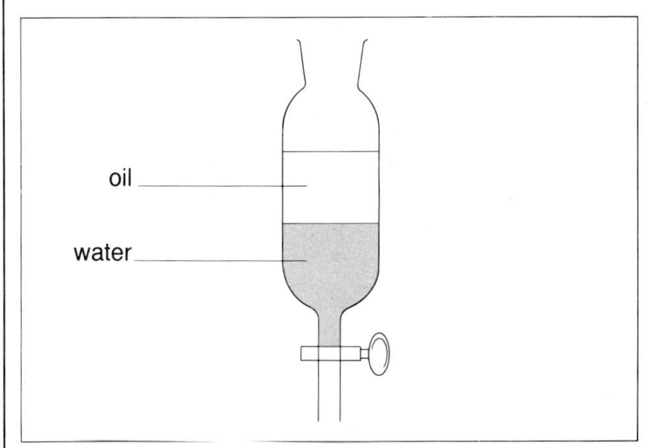

figure 11.11 *Using a separating funnel to separate oil and water*

ATOMIC STRUCTURE AND BONDING

You have already seen in unit 1 that all substances are made up of tiny particles. In solids these particles are packed closely together and are not free to move about. In liquids the particles are still close together but they are able to move. In gases the particles are widely spaced and they are free and constantly moving about in every direction. In this section you are going to find out more about these particles and what makes them join together.

Evidence for particles

For thousands of years people have asked the question "What is matter made from?" Some argued that it was continuous, i.e. the same all the way through without any spaces in it. Others said it was made from particles.

In 1810 John Dalton, an English chemist, proposed a theory that all substances were made up from very small particles. He called these particles **atoms**. The people of the time were not very impressed with his theory, mainly because he suggested that these particles were too small to be seen. Today we accept the majority of Dalton's atomic theory even though we still cannot see individual atoms. Figure 11.12 shows an electron microscope photograph of groups of atoms in a silver compound. If we cannot see atoms, how do we know that they exist? Diffusion experiments give some of the best evidence (see unit 1).

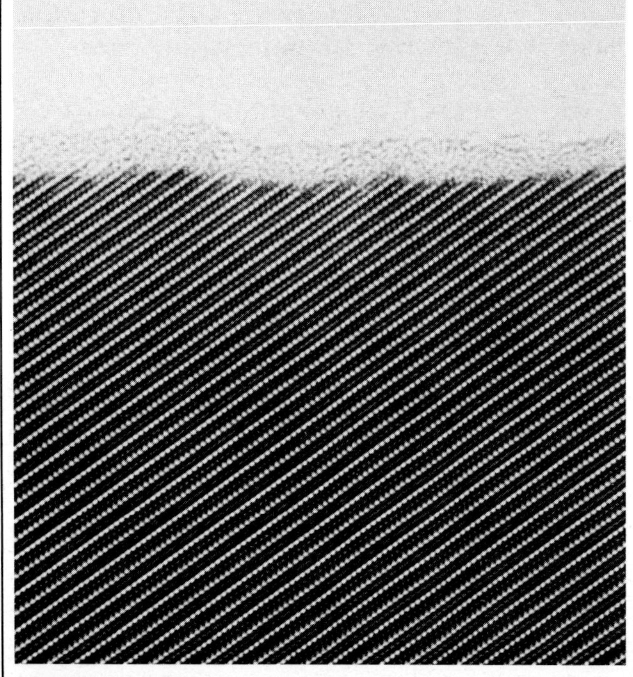

figure 11.12 Electron micrograph showing individual particles in a thin crystal of sodium beta-aluminium

Atoms and molecules

An **atom** is the smallest part of an element that can exist. All the atoms in a particular element are chemically the same but are different from the atoms of any other element.

A molecule is the smallest part of an element or compound that can exist on its own. A molecule is made up of atoms joined together, e.g. a molecule of sulphur consists of eight sulphur atoms joined in a ring: a molecule of water (H_2O) consists of two atom of hydrogen joined to one atom of oxygen (see figure 11.13).

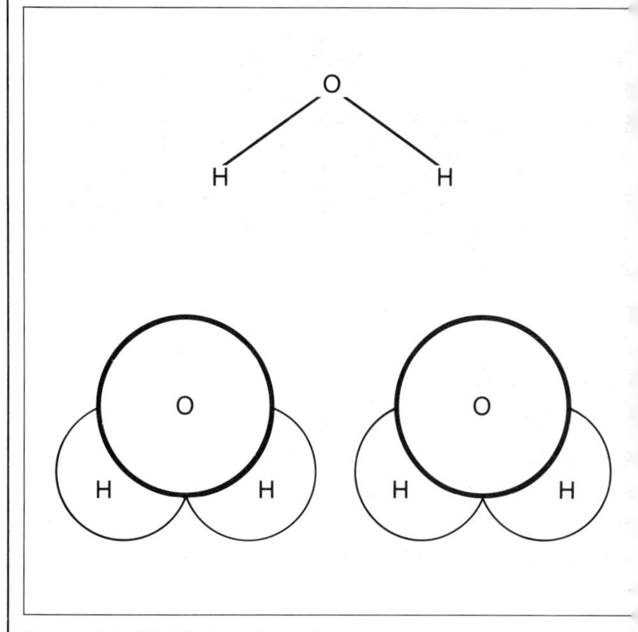

figure 11.13 Molecules of water

Atomic structure

Atoms are made up from three basic particles which are even smaller – **protons**, **neutrons** and **electrons**.

Protons have a positive electric charge, electrons have a negative electric charge and neutrons have no charge.

Protons and neutrons are bound tightly together and form the tiny centre of the atom called the **nucleus**. The electrons orbit the nucleus in particular energy levels called **orbits** or **shells** (see figure 11.14). All atoms are electrically neutral and so must contain equal numbers of protons and electrons.

The **atomic number** of an element is the number of protons in an atom of that element. It is also the number of electrons in an atom of that element, because elements have electrically neutral atoms. The **mass number (nucleon number)** is the total number of protons and neutrons (**nucleons**) in an atom.

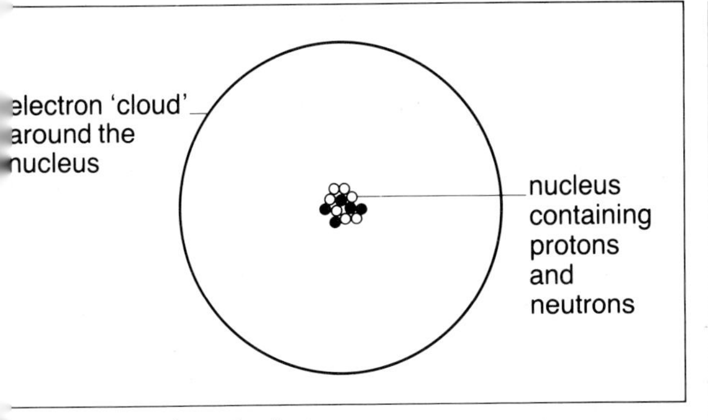

igure 11.14 A typical atom

Electron arrangements

The way in which the electrons are arranged round the nucleus of an atom is very important because it is this that determines the chemical reactions of the atom.

For the first 20 elements (those with atomic numbers 1–20) the following rules tell you which orbit or shell the electrons are in.

The first shell can hold up to 2 electrons.
The second shell can hold up to 8 electrons.
The third shell can hold up to 8 electrons.
The fourth shell can hold up to 8 electrons.

A new shell cannot be started until the previous one is full, for example if an atom has 16 electrons, 2 will go in the first shell and 8 in the second shell, leaving 6 to go in the third shell. The **electron arrangement** would be written 2, 8, 6.

Table 11.1 shows the atomic structure and electron arrangement for the first 20 elements. *p*, *n* and *e* are the numbers of protons, neutrons and electrons in an atom of the element.

Figure 11.15 shows the atomic structures of a sodium atom and an oxygen atom.

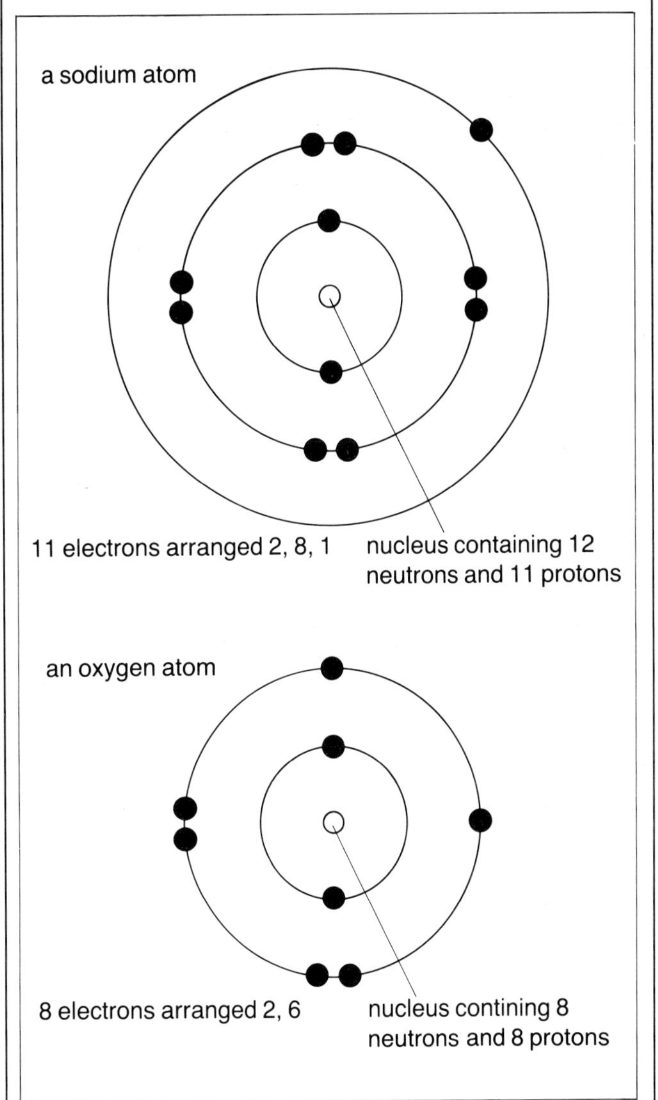

figure 11.15 Atomic structures

element	atomic number	mass number	P	n	e	electron arrangement
hydrogen	1	1	1	0	1	1
helium	2	4	2	2	2	2
lithium	3	7	3	4	3	2, 1
beryllium	4	9	4	5	4	2, 2
boron	5	11	5	6	5	2, 3
carbon	6	12	6	6	6	2, 4
nitrogen	7	14	7	7	7	2, 5
oxygen	8	16	8	8	8	2, 6
fluorine	9	19	9	10	9	2, 7
neon	10	20	10	10	10	2, 8
sodium	11	23	11	12	11	2, 8, 1
magnesium	12	24	12	12	12	2, 8, 2
aluminium	13	27	13	14	13	2, 8, 3
silicon	14	28	14	14	14	2, 8, 4
phosphorus	15	31	15	16	15	2, 8, 5
sulphur	16	32	16	16	16	2, 8, 6
chlorine	17	35*	17	18	17	2, 8, 7
argon	18	40	18	22	18	2, 8, 8
potassium	19	39	19	20	19	2, 8, 8, 1
calcium	20	40	20	20	20	2, 8, 8, 2

table 11.1 Atomic structures and electron arrangements

* Note that chlorine gas is a mixture of two slightly different chlorine atoms. One has a mass number of 35 i.e. has 18 neutrons and the other has a mass number of 37 , i.e. has 20 neutrons.

Atoms of the same element which have a different mass number are called **isotopes**. You will often see the mass number of chlorine given as 35.5.

Chemical reactions

When an atom reacts it will be most stable if it obtains a full outer shell of electrons. The atom achieves this by losing electrons, gaining electrons or sharing electrons.

If the shell only contains a few electrons, the atom will tend to lose them. If the shell is nearly full the atom will tend to gain the extra electrons needed. For example, a sodium atom has 11 electrons arranged 2, 8, 1 and so when it reacts it tends to lose the one electron in the outer shell. An oxygen atom has 8 electrons arranged 2, 6 and so when it reacts it tend to gain 2 electrons.

What do you think an aluminium atom tends to do when it reacts?

Ionic compounds

If an atom completes its outer shell by losing or gaining electrons it will no longer have an equal number of protons and electrons. This means that it will no longer be neutral. If the atom has lost electrons it will now have more protons than electrons and so will have a positive charge.

If the atom has gained electrons it will now have more electrons than protons and so will have a negative charge.

A charged atom or group of atoms is called an **ion**. An ionic compound is formed when one or more electrons are transferred from one atom to another.

Consider the reaction between magnesium and oxygen. A magnesium atom has 12 electrons (2, 8, 2) and so needs to lose the 2 in the outer shell. An oxygen atom has 8 electrons (2, 6) and so needs to gain 2 electrons. Figure 11.16 shows what happens when these two elements react.

Note that after the reaction the magnesium has only 10 electrons but still has 12 protons in the nucleus, so a magnesium ion will have a 2+ charge. The oxygen, on the other hand, will now have 2 more electrons than it has protons and so an oxygen ion (oxide) will have a 2− charge.

This also explains why the chemical formula for magnesium oxide is MgO. Each magnesium atom needs to lose 2 electrons and each oxygen atom needs to gain 2, therefore one atom of each element react together.

Try a similar exercise to show what happens when sodium reacts with chlorine to form sodium chloride.

Properties of ionic compounds

Ionic compounds are formed when atoms which contain only a few electrons in their outer shells have these electrons taken away, by atoms with nearly full

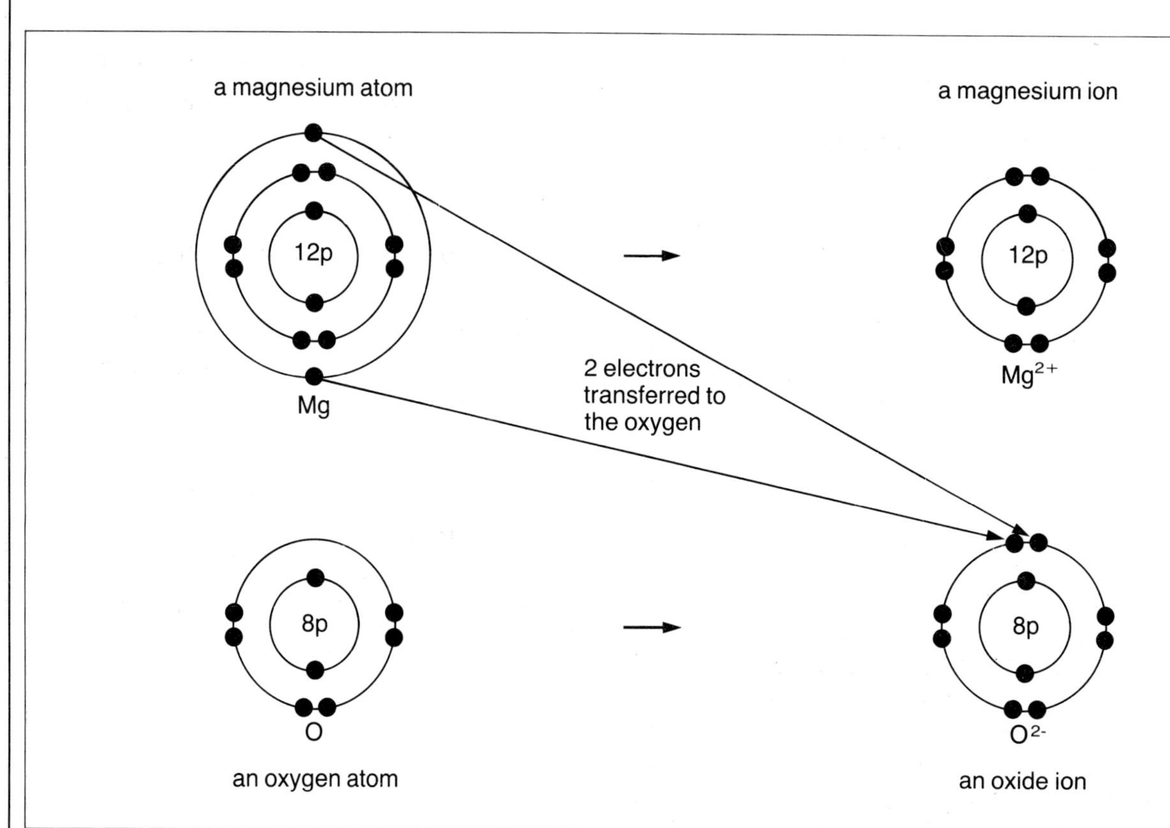

figure 11.16 The formation of magnesium oxide – an ionic compound

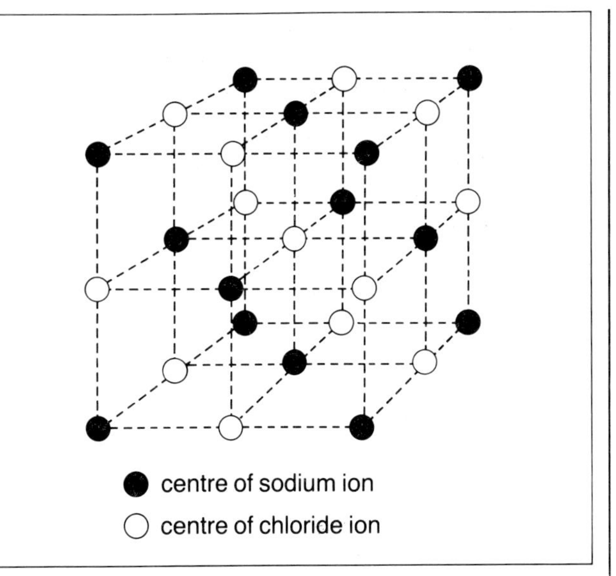

figure 11.17 *Crystal lattice for sodium chloride*

outer shells. Notice that energy is required to pull the electrons away from the positively charged nucleus.

From table 11.1 (page 171) you can see that the elements with only a few electrons in their outer shells are metals and hydrogen. This provides us with the main chemical difference between metals and non-metals.

When metals react they always lose electrons and so form positive ions.
Metals *always* form ionic compounds. If non-metals form ions they *always* form negative ions.

An ionic compound, then, is made up of oppositely charged particles called ions. The oppositely charged particles attract each other and so are arranged closely together in a regular pattern or **lattice**. Figure 11.17 shows the lattice in sodium chloride.

Ionic compounds have the following main properties.

a They are crystalline solids with high melting points.
b They usually dissolve in water but are insoluble in non-aqueous solvents.
c They conduct electricity when molten or when dissolved in water (i.e. they are **electrolytes**).
d They usually contain a metal.

Covalent compounds

When non-metals react with each other they complete their outer shells by **sharing** electrons. In order to do this their outer shells must overlap and each atom must contribute one electron to a shared pair. A shared pair of electrons is called a **covalent bond**. Electrons are shared so that each atom ends up with a full outer shell. The best way to describe the bonding in a covalent compound is to draw a diagram of the outer shells of the atoms involved.

Hydrogen chloride
A hydrogen atom has just one electron and so requires just one more to fill its shell.

A chlorine atom has 17 electrons (2, 8, 7) and so requires one more to fill its shell.

Figure 11.18 shows how the outer shells overlap and how by sharing electrons each atom fills its outer shell. Note that only the outer shells are shown.

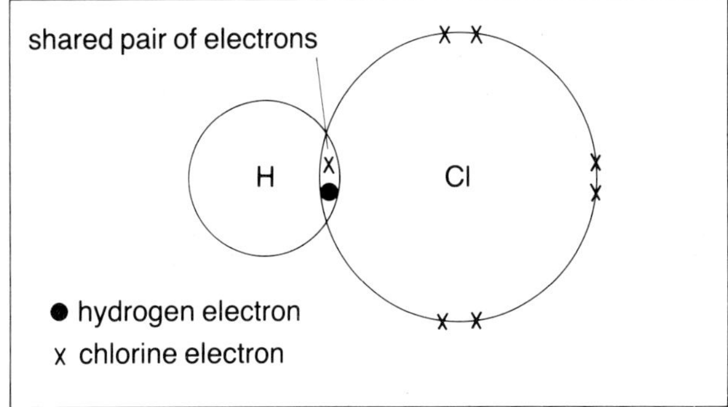

figure 11.18 Hydrogen chloride – a covalent compound

Figure 11.19 shows the covalent diagrams for water, ammonia and methane.

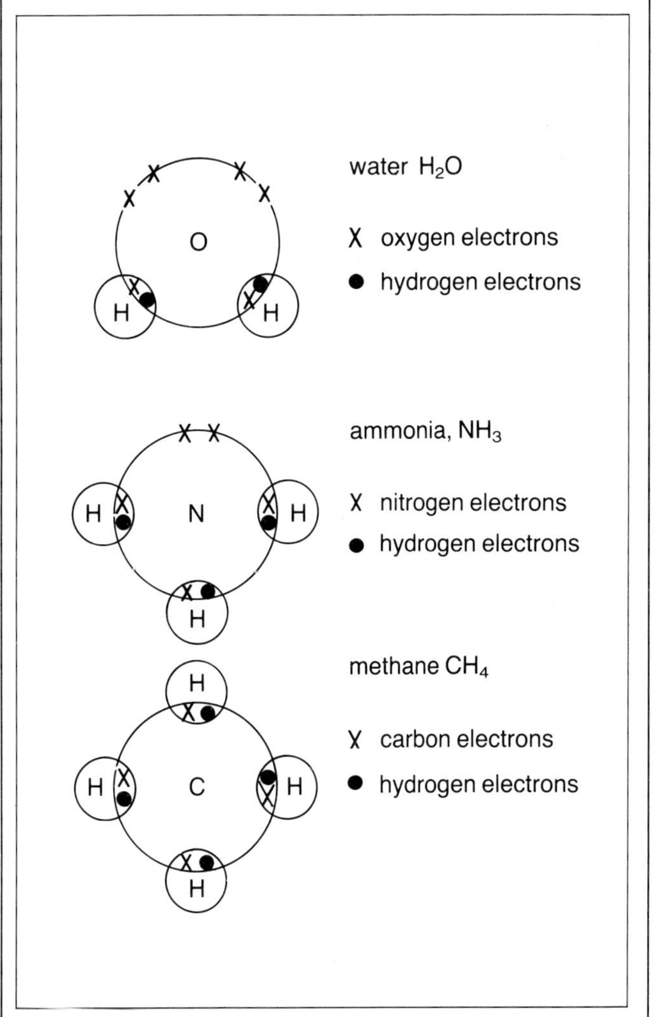

figure 11.19 More covalent compounds

UNIT 11

Properties of covalent compounds

There is no transfer of electrons in covalent compounds and so there are no charged particles. In fact, in a covalent compound the particles are individual molecules with little attraction for each other. (Covalent bonding is sometimes called **molecular bonding**.) This means that the particles are usually widely spaced, or if they are close together they are fairly easy to separate.

Covalent compounds

a are often gases at room temperature. Larger molecules may be liquids or low melting point solids

b are usually insoluble in water but dissolve in organic solvents

c do not conduct electricity at any time

d do not contain a metal.

Giant structures

A giant structure is one where very large numbers of particles are joined together in a giant framework or lattice. It is possible to have a giant structure of atoms, molecules or ions.

Metals

Metals are giant structures of closely packed atoms. The closeness of the atoms accounts for the high density of metals. The atoms are arranged in a regular pattern, and because of this we might expect to be able to see crystals of metals. Normally we would not describe metals as crystalline, but if we look we can readily find examples of metal crystals. One everyday example is the regular pattern on the surface of galvanized iron. The iron is dipped into molten zinc and as the zinc cools it forms crystals (see figure 11.20).

figure 11.20 Zinc crystals on a galvanized iron dustbin

INVESTIGATION 11.4

CRYSTALS OF LEAD

Place a small clean piece of zinc sheet in a test tube half full of lead nitrate solution. After about five minutes you should be able to see the small lead crystals on the surface of the zinc. Figure 11.21 shows silver crystals growing from the surface of some mercury in silver nitrate solution.

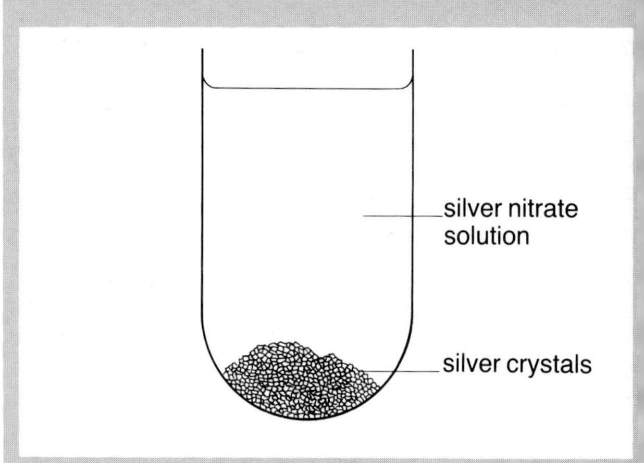

figure 11.21 Crystals of lead on zinc

This closely packed regular arrangement of atoms in a metal can explain the main physical properties of a metal.

Figure 11.22 shows how one layer of atoms can fairly easily be forced over another layer without altering the structure, so accounting for the way in which a metal can be hammered into shape.

Metals are good conductors of heat because the vibrations which caused the heat are easily passed on from one atom to another.

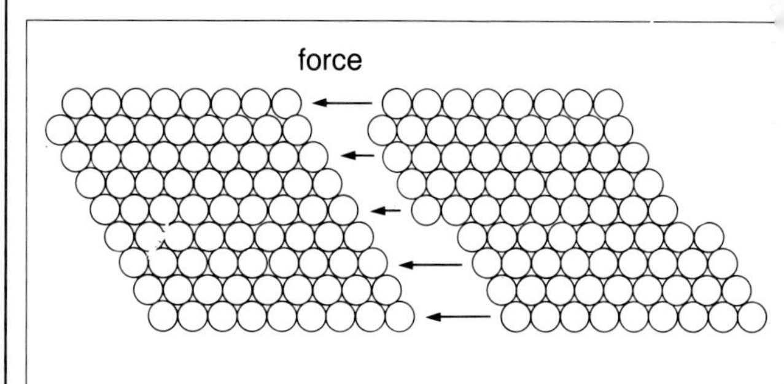

figure 11.22 Atoms in a metal can slip over each other

The good electrical conductivity of metals is also due to the closeness of the atoms. The outer electron shells of the atoms are touching each other and so it is easy for the electrons in these outer shells to move from one atom to another. An electric current is a flow of electrons (see pages 118–9). When extra electrons are fed into one end of the metal wire, electrons are displaced along the wire. A metal structure is sometimes described as a lattice of metal ions set in a sea of electrons.

arbon

arbon is an important non-metallic element. It can xist in two main forms, because the atoms can be rranged in two different ways. Both ways result in e formation of giant structures. When an element an exist in two or more different physical forms it is aid to show **allotropy**. The different forms are called **llotropes**. The two allotropes of carbon are diamond nd graphite (see figures 11.23 and 11.24).

property	diamond	graphite
density	3.5 g/cm^3	2.2 g/cm^3
hardness	very hard	soft
appearance	transparent	black
melting point	sublimes at 3500 °C	3720 °C
electrical conductivity	non-conductor	good conductor

able 11.2 Physical properties of diamond and raphite

he different arrangements give the forms of carbon ifferent physical properties but they have the same hemical properties (see table 11.2). Diamond **ublimes** – this means it turns straight from a solid to gas without forming a liquid.

iamonds are the hardest of all naturally-occuring ubstances and they are used in jewellery, for cutting lass and on the ends of drills.

raphite is a good lubricant because the layers of toms can slide over one another easily. It is also used n bearings as a lubricant and pencil leads and has many important electrical uses. Charcoal and soot are graphite.

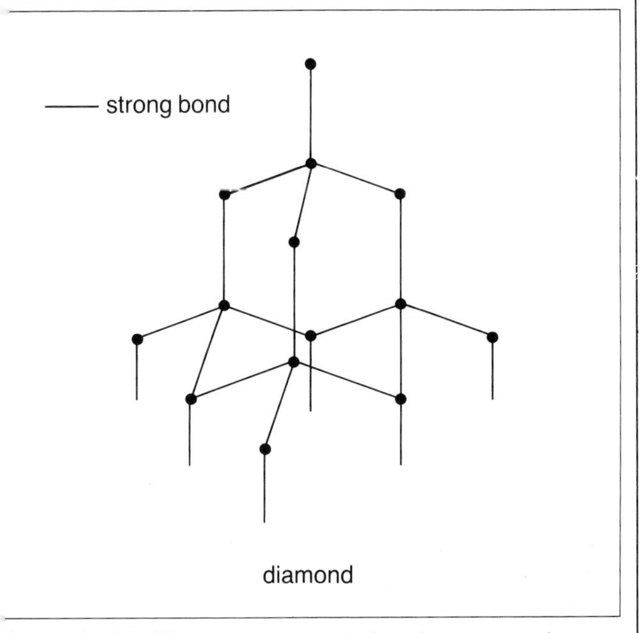

figure 11.23 The arrangement of carbon atoms in diamond

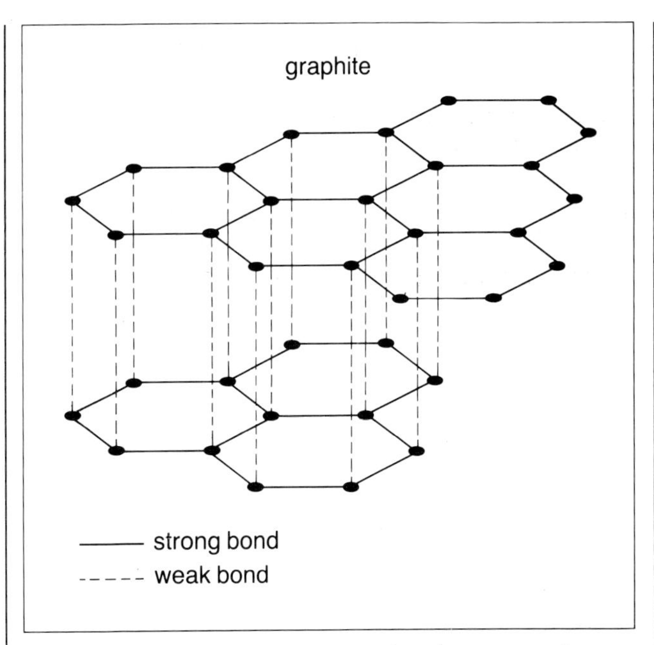

figure 11.24 The arrangement of carbon atoms in graphite

Giant structures of molecules

Elements

Iodine is a solid at room temperature. The crystals are made up from I_2 molecules loosely bound together. When iodine crystals are heated they melt quickly and sometimes sublime as the heat energy soon overcomes the weak intermolecular attractions.

Sulphur, which like carbon has allotropes, is made up from crown-shaped rings of S_8 molecules (see figure 11.25). The relatively low melting point of sulphur (117 °C) indicates the weakness of the attraction between the sulphur molecules.

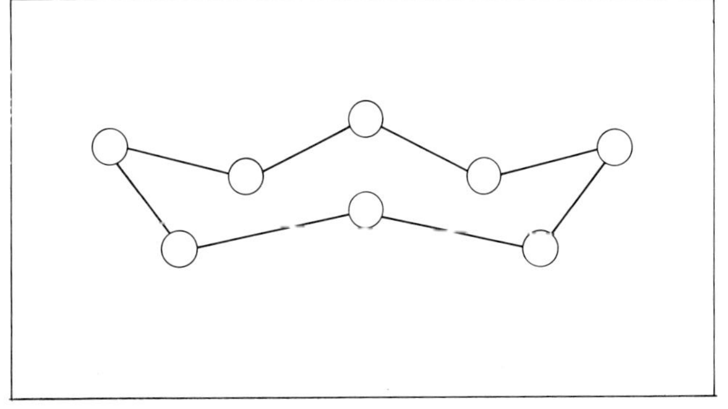

figure 11.25 A sulphur molecule – a crown-shaped ring

Compounds

Polymers are compounds made up from large molecules which themselves are made from many repeating units. The structure of many polymers is rather complicated with cross-linking between molecules. All the modern substances we call plastics are polymers.

Giant structures of ions

All ionic compounds are giant structures of ions. The oppositely charged ions attract each other and form a very strong regular lattice. The strength of the lattice accounts for the high melting point of ionic substances. A large amount of energy is needed to break up the structure.

An ionic compound will not conduct electricity when it is solid, because the ions are not free to move. When the solid is melted or when it is dissolved in water the ions become free to move and the compound will then conduct electricity.

THE PERIODIC TABLE OF ELEMENTS

The periodic table (see figure 11.26) is a list of all the elements arranged in order of their increasing atomic number. Elements with similar chemical properties are arranged in groups. The periodic table as we know it today was largely the work of a Russian chemist called Dmitri Mendeleev in 1869. He developed work which had been done by Johann Döbereiner (1829), John Newlands (1863) and Lothar Meyer (1869), who had all found various patterns in the properties of the elements.

The patterns in the periodic table can be used to explain many practical observations.

General structure of the periodic table

As we have seen the elements are arranged in order of their increasing atomic number. You can see from Figure 11.26 that there are 103 different elements listed. Those in the shaded area are called the **main block** elements and those in the middle or at the bottom are called the **transition elements**.

The main block

As more and more elements were being discovered during the last century it became clear that some sorts of chemical families existed. Cases where two or three elements had similar reactions were investigated and when the results were compared with their atomic numbers a periodic variation could be seen. It was found that every eighth element fitted into a family or group.

The vertical columns in the main block of the table are called **groups**. The group number is shown across the top of the table. Note also how the group number is the number of electrons in the outer shell of the elements in this group. This explains their similar reactions.

The horizontal rows of elements across the table are called **periods**.

The heavy stepped line that goes down the side of aluminium, germanium, antimony and polonium divides the metals from the non-metals. You can see from this then that most elements are metals and that the non-metals are found on the right-hand side of the table.

These general properties of metals and non-metals are shown in table 11.3. See if you can think of any exceptions.

metals	non-metals
silver colour	no general colour
shiny	dull
high melting point solids	solids, liquids and gases
bend easily (malleable and ductile)	brittle
are good conductors of heat and electricity	bad conductors of heat and electricity
form positive ions	form negative ions (if any)
form neutral or alkaline oxides	form acidic oxides
sometimes displace hydrogen from dilute acids	never displace hydrogen from dilute acids

table 11.3 General properties of metals and non-metals

The transition elements

These are the unshaded elements on figure 11.26. They are all metals and fit into a different pattern from the main block elements. It is in this block of elements that we find most of the common useful metals. The elements in this block have a number of similar properties.

a They are typical metals with high melting points and high densities.
b They can often form more than one positive ion, e.g. copper(I) Cu^+ and copper(II) Cu^{2+}; iron(II) Fe^{2+} and iron(III) Fe^{3+}.
This is sometimes described as having a variable valency, see unit 16.
c They usually form coloured compounds.
d They are often used as catalysts (see unit 13).

figure 11.26 The periodic table

KEY:
mass (nucleon) number — 7
Li
atomic number — 3
(242) approximate mass number

I	II											III	IV	V	VI	VII	O
1 H hydrogen 1																	4 He helium 2
7 Li lithium 3	9 Be beryllium 4											11 B boron 5	12 C carbon 6	14 N nitrogen 7	16 O oxygen 8	19 F fluorine 9	20 Ne neon 10
23 Na sodium 11	24 Mg magnesium 12											27 Al aluminium 13	28 Si silicon 14	31 P phosphorus 15	32 S sulphur 16	35 Cl chlorine 17	40 Ar argon 18
39 K potassium 19	40 Ca calcium 20	45 Sc scandium 21	48 Ti titanium 22	51 V vanadium 23	52 Cr chromium 24	55 Mn manganese 25	56 Fe iron 26	59 Co cobalt 27	59 Ni nickel 28	64 Cu copper 29	65 Zn zinc 30	70 Ga gallium 31	73 Ge germanium 32	75 As arsenic 33	79 Se selenium 34	80 Br bromine 35	84 Kr krypton 36
85 Rb rubidium 37	88 Sr strontium 38	89 Y yttrium 39	91 Zr zirconium 40	93 Nb niobium 41	96 Mo molybdenum 42	99 Tc technetium 43	101 Ru ruthenium 44	103 Rh rhodium 45	106 Pd palladium 46	108 Ag silver 47	112 Cd cadmium 48	115 In indium 49	119 Sn tin 50	122 Sb antimony 51	128 Te tellurium 52	127 I iodine 53	131 Xe xenon 54
133 Cs caesium 55	137 Ba barium 56	139 La lanthanum 57 *	178 Hf hafnium 72	181 Ta tantalum 73	184 W tungsten 74	186 Re rhenium 75	190 Os osmium 76	192 Ir iridium 77	195 Pt platinum 78	197 Au gold 79	201 Hg mercury 80	204 Tl thallium 81	207 Pb lead 82	209 Bi bismuth 83	210 Po polonium 84	210 At astatine 85	222 Rn radon 86
223 Fr francium 87	226 Ra radium 88	227 Ac actinium 89 †	? Ku kurchatovium 104														

*58 – 71 lanthanum series	140 Ce cerium 58	141 Pr praseodymium 59	144 Nd neodymium 60	147 Pm promethium 61	150 Sm samarium 62	152 Eu europium 63	157 Gd gadolinium 64	159 Tb terbium 65	162 Dy dysprosium 66	165 Ho holmium 67	167 Er erbium 68	169 Tm thulium 69	173 Yb ytterbium 70	175 Lu lutetium 71
†90 – 103 actinum series	232 Th thorium 90	231 Pa protactinium 91	238 U uranium 92	237 Np neptunium 93	(242) Pu plutonium 94	(243) Am americium 95	(247) Cm curium 96	(249) Bk berkelium 97	(251) Cf californium 98	(254) Es einsteinium 99	(253) Fm fermium 100	(256) Md mendelevium 101	(254) No nobelium 102	(257) Lr lawrencium 103

UNIT 11

AN INTERESTING FAMILY – GROUP I, THE ALKALI METAL FAMILY

This is a family of very reactive metals. The first three members of the group are lithium, sodium and potassium.

Table 11.4 shows some of the physical properties of lithium, sodium and potassium.

element	appearance	ease of cutting	melting point	density
lithium	dark grey lumps	quite hard	181 °C	0.54 g/cm^3
sodium	light grey lumps	easy	98 °C	0.97 g/cm^3
potassium	blue/grey lumps	very soft	63 °C	0.86 g/cm^3

table 11.4 *Physical properties of some alkali metals*

These elements are all stored in oil to keep them away from the water vapour in the air. The very low density of lithium means that it often floats in the oil.

Table 11.5 summarizes the reaction of these elements with water.

element	observations
lithium	Floats on the water slowly bubbling and giving off hydrogen. Moves gently on the surface until it has all dissolved. The solution left is a strong alkali.
sodium	Floats on the surface of the water. Quickly melts into a ball which skims from side to side of the container. Sizzles as it gives off hydrogen. The solution left is a strong alkali.
potassium	Floats on the surface of the water. Very quickly melts into a ball which races across the surface. The hydrogen given off is set alight by the heat of the reaction and burns with a lilac flame. The solution left is a strong alkali (see page 191).

table 11.5 *Reaction of some alkali metals with water*

INVESTIGATION 11.5

THE REACTION OF LITHIUM WITH WATER

Warning Lithium must not be touched with your fingers and you should wear safety goggles at all times during the experiment.

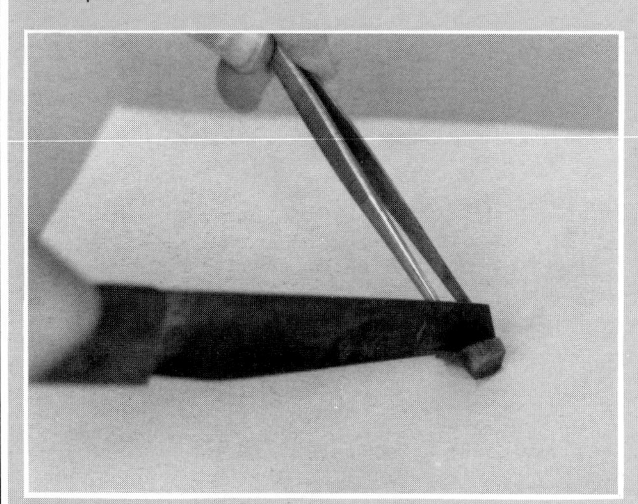

figure 11.27 *Cutting an alkali metal*

a Carefully examine a piece of lithium about the size of a grain of rice. If possible, observe it being cut up (see figure 11.27).
b Tip the piece of lithium into a small beaker about half full of cold water. Note carefully everything that happens. At the end of the reaction, test the water with a piece of universal indicator paper (see pages 191–2).
c Add another similar sized piece of lithium to a test tube half full of water. Test the gas given off by putting a lighted spill to the end of the test tube. The gas has come from water so must be hydrogen or oxygen. If the gas is hydrogen it should burn with a 'pop'. If it is oxygen the spill will burn brighter.

Questions

1 Which gas is given off in the reaction?
2 What colour did the indicator go and what does that mean?
3 In what ways does lithium differ from a typical metal?

Your teacher will probably show you how sodium and potassium react with water (see figure 11.28). Look carefully for all the similarities but also be on the look out for some differences.

figure 11.28 *Sodium reacting with water*

The chemical equations for the reactions in table 11.5 are

lithium + water → lithium hydroxide + hydrogen
$2Li + 2H_2O \rightarrow 2LiOH + H_2$
sodium + water → sodium hydroxide + hydrogen
$2Na + 2H_2O \rightarrow 2NaOH + H_2$
potassium + water → potassium hydroxide + hydrogen
$2K + 2H_2O \rightarrow 2KOH + H_2$

Order of reactivity

From these reactions with water we can see that there is a gradual increase in reactivity from lithium to sodium to potassium. How can we explain this? We need to think about the electron arrangements of all three elements.

lithium has 3 electrons arranged 2, 1
sodium has 11 electrons arranged 2, 8, 1
potassium has 19 electrons arranged 2,8, 8, 1

All three have just one electron in the outer shell and so need to lose one electron when they react. Remember that the electrons are held in position by the positive attraction of the nucleus. The further an electron is away from the nucleus the less attraction the nucleus will have for that electron and so the outer electron in a potassium atom is held less tightly than the outer electron in a sodium atom. This makes it easier for it to be taken away in a reaction, so potassium is more reactive. In a similar way sodium is more reactive than lithium.

Look at the periodic table (page 177) and decide which member of group I will be the most reactive.

ANOTHER FAMILY – GROUP VII, THE HALOGENS

This is a chemical family of reactive non-metals. They are in group VII so they have 7 electrons in the outer shell. They all need to gain one electron when they react. They can do this by complete transfer of an electron, so forming ionic compounds, or by sharing electrons, so forming covalent compounds. The first four members of the group are

flourine	a pale yellow gas
chlorine	a green gas
bromine	a red liquid
iodine	a dark grey crystalline solid

From these appearances at room temperature it might seem strange that they are in the same family, but when we look at their chemical properties we can see that they have quite a lot in common.

Fluorine is too reactive to experiment with in a school laboratory so we shall look at some of the reactions of chlorine and iodine.
Warning Chlorine is poisonous and has an irritating smell.

Reaction with water
Chlorine gas dissolves slightly in cold water to form a pale green solution. When the solution is tested with universal indicator paper, the paper first turns red and then loses its colour as the chlorine bleaches the paper.

Iodine is hardly soluble in cold water but will dissolve slightly in hot water to form a pale orange/brown solution. When the solution is tested with universal indicator paper (see pages 191 – 2), the paper turns yellow and is slightly bleached.

Reaction with sodium hydroxide (a strong alkali)
When a few drops of sodium hydroxide are put into a gas jar of chlorine the green colour quickly disappears showing that the chlorine has reacted.

When a few small crystals of iodine are shaken in sodium hydroxide solution, the crystals dissolve easily to form a colourless solution.

Reaction with iron
When a piece of hot iron wool is dropped into a gas jar of chlorine, the iron wool glows brightly and sparks. Clouds of brown smoke are given off (figure 11.29).

iron + chlorine → iron(III) chloride
$2Fe + 3Cl_2 \rightarrow 2FeCl_3$

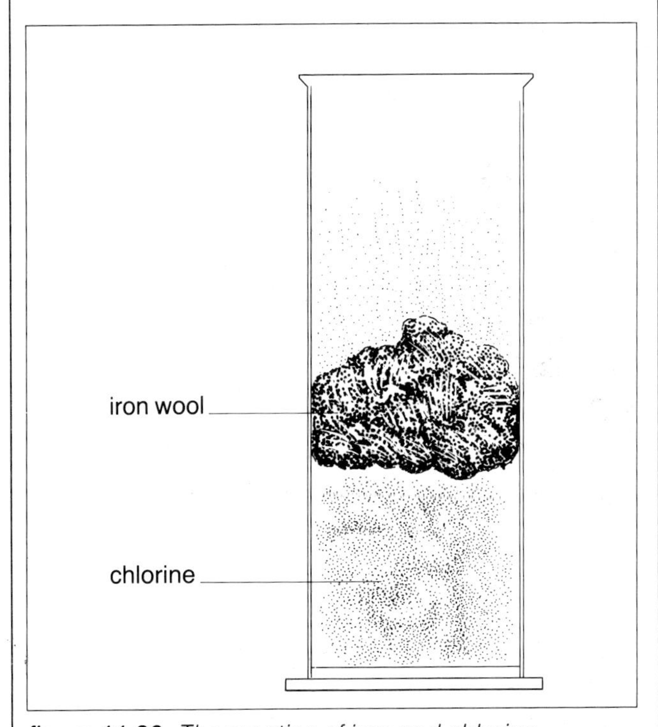

figure 11.29 The reaction of iron and chlorine

When iodine vapour is passed over heated iron wool, the purple colour of the iodine vapour disappears and a rusty brown solid forms in the test tube (see figure 11.30).

iron + iodine → iron(III) iodide
$2Fe + 3I_2 \rightarrow 2FeI_3$

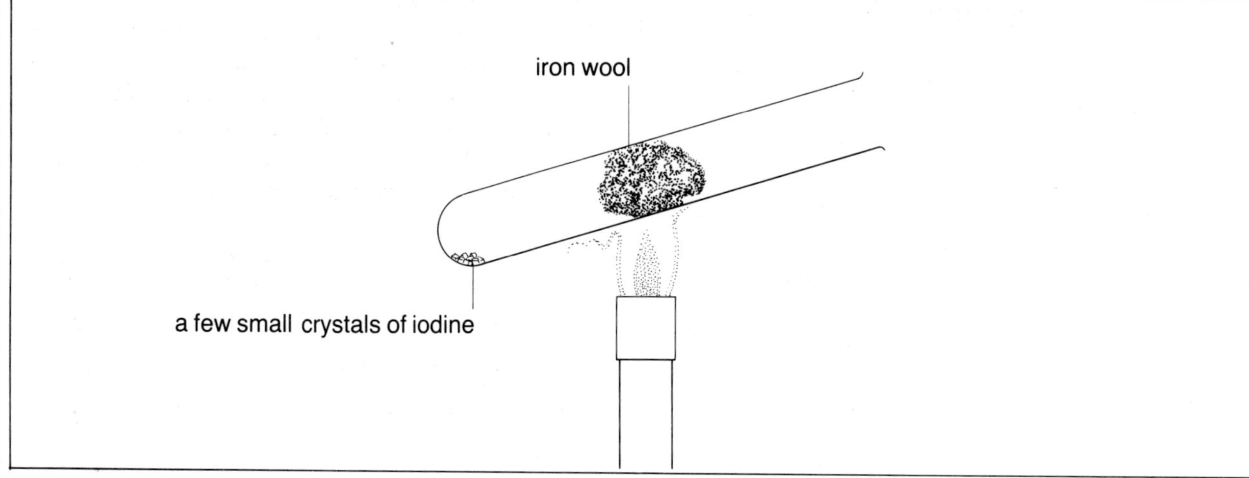

figure 11.30 *The reaction of iron and iodine*

Order of reactivity

These reactions show some of the properties that chlorine and iodine have in common, and that chlorine is more reactive than iodine. We can explain the order of reactivity by looking at the electron arrangements of the first two halogens.

fluorine has 9 electrons arranged 2, 7
chlorine has 17 electrons arranged 2, 8, 7

When they react they need to obtain another electron to complete their outer shells. This extra electron will be attracted to the atom by the positive attraction of the nucleus. This means that the closer the electron can get to the nucleus the greater will be the attraction.

This makes fluorine the most reactive halogen with the reactivity decreasing down the group. This order can be demonstrated using **displacement reactions** in which one halogen takes the place of another.

A reaction will only take place if a more reactive halogen is put into a solution containing a halogen ion (halide) of a less reactive member of the group. Chlorine can 'push out' bromine from a bromide and iodine from an iodide.

Bromine can 'push out' iodine from an iodide.

Iodine *cannot* 'push out' chlorine or bromine.

sodium bromide + chlorine → sodium chloride + bromine

$$2NaBr + Cl_2 \rightarrow 2NaCl + Br_2$$

See if you can write the other displacement equations.

INVESTIGATION 11.6

DISPLACEMENT REACTIONS OF THE HALOGENS

a Using a dropping pipette add a few drops of chlorine water (a solution of chlorine in water) to a test tube a quarter full of
  i sodium chloride solution
  ii sodium bromide solution
  iii sodium iodide solution.

b Using a dropping pipette add a few drops of bromine water to a test tube a quarter full of
  i sodium chloride solution
  ii sodium bromide solution
  iii sodium iodide solution.

c Using a dropping pipette and add a few drops of iodine water to test tube a quarter full of
  i sodium chloride solution
  ii sodium bromide solution
  iii sodium iodide solution

Questions

1 In which cases was there a colour change?
2 Which is the most reactive halogen of these three?
3 What effect would fluorine have on sodium bromide?

Application 11.4

USES OF THE HALOGENS

fluorine making the propellant gas for aerosols (see page 8),
making the non-stick coating for frying pans (PTFE – polytetrafluoroethylene),
fluoride added to drinking water to prevent tooth decay (see page 50).

chlorine making plastics, e.g. PVC polyvinyl chloride,
as a bleach,
disinfecting water (tap and swimming baths) (it kills bacteria and other microorganisms),
making insecticides,
making medicines and anaesthetics,
making dry cleaning liquids,

bromine making photographic film,
making petrol additives,
in medicine to treat some forms of epilepsy,

iodine making photographic film,
making antiseptics and other medicines,
making car headlights
in medicine to study the thyroid gland (see page 190).

ALL YOUR OWN WORK

The basic substances from which all chemicals are made are called ________.
A compound is formed when two or more ________ ________ together.
Atoms are made up from three basic particles called ________, ________ and ________.
If electrons are transferred from one atom to another an ________ bond is formed. If atoms complete their outer shells by sharing electrons a ________ bond is formed.
The periodic table is a list of all the ________ in order of increasing ________ ________.

Why are the noble gases (group 0) so unreactive?

Explain why simple covalent compounds are usually gases at room temperature.

a Table 11.6 compares the numbers of protons and electrons in sodium and chlorine atoms and gives the arrangement of electrons in each atom.

	sodium Na	chlorine Cl
number of protons	________	17
number of electrons	11	________
arrangement of electrons	2, 8, 1	2, 8, 7

table 11.6

i Complete table 11.6.
ii Which particles, apart from protons, are found in the nucleus of a sodium or chlorine atom?
iii In which group of the periodic table are the elements sodium and chlorine placed?
b Table 11.7 compares the numbers of protons and electrons in sodium and chloride *ions*.

	sodium Na^+	chlorine Cl^-
number of protons	11	17
number of electrons	10	18
arrangement of electrons	2, 8	2, 8, 8

table 11.7

i What change takes place when a sodium ion is formed from a sodium atom?
ii What change takes place when a chloride ion is formed from a chlorine atom?
c Figure 11.17 shows the arrangement of sodium and chloride *ions* in a crystal of sodium chloride.
i State *two* changes which take place in the crystal when melting occurs.
ii Why does sodium chloride have a high melting point?

[LEAG]

RADIOACTIVITY

HUMAN CREATION?

When people hear the word radioactivity they are often a little fearful. They think of nuclear power stations and atomic bombs. Radioactivity is something most of us are ignorant about, and we may feel it is best avoided. However, radioactivity is all around us and has been since the beginning of time. It is not new and is not a creation of the scientific age!

In the background

Throughout its history, the Earth and everything on it has been constantly exposed to radiations. Most of these **natural radiations** come from outer space in the form of cosmic rays. The upper atmosphere protects us from the harmful effects of these rays by absorbing some of them as they pass through it. The higher up in the atmosphere we go, the more radiation we receive. A city 2000 m (6000 ft) above sea level receives twice as much cosmic radiation as a city at sea level. The total amount of radiation you receive from the environment depends on many factors including the area in which you live. Aberdeen in Scotland, for example, has one of the highest values of background radiation in the UK. This is due to radiations from the granite rock formations on which the city is built. Some parts of Cornwall where villages have been built on granite also have a very high background radiation. You will continue to be bombarded by natural radiations all your life, wherever you live.

No sense!

One of the problems associated with radioactive materials is that the radiations they emit cannot be *tasted, touched, heard, smelled* or *seen*.

Your senses do not tell you that radiation is present. Your senses let you down as far as radiation is concerned. This can be a real hazard. All the time radiations pass through you there is the very real possibility they are doing damage to the cells of your body (see page 186). The body repair mechanisms can only cope with a certain amount of damage before they become overloaded. The process of self repair, by the cells of your body is an important one because there are radiations passing through you all the time – even now. Fortunately the naturally occurring radiations are small in number.

Discovery of radioactivity

Soon after Willhelm Roentgen had discovered X-rays in 1895, Henri Becquerel noticed that certain uranium salts emitted radiation. In 1896 he found that a sealed photographic plate which had been placed near a uranium compound would be 'foggy' when developed, even though the plate had not been exposed to light. Pierre and Marie Curie were also interested in radioactivity and they isolated radium in 1898. They both died from radiation related diseases.

THE RADIATIONS

There are four main types of radiation given out by radioactive materials: **alpha (α) particles, beta (β) particles, neutrons (n)** and **gamma (γ) rays**. The first three are particles and the fourth is a member of the electromagnetic spectrum. Each time an atom in a radioactive material **decays** it gives off one or more of the four radiations.

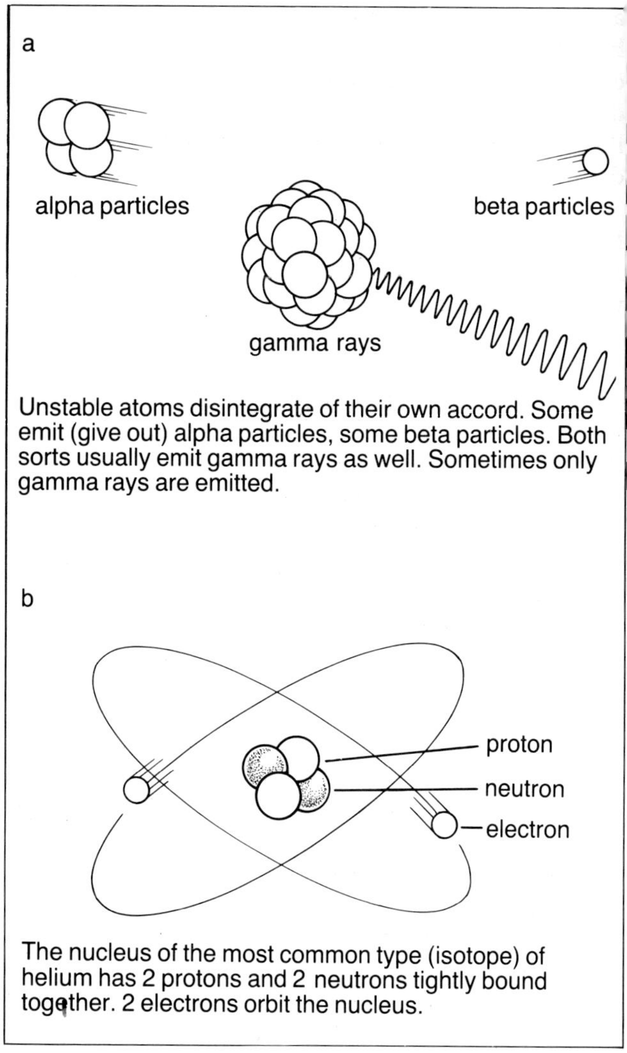

a

alpha particles

beta particles

gamma rays

Unstable atoms disintegrate of their own accord. Some emit (give out) alpha particles, some beta particles. Both sorts usually emit gamma rays as well. Sometimes only gamma rays are emitted.

b

proton

neutron

electron

The nucleus of the most common type (isotope) of helium has 2 protons and 2 neutrons tightly bound together. 2 electrons orbit the nucleus.

figure 12.1 a The three most common radiations
b The helium nucleus is made up in exactly the same way as an alpha particle

Alpha radiation

This is the most easily stopped of the set of four radiations. Alpha particles are the heaviest particles. Alpha particles consist of two neutrons and two protons. The total mass (nucleon number) of the particles is 4. Each of the protons has a positive charge of +1 so the total charge (atomic number) of the alpha particle is +2. By chance, there is an atomic nucleus that has the same structure as an alpha particle (see figure 12.1). The nucleus of a helium atom (symbol He) has four nucleons (2 protons and 2 neutrons). Because it has 2 protons it has 2 positive charges. An alpha particle and a helium nucleus are identical. Quite often you will hear of an alpha particle being called a helium nucleus. The symbol below is used to represent an alpha particle.

mass (nucleon) number —— 4
atomic number —— 2 He —— element symbol

$${}^{4}_{2}\mathrm{He}$$

Penetrating power (see figure 12.2)
Alpha particles will pass through
a a few centimetres of air
b a thin sheet of paper
c clothing (including gloves)
d skin.
Because alpha particles are charged (+2) they are deflected by magnetic fields.

Detection
All radiation detection methods (described overleaf) will record alpha particles.

Beta radiation

Beta particles are electrons that have come from the nucleus of an atom that is undergoing radioactive decay. Beta particles are light, fast and quite energetic. They are potentially much more dangerous than alpha particles. (Note: a stream of electrons is known by several different names depending where it originates from, e.g. cathode rays, electron beam, beta particles, beta rays, beta radiation.) An electron has a single negative charge (−1).
This symbol is used for a beta particle.

$${}^{0}_{-1}\mathrm{e}$$

Penetrating power (see figure 12.2)
Beta particles will pass through
a many centimetres of air
b human tissues
c several millimetres of aluminium
d thin lead sheet
Because beta particles are charged (−1) they are deflected by magnetic fields. They are more easily deflected than alpha particles.

Detection
All radiation detection methods (described overleaf) will record beta particles.

Neutron radiation

This is the rarest form of radiation. Neutrons are particles and as their name suggests they are neutral and have no charge. They have a mass of 1 and are small compared with an alpha particle, but very large compared with a beta particle. They are very fast and extremely energetic.
This symbol is used for a neutron.

$${}^{1}_{0}\mathrm{n}$$

Penetrating power (see figure 12.2)
Neutrons will pass through
a several centimetres (or more) of lead
b 2 metres (or more) of concrete
They cannot be deflected by magnetic fields.

Detection
The usual radiation detection methods do not record neutrons so it is difficult to detect them. Photographic emulsions will detect neutrons.

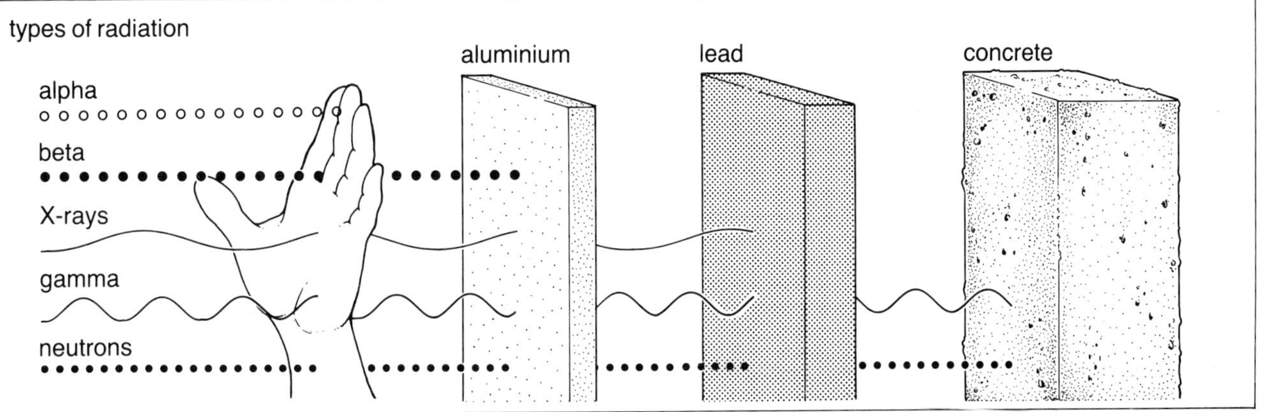

figure 12.2 Radiations are penetrating – some more than others

Gamma radiation

This 'label' is used to cover a wide range of energy waves belonging to the electromagnetic spectrum. Gamma radiation waves have very, very short wavelengths and very, very high frequencies. It is the high frequency that gives the wave its high energy value. The higher the energy value the greater the penetrating power of the wave.

Penetrating power (see figure 12.2)
a several centimetres (or more) of lead
b 2 metres (or more) of concrete
They cannot be deflected by magnetic fields.

Detection
All radiation detection methods will record gamma radiation.

Ionizing radiations

Alpha, beta and gamma radiations are known as **ionizing radiations**. They can cause atoms and molecules to become ionized (lose or gain electrons). In effect this causes the atoms or molecules to become chemically active. It is particularly dangerous if this happens in living tissues.

RADIATION DETECTORS (see figure 12.3)

Geiger–Muller (G–M) tubes

When a G–M tube is connected to a ratemeter or counter then the assembled device is called a Geiger counter (see figure 12.3a). Each time a piece of ionizing radiation enters a Geiger counter it causes an electrical pulse to be generated. This pulse triggers the ratemeter and the radiation is 'counted'. By this method it is possible to count up to 10 000 particles a second!

Though they register the presence of radiations they are unable to distinguish between the various types. Geiger counters are popular because they are relatively inexpensive and robust. They are easily used indoors and outdoors.

Photographic emulsions

Radiations have the same effect on special photographic emulsions as light has on film negative or photographic paper. They cause the material to change chemically. The more radiation the greater the change, and the more energy the radiation has the greater the change. Unlike many detectors it is not possible to obtain an instantaneous count as the emulsions have to be developed after exposure to

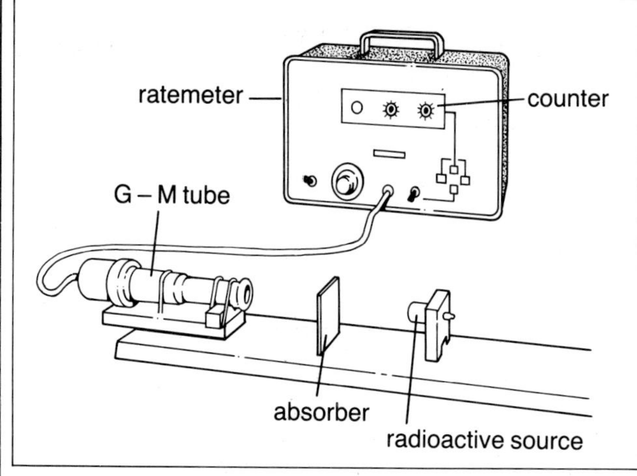

figure 12.3 **a** *A Geiger counter used to detect ionizing radiations*

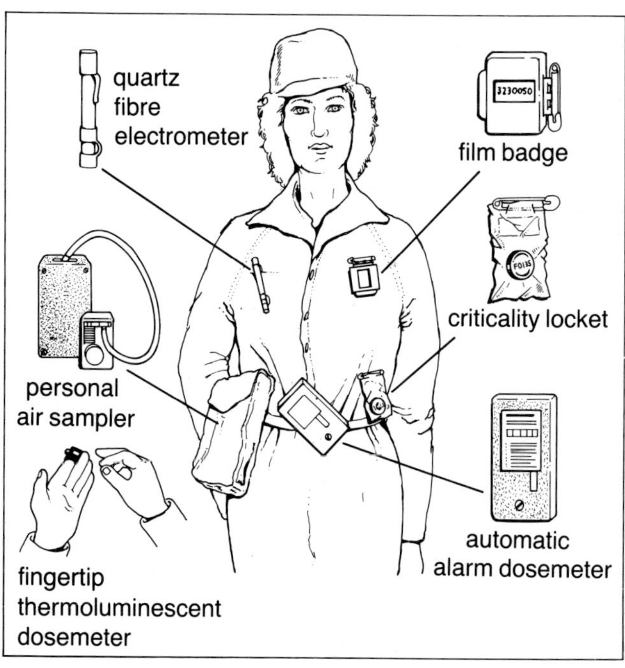

b *Methods used to monitor the amount of radiation a person is exposed to*

radiation. Various filters are placed in front of the emulsions to identify different radiations.

Cloud chambers

These are really only useful for the detection of alpha and beta radiations. They usually contain saturated methylated spirit vapour. This is kept very cold using dry ice. When ionizing radiation passes through the chamber it creates a 'vapour trail', rather like that of a high-flying aircraft. The trail is due to the vapour condensing around the ions which the radiation creates. The pathways formed by beta particles tend to wander far more than the pathways of alpha particles because beta particles are much lighter and more easily knocked off their course by the molecules in the chamber.

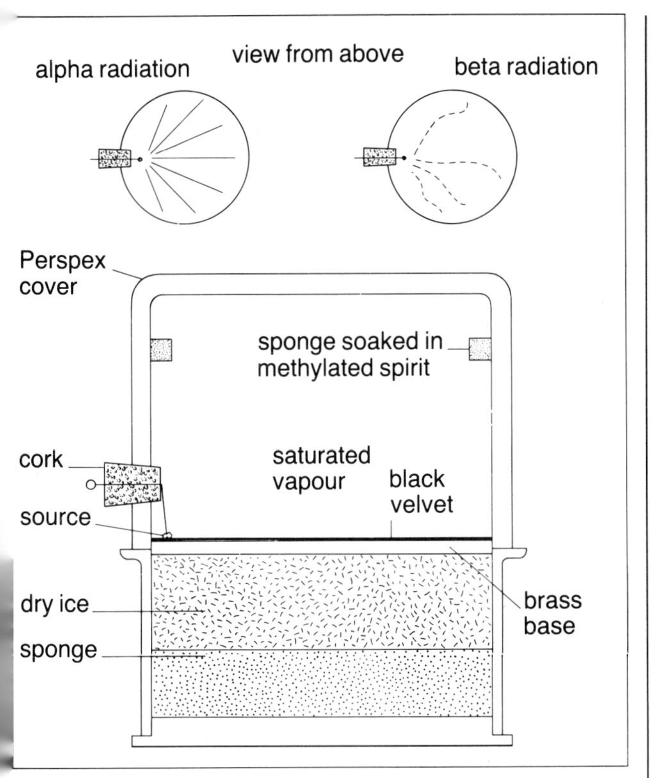

figure 12.4 The Wilson cloud chamber

DECAY SERIES

The nucleus of a decaying atom will continue to emit one or more of the four radiations until it reaches a **stable state**. To arrive at this stable state may involve several stages and different types of radiation. These changes can be written down as a 'decay series'. Elements in a decay series are shown with their mass (nucleon) number and atomic number in the same way as radiations, e.g.

mass (nucleon) number — atomic number — $^{238}_{92}U - ^{4}_{2}He = ^{234}_{90}Th$

After emitting an alpha particle an atom of uranium-238 becomes an atom of thorium-234.

Isotopes

In the decay series for naturally occurring uranium there are three references to lead (Pb).

$$^{214}_{82}Pb \rightarrow ^{210}_{82}Pb \rightarrow ^{206}_{82}Pb$$

How is it possible to have three different varieties of lead? The differences are in the mass (nucleon) numbers (see page 171). The atomic number remains the same each time. The third symbol is the naturally occurring stable isotope of lead. The others are radioactive isotopes. They have additional neutrons in the nucleus which makes them unstable. All there varieties have the same chemical properties.

Half-life

Some changes in the decay series take place very quickly whereas others take a long time. The rate of decay of radioactive isotopes is expressed by their **half-life** ($t_{\frac{1}{2}}$). The half-life of a substance is the time taken for its radioactivity to **decrease to half its value**: i.e. for half the radioactive nuclei to decay. The time for which a material is radioactive is many times its half-life. (In fact the material will be radioactive until the last nucleus splits up.) Each radioactive isotope has its own half-life value. This can range from fractions of a second to tens of thousands of years. The half-life value is **totally independent** of temperature, concentration, chemical composition etc.

material	half-life
uranium-235	700 000 000 years
radium-226	1600 years
iodine-131	8 days
protactinium-234	72 seconds

table 12.1

Figure 12.5 shows a graph of radioactivity in counts per second against time in minutes. The half-life of this substance is 5 minutes – it took 5 minutes for the radioactivity to drop from 50 counts per second to 25, another 5 minutes to 12.5, and so on.

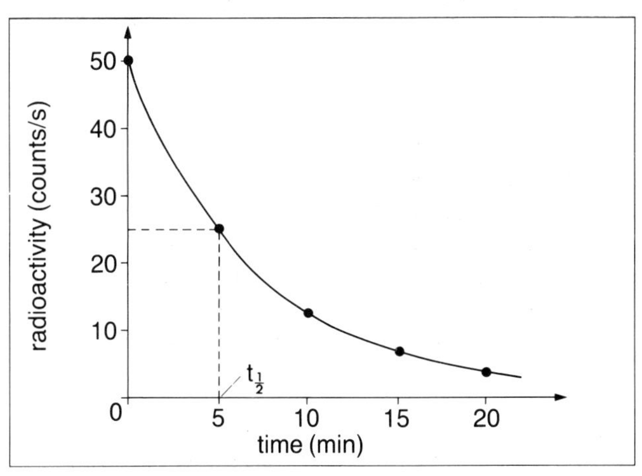

figure 12.5 A typical decay curve

RADIATION AND LEUKAEMIA

Leukaemia is a **cancer** affecting the bone marrow or lymph tissue. As a result of this disease a very large number of white blood cells is produced, preventing red blood cell production. Survivors of the atom bomb attacks on Hiroshima and Nagasaki (9 June 1945) suffered an increased risk of leukaemia.

For those who received a dose of radiation of more than one gray (Gy) the risk of leukaemia was seven times greater than the national average (see table 12.2).

radiation dose	death rate per 100 000
high	18.8
moderate	11.7
low	6.7

table 12.2

Although radiation can cause leukaemia, not everyone who receives a high dose of radiation develops the illness. Scientists are carrying out research to discover what causes this disease to attack some people and not others.

RADIATION AND MEDICINE

X-ray radiation can cause ill health if we have too much but it is also a vital tool in medicine. Used with care, it can be beneficial to health. X-rays pass through living tissue. If they then fall on a sensitive film an image of the inside of our bodies is produced. The image is formed from the shadows of the internal organs and structures that do not let the X-rays pass through so easily. Sometimes the image is made clearer by giving the patient a substance that is opaque to X-rays, for example, barium sulphate.

Application 12.1

ATOMIC BOMBS AND THE LESSON OF HIROSHIMA

The first atomic bombs used uranium-235 as their fuel. In a small amount of radioactive uranium-235, most neutrons released by the random decay of a nucleus escape and do not trigger off the fission (splitting up) of other nuclei. If, however, a large block is used, neutrons produced by one decaying nucleus collide with other active nuclei and cause a chain reaction. The amount of material required for a chain reaction is called the **critical mass**.

In a fusion bomb, amounts less than the critical mass are forced together to make an amount large enough to cause a chain reaction. This results in the release of nuclear energy, causing a tremendous explosion.

figure 12.6 *Atomic explosion over Nagasaki*

The hydrogen bomb is even more powerful and considerably more destructive. Here a fission explosion is used to generate enough heat for a fusion (forcing together) reaction to take place. The fusion reaction is between two atoms of deuterium. (Deuterium is an isotope of hydrogen.)

$$^{2}_{1}H + ^{2}_{1}H = ^{3}_{2}He + ^{1}_{0}n$$

The energy produced by a fusion explosion is many thousands of times greater than that caused by a fission explosion, because a great deal more nuclear energy is released by the fusion reaction.

At 8.15 a.m. on 6 August 1945 a United States B-29 bomber plane released an atomic bomb, nicknamed 'Little Boy', over the Japanese city of Hiroshima. The force of the explosion flattened buildings over an area of 20 square kilometres. Within seconds a fireball, at a temperature of 3000 °C. had spread over the city. One hundred thousand people were killed instantly. Over the following months many more people suffered from the high dose of radiation they received from the explosion. Radiation affects dividing cells particularly, so the first symptoms of radiation sickness were loss of hair, skin problems and weakness and vomiting due to damage to the gut lining. The bone marrow was also affected, and stopped producing red and white blood cells. When the existing red blood cells were all worn out, after about six weeks, there were no new ones to take their place so oxygen could not be carried around the body. Lack of white blood cells made the victims prone to infections. Those who survived radiation sickness suffered long-term affects after many years. These included leukaemia and cancer of the thyroid gland, salivary gland and breast, and deformities in babies born to victims.

Why did radiation from the atomic bombs have such a disastrous effect on human life? When radiation passes through cells, it causes the molecules to break up. If radiation passes through the nucleus it damages the vital DNA molecules which store the genetic information for building the cell. As a result the cell may either die or fail to divide. In some cases the damaged DNA molecules may carry the faulty information into future generations. Cells that are rapidly dividing are particularly vulnerable to the effects of radiation. They are found in the skin, hair follicles, blood vessels, intestine and bone marrow.

Radiotherapy

Radiation can kill cells inside living organisms. This is useful when treating certain types of cancer. A beam of intense radiation can be focused on the cancer cells. Since cancer cells are dividing, they are more sensitive to X-radiation than normal cells, so they are destroyed more easily. Neutron radiation can also be used in radiotherapy.

Safety precautions

In the early part of this century the first radiographers were not fully aware of the potential hazards of the X-rays they were using, and many received dangerous doses of radiation. Since then safety precautions have been introduced for both the radiographers and the patients being X-rayed. Protective 'lead rubber' can be placed over a person's body to prevent stray X-rays passing through. This ensures that the smallest area possible is exposed to radiation.

The gonads (ovaries and testes) are protected against X-radiation where possible. This is because the dividing cells producing sperm and ova are particularly sensitive to damage by X-rays. Also any alteration in their DNA would be copied in the genes of any offspring. This might result in a child being handicapped in some way.

Very low doses of radiation received during an X-ray will not affect adult cells which are not dividing. However, they may harm the sensitive cells of an embryo developing in a mother's uterus. It is therefore important that pregnant women should not be exposed to X-rays.

NUCLEAR POWER

An increasing amount of the world's energy is being supplied by nuclear power. In Britain in 1984 11% of the electricity produced was from nuclear power stations. The energy is produced by **nuclear fission** reactions (see figure 12.7a) As the large nuclei split they release large amounts of energy. The most common fuel is uranium. When a neutron collides with a uranium-235 nucleus it causes it to split forming two smaller nuclei. As well as energy, more neutrons are released. These then collide with other uranium nuclei and so set off a chain reaction (see figure 12.7b).

The uranium is loaded in fuel rods which are lowered into the reactor core. Carbon **moderators** (graphite rods) are used to **slow down** the neutrons. A slow-moving neutron is much more effective in causing fission in the remaining uranium nuclei. Central rods of cadmium, or steel with a high boron content, are used to absorb neutrons. If the reactor becomes too hot these control rods are lowered into it. In the case of an emergency these rods can be lowered completely and the reactor shut down in a matter of seconds. The heat produced in the core is transferred

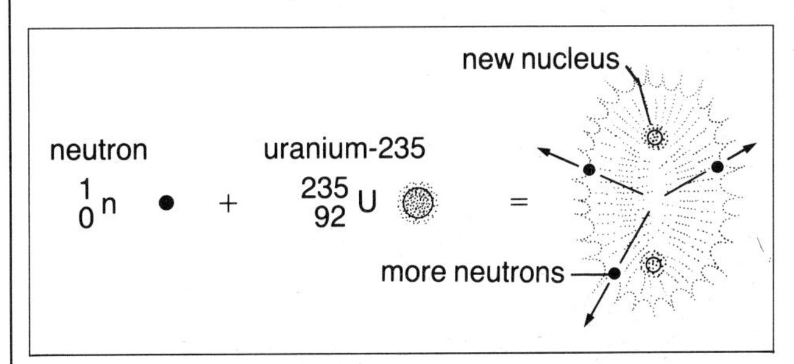

figure 12.7 a Nuclear fission

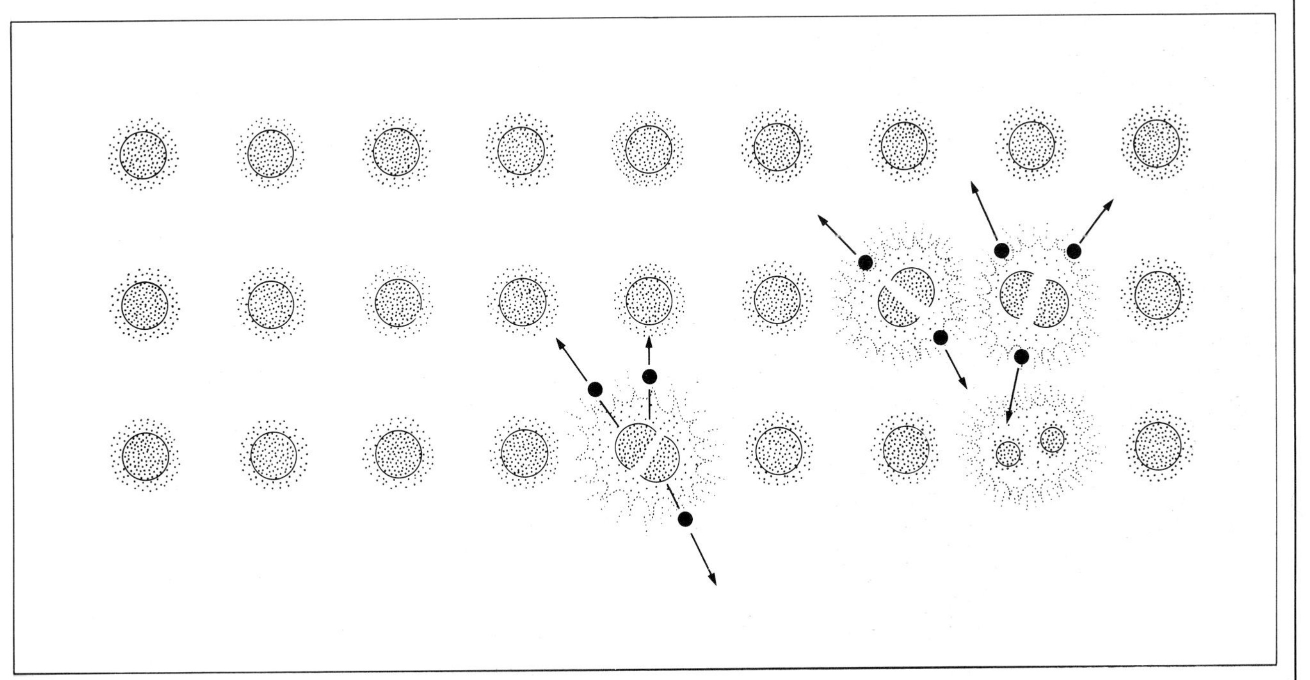

b Chain reactions

from the reactor by a heat exchanger. The steam produced drives conventional turbines. Water, carbon dioxide and liquid sodium are all used as cooling fluids (see figure 12.8).

The main problems with this type of nuclear power station are

a the uranium is difficult to obtain and concentrate from its ore, making the process expensive

b large amounts of highly radioactive waste are produced. This cannot be destroyed and its disposal is a major environmental issue.

Time for thought

"If the fission type of nuclear power station is to be developed further the problem of disposal of nuclear waste must be overcome to everyone's satisfaction." How might this be done?

INVESTIGATION 12.1

THE EFFECTS OF RADIATION

Take two groups of barley seeds, one of them treated with ionizing radition, the other to act as a control group.

a Prepare two dishes by placing a layer of wet cotton wool in each.

b Place each of the two groups of seeds on a separate dish and keep them well lit.

c Observe the two groups of seeds at regular intervals over a period of several weeks. Ensure that the cotton wool does not dry out.

Questions

1 What differences do you notice between the two groups of seeds and the seedlings they produce?

2 How do you explain these differences?

NUCLEAR FUSION

In a nuclear fusion reaction, energy is produced, not by splitting larger nuclei but by fusing smaller nuclei together e.g. the fusion of two nuclei of deuterium (an isotope of hydrogen).

$$^{2}_{1}H + ^{2}_{1}H = ^{3}_{2}He + ^{1}_{0}n$$

This type of reaction produces vast amounts of energy. The only problem is that fusion reactions take place at temperatures in excess of 10 000 000 °C. It is by nuclear fusion that the sun and all the other stars produce their energy. How can we cope with such high temperatures here on Earth and how could the substances produced be contained? Considerable research is being carried out to find out how a controlled fusion reaction could be achieved and the energy harnessed. The raw materials are readily available in vast quantities. Perhaps our energy of the future will come from nuclear fusion reactors?

RADIOACTIVE WASTE

Classification

Radioactive waste is classified according to how active it is. The three classifications or levels are **low, intermediate (medium)** and **high**.

Low level wastes contain mostly short-lived radioactive nuclei with only trace (very small) quantities of materials with long half-lives. They are generally of such low radioactivity that they can be released directly into the environment or, after

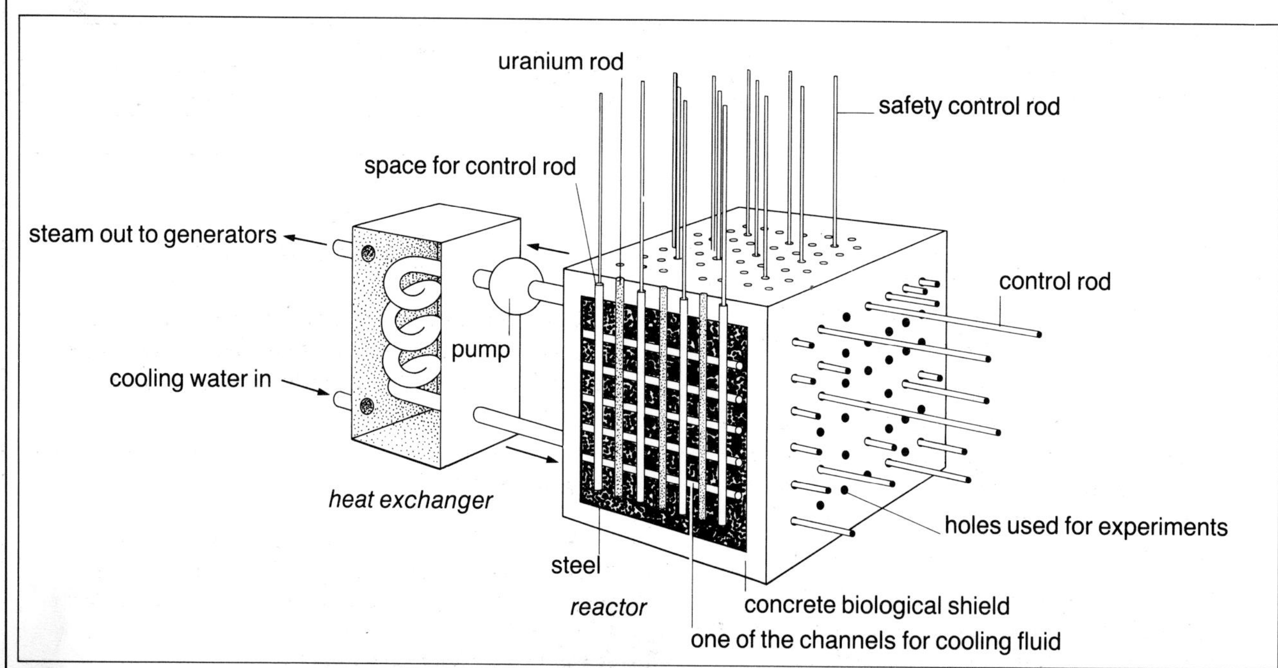

figure 12.8 Nuclear power generation

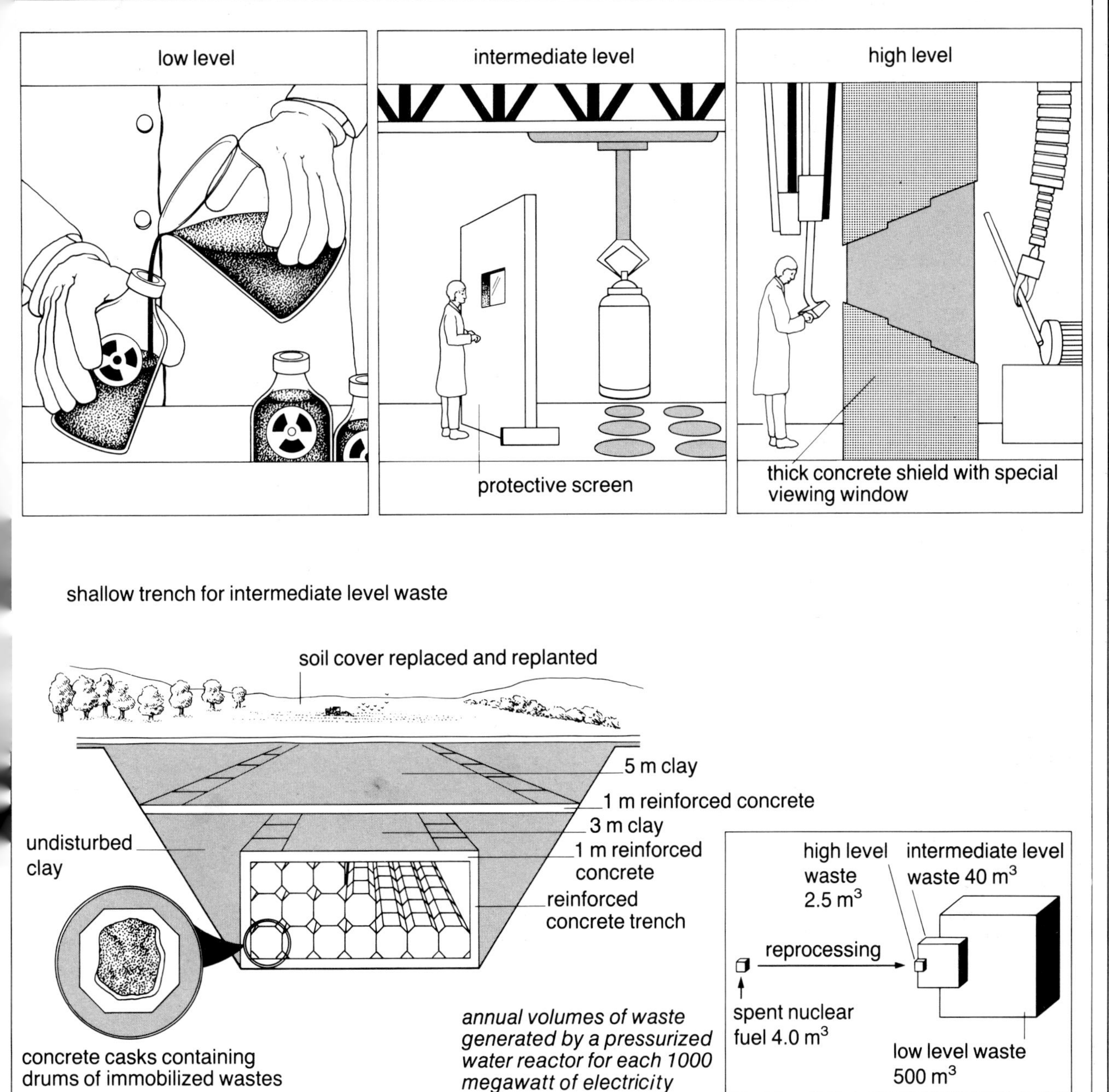

figure 12.9 Keeping waste under control

packaging, disposed of immediately by shallow land burial or at sea. They include almost all the wastes from medical and industrial uses of radioactive materials as well as some of that from the nuclear industry.

Intermediate level wastes contain larger quantities of fission and activation products, and sometimes compounds called **actinides** which have long half-lives. They are characterized by a low rate of radioactive emission but considerable bulk.

High level wastes contain most of the fission products and actinides (except plutonium) from the nuclear fuel cycle. They have a high rate of radioactive emission but low bulk (see figure 12.9).

Storing radioactive waste

Most intermediate level wastes and all high level wastes are currently stored at various nuclear sites operated by the United Kingdom Atomic Energy Authority.

Time for thought

In 1980 the Mayor of Hiroshima made the following plea: "It is now high time for us all to call for the solidarity of all mankind and to shift our common path from self-destruction towards survival."
Do you agree? If so, what can be done?

UNIT 12

Application 12.2

RADIATION – PROBLEMS AND SOLUTIONS

Problem How do you check that the support piles for North Sea oil rigs have been 'planted' properly in the sea bed?
Solution Use radioactive cement around the bottom of the piles and check the pathway of the radiations.

Problem How do you tell the date of wooden archaeological remains?
Solution Radiocarbon dating is a well accepted dating technique. A very tiny proportion of all carbon is radioactive. Measure the activity of an old sample and compare it with the activity of a modern sample of the same wood. The half-life of the carbon is known, so from the amount of radiation the old sample has lost, you can work out how old it is. This technique can be used with almost any 'artefact' that takes in carbon during its lifetime.

Problem Plastic syringes are much cheaper to manufacture than glass ones. How do you sterilize plastic?
Solution Seal the syringes in plastic bags and then subject them to intense gamma radiation. The radiation kills all organisms in the bags (without making the syringes radioactive) leaving the contents completely sterile.

Problem How do you test for even thicknesses in products such as tin foil?
Solution As the foil comes off the production line bombard it with radiation and measure how much is transmitted through the material. If the amount of transmitted radiation varies then the thickness of the foil must be changing.

Problem Doctors want to measure how much iodine in a patient's diet is being taken up by the thyroid gland.
Solution Give the patient a known quantity of mildly radioactive iodine in their diet. After a few days an amount of radiactivity given off by the thyroid gland is measured. This allows the amount of iodine taken up by the thyroid gland to be measured.

Problem Farmers want to get rid of fruit flies that are ruining their crops. They do not want to use pesticides that might pollute the environment.
Solution Zoologists produce a population of sterile male fruit flies by bombarding them with gamma radiation. When these males are released they mate with normal females but no offspring are produced. The fruit fly population will gradually die out and the crops will be protected.

Problem Doctors want to remove a cancer tumour that has developed in a patient's neck. X-ray treatment has not worked and surgery is impossible because of where the tumour is.
Solution A fine beam of fast moving neutrons can be focused accurately on the tumour cells. This kills them rapidly and helps to cure the patient.

ALL YOUR OWN WORK

1 The graph in figure 12.10 shows the variation in activity recorded by a Geiger counter used to measure the thickness of newsprint paper. Describe how the thickness varies along the length of the newsprint.

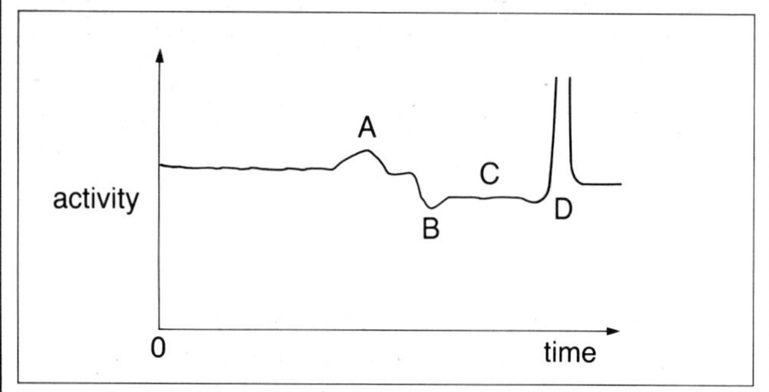

figure 12.10 Measuring paper thickness with radiation

2 Survivors of the atomic bomb explosion at Nagasaki included pregnant women. Table 12.3 separates these women into three groups and shows what happened to their children.

(The hypocentre is the exact point on the ground above which the bomb exploded.)

a Which group of mothers received
i the highest radiation dose
ii the lowest radiation dose?

b Give reasons for your answer to **a**.

c Calculate the following for each of the three groups
i the percentage of miscarriages among mothers
ii the number of mothers whose children survived to reach 1 year old
iii the percentage of mothers whose children survived to reach 1 year old. What explanation can you suggest for this?

group	number of pregnancies	number of miscarriages	number of still-births	number of infant deaths 0–1 year	children with mental retardation
mothers within 2000 m of hypocentre who suffered radiation sickness	30	3	4	6	4
mothers within 2000 m of hypocentre who did not suffer radiation sickness	68	1	2	3	1
mothers who were exposed within 4–5000 m of hypocentre	113	2	1	4	0

table 12.3

CHEMICALS IN ACTION

ACIDS

When we hear the word **acid**, we usually think of a corrosive liquid that eats things away, substances to be avoided.

While it is true that some acids will burn your skin and rot your clothes, others like the acids in fruit juice are quite safe to drink. Acids that we drink usually have a sharp or sour taste.

What do acids have in common, and how can we detect them?

Table 13.1 lists some common acids.

acid	where it is found
sulphuric acid nitric acid	the common laboratory acids
hydrochloric acid	occurs in the stomach
citric acid ascorbic acid (vitamin C) tartaric acid	oranges and lemons
ethanoic acid (acetic acid)	vinegar

table 13.1

Indicators

An indicator is a substance that gives different colours in acids and in **alkalis** (an alkali is the chemical opposite of acid). Many of the dyes obtained from plants can be used as indicators.

Litmus

Litmus, a dye obtained from plants, is often used as an indicator. You may have already come across litmus paper. This is paper rather like filter paper or blotting paper that has been soaked in a solution of the dye. When you want to test a substance to see if it is acid, put a drop on a piece of blue litmus paper. If the paper turns red the substance is an acid.

If you want to test a substance to see if it is an alkali, put a drop on a piece of red litmus paper. If the paper turns blue the substance is an alkali.

Neutral substances (substances that are neither acid nor alkali) do not alter the colour of litmus.

INVESTIGATION 13.1

EXTRACTING AND TESTING THE COLOUR FROM RED CABBAGE

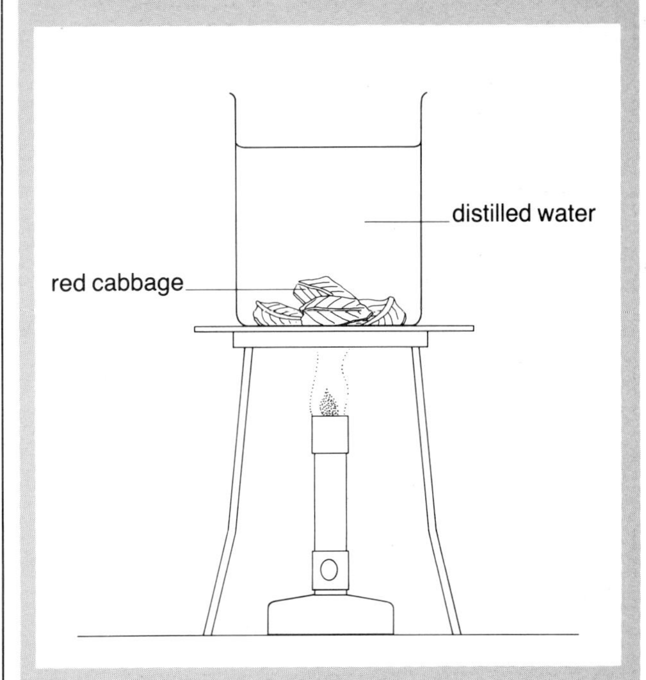

figure 13.1 Extracting the dye from red cabbage

1. Tear up a square of red cabbage leaf about 5 cm^2 into small pieces.
2. Place the pieces of red cabbage into a 250 cm^3 beaker half full of distilled water.
3. Boil the water by placing the beaker on a tripod and gauze and heat it using a medium bunsen flame (see figure 13.1). Boil the water for about 15 minutes. (Do not let all the water boil away.)
4. As soon as the beaker is cool enough to handle safely, quarter fill three test tubes with the coloured water.
5. To test tube 1, add a few drops of ethanoic acid. To test tube 2, add a few drops of distilled water. To test tube 3, add a few drops of dilute alkali (*Care* – use very dilute sodium hydroxide).
   Note the colour in each case.

Universal indicator

Universal indicator is a mixture of several different indicators. This means that it can be one of many colours. It is able to indicate not only whether a substance is acid, alkaline or neutral, but also how strong an acid or alkali it is. Figure 13.2 (overleaf) shows the colours of universal indicator.

You will notice that corresponding to each colour there is a number. This is called the **pH**. pH is a way of describing how acid or alkaline a substance is. The pH scale is from 1 to 14. Numbers less than 7 indicate acids, numbers greater than 7 indicate alkalis and 7 is neutral.

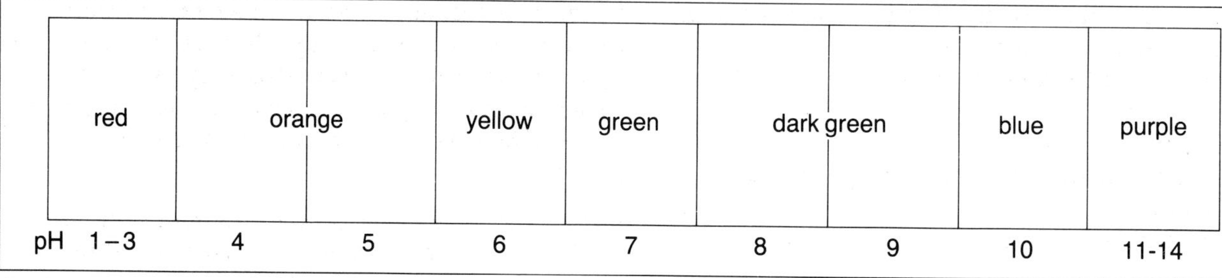

figure 13.2 Colours of universal indicator

acids neutral alkalis
1 2 3 4 5 6 7 8 9 10 11 12 13 14

The further away from 7 the stronger the acid or alkali is, for example pH 1 will be a strong acid, pH 5 a weak acid, pH 9 a weak alkali and pH 14 a strong alkali.

INVESTIGATION 13.2

USING UNIVERSAL INDICATOR

Test the following substances with universal indicator.

sodium chloride	sugar
soap	dilute sulphuric acid*
washing up liquid	distilled water
sodium carbonate	ethanoic acid
citric acid	sodium hydrogencarbonate
ammonia solution*	dilute sodium hydroxide*

Those marked* need special care

1. Cut up the indicator paper into pieces about 1 cm long.
2. Spread 12 pieces out on a piece of scrap paper and test each substance in turn. If the substance is a liquid, dip a *clean* glass rod into the liquid and then touch the rod on to the indicator paper. Wash the rod straight away.
3. If the substance is a solid, make a solution first by adding a small amount of the solid to a test tube one-quarter full of distilled water. Carefully shake or stir the mixture until the solid has dissolved. Test a drop of the liquid as before.
4. Make a table of your results like table 13.2.

Table 13.2 shows a set of results for some of the substances tested in investigation 13.2.

substance	colour of indicator	pH	type of substance
salt	green	7	neutral
ammonia solution	blue	11	alkali
citric acid	orange	5	weak acid
sodium hydroxide	purple	14	strong alkali
distilled water	green	7	neutral

table 13.2

What makes a substance an acid?

By using an indicator we can soon find out whether a substance is an acid or not, but this does not tell us why some things are acids, or what acids have in common.

Apart from their effect on indicators, acids have four main common reactions. They react

a with a fairly reactive metal (e.g. magnesium, zinc, iron)
b with a metal oxide (a base)
c with a metal carbonate
d with an alkali (a soluble base).

INVESTIGATION 13.3

INVESTIGATING THE REACTIONS OF A DILUTE ACID

Warning great care must be taken when using acids and alkalis. Goggles must be worn at all times and if any acid is spilled it must be wiped up at once with a wet cloth. If you get any acid on you or your clothes, wash it off at once with plenty of water and tell your teacher.

1. Place a piece of magnesium ribbon 2 cm long in a test tube one-quarter full of dilute sulphuric acid. Note what happens. Test the gas given off with a lighted spill (see page 178).
2. Add a small measure of copper(II) oxide to a test tube one-quarter full of dilute sulphuric acid. Stand the test tube in a beaker of boiling water. Note any reaction.
3. Add a small measure of copper(II) carbonate to a test tube one-quarter full of dilute sulphuric acid. Note any reaction. Test the gas given off with a lighted spill.
4. Add 10 drops of dilute sulphuric acid to a test tube and then add about 5 drops of universal indicator solution. Add sodium hydroxide solution a drop at a time to the test tube and see whether you can make a neutral solution.
5. Repeat these four reactions using dilute hydrochloric acid instead of dilute sulphuric acid.
6. Repeat the four reactions using dilute nitric acid.

Question

What do the three acids have in common?

The results of the investigation 13.3 show that these three common laboratory acids

a react with a fairly reactive metal and give off hydrogen.
b dissolve metal oxides
c react with metal carbonates and give off carbon dioxide.
d are made neutral (neutralized) by alkalis.

How can we account for these common properties? Before we can really answer this question we must look at the part water plays in the reaction of acids.

Hydrogen chloride is a covalent compound. It is a steamy gas which will dissolve in both water and methyl benzene (a liquid similar to petrol). The two solutions, however, have very different properties.

Carry out investigation 13.4 (below), then look at table 13.3 which shows some typical results.

test	result in methyl benzene	result in water
thermometer	no change	temperature rises
universal indicator	no change	turns red
magnesium	no reaction	gives off bubbles of gas
marble	no reaction	gives off bubbles of gas
electrical conductivity	does not conduct	conducts

table 13.3

From investigation 13.4 we can see that hydrogen chloride acts as an acid only when it is in water. The solution in water conducts electricity so it must contain ions (see page 172) whereas the solution in methyl benzene does not conduct so contains no ions.

The temperature change on the formation of the solution in water indicates a change in bonding. When the hydrogen chloride dissolves in water it changes into **hydrogen ions** and **chloride ions**.

$$\text{hydrogen chloride} \xrightarrow{\text{water}} \text{hydrogen ions} + \text{chloride ions}$$
$$HCl \rightarrow H^+ + Cl^-$$

It is the hydrogen ions that are responsible for acid reactions. Water has to be present before anything can act as an acid because water allows the acid to form ions.

The more hydrogen ions there are the stronger the acid will be. pH is a measure of the hydrogen ion concentration.

Why acids have four common reactions

Dilute acid + fairly reactive metal
This reaction works with magnesium, zinc and iron. Magnesium reacts quickest and iron slowest. In each case hydrogen gas is given off. The metal takes the place of the hydrogen in the acid. The compound formed when the hydrogen of an acid is replaced by a metal is called a **salt**.

INVESTIGATION 13.4

WATER IN ACID REACTIONS
This experiment should be demonstrated by your teacher.

Hydrogen chloride gas can be made by reacting common salt (sodium chloride) with concentrated sulphuric acid, and dried by passing through concentrated sulphuric acid (see figure 13.3) Two *dry* gas jars full of hydrogen chloride gas are needed for this experiment.

1 A thermometer is dipped in some methyl benzene, the temperature noted and then the same thermometer held without wiping it in one of the gas jars of hydrogen chloride. Any change in temperature is noted. (Methyl benzene is very inflammable and is also a very good solvent for plastics and synthetic fibres.) A second thermometer is dipped in water, the temperature noted and then the thermometer held without wiping it in the gas jar of hydrogen chloride. Any temperature change is noted.
2 About 1 cm depth of methyl benzene is placed in the first jar and about 1 cm depth of water in the second gas jar. The coverslips are replaced and the gas jars gently shaken to dissolve the hydrogen chloride. The following tests are carried out on each solution.

a Put a spot of the solution on a piece of universal indicator paper that has been dried by warming it.
b Test a small amount of each solution to see if it conducts electricity (use a watch glass).
c Place a small piece of magnesium ribbon in each gas jar.
d Place a small piece of marble (calcium carbonate) in each gas jar.

Question
Make a table to show the results of each test in the methyl benzene solution and the water solution.

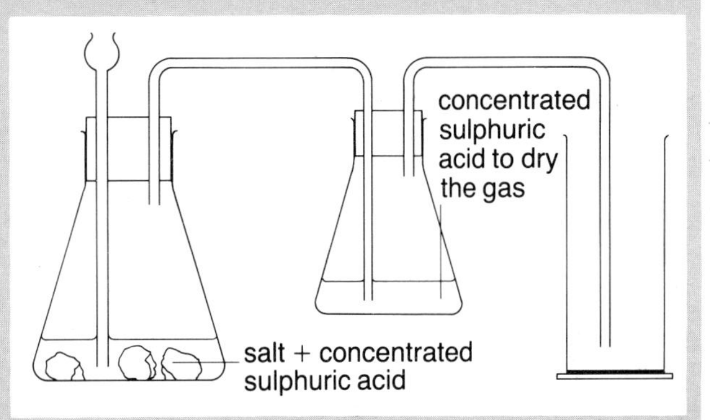

figure 13.3 Making hydrogen chloride

UNIT 13

Sulphuric acid forms **sulphates**, hydrochloric acid forms **chlorides** and nitric acid forms **nitrates**, for example

magnesium + dilute sulphuric acid → magnesium sulphate + hydrogen

$$Mg + H_2SO_4 \rightarrow MgSO_4 + H_2$$

This reaction does not work very well with dilute nitric acid.

Dilute acid + metal oxide (base)

This reaction works well for nearly all metal oxides. Usually the acid has to be warmed. In this reaction a salt and water are formed, for example

copper oxide + dilute nitric acid → copper nitrate + water

$$CuO + 2HNO_3 \rightarrow Cu(NO_3)_2 + H_2O$$

Dilute acid + metal carbonate

Most metals carbonates dissolve easily in cold dilute acids. Carbon dioxide gas is given off and a salt and water are formed, for example

calcium carbonate + dilute hydrochloric acid → calcium chloride + carbon dioxide + water

$$CaCO_3 + 2HCl \rightarrow CaCl_2 + CO_2 + H_2O$$

Dilute acid + alkali (a soluble base)

Dilute acids are neutralized by alkalis to form a salt and water, for example

sodium hydroxide + dilute hydrochloric acid → sodium chloride + water

$$NaOH + HCl \rightarrow NaCl + H_2O$$

The above reactions can be summarized as follows.

1 acid + fairly reactive metal → salt + hydrogen
2 acid + metal oxide → salt + water
3 acid + metal carbonate → salt + water + carbon dioxide
4 acid + alkali → salt + water

These four reactions work providing that the acid is dilute and the salt formed is soluble.

Remember that the reactions of a dilute acid are really the reactions of hydrogen ions.

Ionic equations show how the hydrogen ions react.

1 $Mg + 2H^+ \rightarrow Mg^{2+} + H_2$
2 $CuO + 2H^+ \rightarrow Cu^{2+} + H_2O$
3 $CO_3^{2-} + 2H^+ \rightarrow CO_2 + H_2O$
4 $H^+ + OH^- \rightarrow H_2O$

Note this simplest equation for the reaction between an acid and an alkali.

Having seen what makes a substance an acid we can now give more detailed definitions of an acid. An acid is a compound containing hydrogen that can be replaced by metal to form a salt. It has a pH of less than 7. It is a solution containing a high concentration of hydrogen ions. An acid is a proton donor (a hydrogen ion, H^+, is a proton).

INVESTIGATION 13.5

PREPARING SALTS

These experiments may be used for practical assessment.

Acid + fairly reactive metal: making magnesium sulphate

Remember all the normal safety precautions when using acids, especially to wear goggles all the time. Make sure that there are no Bunsens lit or other flames nearby during the first stage of the experiment, i.e. while hydrogen is being produced.

1 Measure about 25 cm^3 of dilute sulphuric acid into a 100 cm^3 glass beaker.
2 Add three pieces of magnesium ribbon about 2 cm long to the acid. Take care not to breathe in the acid spray given off with the hydrogen and make sure that the laboratory is well ventilated.
3 When all the magnesium has dissolved add two or three more similar sized pieces of magnesium.
4 Continue adding magnesium until no more will react. Do not have more than three pieces of magnesium in the beaker at any time because this will cause the acid to become too hot and give off too many unpleasant fumes.
5 When the reaction has stopped, filter off the excess magnesium and evaporate the solution to about half its volume.
6 Leave the solution to crystallize.

magnesium + dilute sulphuric acid → magnesium sulphate + hydrogen

$$Mg + H_2SO_4 \rightarrow MgSO_4 + H_2$$

Acid + metal oxide: making copper sulphate

The basic method is the same as in the last experiment, i.e. adding excess solid to the acid to ensure that all the acid is used up, and then filtering the mixture to remove the excess solid. This reaction usually needs warming.

1 Place about 25 cm^3 of dilute sulphuric acid in a 100 cm^3 glass beaker.
2 Carefully warm the acid using a Bunsen burner, tripod and gauze until the acid just starts to boil. Turn off the Bunsen burner but leave the beaker on the tripod and gauze. Take great care not to knock the beaker over as hot acids are very corrosive.
3 Add a spatula measure of copper(II) oxide to the hot acid.
4 When all the copper(II) oxide has dissolved, add another measure of copper(II) oxide.
5 Continue adding copper(II) oxide a measure at a time until no more dissolves, even with stirring.
6 Filter the mixture and leave the solution to crystallize.

copper oxide + dilute sulphuric acid → copper sulphate + water

$$CuO + H_2SO_4 \rightarrow CuSO_4 + H_2O$$

Acid + metal carbonate: making lead nitrate

This reaction works well at room temperature. Again excess solid is added to make sure that all the acid is used up.

1 Place about 25 cm^3 of dilute nitric acid in a 100 cm^3 glass beaker.
2 Add a spatula measure of lead carbonate to the acid.
3 When the lead carbonate has all reacted, add another measure of lead carbonate. Continue to add lead carbonate a measure at a time until no more will react.
4 Filter the mixture into a clean evaporating basin.
5 Evaporate the solution over a water bath (see figure 13.4) until all the water has gone.

Lead nitrate crystals do not contain any water of crystallization so the solution can be evaporated to dryness. However, lead nitrate decomposes when it is heated and so the evaporation needs to be at a relatively low temperature, hence the water bath.

lead carbonate	+	dilute nitric acid	→	lead nitrate	+	carbon dioxide	+	water
$PbCO_3$	+	$2HNO_3$	→	$Pb(NO_3)_2$	+	CO_2	+	H_2O

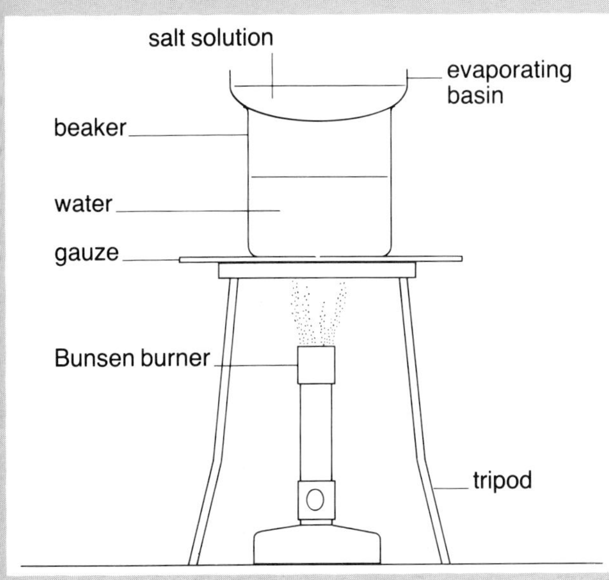

figure 13.4 *Evaporating over a water bath*

Acid + alkali: making sodium chloride
This experiment cannot be done in the same way as the other three salt preparations because both the acid and the alkali are solutions. This means that you cannot add an excess of alkali and then filter it off. A totally different technique has to be used, called **titration**. The experiment uses an indicator to show when the acid and alkali have neutralized each other. The experiment is then repeated with exactly the same quantities but without the indicator. Repeating the experiment accurately makes this quite a difficult experiment to do.

Before you start the experiment make sure that all the apparatus is clean. Remember that alkalis can be as corrosive as acids so take care.

1 Using a pipette and a pipette filler (see figure 13.5) place 25 cm^3 of dilute sodium hydroxide solution in a 100 cm^3 conical flask.
2 Add 5 drops of phenolphthalein indicator to the sodium hydroxide.

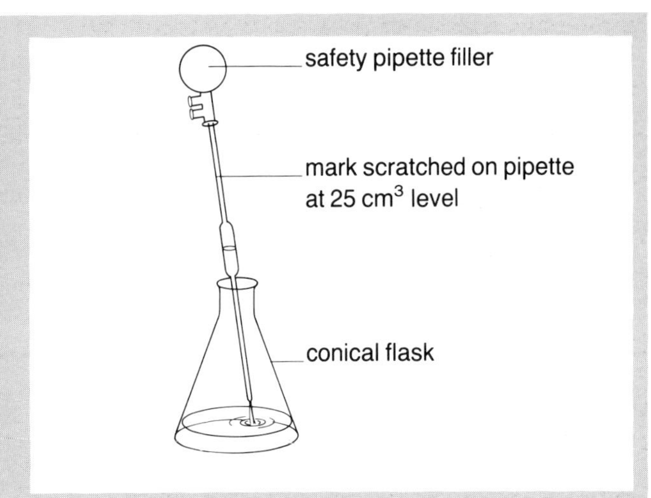

figure 13.5 *Using a pipette*

3 Fill a burette with dilute hydrochloric acid.
4 Add acid from the burette to the alkali until the indicator changes from crimson to colourless (see figure 13.6). Swirl the flask all the time. Note the volume of acid needed.
5 Wash out the flask and then add another 25 cm^3 of sodium hydroxide using the pipette. Add 5 drops of indicator as before.
6 Refill the burette with acid and repeat the titration. This time, when you are within 2 cm^3 of your previous reading turn the tap so that you add the acid a drop at a time. Find the exact volume that changes the indicator.
7 Thoroughly wash out the flask to remove all traces of indicator. Place 25 cm^3 of sodium hydroxide in the flask as before. Do *not* add any indicator.
8 Refill the burette and then add exactly the same volume of acid to the alkali as you needed in the last experiment. The solution should now be neutral.
9 Evaporate the solution to obtain a sample of pure salt.

sodium hydroxide	+	dilute hydrochloric acid	→	sodium chloride	+	water
NaOH	+	HCl	→	NaCl	+	H_2O

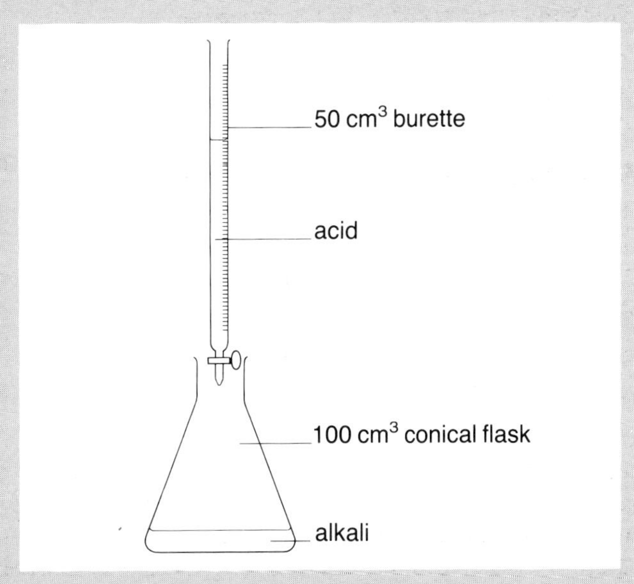

figure 13.6 *Titrating acid and alkali*

Using acids to prepare salts

The four main reactions of dilute acids can be used to make soluble salts (see investigation 13.5, previous page). Before you can decide which method to use you must first find out whether the salts are soluble or not. Most common salts are soluble. The list below gives the rules of solubility.

- **a** All **nitrates** are soluble.
- **b** All salts of **sodium**, **potassium** and **ammonia** are soluble.
- **c** Chlorides are soluble except **silver**, **lead** and **mercury(I)**.
- **d** Sulphates are soluble except **lead** and **barium** (**calcium** is slightly soluble).
- **e** Carbonates are insoluble except those of **sodium**, **potassium** and **ammonium**.

Decide which of the following salts are soluble and which are insoluble: copper sulphate, zinc carbonate, potassium carbonate, lead nitrate, silver chloride, barium chloride, lead sulphate, ammonium chloride, barium sulphate, silver nitrate.

Making insoluble salts

There is only one way to make an insoluble salt, and that is by mixing together two soluble salts. Each soluble salt needs to contain one part of the required insoluble salt. For example, to make a sample of lead chloride, we need a soluble lead salt (e.g. lead nitrate) and a soluble chloride (e.g. sodium chloride). When the two salts are mixed together a white solid (precipitate) of lead chloride is formed, which can be filtered out and washed.

lead nitrate + sodium chloride → lead chloride + sodium nitrate

$$Pb(NO_3)_2 + 2NaCl \rightarrow PbCl_2 + 2NaNO_3$$

Application 13.1

ACID RAIN

When water falls through the air small amounts of the gases in the air dissolve in the water. The most soluble of the gases in clean air is carbon dioxide. When this dissolves in water it produces a very weak acid called **carbonic acid**. This acid is not very corrosive, although it will attack limestone very slowly. However, in areas where fossil fuels (coal, oil and natural gas) are burned quite large amounts of **sulphur dioxide** are produced. The sulphur present in these fuels combines with oxygen when the fuel burns.

Sulphur dioxide dissolves in water to form **sulphurous acid** which is a much stronger acid than carbonic acid, and can cause considerable damage.

In Norway and Sweden there is a great deal of concern over thousands of trees that are dying from what is thought to be the effect of acid rain. It is not the rain itself that damages the plants but the increased acidity which causes changes in the chemicals in the soil. This in turn harms the plants.

figure 13.7 Acid rain can have disastrous effects on trees

figure 13.8 Acid rain in some areas can cause severe damage to stonework

It is claimed that polluted air from coal power stations in the UK is blown over to Scandinavia by the prevailing winds, causing acid rain there.

The government in the UK together with the Central Electricity Generating Board have recently started a programme to extract the sulphur dioxide from the fumes produced by the power stations. This will be a long and expensive task, but the effects of chemicals on our environment must be considered carefully.

Figures 13.7 and 13.8 show some of the effects of acid rain.

Application 13.2

USING ACIDS

Acids are very useful industrial chemicals. Sulphuric acid is especially useful and over 4 million tonnes of it are produced in this country every year. It is used in car batteries, fertilizers, man-made fibres and detergents, for example. Nitric acid is used in fertilizers and explosives. Vinegar is a solution of enthanoic acid (acetic acid) and is used to preserve foods. The bacteria which cause food to decay cannot survive in vinegar and so the 'pickled' food does not go bad.

Application 13.3

CONTROLLING pH

Acid rain has also caused the pH of some lakes to become far too low for fish and other organisms to survive. In order to counteract this, experiments have been taking place in the Lake District. Controlled amounts of lime (calcium hydroxide, a base) have been added to the water in an attempt to reduce the acidity. Great care has to be taken and the lime added in small amounts so that the lime itself does not damage the living things.

Increased use of fertilizers can cause the pH of the soil to become too low for plants. When this happens farmers put lime on the soil to restore the best pH value for the crops.

Indigestion is usually caused by too much hydrochloric acid being produced by the stomach. This can be neutralized by taking a weak alkali. This is the main ingredient of the various indigestion tablets you can buy.

HOW FAST?

It is useful to be able to measure the rates of chemical reactions, and to know how we can alter the rates. It is very important in industrial chemistry to be able to control the rate of a reaction, for reasons of both safety and economics. Industry is usually concerned with speeding up a reaction in order to get a greater amount of product each day.

In the laboratory we can study what changes affect the rate of a reaction.

Studying rates

Some reactions you have already come across are so fast that they appear to be instantaneous, for example when an insoluble salt is prepared by mixing together two soluble salts, the solid appears immediately. Other reactions take a few seconds, for example burning a small length of magnesium ribbon, and some reactions seem to take a very long time, like the tarnishing of a silver tray or the rusting of iron.

In order to follow a reaction in the laboratory we need an observable or measureable change. Changes we can measure include

a a change in pH
b a change in temperature
c a change in colour
d a change in mass
e the disappearance of a reactant
f the appearance of a precipitate
g a volume of gas given off.

Some of the easiest reactions to follow in the laboratory are those in which a gas is given off.

Explaining rates

Carry out investigation 13.6, overleaf. You can see from figure 13.12 that the slope is steepest at the beginning, and gradually curves until it is a horizontal line. When the graph is steepest it means that the reaction is at its fastest. When the graph is horizontal it means that no more gas is being given off, i.e. the reaction has finished. Note that a reaction will finish when *one* of the reactants is used up.

The big collision

In order for substances to react they must come into contact with each other. In the case of magnesium and sulphuric acid the particles in the acid (the hydrogen ions) must **collide** with the surface of the magnesium.

In a liquid the particles are always moving around, and so will collide with each other and anything else in the liquid (see page 7).

If a collision is to cause a reaction to take place the collision must have enough energy and be in the right direction. If a tennis ball accidentally hits a window, you hope that it will not break the window but will bounce back. If the ball is travelling fast enough, however, it will go through the window and break the glass (see figure 13.9).

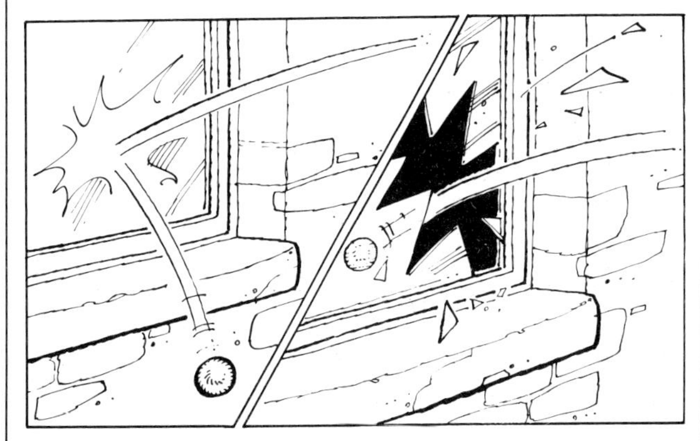

figure 13.9 Activation energy

INVESTIGATION 13.6

THE RATE OF REACTION BETWEEN MAGNESIUM AND DILUTE SULPHURIC ACID

This is one of the typical reactions of a dilute acid.

dilute acid + fairly reactive metal → salt + hydrogen

dilute sulphuric acid + magnesium → magnesium sulphate + hydrogen

We can follow the reaction by measuring the volume of hydrogen produced every five seconds. The volume of gas is measured using a gas syringe. This is a very expensive piece of apparatus and so needs special care. Figure 13.10 shows the apparatus arrangement. Study the diagram carefully before you start the experiment. The purpose of the plastic tube is to keep the reactants apart.

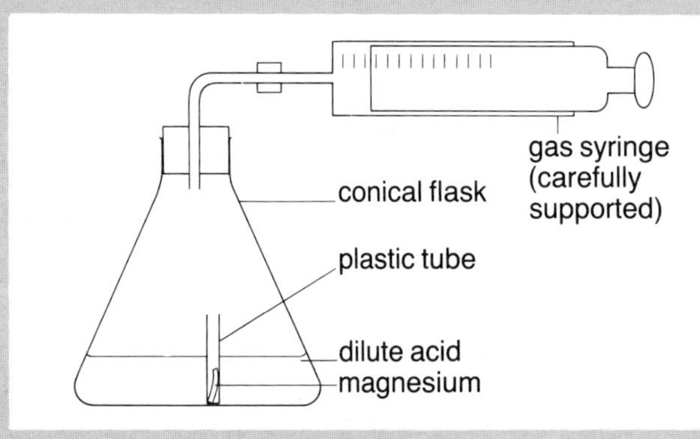

figure 13.10 *Following the rate of a reaction*

1 Make sure that your syringe is free running and then clamp it horizontally at the correct height. Do not clamp it too tightly. Make sure that you can see the scale.
2 Carefully place 25 cm^3 of dilute sulphuric acid in a 100 cm^3 conical flask.
3 Cut about 7 cm of magnesium ribbon into three pieces and then place them in the clean dry plastic tube.
4 Lower the plastic tube into the conical flask containing the acid using a dry glass rod (see figure 13.11).
5 Connect the rubber bung and delivery tube to the flask, taking care not to tip over the plastic tube.
6 Connect the delivery tube to the syringe.

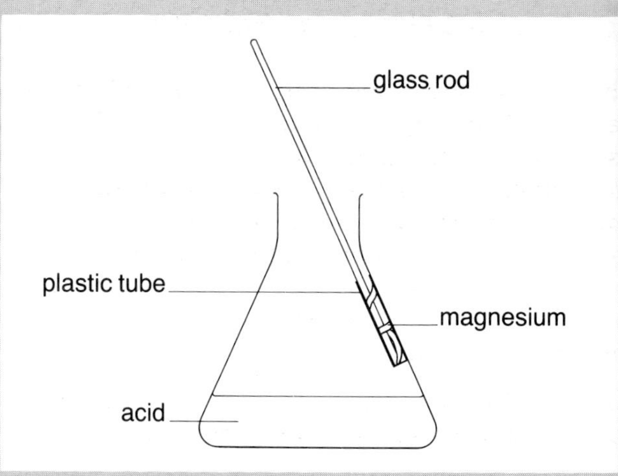

figure 13.11 *Keeping the reactants apart*

7 Shake the flask so that the tube falls over and all the magnesium comes out of the tube. At the same time start a stop clock.
8 Note the volume of gas given off every 5 seconds until the reaction has finished. Make a table of your results.

Question

Plot a graph of your results with volume on the *y*-axis and time on the x-axis.

Figure 13.12 shows a typical graph.

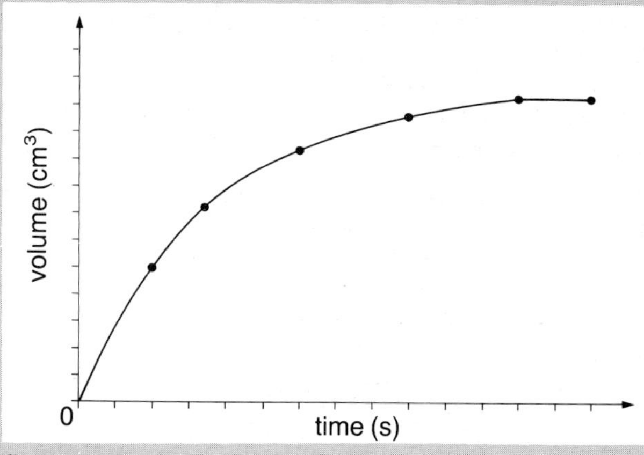

figure 13.12 *A rate of reaction graph*

In a similar way, not all collisions result in reaction. A certain energy barrier (threshold) has to be reached. This is called the **activation energy**. Each different reaction has its own activation energy.

At the start of a reaction there is the greatest concentration of particles of reactants present, and therefore there is a much greater chance of the reactant particles colliding with each other. The more collisions there are the faster the reaction will be.

As the reaction proceeds the reactants are being used up and so there will be fewer collisions, which in turn lead to a slower reaction.

Eventually one of the reactants will be used up completely. When this happens there will be no more collisions between the reactants and so no more reaction. Look back at figure 13.12 and see how you can account for its shape.

Altering the rate of a reaction

In order to alter the rate of a reaction either the **number** or the **energy** of collisions must be changed. The more collisions there are, the faster the reaction. The harder the collisions are, the greater the proportion of collisions that will reach the activation energy. Both these facts lead to a faster reaction.

The effect of concentration

'he effect of concentration on a reaction could easily ·e demonstrated by repeating investigation 13.6, ısing dilute sulphuric acid of different ·oncentrations. You would need to ensure that ·ou used the same amount of magnesium ·ach time. This highlights an important principle in ·xperimentation. In order to investigate the effect of ·ne change in conditions all the other conditions ·nust be kept the same.

·n alternative experiment involving the effect of ·oncentration is given in investigation 13.7.

The effect of temperature

When a liquid is heated, energy is given to the particles in the liquid. This energy will make the particles move faster. What effect will this have on the rate of the reaction? Carry out investigation 13.8, overleaf.

You probably found in that the higher the temperature, the quicker the reaction.

When a solution is heated, the particles move faster. This has two effects on the collisions between the reacting substances. As the particles are moving

INVESTIGATION 13.7

THE EFFECT OF CONCENTRATION ON THE REACTION BETWEEN SODIUM THIOSULPHATE AND DILUTE ACID

The reaction between sodium thiosulphate and a dilute acid produces a precipitate of sulphur. The actual reaction is between the thiosulphate ions and the hydrogen ions from the acid.

thiosulphate ions + hydrogen ions → sulphur dioxide + water + sulphur

$$S_2O_3^{2-} + 2H^+ \rightarrow SO_2 + H_2O + S$$

You can measure the time taken for the solution to become so cloudy that you cannot see through it. This is done by placing the conical flask containing the reactants on top of a marked piece of paper (see figure 13.13). The mark is observed by looking down through the solution. It is important to have the same depth of liquid in the flask each time.

Before you start the actual experiment you must thoroughly wash all the apparatus.

a Pour 50 cm^3 of sodium thiosulphate solution (concentration 50 g/dm^3) into a 100 cm^3 conical flask.
b Make an 'X' on a piece of paper and then stand the flask on the paper (see figure 13.13).
c Add 5 cm^3 dilute sulphuric acid to the flask, starting the stop clock at the same time. Swirl the flask to mix the chemicals.
d Note the time taken for the solution to become so cloudy that you cannot see the 'X'.
e Wash out the flask several times and then repeat the experiment using 40 cm^3 sodium thiosulphate solution with 10 cm^3 distilled water. Still use 5 cm^3 dilute sulphuric acid.
f Repeat the experiment three more times using
  i 30 cm^3 sodium thiosulphate solution + 20 cm^3 water
  ii 20 cm^3 sodium thiosulphate solution + 30 cm^3 water.
  iii 10 cm^3 sodium thiosulphate solution + 40 cm^3 water.

In each case 5 cm^3 dilute sulphuric acid is used. Note how the total volume of liquid in the flask is kept the same.

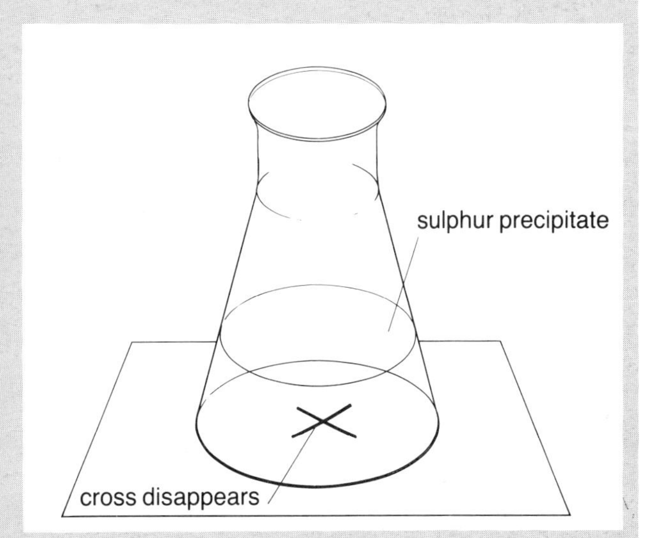

figure 13.13 The thiosulphate reaction

Questions

1 Make a results table similar to table 13.4.

volume of solution (cm^3)	volume of water (cm^3)	concentration (g/dm^3)	time (s)
50	0	50	
40	10	40	
30	20	30	
20	30	20	
10	40	10	

table 13.4

2 Plot a graph of concentration of sodium thiosulphate against the time taken for the 'X' to disappear.
3 You should find that the greater the concentration the faster the reaction. The higher the concentration of the solution the more particles there are present. This leads to a greater number of collisions and hence a faster reaction. From your graph, work out how long you think it would have taken for the 'X' to disappear if 25 cm^3 of solution and 25 cm^3 of water had been used.

faster they travel a greater distance in a given time, and so are involved in a greater number of collisions. Also, because the particles are moving faster, they hit each other harder. This means that a greater number of collisions reach the activation energy and result in a reaction. The higher the temperature the greater the rate of reaction.

INVESTIGATION 13.8

THE EFFECT OF TEMPERATURE ON THE RATE OF A REACTION

Repeat investigation 13.7 using 25 cm^3 sodium thiosulphate solution and 25 cm^3 distilled water. Each time use 5 cm^3 dilute sulphuric acid. Carry out the reaction at different temperatures.

1. The first experiment should be carried out at room temperature. Measure the sodium thiosulphate solution and the water into the conical flask.
2. Add the 5 cm^3 of acid as before and start the stop clock.
3. Note the temperature of the reaction mixture *during the reaction* and also note the time taken for the 'X' to disappear.
4. Wash out the flask and the thermometer and repeat the experiment. This time, after you have mixed the solution and the water, gently warm the mixture until it is at about 30 °C.
5. Stand the flask over the 'X' and leave it for about 30 seconds. Add the 5 cm^3 of acid and start the stop clock. Note accurately the temperature during the reaction and the time taken for the 'X' to disappear.
6. Repeat the experiment several times, warming the mixture of the solution and water (before you add the acid) to temperatures of about 40 °C, 50 °C, 60 °C and 70 °C.

Remember that the temperature you need to record is the temperature during the reaction.

The effect of pressure

Pressure will have an effect only if the reaction involves gases. In a reaction between two gases an increased pressure will have the effect of forcing the gas particles closer together, i.e. increasing the concentration. This will lead to an increased rate of reaction.

The effect of particle size

In a reaction involving a solid and a liquid or a solid and a gas the reaction takes place on the surface of the solid. The particles of the liquid or gas constantly collide with the surface of the solid. If these collisions have enough energy a reaction will take place. The amount of surface there is will affect the number of collisions and alter the rate of the reaction.

INVESTIGATION 13.9

THE EFFECT OF PARTICLE SIZE ON THE RATE OF A REACTION

The reaction here is that in investigation 13.6 (page 198), between magnesium and dilute sulphuric acid. The reaction rate is again following by measuring the volume of hydrogen given off using a gas syringe.

In this experiment the amount of each chemical used is kept the same. The only difference is the form of the magnesium.

a. Check that the gas syringe is free moving and then clamp it horizontally at the correct height and so that you can see the scale (see figure 13.10).
b. Place 25 cm^3 dilute sulphuric acid in a 100 cm^3 conical flask.
c. Weigh out exactly 0.08 g magnesium ribbon (this should be a piece about 8 cm long) and cut it into three or four smaller pieces. Put the pieces into a small dry plastic tube.
d. Carefully lower the tube into the conical flask using a glass rod (see figure 13.11).
e. Push the bung and delivery tube into the flask and connect the delivery tube to the gas syringe.
f. Swirl the flask to tip over the tube and ensure that all the magnesium comes out of the tube. At the same time start the stop clock.
g. Note the volume of hydrogen given off every 5 seconds until the reaction has stopped.
h. Repeat the experiment using 0.08 g magnesium turnings.
i. Repeat the experiment using 0.08 g magnesium powder.

Questions

1. Plot all three sets of results on the same graph. (See figure 13.14)
2. From your graph, which reacts faster, the magnesium powder, the magnesium ribbon or the magnesium turnings? Why is this the case?

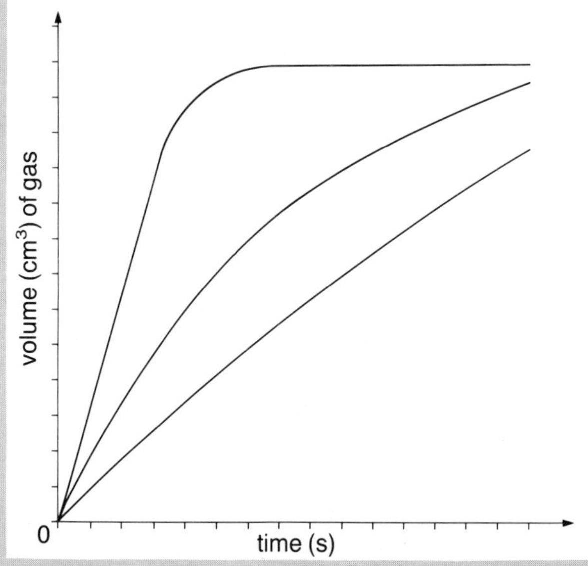

figure 13.14 Which graphs represent powder, turnings and ribbon?

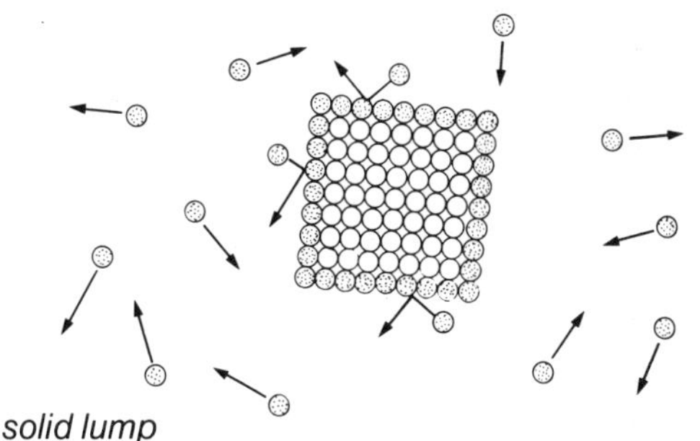

solid lump
32 particles of solid exposed to bombardment by liquid or gas particles

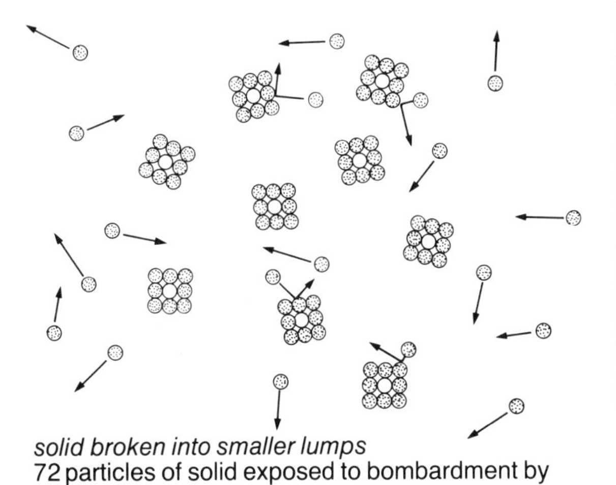

solid broken into smaller lumps
72 particles of solid exposed to bombardment by liquid or gas particles

igure 13.15 Why particle size makes a difference

.ook at figure 13.15. When the block is in one piece, he reaction can only take place round the outside of he solid, i.e. on its surface. When the block is broken n two, two additional surfaces are created. You can ee that the more the solid is broken up the more urface is produced. This means there is a greater urface available for collisions to take place, and so he rate of reaction increases.

n the reaction between a solid and a liquid or a solid ınd a gas, the smaller the pieces the solid is broken lown into, the faster the rate of the reaction.

Using a catalyst

The rates of some reactions can be altered by adding other chemicals to the reaction mixture. A substance that can alter the rate of a reaction, without altering the reaction in any other way and without being used up during the reaction, is called a **catalyst**. Catalysts are very important in industry. Carry out nvestigation 13.10.

You may well have found that manganese (IV) oxide s the best catalyst for this reaction. This reaction is often used to prepare oxygen in the laboratory. Chemicals in your blood and liver will also catalyze this reaction.

INVESTIGATION 13.10

THE EFFECT OF A CATALYST ON THE DECOMPOSITION OF HYDROGEN PEROXIDE SOLUTION

Hydrogen peroxide solution decomposes to give water and oxygen.

hydrogen peroxide → water + oxygen

$$2H_2O_2 \rightarrow 2H_2O + O_2$$

The reaction could be followed using a method similar to that in investigation 13.6, i.e. using a gas syringe to measure the volume of oxygen produced.

In this experiment, however, you can find out which substances speed up the reaction sufficiently to produce enough oxygen to relight a glowing spill.

1 Quarter fill four test tubes with 10 volume hydrogen peroxide solution. (Take care – hydrogen peroxide is acidic and is a bleach.)
2 To test tube 1 add one spatula measure of copper(II) oxide.
  To test tube 2 add one spatula measure of magnesium oxide.
  To test tube 3 add one spatula measure of zinc oxide.
  To test tube 4 add one spatula measure of manganese(IV) oxide.
3 Notice whether any gas is given off and test each tube with a glowing spill (see figure 13.16). See if the spill relights. Record your results.
4 If the spill does not relight, gently warm the mixture until it boils, return the test tube to the stand and then re-test for oxygen with a glowing spill. Record your results.

figure 13.16 Testing for oxygen

Remember that in order to show that a substance acts as a catalyst you must show that

a the rate is altered
b the reaction is not altered in any other way
c the substance itself is not used up.

Biological catalysts — enzymes

A large number of catalysts occurs in living cells. These are called **enzymes**. The enzyme which speeds up the decompostion of hydrogen peroxide is called **catalase**. As catalysts are not used up in a reaction only a small amount of the substance is needed.

Many different chemical reactions happen inside each living cell. Without enzymes these reactions would happen so slowly the cells would die. Enzymes also control these reactions so they happen in an orderly way. An enzyme works by bringing the chemicals together at a particular location on its surface – the **active site**. Figure 13.17 shows how the active site works. The active site of each enzyme has a different shape. One particular shape will fit a specific combination of molecules. This means each chemical reaction in the body requires a different type of enzyme. Enzymes are therefore **specific** to certain chemical reactions.

Unlike metal catalysts, enzymes are destroyed by high temperatures. They work best between 37 °C and 40 °C. Enzymes are also sensitive to acidity and alkalinity (see investigation 13.11).

INVESTIGATION 13.11

HOW SENSITIVE ARE ENZYMES?

Protease is an enzyme that breaks down protein into amino acids (see figure 3.9, page 48). Amino acids are soluble in water. The small pieces of boiled egg white in this investigation consist of the protein called albumin.

a Label four test tubes A to D. Place a small cube of egg white into each test tube.

b Add 20 cm^3 of the following solutions to the different test tubes.

A – acid protease solution
B – acid solution only
C – alkali protease solution
D – alkali solution only

c Leave the test tubes until your next science lesson. Ideally they should be kept at 37 °C.

Questions

1 What happened to the egg white in each test tube?
2 Why was it necessary to include tests without the enzyme (test tubes B and D)?
3 How does acidity and alkalinity affect the action of protease?
4 How could you extend this investigation to find out the exact pH at which protease works best (the optimum pH)?
5 Design an investigation to discover how enzymes are affected by different temperatures.

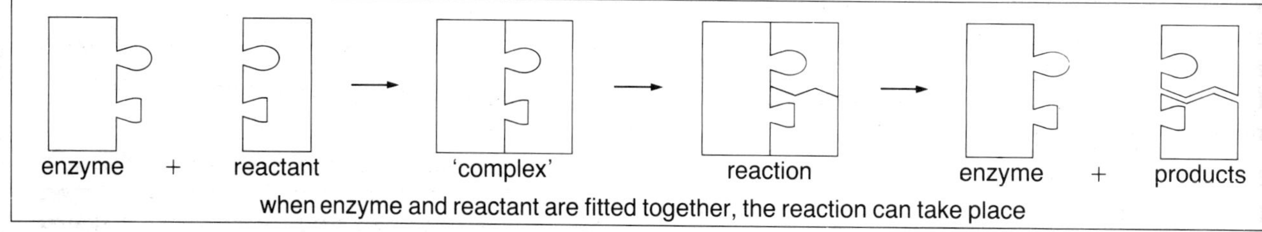

figure 13.17 How enzymes work

Application 13.4

ENZYMES AT WORK

For thousands of years the enzymes contained in yeast cells have been used to make bread dough rise and to produce alchoholic drinks such as wine and beer. Even when enzymes are removed from living cells they continue to work. In recent years a wide variety of enzymes has been extracted and used in industry. Table 13.5 summarizes some of these industrial applications.

application	enzymes used	benefits
biological washing powders	proteases	digest protein stains such as blood, grass or food on clothes
cheese making	rennin (extracted from the stomachs of young calves and lambs)	digests milk protein
baking bread	amylase	starch in the flour is broken down to sugar which uses the yeast to produce carbon dioxide, which makes the dough rise
textile manufacture	amylase	breaks down starch that is added as an adhesive to prevent damage to the threads during weaving
leather industry	protease	removes certain proteins in leather to make it more supple
brewing industry	protease	breaks down starch and proteins to sugars and amino acids which are used by yeast to increase alcohol production

table 13.5

Application 13.5

INDUSTRIAL REACTIONS INVOLVING CATALYSTS

Table 13.6 summarizes the use of other catalysts in industry.

process	catalyst used	main product
Haber process	iron	ammonia
Contact process	platinum	sulphuric acid
oxidation of ammonia	platinum	nitric acid

table 13.6

Reversible reactions

Some reactions can go in both directions, depending on the conditions. If some blue copper sulphate crystals are heated they give off water of crystallization to become anhydrous copper sulphate. This is a white powder. If water is added to this white powder it becomes very hot and turns blue again. This is an example of a **reversible** change. In an equation a reversible change is shown by an arrow going each way.

copper sulphate crystals ⇌ anhydrous copper sulphate + water

$$CuSO_4.5H_2O \rightleftharpoons CuSO_4 + 5H_2O$$

When ammonium chloride crystals are heated they appear to sublime, i.e. change directly from a solid to a gas without changing to a liquid. At the top of the test tube a ring of solid ammonium chloride reappears on cooling.

In fact, the ammonium chloride splits up (decomposes) into ammonia and hydrogen chloride, which recombine when they cool down.

ammonium chloride ⇌ ammonia + hydrogen chloride

$$NH_4Cl \rightleftharpoons NH_3 + HCl$$

If a reversible reaction is enclosed in a sealed system so that no chemicals can enter or leave, a situation can arise where both the forward and the backward reaction are taking place at the same time. A balance is set up where the rate of the forward reaction is equal to the rate of the backward reaction. This is called a **chemical equilibrium**.

When an equilibrium is set up the overall reaction appears to have stopped. A model to help you understand what is happening in a chemical equilibrium is someone walking up a down escalator. If the person keeps walking at exactly the speed of the escalator he or she will appear to be standing still (see figure 13.18).

If the conditions of a chemical equilibrium reaction are altered, the position of the equilibrium alters to cancel out the change. For example, if more of one chemical is added, the reaction will move in a direction to use up that chemical. If a chemical is removed from the mixture, the reaction will go in the direction that makes more of that chemical.

figure 13.18 *An equilibrium position*

Application 13.6

THE HABER PROCESS

This is an important industrial process involving a reversible reaction. The product of the reaction is ammonia. It is made by reacting nitrogen and hydrogen.

nitrogen + hydrogen ⇌ ammonia (+ HEAT)

$$N_2 + 3H_2 \rightleftharpoons 2NH_3$$

Under normal laboratory conditions the equilibrium is very much on the left hand side , i.e. hardly any nitrogen and hydrogen react together to form ammonia. Extreme conditions are used in industry to force the equilibrium towards the right-hand side and so produce ammonia.

The reacting gases are subjected to very high pressure (about 200 atmospheres, 20 000 kPa). Apart from forcing the particles closer together, this makes the reaction tend to reduce the pressure.It can do this by moving to the right-hand side. A total of 4 molecules are present on the left-hand side, one of nitrogen and three of hydrogen. On the right-hand side of the equation there are only two molecules of ammonia. Fewer molecules means less pressure.

Iron is used as a catalyst to speed the reaction up, but this cannot alter the position of the equilibrium.

The forward reaction produces heat, i.e. is **exothermic**, and so cooling the reaction down should encourage the reaction to make more heat and ammonia. However, cooling the reaction slows it down and the lowest temperature at which the reaction goes at a reasonable rate is about 450 °C. Even under these conditions only about 15 per cent of the reacting gases combine.

THE REACTIVITY SERIES OF METALS

The **reactivity series** is a list of the common metals, placed in order of their chemical reactivity. The most reactive metal is at the top. The reactivity series is useful because it can be used to predict properties of metals and so help us to decide which metal to use for a particular job.

The reactivity series can be built up by looking at a few simple experiments. It is helpful to divide the metals up into four main groups.

The very reactive metals

These metals all react with cold water. We have already seen in unit 11 how the group I metals lithium, sodium, and potassium react with water. The most reactive of these is potassium, then sodium, then lithium. Another metal that will react with cold water is calcium. See investigation 13.12.

From the reaction with cold water we can place these four metals in order of reactivity: potassium, sodium, lithium and calcium.

The fairly reactive metals

These metals hardly react at all with cold water but they will react with steam. Figure 13.19 shows the apparatus used to demonstrate the reaction of steam with magnesium. The heated magnesium burns in the steam and the hydrogen produced ignites at the end of the tube.

INVESTIGATION 13.12

THE REACTION OF CALCIUM WITH COLD WATER

a Place a small piece of freshly cut calcium into a beaker half full of cold water.
b When it has finished reacting, test the solution with universal indicator paper.
c Add another piece of calcium to a large test tube one-quarter full of cold water. Test the gas given off with a lighted spill. Notice carefully what happens to the solution.

Questions

1 Complete the equation
calcium + water → ______ + ______
2 What is the pH of the solution at the end of the reaction?
3 What is the common name for calcium hydroxide solution?
4 Why does the mixture in the test tube go cloudy after the reaction has been going a short time?
5 Give three reasons why calcium is not placed in the same group or chemical family as sodium and potassium.

magnesium + water → magnesium oxide + hydrogen

$$Mg + H_2O \rightarrow MgO + H_2$$

Zinc will react in a similar way, and iron wool will react slowly. (This reaction is reversible).

Another reaction that these metals have in common i their reaction with dilute acids. (See page 193.)

dilute acid + a fairly reactive metal → a salt + hydrogen

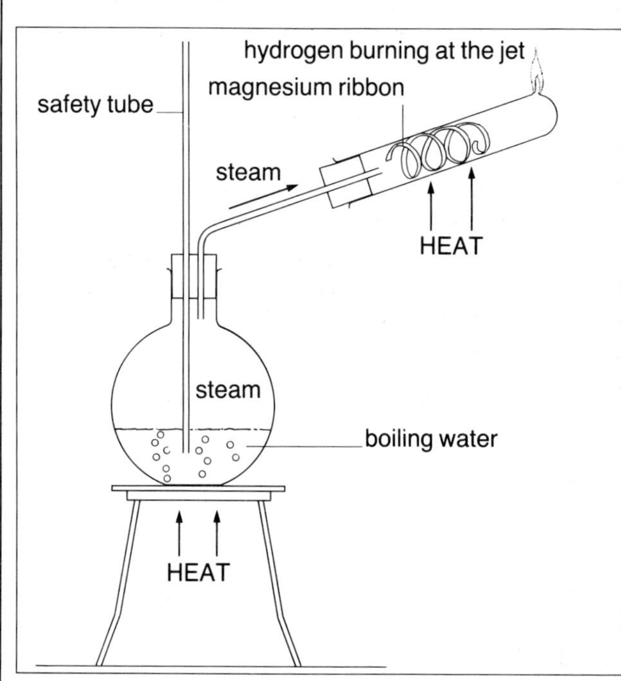

figure 13.19 Reaction of magnesium and steam

INVESTIGATION 13.13

WHICH METALS REACT WITH DILUTE SULPHURIC ACID?

Remember to wear goggles and to take special care when using dilute acids.

a Quarter fill six test tubes with dilute sulphuric acid.
b Add one of the following metals to each test tube of dilute acid.
  i 2 cm magnesium ribbon
  ii a small piece of lead foil
  iii a small piece of aluminium foil
  iv a small ball of iron wool
  v a small piece of copper foil
  vi a small piece of granulated zinc
c Test any gas given off with a lighted spill.

Questions

1 Which metal reacts the best?
2 Which other metals react with the acid?

Time for thought

You might expect aluminium to be about as reactive as magnesium, but it does not appear to react at all. Why?

We can see from these reactions that the fairly reactive metals, in order of reactivity, are: magnesium, (aluminium), zinc, iron.

The fairly unreactive metals

These are metals that do not react with water or dilute acid. They react very slowly with the air to lose their shiny appearance. Lead and copper are in this group.

The very unreactive metals

These include the metals used for jewellery. They are unreactive and keep their shiny appearance. They have been known for a long time because they can either be found uncombined in the earth, or can be easily obtained from their ores. Gold, silver and mercury are in this group.

Table 13.7 shows part of the reactivity series.

potassium sodium lithium calcium	very reactive
magnesium (aluminium) zinc iron	fairly reactive
lead copper	fairly unreactive
mercury silver gold	very unreactive

table 13.7

Displacement reactions

If a more reactive metal is heated with a compound of a less reactive metal, the less reactive metal is pushed out (**displaced**) by the more reactive metal. You can think of it as the more reactive metal being stronger and taking something from the weaker, less reactive metal.

A good example of this type of reaction is that between aluminium powder and iron(III) oxide. The 'stronger' aluminium takes the oxygen from the 'weaker' iron.

aluminium + iron(III) oxide → aluminium oxide + iron

$$2Al + Fe_2O_3 \rightarrow Al_2O_3 + 2Fe$$

This reaction produces a lot of heat, enough to melt the iron formed. This reaction is called the **Thermit reaction** and it is used to fill gaps in railway lines and weld them together (see figure 13.20). Your teacher may demonstrate this reaction to you. It is far safer to do the reaction outside.

Displacement reactions also take place in solution.

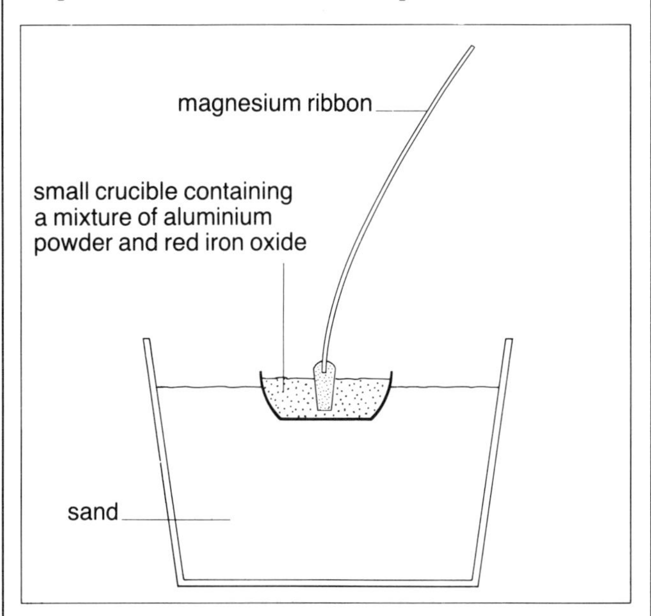

figure 13.20 The Thermit reaction ***a*** *In the laboratory*

figure 13.20 ***b*** *Molten iron from the reaction runs into a mould around the rails. When the iron has cooled the mould is removed and excess iron trimmed off*

INVESTIGATION 13.14

DISPLACEMENT OF METALS IN SOLUTION

a Add a small ball of iron wool to a test tube one-quarter full of copper(II) sulphate solution. Leave it for 2–3 minutes and then observe both the colour of the solution and the iron wool.

b Add a small piece of zinc foil to a test tube one quarter full of lead nitrate solution. Leave it for about 30 minutes and then observe the surface of the zinc.

Questions

1 What happened to the iron wool?

2 Complete the equation
copper sulphate + iron → ________ + ________

3 Describe the appearance of the surface of the zinc. What does this suggest about the structure of metals?

Application 13.7

USEFUL METALS

Titanium is extracted by heating titanium chloride with sodium. It is used for making light alloys for aircraft.

The use of a metal depends on its reactivity. It would be no use making a bucket out of a metal that reacts with water. Table 13.8 shows the uses of some common metals.

metal	use
iron (steel)	girders, cars, ships etc., nails, screws, tools
aluminium	aircraft, window frames, greenhouses, light alloys for racing cars and bikes, cooking foil
copper	electric wires, water pipes, coins, roofs
gold	jewellery and ornaments
lead	church roofs, flashing on house roofs, car batteries
zinc	galvanizing iron, dry batteries

table 13.8

The extraction of metals

If you find out the dates of discovery of various metals, you will find that some of them have been known for thousands of years, e.g. gold, whereas others, like potassium, have been known for only a few hundred years. When you relate this to the reactivity series you should find a fairly regular pattern. The metals near the bottom of the series have been known longest and those near the top of the series are relatively recent discoveries.

Ease of extraction is also related to the reactivity series. The lower a metal is down the series the easier it is to extract.

The extraction of iron

The high strength and relative abundance of iron make it a very useful metal. Iron is extracted in a **blast furnace** (see figures 13.21 and 13.22).
There are five main reactions that take place.

a The carbon burns near the bottom of the furnace to form carbon dioxide.

carbon + oxygen → carbon dioxide

$$C + O_2 \rightarrow CO_2$$

b As the carbon dioxide rises up the furnace it reacts with more carbon to form carbon monoxide.

carbon dioxide + carbon → carbon monoxide

$$CO_2 + C \rightarrow 2CO$$

c The carbon monoxide reacts with the iron oxide. It takes the oxygen away from the iron oxide (**reduces** it) to leave iron and carbon dioxide.

iron oxide + carbon monoxide → iron + carbon dioxide

$$Fe_2O_3 + 3CO \rightarrow 2Fe + 3CO_2$$

The molten iron runs to the bottom of the furnace.

Two other reactions are concerned with getting rid of the impurities present in the iron ore.

d The limestone (calcium carbonate) decomposes to calcium oxide (quicklime) and carbon dioxide.

calcium carbonate → calcium oxide + carbon dioxide

$$CaCO_3 \rightarrow CaO + CO_2$$

e The calcium oxide then reacts with silicon dioxide (sand or quartz) present in the iron ore.

calcium oxide + silicon dioxide → calcium silicate

$$CaO + SiO_2 \rightarrow CaSiO_3$$

The molten calcium silicate sinks to the bottom of the furnace where it floats on top of the molten iron.

The iron produced by the blast furnace contains some carbon together with other impurities. This form of iron is rather brittle and has limited use (e.g. road grids and man-hole covers). Most iron produced is changed into steel.

The conversion of iron to steel

First, all the impurities are burnt out of the iron by passing oxygen gas through or over the molten iron (see figure 13.23). This makes pure iron. Carefully controlled amounts of carbon are then added to the molten iron. This changes the iron to steel. Stainless steel contains very small amounts of some other metals like chromium and manganese.

Table 13.9 shows the composition and uses of some typical steels.

gure 13.21 A blast furnace

iron ore + limestone + coke
double cone for charging furnace
hot waste gases
iron oxide reduced to iron by carbon monoxide
250 °C
600 °C
1200 °C
1800 °C
firebrick lining
limestone decomposed to quicklime
sand impurity fused with quicklime to form slag
hot air blast through tuyères
slag
coke burned in air blast to give carbon monoxide and heat
molten iron

figure 13.22 The reactions in a blast furnace

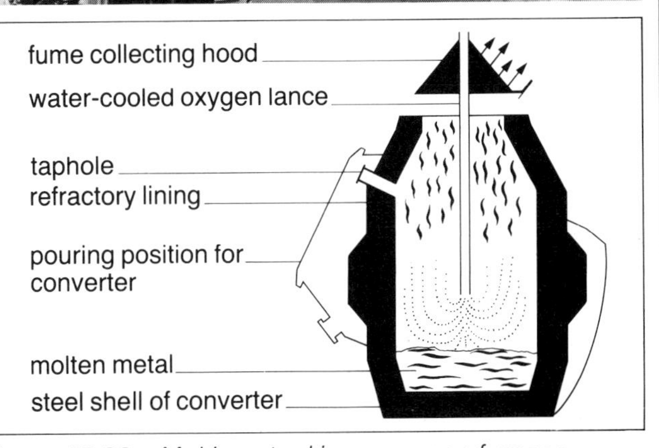

igure 13.23 Making steel in an oxygen furnace

type of steel	percentage of carbon	use
mild steel	0.1 – 0.25	car bodies
medium carbon steel	0.25 – 0.5	girders
high carbon steel	0.5 – 1.5	tools and drills

table 13.9

The rusting of iron

We have seen that iron and steel are very useful metals. This is mainly because of their high strength. Iron and steel, however, have one very serious defect – they rust. Most metals corrode when exposed to the air. The more reactive the metal the faster it corrodes. In some cases this corrosion does not cause much of a problem. You have already come across the thin layer of aluminium oxide on the surface of aluminium. This is a thin, tightly held layer which then protects the aluminium against further attack.

Unfortunately this is not the case with iron. The layer of corrosion is porous and not tightly held. The rust comes away from the surface of the iron and flakes off. This exposes new iron to the air. Before we can look at the prevention of rusting we need to find out what causes rusting.

INVESTIGATION 13.15

WHAT CONDITIONS ARE NEEDED FOR IRON TO RUST?

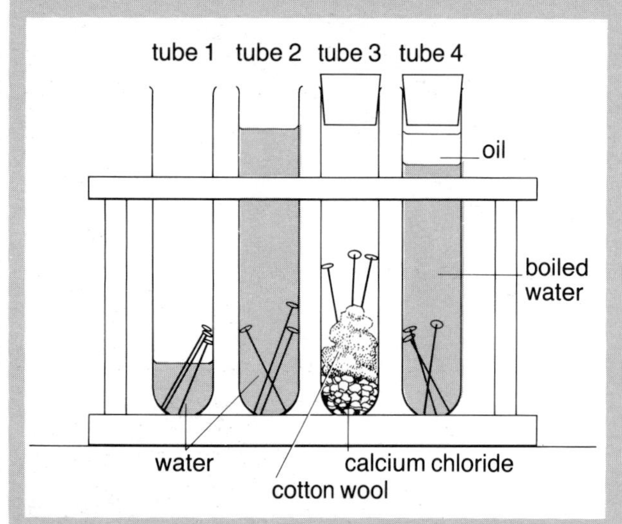

figure 13.24 Investigating the rusting of iron

Set up the apparatus shown in figure 13.24.
Test tube 1 Make sure that the iron nail is partly in the water and partly in the air.
Test tube 2 The nail is in a test tube full of tap water.
Test tube 3 Make sure that the test tube is dry to start with.
The lumps of calcium chloride will absorb any water present. Add a small piece of cotton wool to keep the nail away from the calcium chloride. Do not forget to put a bung in the tube.
Test tube 4 Boil about 50 cm^3 of water for about 5 minutes. Pour some of the boiling water onto the nail in the test tube and quickly add a layer of oil on top of the hot water. Again remember to put a bung in the tube.

Leave the test tubes for a week and then look for signs of rusting.

Questions

1. In which test tube are there most signs of rust?
2. In which test tube are there least signs of rust?
3. Why was the water boiled in test tube 3, and what was the purpose of the layer of oil?
4. What conditions are needed for iron to rust?

Time for thought

1. *How would you design experiments to show that the following can prevent rusting?* ***a*** *painting a piece of iron* ***b*** *keeping a piece of iron in contact with magnesium.*
2. *The iron ore in the ground is being used up and one day will run out. One possible solution to this problem is recycling. Find out what is meant by recycling. What other materials can be recycled?*

Application 13.8

THE COST OF RUSTING AND ITS PREVENTION

Rusting is a serious problem. If iron pipes rust away they will burst. If girders rust away they will lose their strength and structures will fall down. Unfortunately, it is very difficult to stop rusting completely, but it is often possible to slow it down considerably.

figure 13.25 It can cost £200 000 a year to maintain a pier's paintwork and prevent it rusting

Painting

This is the commonest method of trying to prevent rusting. It is relatively cheap and also improves the appearance of the metal. It is not a permanent treatment and has to be repeated at intervals. You might have seen pictures of the painting of the Forth Bridge (see figure 13.26). It takes about four years to paint it from one end to the other. As soon as they have finished, the painters go back and start again. Painting is no good where moving parts rub together.

figure 13.26 The Forth bridge takes four years to paint

Grease and oil

This method is used for protecting moving parts, where the oil also acts as a lubricant. Many new machinery parts are coated with a layer of grease to protect them while they are stored.

Galvanizing

You might have noticed sometimes a pattern on the surface of metal dustbins, buckets, watering cans etc. This is caused by the crystal structure of the thin layer of zinc that covers the iron. The iron is dipped into molten zinc and a thin protective layer of zinc sticks to the iron. This process is called galvanizing.

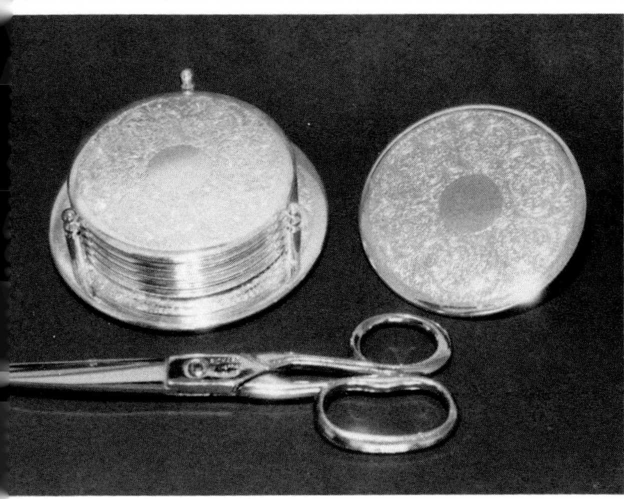

figure 13.27 *Products using electroplating*

Plating

Here the iron is covered with a thin layer of another metal. In addition to protecting the iron, it can also improve the appearance of the object, e.g. chrome plating on kettles, taps and bicycles. This layer is put on the iron by a process called electrolysis (see figure 13.27).

figure 13.28 *Potentially vulnerable areas of a car body are treated with sealants*

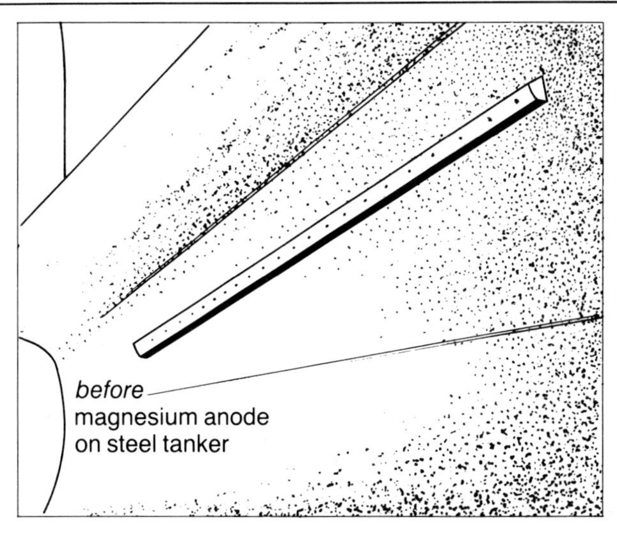

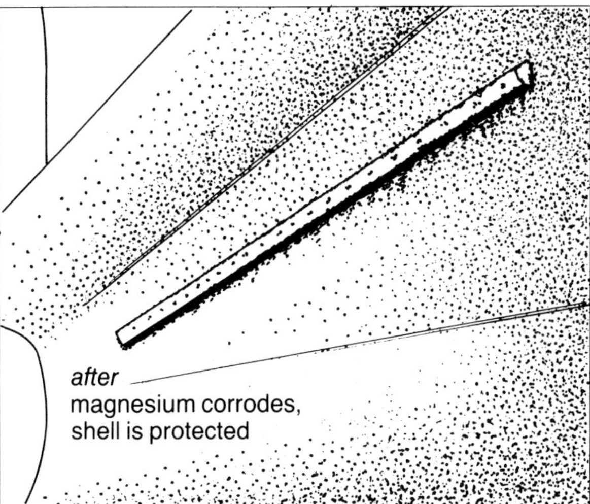

figure 13.29 *Sacrificial corrosion – the magnesium protects the stern of the tanker*

Special treatments

The iron can be treated with special chemicals and protective paints to prevent rusting. You will have seen adverts for cars saying that the body is guaranteed not to rust for six to ten years because of the anti-rust treatment (see figure 13.28).

Sacrificial protection

It is possible to protect some products made of iron by placing a more reactive metal in contact with the iron. The more reactive metal corrodes away instead.

This is a good way of protecting iron where it is not easy to get at e.g. an underground pipe or a ship's hull. Large blocks of magnesium are attached to the iron at easily accessible places. The iron will not rust as long as there is any magnesium left. The blocks of magnesium have to be replaced at regular intervals (see figure 13.29).

Alloys

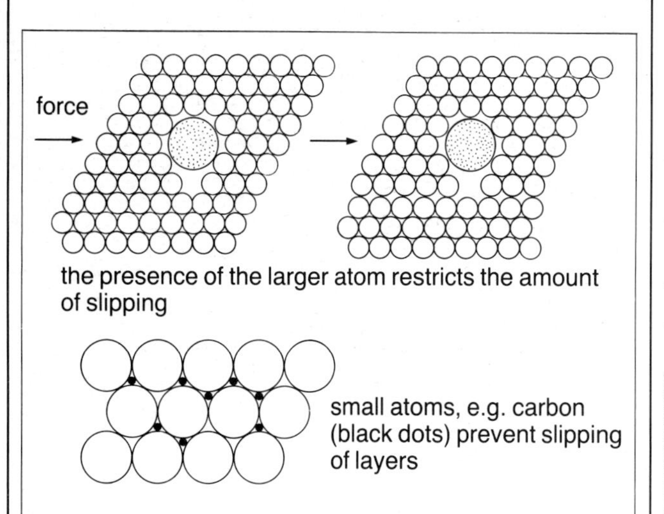

force
the presence of the larger atom restricts the amount of slipping
small atoms, e.g. carbon (black dots) prevent slipping of layers

figure 13.30 The structure of alloys

An **alloy** is a metal mixture. Alloys are usually made by mixing together the molten metals in certain proportions.

By introducing different sized atoms it is possible to alter the structure and therefore the properties of the metal. Figure 13.30 shows how different sized atoms alter the metal structure.

Application 13.9

ALLOYS

Some common alloys and their uses are shown in table 13.10.

alloy	compostion (%)	use
brass	copper 70 zinc 30	screws, electric plugs and ornaments
bronze	copper 78 tin 22	statues and bells
coinage bronze	copper 97 zinc 2.5 tin 0.5	1p and 2p coins
solder	tin 60 lead 40	electrical connections and plumbing
pewter	tin 80 lead 20	ornaments
duralumin	aluminium 95 copper 4 manganese 0.5 magnesium 0.5	sheet metal work

table 13.10

ELECTROLYSIS

One of the properties of metals that we have seen is that they are good conductors of electricity. When a metal conducts electricity it is not permanently changed. A wire might get hot while the electricity is actually flowing, but goes cold when the electricity is switched off. Carbon in the form of **graphite** also conducts.

Some chemical compounds will conduct electricity but only when they are molten or in solution in water.

INVESTIGATION 13.16

PASSING ELECTRICITY THROUGH LEAD BROMIDE

This investigation will be demonstrated by your teacher.

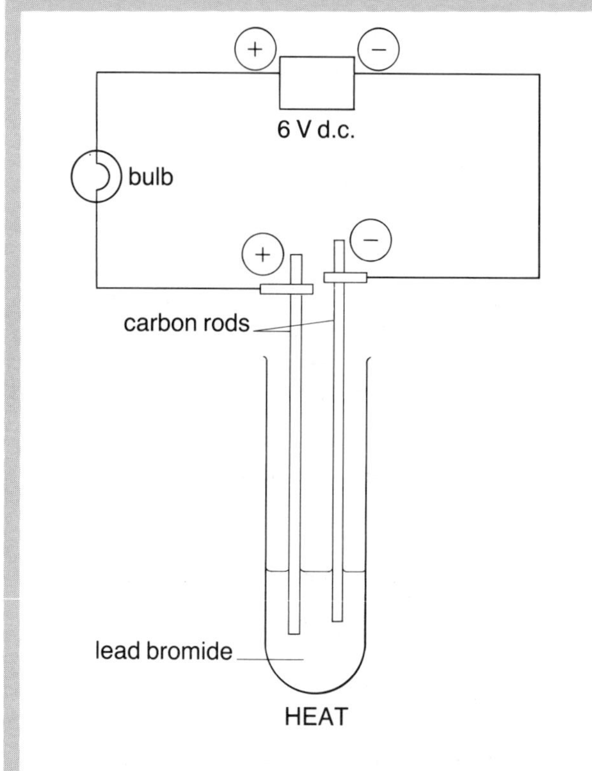

figure 13.31 The electrolysis of lead bromide

A simple circuit is connected as shown in figure 13.31. A 6 V d.c. supply is used. The bulb indicates when the circuit is complete.

a The circuit is checked by touching the long carbon rods together.
b The solid lead bromide is tested to see whether it conducts electricity, by dipping the rods into the powder.
c The boiling tube is heated until the lead bromide melts. Any changes are carefully noted.

Questions

1 When does the lead bromide conduct electricity?
2 What do you see at each rod, while the bulb is lit?

Lead bromide is an ionic substance, i.e. it is made up from positively charged lead ions (Pb^{2+}) and negatively charged bromide ions (Br^-). In the solid the ions are fixed in position and cannot move. Positively charged ions are called **cations** and negatively charged ions **anions**.

When the lead bromide melts, the ions become free to move. While the electricity is switched on the negative rod (**cathode**) attracts the positive ions and the positive rod (**anode**) attracts the negative ions.

At the cathode the lead ions receive electrons and become lead atoms, i.e. lead metal.

At the anode the bromide ions lose electrons and become bromine molecules (bromine atoms are always found in pairs).

The reaction at the cathode is

$$\text{lead ions} + \text{electrons} \rightarrow \text{lead atoms}$$
$$Pb^{2+} + 2e^- \rightarrow Pb$$

The reaction at the anode is

$$\text{bromide ions} - \text{electrons} \rightarrow \text{bromine molecules}$$
$$2Br^- - 2e^- \rightarrow Br_2$$

The process of splitting up a compound in this way is called **electrolysis**. Electrolysis is the passing of a direct electric current through a compound in solution in water or when molten, resulting in the decomposition of the compound. Figure 13.32 shows what happens during electrolysis.

Human beings conduct electricity because our bodies contain ions in solution. We conduct electricity by electrolysis.

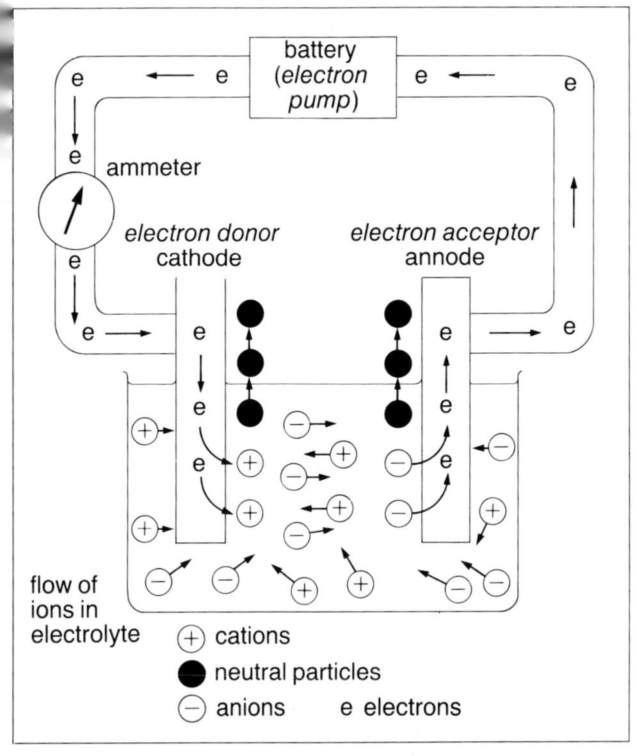

figure 13.32 *How electrolysis works*

INVESTIGATION 13.17

ELECTROLYSIS OF SOLUTIONS

a Connect an electrolysis cell (see figure 13.33) to a 6 V d.c. supply. Include a 6 V bulb in the circuit.
b Fill the cell to the top of the carbon rods with distilled water.
c Switch on and note whether the bulb lights.
d Empty the cell and refill it with strong salt solution. Note whether the bulb lights now.
e If any gas appears at the carbon rods, collect the gas in an inverted test tube and identify it.
f Repeat the experiment with
  i copper sulphate solution
  ii very dilute sulphuric acid (*care*).

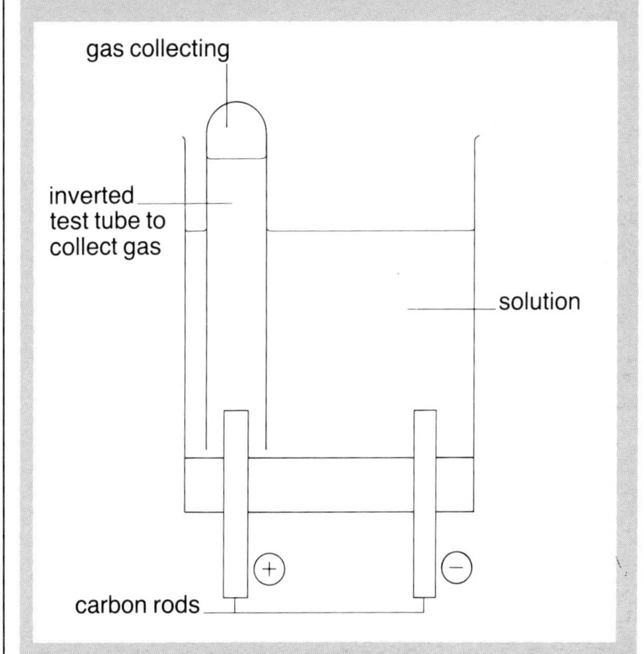

figure 13.33 *An electrolysis cell*

Questions

1 Why is pure water a poor conductor of electricity?
2 Which gas is produced at the cathode?
3 Devise an experiment to plate a metal object with a layer of copper.

Table 13.11 shows a summary of the electrolysis of various solutions.

solution	anode product	cathode product
sodium chloride	chlorine	hydrogen *
lead nitrate	oxygen*	lead
dilute sulphuric acid	oxygen*	hydrogen
copper sulphate (carbon electrodes)	oxygen*	copper
copper sulphate (copper electrodes)	anode dissolves	copper

*These products come from the water in the solutions

table 13.11

Application 13.10

USING ELECTROLYSIS

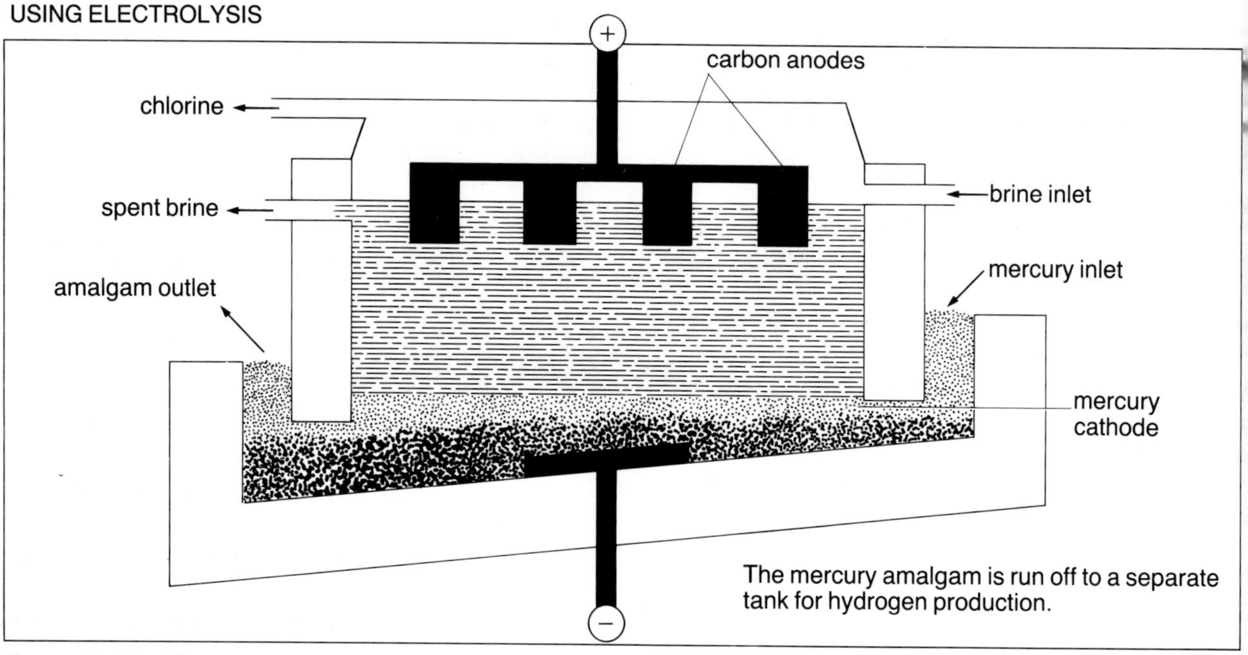

+
carbon anodes
chlorine
spent brine
brine inlet
mercury inlet
amalgam outlet
mercury cathode
The mercury amalgam is run off to a separate tank for hydrogen production.
–

figure 13.34 The mercury cathode cell

Electrolysis of salt solution

This is a very important industrial process. Salt is a readily available raw material and all three of the products, sodium, chlorine and hydrogen, are very useful. The cell is designed so that the process can run continuously and so that the three products are produced separately (see figure 13.34).

Hydrogen is used to make ammonia, margarine and hydrochloric acid. Chlorine is involved in purifying water, making plastics, disinfectants and hydrochloric acid and used as a bleach. Sodium hydroxide is used in the manufacture of paper, rayon and soap.

Extraction of aluminium

Aluminium is the commonest metal in the Earth's crust. It is a very useful, light structured metal. Aluminium is extracted from molten aluminium oxide. The ore is called **bauxite**. Aluminium oxide has a very high melting point and so it is mixed with another substance called cryolite. This reduces the melting point to about 800 °C. The aluminium is produced at the cathode and collects as the molten metal.

$$Al^{3+} + 3e^- \rightarrow Al$$

aluminium ions → aluminium metal

Oxygen is produced at the anode which is made of huge blocks of carbon.

$$2O^{2-} - 4e^- \rightarrow O_2$$

oxide ions → oxygen gas

These carbon anodes have to be replaced frequently because the oxygen combines with the hot carbon to form carbon dioxide (see figure 13.35).

The main uses of aluminium are shown in table 13.8, page 206.

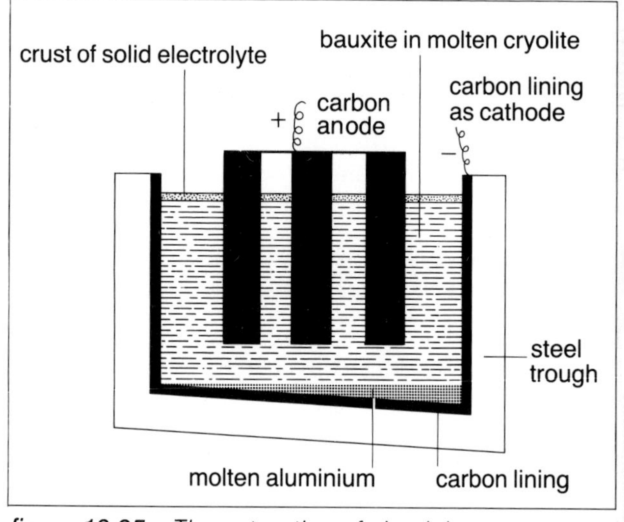

figure 13.35 The extraction of aluminium

figure 13.36 Another use of aluminium

Time for thought

1. *There is a large aluminium producing factory near Fort William in Scotland. Why do you think it is sited there?*
2. *Many drinks cans are made of aluminium. What do you think would be the best thing to do with the empty cans?*

Electroplating
We have already seen how a thin layer of one metal is plated on top of another to prevent corrosion. Another reason for plating is to improve the appearance of the metal. Sometimes copper or iron is coated with a thin layer of an expensive metal like gold and silver. This gives the object the appearance of the expensive metal although sometimes this layer is very thin and wears off. The object to be plated is connected to the cathode and the solution contains ions of the expensive metal.

Purification of copper
During the electrolysis of copper(II) sulphate solution using copper electrodes, copper is transferred from the anode to the cathode. In the purification of copper, the anode is made of a large block of impure copper, and the cathode is made of a thin sheet of pure copper. As the copper is transferred during electrolysis, the impurities sink to the bottom. These impurities often contain silver and gold which can be recovered later.

ALL YOUR OWN WORK

1. All acids contain the element ________.
2. Before anything can act as an acid ________ must also be present. This allows the acid to form ________ ________.
3. ________ and ________ are examples of fairly reactive metals.
4. The ________ is a measure of the strength of an acid. This is also a measure of the ________ ________ concentration.
5. Why are most reactions fastest at the beginning?
6. Why is the recycling of metals so important?
7. Magnesium reacts with dilute hydrochloric acid as follows. (For an explanation of (s), (aq) and (g) see page 254.)

$$Mg(s) + 2HCl(aq) \rightarrow MgCl_2(aq) + H_2(g)$$

a A pupil carried out an experiment to investigate the rate of reaction between a piece of magnesium and 40 cm^3 of dilute hydrochloric acid at 20 °C. The acid was in excess.
The results are shown in figure 13.37.

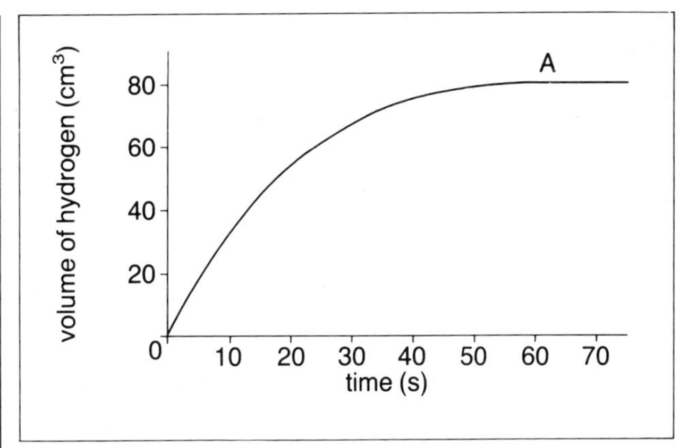

figure 13.37

i What is the gas produced when magnesium reacts with hydrochloric acid?
ii What volume of gas was collected after 30 seconds?
iii After how many seconds was the reaction just completed?

b The experiment was repeated using an equal mass of magnesium cut into very small pieces and a fresh 40 cm^3 of hydrochloric acid at 20 °C. The results are shown in table 13.12.

time/s	0	5	10	15	20	25	30	35	40	45	50
volume of gas/cm^3	0	34	50	62	70	76	79	80	80	80	80

table 13.12

i Roughly sketch out the graph in figure 13.37, then plot these results on the graph and label this line B.
ii Which of the two lines on the graph refers to the faster reaction?
iii Explain why the two reactions do not take place at the same time.

c Which one of the following methods would increase the **final** volume of gas collected? Write the letter of the method chosen and explain your reasoning.
A use of more magnesium
B use of a larger volume of dilute hydrochloric acid
C use of a more concentrated solution of hydrochloric acid

d Great care is taken in flour mills to prevent sparks or flames.
Why is the risk of explosion high in a flour mill?
[LEAG]

UNIT 14

CHEMICALS AROUND US

THE AIR

Air is all around us, and because we cannot see it, we often take it for granted. At times we may forget it is there. Air is, however, needed by every living thing.

Composition of the air

Air is a mixture of several different gases. The amount of each gas varies very slightly. Table 14.1 shows the composition of a typical sample of dry air.

gas	percentage of gas in air
nitrogen	78
oxygen	21
argon	0.9
carbon dioxide	0.04
neon, krypton, xenon	0.06

table 14.1 *Compostion of dry air*

You will see that 99 per cent of the air is made up from just two gases, nitrogen and oxygen.

Air also contains a variable amount of water vapour. The **humidity** is a measure of how much water vapour there is in the air.

Obtaining oxygen and nitrogen from the air

As air is a mixture it is possible to separate it. A mixture of gases, however, is fairly difficult to separate. A mixture of liquids is much easier to separate. The air is liquefied. First the water vapour and carbon dioxide must be removed. These two compounds have much higher freezing points so would soon form solids on cooling and block up the pipes.

In order to return the air into a liquid it must be cooled to a very low temperature, about −200 °C. Temperatures of this sort are not easy to reach. The air is compressed and cooled. When it is suddenly allowed to expand again the molecules spread out and slow down, so it gets even colder. This process is repeated until it turns into a liquid (see figure 14.2).

The liquid air can then be fractionally distilled (see page 168). The nitrogen has the lower boiling point, −196 °C, and so distils off first, leaving the liquid oxygen, boiling point −182 °C behind. The noble gases can also be obtained from the liquid air by fractional distillation.

INVESTIGATION 14.1

HOW MUCH OXYGEN IS IN A SAMPLE OF AIR?

a Connect up two gas syringes as shown in figure 14.1. Make sure that one syringe is completely empty and that the other syringe contains 100 cm^3 of air.
b Pass the air over the copper to check that there are no leaks in the apparatus.
c Heat the copper strongly and then pass the air over the hot copper. Note any change in volume.
d Move the Bunsen along the tube in order to heat a different sample of copper. Pass the air over the copper again.
e Repeatedly pass the air over the copper until there is no further decrease in volume.

Questions

1 Why is the copper in the form of many tiny pieces?
2 What do you notice happening to the hot copper when the air is passed over it?
3 What was the final volume in the gas syringe?
4 How could you be sure that all the oxygen had been used up?

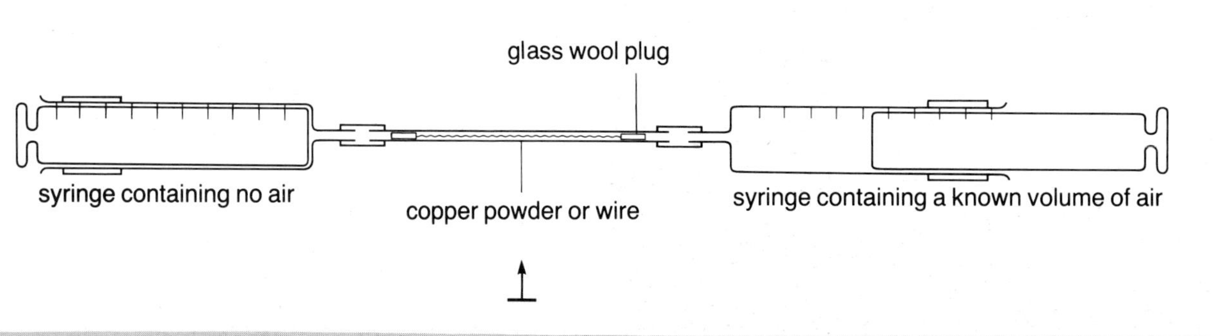

figure 14.1 *Finding the percentage of oxygen in the air*

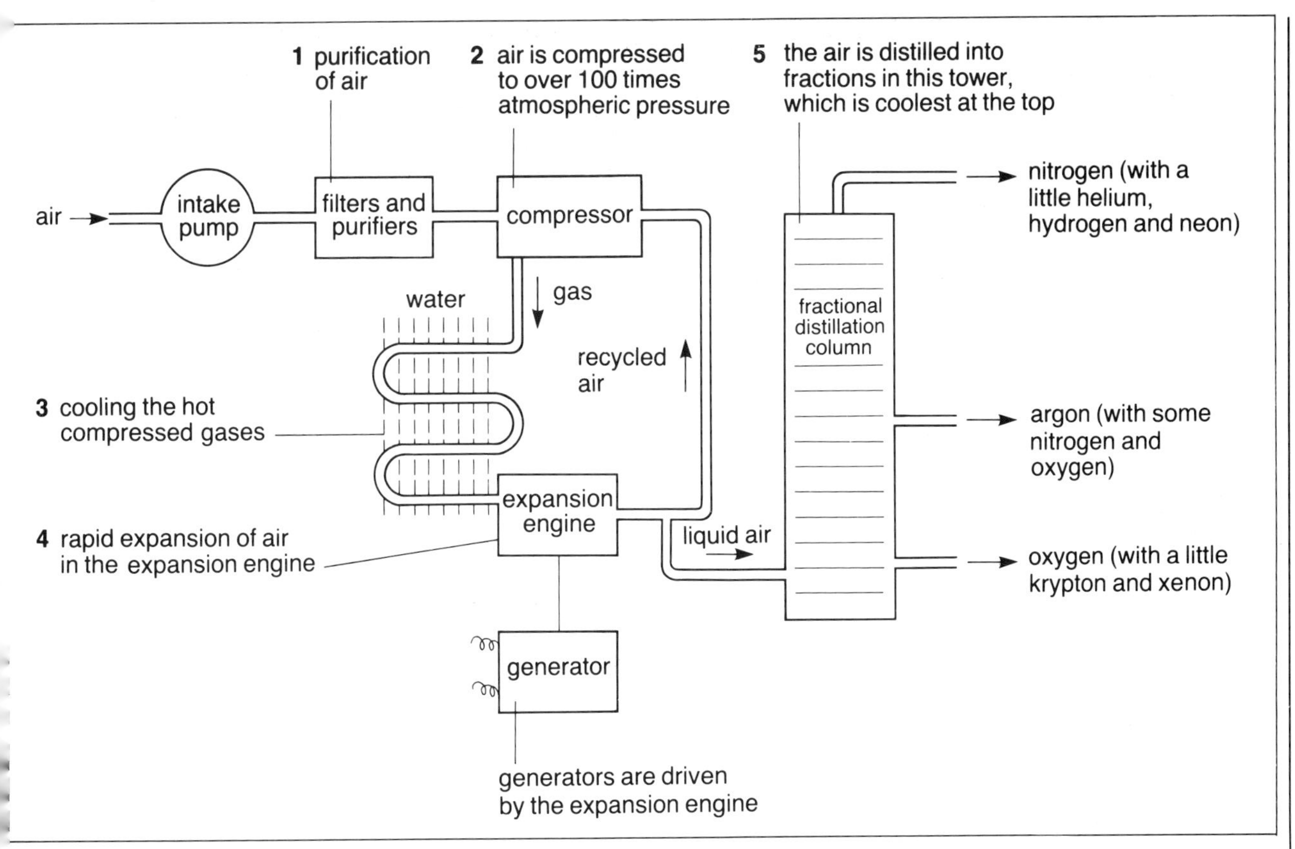

figure 14.2 *Separating the gases in the air*

OXYGEN

Application 14.1

USES OF OXYGEN

Breathing

Oxygen is needed by all living things. Astronauts, pilots of high flying aircraft and climbers of very high mountains need to take oxygen with them because there is less oxygen high up in the atmosphere.

The oxygen is often mixed with an unreactive gas like helium to dilute it, just as nitrogen dilutes the oxygen in the air. It is not good for us to breathe pure oxygen for very long.

Oxygen enriched air is used in hospitals for people who have difficulty absorbing enough oxygen into the their bloodstreams.

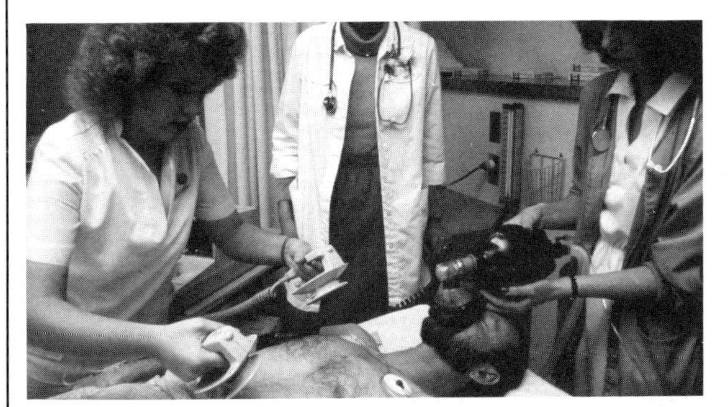

figure 14.3 *Oxygen assists breathing*

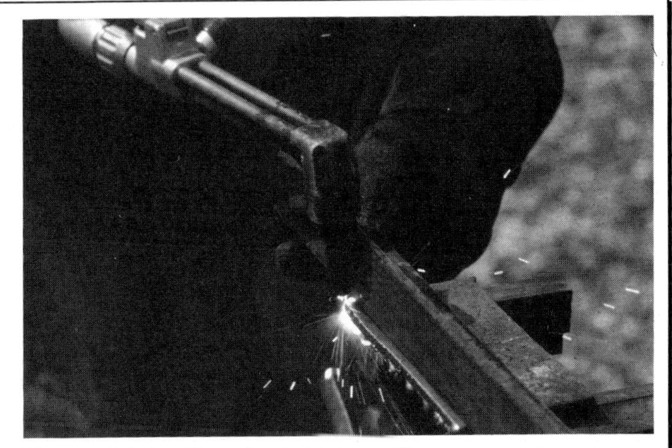

figure 14.4 *Oxy-acetylene cutting*

Making steel

Large amounts of oxygen are needed to "burn out" the impurities in the iron from the blast furnace before carefully controlled amounts of carbon and other metals can be added to the molten iron (see figure 13.23, page 207).

Oxy-acetylene welding

Acetylene (ethyne) burns at a very high temperature when it is mixed with the correct amount of oxygen. The flame is hot enough to melt iron and can be used for cutting through metals or welding them together (see figure 14.4).

INVESTIGATION 14.2

MAKING OXYGEN

One of the easiest ways of making oxygen in the laboratory is by the decomposition of hydrogen peroxide. Hydrogen peroxide splits up into water and oxygen. The reaction is very slow at room temperature but can be speeded up by adding some manganese(IV) oxide as a catalyst (see page 201).

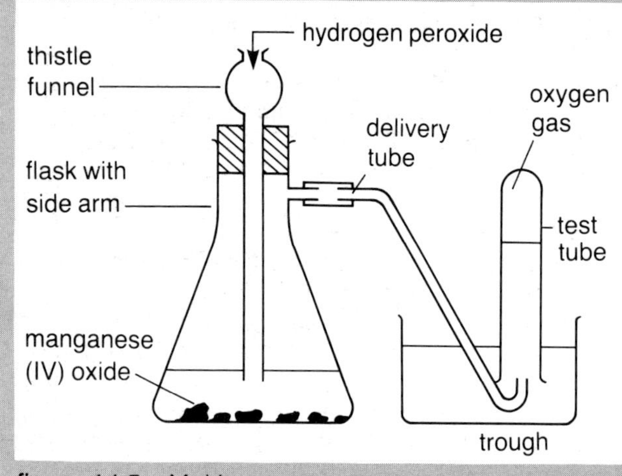

figure 14.5 Making oxygen

Warning Hydrogen peroxide is an acid and a bleach.

a Connect up the apparatus as shown in figure 14.5.
b Place a small amount of manganese (IV) oxide in the conical flask.
c Add a few cm^3 of hydrogen peroxide solution to the flask through the thistle funnel. Do not collect the bubbles of gas.
d When the bubbles stop, place a test tube full of water over the delivery tube and add a few more cm^3 of hydrogen peroxide to the flask. Collect a test tube full of oxygen.
e Test the gas with a glowing spill.

Questions

1 Why did you not collect the first lot of bubbles?
2 What happened to the glowing splint?
3 What does the method of collection tell you about oxygen?
4 How could you show that the manganese(IV) oxide had not been used up in the reaction?

Oxides

An **oxide** is a compound of another element with oxygen. Most of the other elements will react with oxygen to form oxides. You will already have come across several different oxides e.g. copper oxide, carbon dioxide and water (hydrogen oxide). From these examples you can see that oxides can look and act very differently.

Types of oxides

Oxides can be divided into four main groups.

Basic oxides
These are the usual oxides of metals. They are solids at room temperature and react with acids to form a salt and water, e.g. magnesium oxide, copper oxide.

Acidic oxides
These are the usual oxides of non-metals. They are often gases at room temperature and dissolve in water to form acids, e.g. carbon dioxide, sulphur dioxide.

Neutral oxides
These are also oxides of non-metals, but they do not dissolve in water to form acids. The most common examples are carbon monoxide, nitrogen monoxide and, of course, water.

Amphoteric oxides
This is a group of rather unusual metal oxides that will react with both acids and alkalis. The most common examples are zinc oxide and aluminium oxide.

Combustion

Combustion is another name for burning. We have seen that when an element burns in oxygen an oxide is formed. Combustion is an **exothermic** reaction, i.e. the reaction gives out heat. We often use combustion to produce heat for ourselves, e.g. when using a gas cooker or coal fire (see figure 14.6).

figure 14.6 Combustion to keep us warm

INVESTIGATION 14.3

MAKING OXIDES

Magnesium

a Wrap the end of a piece of magnesium ribbon round the end of a burning spoon leaving about 2–3 cm ribbon below the spoon.
b Set fire to the magnesium and lower it into a boiling tube full of oxygen (see figure 14.7). (Take special care and do not stare at the burning magnesium.)
c Add 1 cm depth of water and shake the ash in the water. Add a few drops of universal indicator solution.

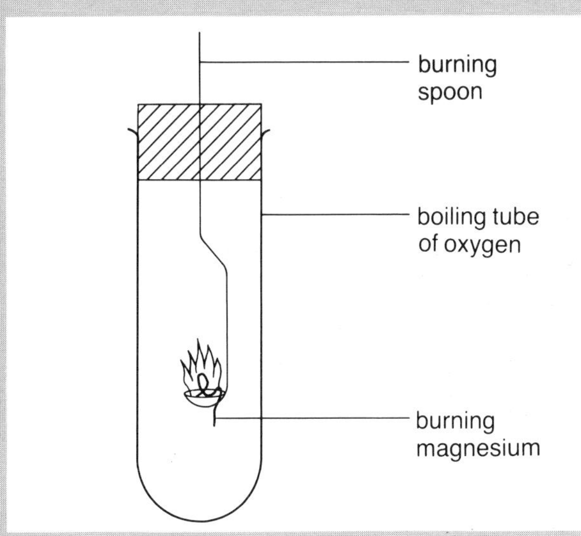

figure 14.7 Making magnesium oxide

Iron

a Wrap a small piece of iron wool loosely round the burning spoon.
b Heat it in the Bunsen flame and quickly lower it into a boiling tube full of oxygen.
c Add a few drops of universal solution.

Carbon

a Carefully balance a small piece of charcoal on the burning spoon and heat it until it is red-hot.
b Lower the red-hot charcoal into a boiling tube full of oxygen.
c When the charcoal has gone out, add 1 cm depth of water to the boiling tube, making sure that the gas does not escape. Shake the boiling tube and then add a few drops of universal indicator.

Sulphur

a Dip the burning spoon into some powdered sulphur to put a small pile of sulphur on the spoon.
b Heat the sulphur in the Bunsen flame until it melts. Lower the burning sulphur into the boiling tube full of oxygen.
c When the sulphur has gone out, add about 1 cm depth of water to the boiling tube taking care not to let the gas escape (the gas has an unpleasant, choking smell). Shake the tube and add a few drops of universal indicator solution.

Questions

1 Complete table 14.2.
2 Are non-metal oxides different from metal oxides?

element	appearance	reaction with oxygen	appearance of oxide	colour of indicator
magnesium				
iron				
carbon				
sulphur				

table 14.2

When a compound burns, the products are usually the oxides of the elements in the compound, e.g. methane is a compound of carbon and hydrogen.

methane + oxygen → carbon dioxide + water
(natural gas)

Oxidation and reduction

When substances combine with oxygen to form oxides we can describe the reaction as an **oxidation** reaction. In the reaction between magnesium and oxygen, we say that the magnesium is being **oxidized**.

magnesium + oxygen → magnesium oxide

$$2Mg + O_2 \rightarrow 2MgO$$

figure 14.8 Saturn rocket lifting off

Oxidation can be defined in three ways. It is
a the addition of oxygen
b the loss of hydrogen
c the loss of electrons.

One or more of these definitions may apply to a particular reaction.

When magnesium reacts with oxygen it loses electrons to become a magnesium ion. Sometimes the oxygen is provided by another compound.

If a mixture of lead(II) oxide and carbon is heated, the carbon takes the oxygen away from the lead(II) oxide.

lead(II) oxide + carbon → carbon dioxide + lead

$$2PbO + C \rightarrow CO_2 + 2Pb$$

The carbon is oxidized by the lead(II) oxide which itself loses oxygen. The taking away of oxygen, i.e. the opposite of oxidation, is called **reduction**.

Reduction is defined as
a the taking away of oxygen
b the adding of hydrogen
c the gaining of electrons.

A reaction that involves both oxidation and reduction is called a **redox** reaction.

Examples of redox reactions are given below.

copper oxide + hydrogen → copper + water

$$CuO + H_2 \rightarrow Cu + H_2O$$

reduced (CuO → Cu); oxidized (H_2 → H_2O)

methane + oxygen → carbon dioxide + water

$$CH_4 + 2O_2 \rightarrow CO_2 + 2H_2O$$

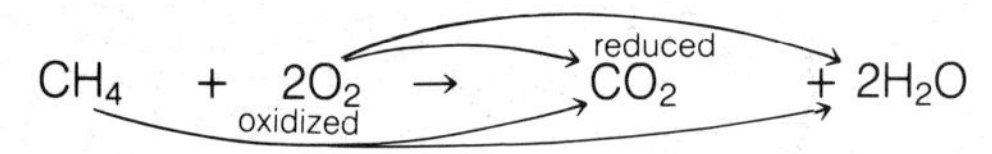

iron(II) chloride + chlorine → iron(III) chloride

$$FeCl_2 + Cl_2 \rightarrow FeCl_3$$

reduced (Cl_2 → $FeCl_3$); oxidized ($FeCl_2$ → $FeCl_3$)

The iron(II) ions have lost an electron, so have been oxidized.

$$Fe^{2+} - 1e^- \rightarrow Fe^{3+}$$

iron(II) → iron(III)

The chlorine atoms gain an electron to become chloride ions, so have been reduced.

$$Cl + 1e \rightarrow Cl^-$$

chlorine atom → chloride ion

concentrated hydrochloric acid + manganese(IV) chloride → managanese(II) chloride + water + chlorine

$$4HCl + MnO_2 \rightarrow MnCl_2 + 2H_2O + Cl_2$$

reduced (MnO_2 → $MnCl_2$); oxidized (HCl → Cl_2)

The substance that brings about the oxidation is called an **oxidizing agent**.

iron(III) oxide + carbon monoxide → iron + carbon dioxide

$$Fe_2O_3 + 3CO \rightarrow 2Fe + 3CO_2$$

reduced (Fe_2O_3 → Fe); oxidized (CO → CO_2)

This is the main reaction in the blast furnace for the production of iron (see page 206).

During electrolysis, electrons are being gained by positive ions at the cathode (see page 211).

copper ions + electrons → copper atoms

$$Cu^{2+} + 2e^- \rightarrow Cu$$

The gaining of electrons is reduction.

At the anode, electrons are being taken away from negative ions.

chloride ions − electrons → chlorine atoms

$$Cl^- - 1e^- \rightarrow Cl$$

The loss of electrons is oxidation.

In a redox reaction the oxidizing agent is itself reduced. A substance that brings about reduction is called a **reducing agent**. In the reaction the reducing agent is itself oxidized. Table 14.3 shows some oxidizing and reducing agents.

oxidizing agents	reducing agents
oxygen	hydrogen
hydrogen peroxide	carbon monoxide
manganese (IV) oxide	sulphur dioxide
potassium permanganate	metals
chlorine	

table 14.3

figure 14.9 Electrolysis used to purify copper

Respiration

Respiration is also an oxidation reaction. It is the oxidation of **carbohydrates** and takes place in all living things. The purpose of respiration is to provide energy.

carbohydrate + oxygen → carbon dioxide + water + ENERGY

Carbohydrates are chemicals made from the elements carbon, hydrogen and oxygen. Starch and sugar are examples of carbohydrates.

figure 14.10 *Oxidation of carbohydrates gives energy*

One of the simplest sugars is called **glucose**, $C_6H_{12}O_6$.

When glucose is oxidized, carbon dioxide and water are produced.

$$C_6H_{12}O_6 + 6O_2 \rightarrow 6CO_2 + 6H_2O + \text{ENERGY}$$

Notice that this equation for the oxidation of glucose in living things is the same as that for burning glucose in oxygen. Table 14.4 shows a comparison of respiration and combustion of glucose.

Photosynthesis

Table 14.4 shows that respiration and combustion use up oxygen and produce carbon dioxide. Does this mean then that the amount of oxygen in the air is getting less and the amount of carbon dioxide increasing?

Table 14.1 (page 214) shows a surprisingly small amount of carbon dioxide in the air. Where does the carbon dioxide produced by respiration and combustion go?

The answer to these questions is **photosynthesis** (see pages 82—3). This is a process that takes place in all green plants. In the presence of sunlight, green plants convert carbon dioxide and water into carbohydrates and oxygen. The reaction is the reverse of respiration.

$$\text{carbon dioxide} + \text{water} \xrightarrow[\text{chlorophyll}]{\text{sunlight}} \text{carbohydrate} + \text{oxygen}$$

Chlorophyll is a catalyst. It is the green colour in plants. So while it is light, all green plants are using up carbon dioxide and making oxygen.

Respiration and photosynthesis maintain fairly constant percentages of oxygen and carbon dioxide in the air.

figure 14.11 *Trees maintain a supply of oxygen*

Time for thought

What effect could it have on us all if the Amazonion rain forest was cut down or destroyed (see page 82)?

respiration	combustion
takes place in living cells, needs oxygen	takes place in air or oxygen
produces energy (heat and chemical)	produces energy (heat and light)
produces carbon dioxide and water	produces carbon dioxide and water
complicated series of reactions helped by enzyme catalysts	relatively simple one stage reaction, no other chemicals involved

table 14.4

CARBON DIOXIDE

The balance of carbon dioxide and carbon compounds was shown in figure 5.1 (page 82).

INVESTIGATION 14.4

MAKING CARBON DIOXIDE

Carbon dioxide can be made by reacting a metal carbonate with a dilute acid.

- a Connect up the apparatus as shown in figure 14.12. Five or six pieces of marble will be enough.
- b Carefully add dilute hydrochloric acid down the thistle funnel.
- c Add a few cm^3 of lime water to the boiling tube and bubble the carbon dioxide through the lime water.
- d In another boiling tube, bubble the carbon dioxide through some universal indicator solution.
- e Fill another boiling tube with carbon dioxide and then put a lighted spill into the boiling tube.

Questions

1. This method of collection is called downward delivery. What does this method tell you about carbon dioxide?
2. What is the chemical name for lime water?
3. What happens to the lime water? This reaction is used as a test for carbon dioxide.
4. What effect does carbon dioxide have on universal indicator and on a lighted spill?

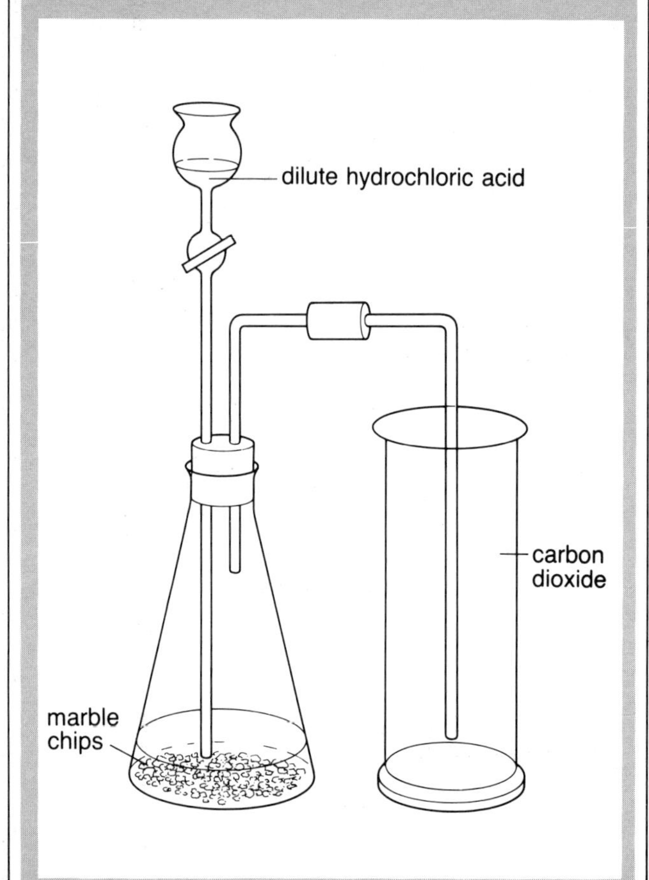

figure 14.12 *Making carbon dioxide*

Application 14.2

USING CARBON DIOXIDE

Fire extinguishers

Carbon dioxide is used in 'foam' fire extinguishers and in carbon dioxide gas extinguishers (see pages 234-5). It is most likely that the extinguisher in your laboratory is a carbon dioxide gas extinguisher.

figure 14.13 *A carbon dioxide fire extinguisher*

The carbon dioxide gas will not support burning, and because it is more dense than air, it forms a 'blanket' over the fire and stops air getting to it.

Fizzy drinks

The 'fizz' in fizzy drinks is carbon dioxide gas. It is dissolved in the drink under quite high pressure. When the drink is opened the carbon dioxide comes out of the solution and streams of bubbles escape to the air.

figure 14.14 *Carbon dioxide in a fizzy drink*

Dry ice

Carbon dioxide sublimes when it is cooled, i.e. it changes straight from the gas to a solid. This solid carbon dioxide can be used for refrigerating ice cream and other foods, especially during transit.

NITROGEN

Nitrogen makes up about 80 per cent of the air. It is essential to all living things as it is in all **proteins**. Proteins are body building materials. As an element, however, nitrogen is very unreactive and only a few other elements will combine with it directly. We get our nitrogen compounds by eating animal or plant protein.

Figure 14.15 outlines the nitrogen cycle.

A plant takes in **nitrates** and some **ammonium compounds** from the soil. These compounds are in solution and are absorbed by the roots of the plant. The plant uses them to make proteins. Some plants have **nitrogen-fixing bacteria** in their roots which enable them to use nitrogen gas from the air. The bacteria convert the atmospheric nitrogen into compounds that the plant can make use of. This family of plants, called **legumes** includes peas, beans and clover.

Plants are either eaten by animals or die and decay in the soil. Animals excrete waste material containing nitrogen compounds and then die and decay. These processes all return nitrogen compounds to the soil. In the soil, bacteria convert some of these compounds to ammonium compounds and nitrates. Other bacteria (**denitrifying bacteria**) change the nitrogen compounds back to nitrogen gas.

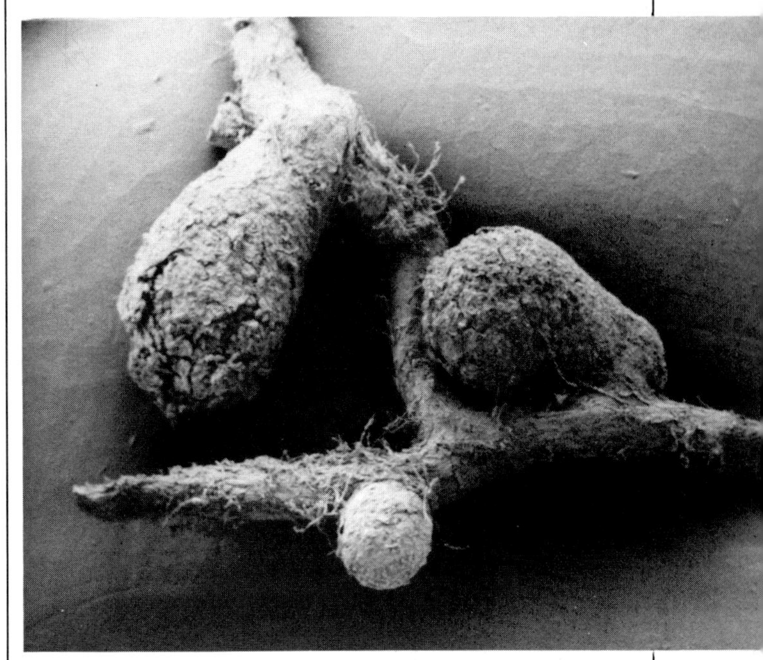

figure 14.16 *Nodules containing nitrogen-fixing bacteria on legume roots*

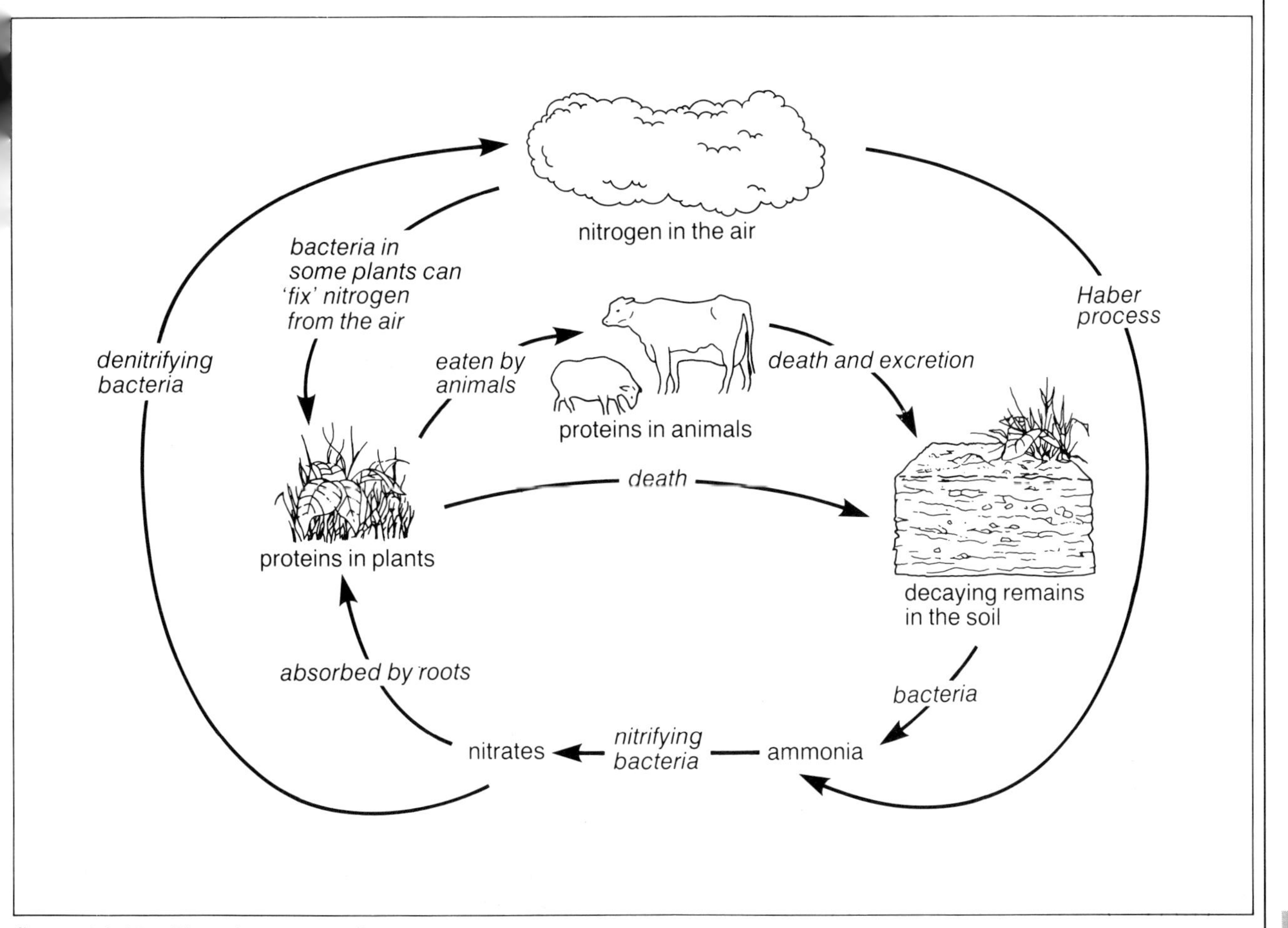

figure 14.15 *The nitrogen cycle*

figure 14.17 *Lightning returns nitrogen to the soil*

During a thunderstorm lightning cause some of the nitrogen in the air to react with oxygen to form nitrogen dioxide. This dissolves in the rain and forms nitrates in the soil.

Extensive farming has meant that extra nitrogen compounds have had to be added to the soil. In the early 1900s a German chemist, Fritz Haber, found a way of making ammonia from nitrogen and hydrogen. The process is still called the Haber process and it is very important in industry today (see page 203). Large amounts of fertilizers are made using this process.

The conversion of nitrogen into its compounds is called the **fixation** of nitrogen.

Application 14.3

USING NITROGEN AND ITS COMPOUNDS

Making fertilizers
Nitrogen is essential to all living things. All proteins contain nitrogen. Nitrogen from the air is converted into ammonia by the Haber process. Ammonia itself can be used as a fertilizer. Usually, however, it is converted into nitric acid and then made into nitrates.

Making explosives
Nitrogen from the air is again converted to nitric acid which can then be used to make many different explosives, e.g. TNT (trinitrotoluene).

Making dyes and drugs
Many dyes and drugs are made using nitrogen compounds.

Fertilizers

Plants need a number of chemicals for healthy growth. Most of these they obtain from the soil. Plants take in chemicals in solution through their roots and so the chemicals they use must be soluble in water.

The three most important elements needed by the plant are **nitrogen**, **phosphorus** and **potassium**. Other elements, called **trace elements**, are needed in small amounts. These include magnesium and iron.

If the same crops are grown on the same piece of land year after year, the soil becomes exhausted of these necessary nutrients and so the plants do not grow so well. **Crop rotation** is one way of trying to avoid this. Three different crops are grown on the area for three years and then the ground is left uncultivated for the fourth year.

In these days of intensive farming, high yields are required and crop rotation is not sufficient to maintain the fertility of the soil. Large amounts of chemical fertilizers are often added.

Some common fertilizers are described below.

NPK fertilizer
This is a general fertilizer containing nitrogen, phosphorus and potassium. The actual composition varies according to the farmer's need and is shown on the label (see figure 14.18).

Nitrochalk
This is a mixture of ammonium nitrate and calcium carbonate.

'Nitram'
This is ammonium nitrate. It contains quite a high percentage of nitrogen and is a quick-acting fertilizer because of its high solubility.

Super phosphate
This is mainly calcium dihydrogen phosphate which is made from rock containing the phosphate.

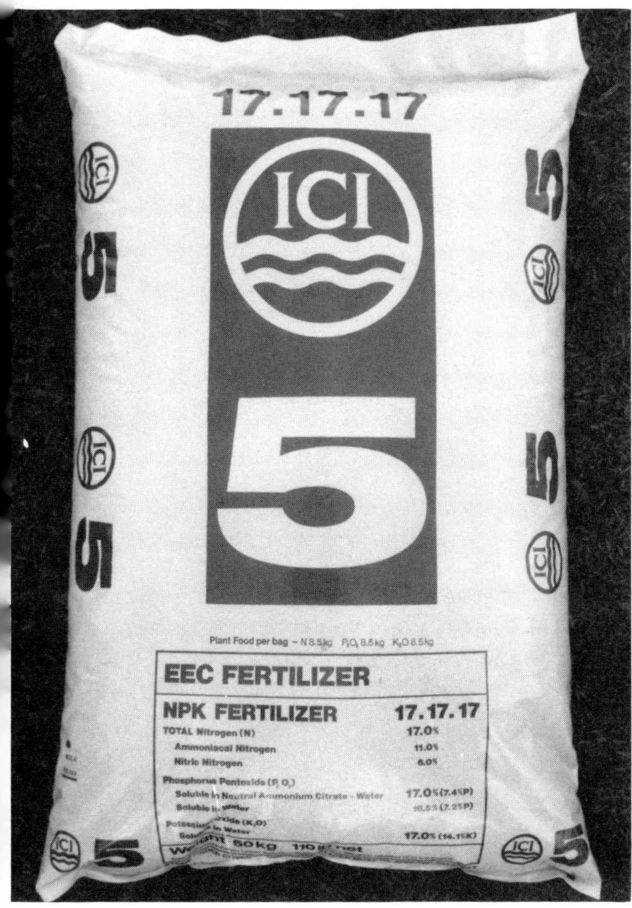

figure 14.18 *NPK fertilizer – the percentages of N, P and K are shown at the top of the bag (17% N, 17% P and 17% K)*

Upsetting the balance

Adding too much fertilizer can cause serious problems. Recent reports have shown that there is an increasingly high concentration of nitrate in our own drinking water. In some areas this is reaching unacceptable levels and there is concern over the effect of high nitrate concentration on our health. As nitrates are so soluble they are easily washed out of the soil into rivers and streams. This can cause another problem. The high nitrogen levels in the ponds and streams encourage rapid growth of bacteria and other microorganisms. This in turn greatly increases the demand for oxygen. Eventually there is not enough oxygen to go round. Fish and plants soon die, as do the microorganisms. This process is called **eutrophication**.

figure 14.19 *Nitrates can cause a lack of oxygen in ponds*

When fertilizers are added, the pH of the soil may alter. Some fertilizers such as ammonia are alkaline whereas others like ammonium sulphate are acidic. To counteract this increased acidity lime (calcium hydroxide) is added to the soil.

It is mainly because of these problems that some farmers are returning to 'organic' farming. On 'organic' farms the farmer uses only natural fertilizers, often without any synthetic chemicals at all.

figure 14.20 *Organically grown foods are gaining popularity*

Application 14.4

AIR POLLUTION

During the last few years we have all become far more aware of the need to look after our environment. The various Clean Air Acts, introduced since 1956, have done a great deal towards cleaning the air in our cities (see figure 14.21).

In the early 1950s, smog, a mixture of smoke and fog, caused serious health problems (see pages 62 and 76). Los Angeles still suffers from smog. The chief problem is the burning of fuels. The **fossil fuels**, coal, oil and natural gas, all contain carbon. In a plentiful supply of air they produce carbon dioxide. If, however, there is not enough air present, carbon (in the form of soot) and carbon monoxide are formed. Carbon monoxide is also produced by cars when petrol is not completely burned. Carbon monoxide is very poisonous. It combines with the red blood cells and prevents them from carrying oxygen.

figure 14.21 *Air pollution in a heavily industrialized area*

figure 14.22 *Unleaded petrol — available now*

Most fossil fuels also contain small amounts of sulphur. When this burns, it produces the gas sulphur dioxide, which dissolves in water to produce sulphurous acid. It is the sulphur dioxide in the air that causes **acid raid** (see page 196). Acid rain can damage buildings and is also thought to be responsible for the deaths of large numbers of trees in Scandinavia. Cars are the cause of another serious pollutant in the air. **Lead compounds** are added to petrol to make it burn better. The waste containing these lead compounds is released into the air by the car exhausts. It has been shown that these lead compounds can cause brain damage, especially in young children. Lead-free petrol is already available (see figure 14.22) and the amount of lead in other petrols has been reduced. By the early 1990s there should be no lead used in petrol in this country.

The banning of lead from petrol and the establishment of 'smokeless zones' go a long way towards making the air cleaner and healthier. We must always be conscious of the effect of our actions on our environment and seek to reduce pollution.

WATER

Water is the most common of all liquids. About two-thirds of the Earth's surface is covered with water. In this country we take it for granted that when we turn on the tap, fresh clean water will come out. In fact, we often complain that we have too much water.

Water is a very special liquid. It is needed by all living things. We can manage without food for several weeks, but we cannot last many days without water. It exists as a solid, liquid or gas over a relatively short range of temperature.

Chemically, water is a compound of hydrogen and oxygen (formula H_2O).

The water cycle

The **water cycle** describes the changes that happen to water in nature. Water is evaporated by the heat from the sun. The water vapour rises into the atmosphere where it cools down and condenses into cloud formations. As the clouds get higher and colder, the droplets of water get bigger and eventually fall out of the sky as rain.

figure 14.23 Fresh water

The rain either soaks into the ground or runs into streams and rivers. Eventually the water makes its way back to the sea, and the cycle is repeated (see figure 14.24).

Time for thought

1. *Can you explain why rain water, although it has come from the sea, does not taste salty?*
2. *Rivers which carry salt to the sea do not taste salty. Why?*
3. *What process which you can carry out in the laboratory is a very similar process to the water cycle? What type of water is produced (see unit 11)?*

Obtaining water for our use

Do you know where your domestic water comes from? In most places in Great Britain the domestic water supply comes from underground sources or reservoirs. Dams are often built in hilly areas to form large lakes to provide water for big cities. The reservoirs in the Elan valley in Wales (see figure 14.25, overleaf) supply much of the water for Birmingham. Each area has its own water authority. They are responsible for ensuring a pure and regular supply of water to homes and industry.

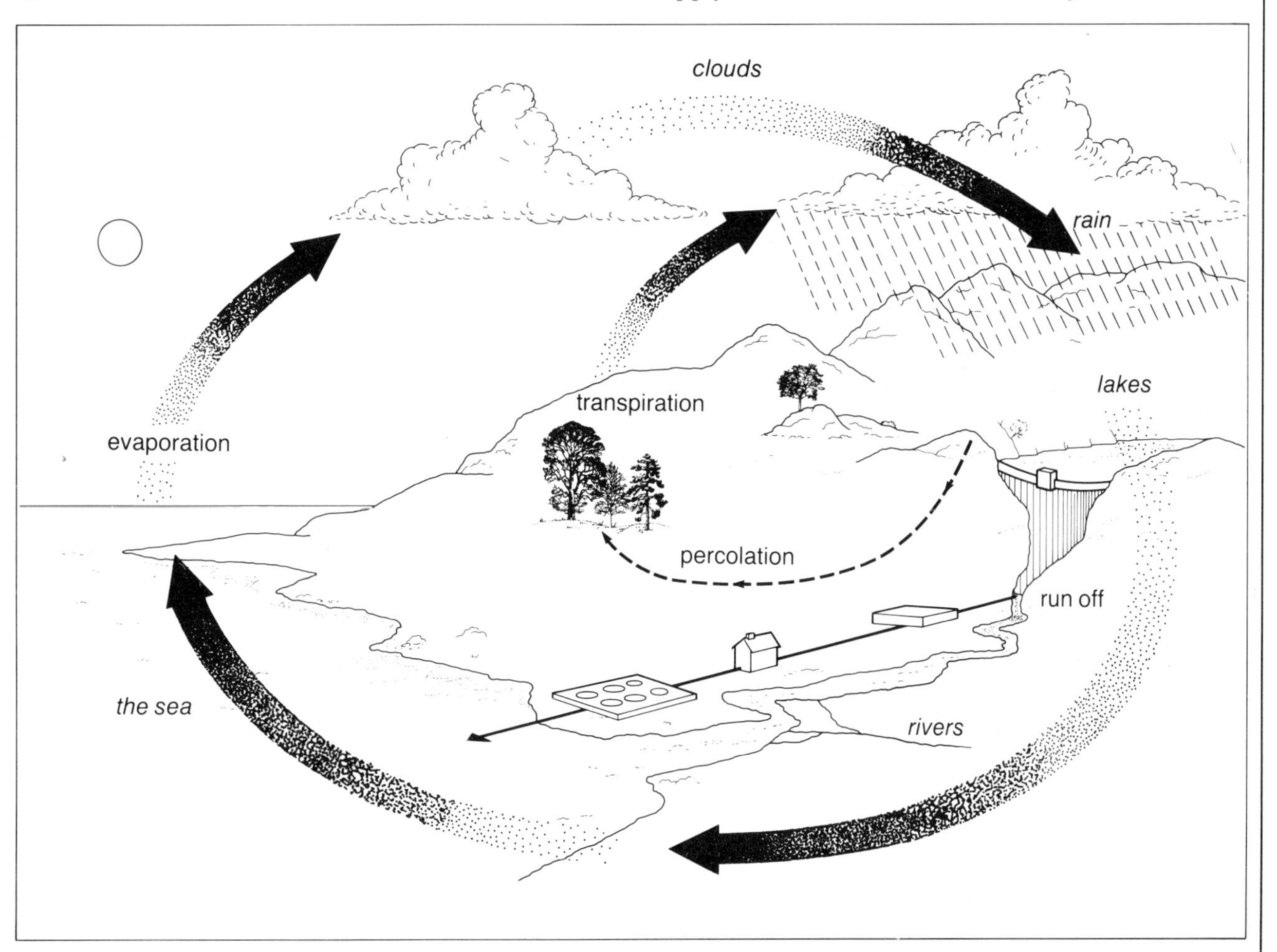

figure 14.24 The water cycle

figure 14.25 *The Elan Valley reservoir*

The water from reservoirs is first filtered to remove solid particles. The filtering is done by letting the water pass through filter beds of sand and gravel. Small amounts of **chlorine** are then added to the water to kill bacteria. Sometimes other chemicals are added, such as **lime** or **fluoride** (see page 50).

The purified water is stored high up in underground reservoirs, where gravity can be used to create the necessary water pressure to supply surrounding towns and villages (see figure 14.26).

Using water

We each use nearly 200 litres of water a day. About 70 per cent of your body is water and each day you need to drink about 2 litres of water.

Figure 14.27 shows the daily intake and excretion of water for an average man or woman.

Think of some of the ways you use water. Figure 14.28 shows an average day's use of water by one person.

In addition to domestic use, industry uses large amounts of water. Nearly all of this water finds its way back into streams and rivers. Unfortunately it is not always very clean when it is discharged into them. This can cause serious problems to life in our river systems.

Water pollution

Water pollution is a serious problem. When we have finished using water we just let it go down the drain. If that unclean water was allowed to run into streams and rivers they would soon become unfit for life, dirty and smelly. At one time untreated **sewage** was simple discharged into the sea. Nowadays there is much stricter control over the disposal of sewage.

Sewage treatment

Towns in the UK all have their own sewage works where the sewage is processed before it is allowed to be discharged into the environment (see figure 14.29).
Figure 14.30 shows a diagramatic representation of the processes involved.

1 Filters (strainers) remove large objects.
2 Non-organic waste settles out.

(*contd. page 228*)

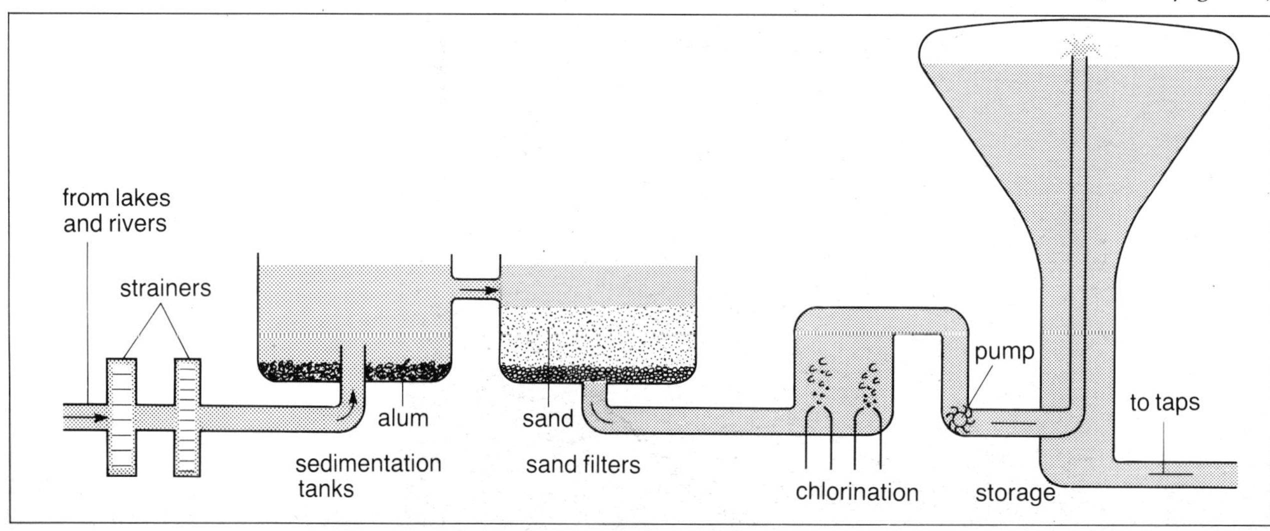

figure 14.26 *Processing water*

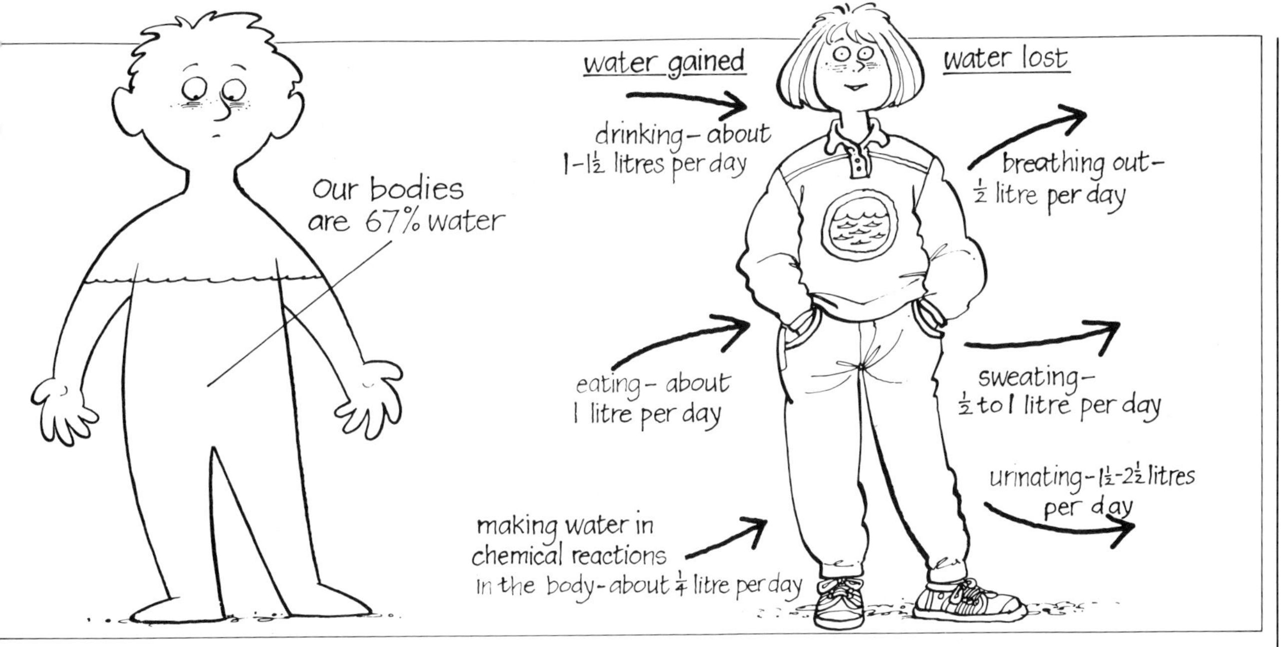

figure 14.27 *Water and us*

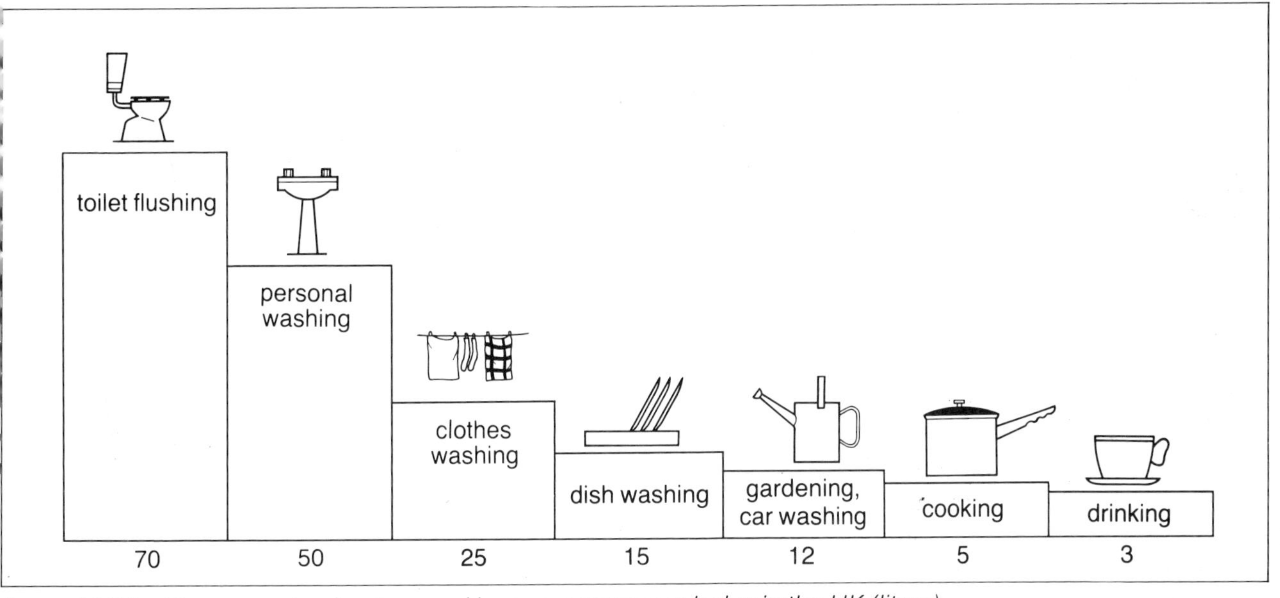

figure 14.28 *The amounts of water used by one person each day in the UK (litres)*

figure 14.29 *Sedimentation tanks at a sewage works*

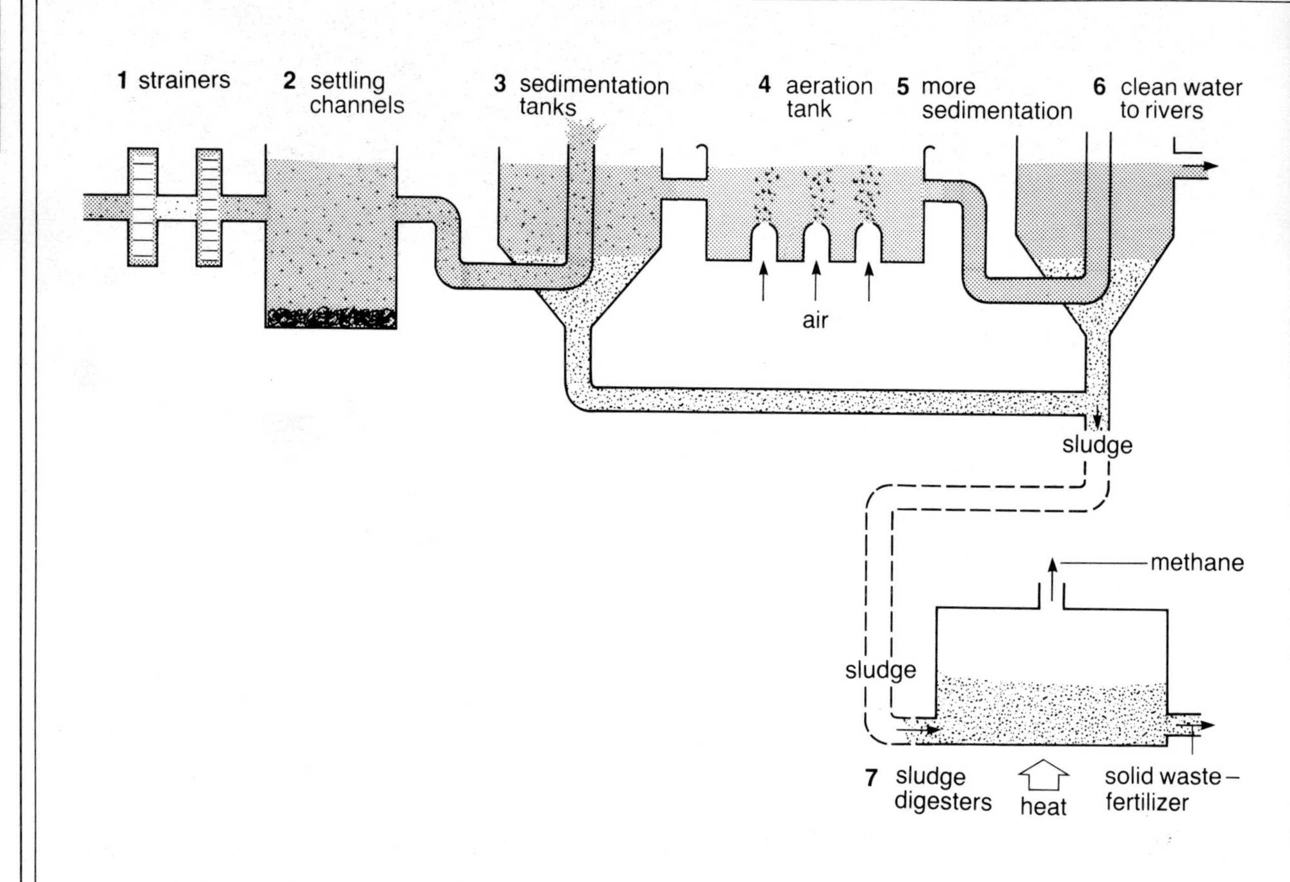

figure 14.30 *Treatment of sewage*

3 Solid organic waste settles out as a sludge in the sedimentation tanks.
4 Air is bubbled through the sewage. This provides extra oxygen for bacteria which break down the sewage.
5 Further solids are removed.
6 The purified water is discharged to nearby rivers.
7 The sludge is treated by bacteria and used to make organic fertilizers and the methane collected and used as fuel.

Industrial waste

Waste water from industry can contain all sorts of harmful chemicals. Some of those most dangerous to animal life contain metals such as mercury and lead.

People have died after eating fish contaminated with mercury compounds. Careless discharge of industrial waste has killed nearly all living things in the water of the Great Lakes near Chicago and Michigan in the USA. Recent reports have shown concern over the pollution levels in the Rhine in Germany. In the UK some of the streams and rivers in our industrial cities have been heavily polluted.

The water authorities have been working hard over the last twenty years or so to clean up our waterways. There has certainly been a great improvement. Heavy fines can be imposed on firms found guilty of discharging chemicals into our rivers, though some still persist.

Farm chemicals

On page 223 we saw how too much nitrogen being washed into ponds and streams causes problems.

Another source of pollution from farms is the use of pesticides. These are chemicals sprayed on crops to kill insects or other pests that are attacking the crop. Unfortunately, these can also get washed into nearby ponds and streams where they can harm plant and animal life. Some of these chemicals have a cumulative effect, e.g. DDT, which can build up in an animal eventually getting to a dangerous level (see page 85).

Detergents

We all use detergents for cleaning. If large amounts of these detergents are released into rivers, the river may become covered with foam (see figure 14.31). Some detergents are harmful to river life. It is possible now to obtain **biodegradable** detergents that are broken down by naturally occurring bacteria.

igure 14.31 Foaming detergents on rivers are rarely seen since the introduction of biodegradable detergents

Application 14.5

OIL POLLUTION

In the last 20 years there have been many cases of serious oil pollution around our coasts. The worst involved accidents with large oil tankers. In 1967 the tanker *Torrey Canyon* ran onto rocks off Land's End in Cornwall. Most of the cargo of crude oil ran out into the sea and formed a huge oil slick. Thousands of sea birds were killed and miles of UK beaches were affected when the oil was washed ashore (see figure 14.32). Since then there have been several other major oil pollution problems. Oil slicks are treated with detergents to try and break them up. Sometimes giant booms are used to prevent the slicks approaching the coast.

As with air pollution, we need to see water pollution as something we must try to keep to a minimum. We all have our part to play in protecting our environment.

figure 14.32 Oil slick from the Torrey Canyon off the Cornish coast

Water as a solvent

Water is a very good solvent, i.e. it is very good at dissolving things. The sea tastes salty because of the dissolved sodium chloride (salt) in it. It also contains several other dissolved substances and it can be used as a valuable source of chemicals (see figure 14.33).

Some gases also dissolve easily in water. Ammonia and sulphur dioxide are very soluble in water. Carbon dioxide is the most soluble of the gases in the air. Oxygen does not dissolve very well, but there is enough for fish and water plants to be able to obtain oxygen by diffusion (see page 6).

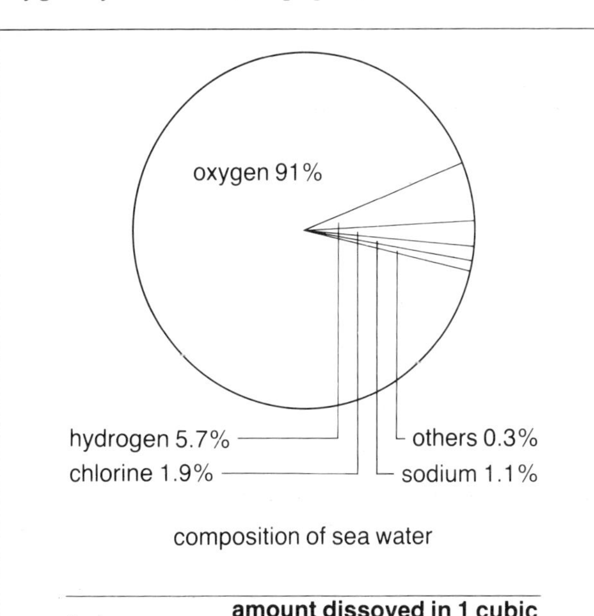

halogen	amount dissoved in 1 cubic mile sea water (tonnes)
fluorine	7000
chlorine	100 000 000
bromine	300 000
iodine	250

figure 14.33 Chemicals from the sea

INVESTIGATION 14.5

HOW MUCH AIR IS DISSOLVED IN A SAMPLE OF TAP WATER?

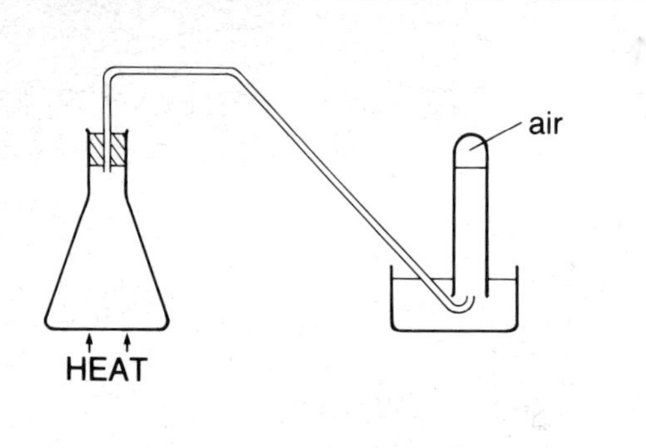

figure 14.34 *Obtaining the dissolved air from tap water*

- a Fill a 100 cm^3 conical flask completely to the top, and then measure how much water it contains using a measuring cylinder.
- b Refill the flask and connect up the apparatus as shown in figure 14.34.
  You must ensure that all the apparatus including the delivery tube is completely full of water.
- c Carefully heat the flask until it starts to boil, and keep it boiling for several seconds.
- d Allow about one-third of the flask to become filled with water vapour and then turn off the Bunsen.
- e As the water vapour condenses the flask will refill with water. Make a mark on the test tube where the water level settles.
- f Remove the test tube and fill it with water up to the line you have made.
- g Measure the volume of water (equal to the volume of air from the water in the flask) using a small measuring cylinder.

Questions

1. What does this experiment tell you about the solubility of gases in hot water?
2. Calculate the percentage of air that was dissolved in 100 cm^3 of water.

$$\text{percentage of air in 100 cm}^3\text{ water} = \frac{\text{volume of air}}{\text{volume of water}} \times 100\%$$

3. Would you expect the composition of the 'boiled out' air to be the same as that of normal air?

Solubility of solids

Investigation 14.5 suggested that the solubility of gases decreases as the temperature increases. With solids, it is the other way round. You might have noticed how sugar dissolves much better in hot tea than in cold tea. There is a maximum amount of solid that will dissolve. You cannot keep putting sugar into a cup of tea and expect it all to dissolve. There comes a time when no more will dissolve. When this happens a **saturated solution** has been formed.

The solubility of a substance is usually defined as the amount of the substance that will dissolve in 100 g of water to form a saturated solution at a given temperature. Table 14.5 shows the solubility of some common chemicals.

chemical	solubility at 20 °C (g)	solubility at 100 °C (g)
sodium chloride	36	39
sucrose	204	487
potassium nitrate	32	245

table 14.5

INVESTIGATION 14.6

THE SOLUBILITY OF SALT AT ROOM TEMPERATURE

- a Place about 25 cm^3 distilled water in a 100 cm^3 beaker.
- b Make a saturated solution by adding salt and stirring until no more will dissolve. (Be absolutely sure that no more will dissolve.)
- c Weigh a clean, dry evaporating basin.
- d Filter your saturated solution into the evaporating basin.
- e Re-weigh the basin and solution.
- f Carefully evaporate the solution until *all* the water has gone. (If you heat it too quickly it will start to spit.)
- g When the evaporating basin is cool, re-weigh the basin and salt.

Questions

1. What was the mass of your solution?
2. What mass of salt was left?
3. What mass of water was there in the solution? (solution = salt + water)
4. How much salt would have dissolved in 100 g water?

$$\text{solubility in 100 g water} = \frac{\text{mass of salt}}{\text{mass of water}} \times 100$$

Hard water

Hard water is water that will not easily form a lather with soap. It is caused by dissolved chemicals in the water. You may have noticed when you go to a different part of the country that the soap lathers better or worse than it does at home. This is because the hardness of water varies from place to place.

Hard water is caused by dissolved **calcium** and **magnesium** compounds in the water. Soaps are usually the sodium or potassium salts of fatty acids (see unit 15), e.g. sodium stearate. When soap is added to hard water the calcium and magnesium ions react with the stearate to form an insoluble compound, e.g. calcium stearate. These insoluble ompounds cause an unpleasant **scum** and also emove the soap from the water, so stopping it athering. This wastes money and prevents the soap vorking properly. There are two types of hard water, emporary hard water and permanent hard water.

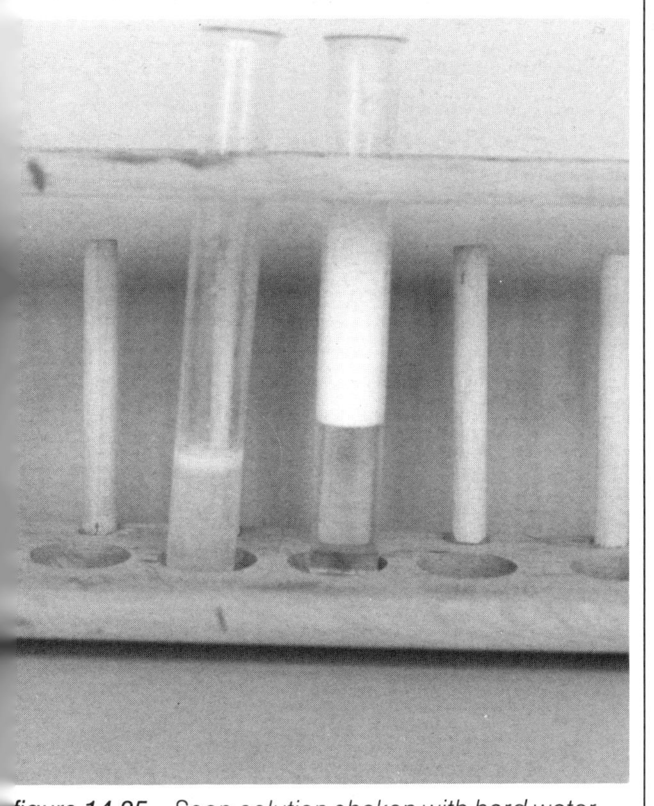

figure 14.35 *Soap solution shaken with hard water (left) and soft water (right)*

Temporary hard water

This type of hard water is caused by **calcium hydrogencarbonate**. Rain water contains a small amount of dissolved carbon dioxide which forms carbonic acid. When this rain falls on calcium carbonate rocks (limestone or chalk) some of the calcium carbonate reacts with the carbonic acid to form a solution of calcium hydrogencarbonate.

calcium carbonate + carbonic acid ⇌ calcium hydrogencarbonate

$$CaCO_3 + H_2CO_3 \rightleftharpoons Ca(HCO_3)_2$$

(The arrows show that the reaction is reversible.) When temporary hard water is boiled the reaction is reversed:

calcium hydrogencarbonate ⇌ calcium carbonate + carbon dioxide + water

$$Ca(HCO_3)_2 \rightleftharpoons CaCO_3 + CO_2 + H_2O$$

The calcium carbonate formed will not dissolve in water and so cannot affect the soap. Temporary hard water is **softened** by boiling.

If you live in an area where there is temporary hard water you will find quite a build up of 'fur' or 'scale' round the element of your electric kettle. Hot water pipes also fur up (see figure 14.36).

figure 14.36 *Furring up of pipes by temporary hard water*

This reversible reaction is also responsible for the formation of stalactites and stalagmites. The solution of calcium hydrogencarbonate drips from the cave roof, and changes back to calcium carbonate. A similar reaction can take place when the drips hit the ground (see figure 14.37).

figure 14.37 *Stalactites and stalagmites – calcium carbonate*

UNIT 14

Permanent hard water
This is caused by other calcium and magnesium compounds in the water. These compounds are unaffected by boiling so permanent hard water is *not* softened by boiling.

INVESTIGATION 14.7

HARD WATER

a About one-quarter fill a test tube with distilled water. Add five drops of soap solution and then shake the test tube.
b Repeat **a** using samples of temporary hard water and then permanent hard water.
c Boil about 30 cm^3 of temporary hard water for about 10 minutes. Repeat the test adding five drops of soap solution.
d Boil about 30 cm^3 of permanent hard water for about 10 minutes. Repeat the test adding five drops of soap solution.
e Add a crystal of sodium carbonate (washing soda) to a test tube one-quarter full of temporary hard water. Shake the tube to dissolve the crystal. Test with soap solution as before.
f Repeat **e** using permanent hard water.
g Test samples of temporary and permanent hard water with a few drops of washing-up liquid.

Questions
1 What did you see while the temporary hard water was being boiled?
2 What effect does sodium carbonate have on
 i temporary hard water
 ii permanent hard water?
3 Does hard water affect the action of a soapless detergent?

figure 14.38 Soap and soapless detergent

The results of investigation 14.7 probably showed that both types of hard water can be softened by adding washing soda (sodium carbonate). The carbonate ions react with the calcium and magnesium ions to produce calcium carbonate and magnesium carbonate. These two carbonates are both insoluble in water and so cannot affect the soap. You should have noticed that when you added the washing soda crystals, the water went a bit cloudy due to the formation of these insoluble substances. Bath salts are mainly sodium carbonate, and many washing powders also contain some sodium carbonate.

Another way of softening both types of hard water is to pass the water through an **ion exchange column**. This is a column packed with small resin beads. The resin is first soaked in strong salt solution. Sodium ions stick on to the surface of the resin. When hard water is passed through the column, the calcium and magnesium ions are exchanged for sodium ions. When all the sodium ions have been replaced the resin can be soaked in salt solution and used again.

Time for thought

Distillation will also soften hard water. Why do you think that it is not used more often?

ALL YOUR OWN WORK

1 The air is a ________ of gases. The most plentiful gas in the air is ________. This makes up about ________ of the air. Nearly all the rest is the gas ________.
2 The amount of oxygen in the air can be found by passing a sample of air over heated ________ using two gas ________. The ________ turns black as it combines with the oxygen to form ________ ________. The air is passed backwards and forwards until the ________ does not go down any more. If you start with 100 cm^3 of air you should end up with about ________, showing that about ________ per cent of the air is oxygen.
3 Oxygen is obtained industrially from the air by ________ ________ of liquid air.
4 Give three similarities between respiration and burning.
5 Hard water is water that will not ________ easily with soap. It is caused by ________ and ________ compounds dissolved in the water. There are two types of hard water, ________ ________ ________ and ________ ________ ________. ________ ________ ________ can be softened by boiling but this has no effect on ________ ________ ________. This water can be softened by adding ________ ________ or by passing the water through an ________ ________ ________.

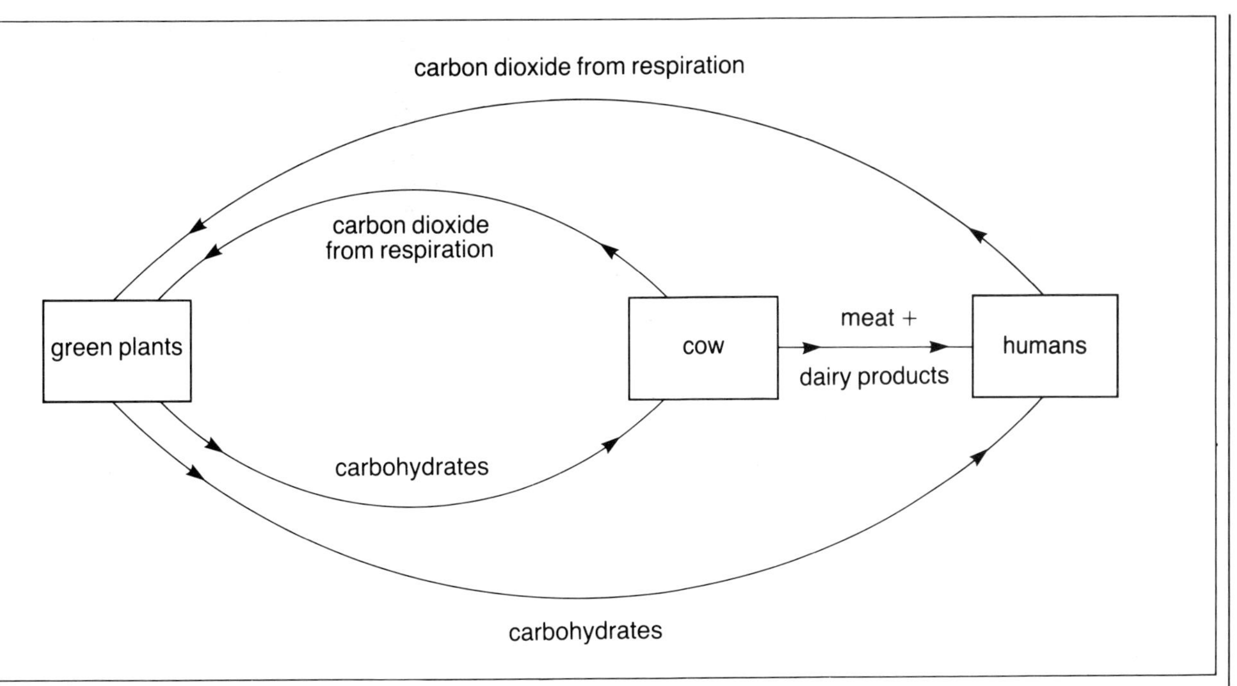

Figure 14.39

6 Figure 14.39 shows a relationship between plants and animals.

a The making of carbohydrates requires two simple compounds and a source of energy.

i Name the two compounds and where they come from.

ii What *form* is the energy in?

iii Where does the energy come from?

b Respiration requires oxygen and produces water. Energy is released in the process.

i Write down a word equation for respiration.

ii The process above uses oxygen; anaerobic respiration does not. Give an example of an organism which respires anaerobically.

c Some people say that a diet consisting only of meat and dairy produce can lead to health problems.

i Name *one* of these health problems and explain how meat and dairy produce may contribute to it.

ii Why might a person be advised to eat less butter and more margarine containing plant fat, e.g. sunflower oil?

[MEG]

CHEMICALS IN USE

FUELS

A fuel is a substance we use to provide us with energy. This energy is usually in the form of heat. It is produced by burning **fossil fuels**. These come from living things that died many millions of years ago. Their remains were buried in the Earth, and were formed into the fuels we use today by the high temperatures and pressures underground. **Coal** was formed mainly from giant fern-like plants that lived in prehistoric forests. **Oil** was formed mainly from marine organisms that lived in the shallow seas which covered large areas of the world at this time. **Natural gas** was also formed from prehistoric remains and gas is often found above oil deposits (see page 134).

figure 15.1 a Coal

b Oil being delivered

Going up in smoke

The burning of a fuel to release energy can happen in one of two ways – with or without flames. Petrol burns with a flame to provide energy. Food is 'burned' without flames during respiration. Three things are necessary for a fire – fuel, oxygen and heat. These three form the **fire triangle**. If any one of these is removed, the fire will go out (see figure 15.2).

A liquid can burn only at its surface where some of the liquid has vaporized. As soon as combustion starts, the liquid begins to warm up, and the rate of vaporization increases. Petrol vaporizes extremely easily, and so burns readily. Lubricating oil vaporizes with difficulty and so is hard to ignite. The **flash point** of a liquid is the lowest temperature at which a liquid will produce enough vapour to burn.

Putting fires out

For fire extinguishers to be effective they must remove one or more of the elements of the fire triangle. Using water is the commonest way of trying to cool the fire, i.e. by removing the heat. Water must not be used on fires involving electrical apparatus.

Carbon dioxide forms a blanket of gas around the fuel and so prevents the oxygen in air getting to the fuel (see page 220 and figure 15.3). Sand acts in a similar way.

c Gasometers which store gas

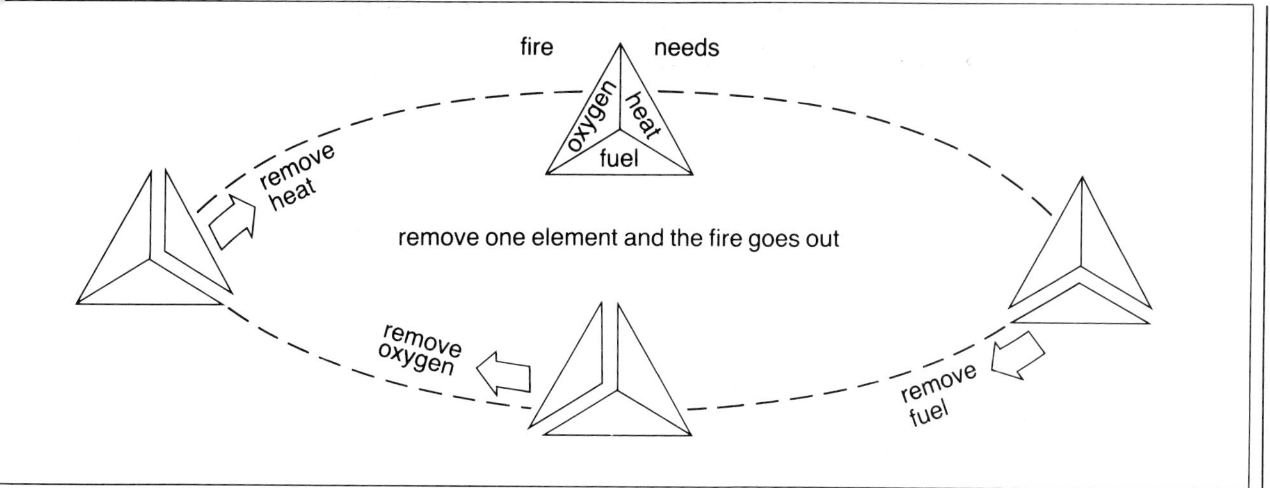
fire
needs
oxygen
heat
fuel
remove heat
remove one element and the fire goes out
remove oxygen
remove fuel

gure 15.2 The fire triangle

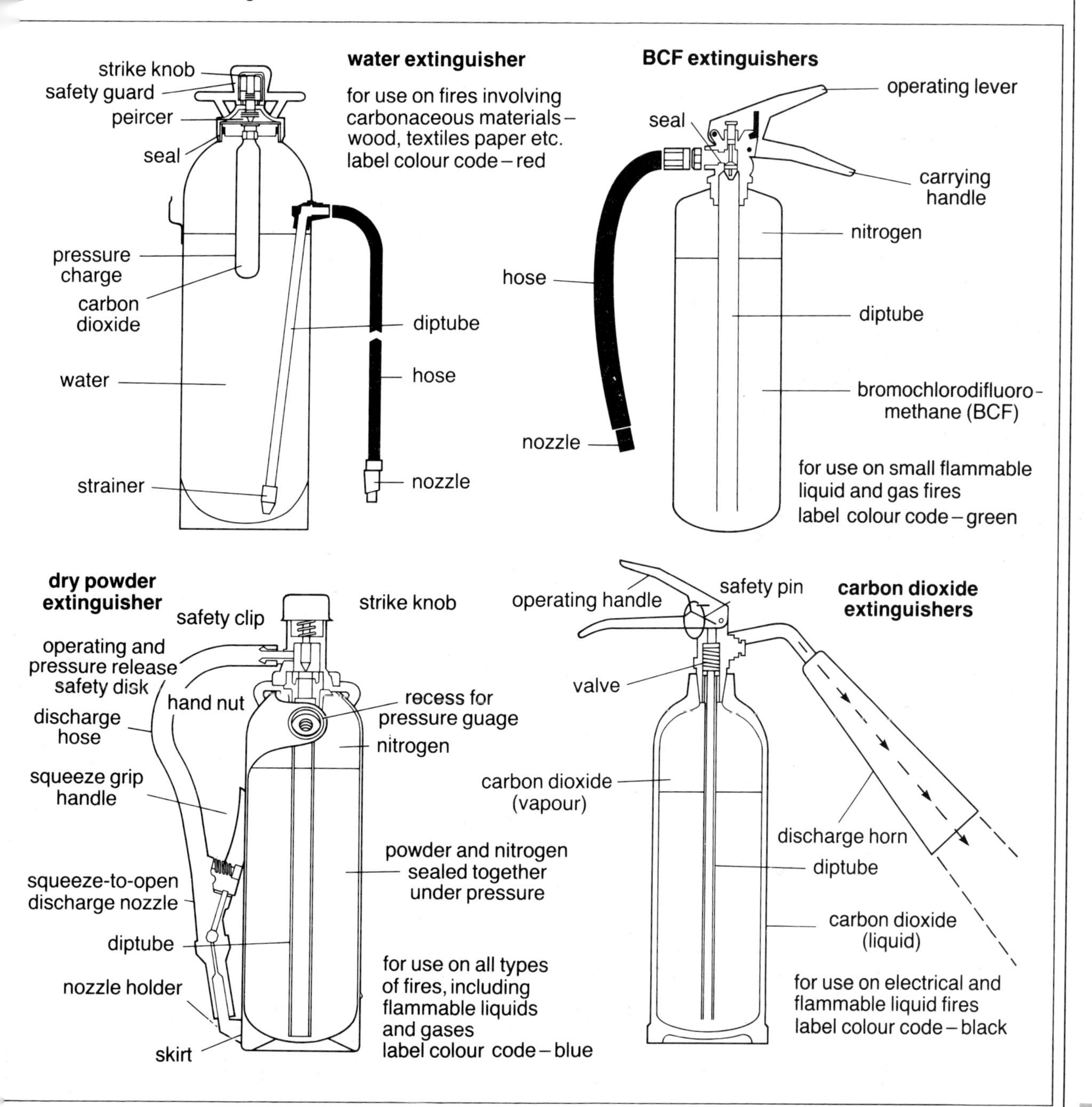
strike knob
safety guard
peircer
seal
pressure charge
carbon dioxide
water
strainer
diptube
hose
nozzle
water extinguisher
for use on fires involving carbonaceous materials – wood, textiles paper etc.
label colour code – red
BCF extinguishers
operating lever
seal
carrying handle
nitrogen
hose
diptube
bromochlorodifluoro-methane (BCF)
nozzle
for use on small flammable liquid and gas fires
label colour code – green
dry powder extinguisher
safety clip
strike knob
operating and pressure release safety disk
hand nut
discharge hose
recess for pressure guage
nitrogen
squeeze grip handle
powder and nitrogen sealed together under pressure
squeeze-to-open discharge nozzle
diptube
nozzle holder
skirt
for use on all types of fires, including flammable liquids and gases
label colour code – blue
operating handle
safety pin
carbon dioxide extinguishers
valve
carbon dioxide (vapour)
discharge horn
diptube
carbon dioxide (liquid)
for use on electrical and flammable liquid fires
label colour code – black

igure 15.3 Types of fire extinguisher

The combustion of fossil fuels

Fossil fuels contain carbon and carbon compounds. Coal is mainly carbon. A **hydrocarbon** is a compound of hydrogen and carbon. Natural gas is mainly the hydrocarbon methane, CH_4. Oil is a rather complicated mixture of hydrocarbons.

When a compound burns the products are usually the oxides of the elements in the compound. So when a hydrocarbon burns we would expect it to produce carbon dioxide and water. Provided there is enough air (oxygen) present, that is what happens.

$$\text{methane} + \text{oxgyen} \rightarrow \text{carbon dioxide} + \text{water}$$
$$CH_4 + 2O_2 \rightarrow CO_2 + 2H_2O$$

If there is not enough air present, carbon monoxide and carbon can be formed. Both these substances are produced by a car engine and cause **pollution** (see page 224). Carbon monoxide is very poisonous and carbon (soot) can pollute the air causing smog (see pages 62 and 76).

Time for thought

Why is it important that you should have a well-ventilated room when using a gas fire?

Energy from fuels

When a fuel reacts with the oxygen in the air, heat is given out. A reaction that produces heat energy is called an **exothermic reaction**. Where does this energy come from?

In any chemical reaction there are changes in **bonding**. Atoms become joined to different atoms and new substances are formed. When bonds are broken and new bonds are formed there is usually an energy change. If more energy is produced by the new bonds forming than is needed to break the old bonds, energy will be given out. When a fuel burns more energy is always produced, and so heat is given out.

The amount of heat produced by a fuel burning is called the **heat of combustion**. It is usually measured in kilojoules (kJ).

What makes a good fuel?

What qualities do you think a good fuel should have? Obviously this depends to some extent on the use of the fuel. The fuel for your open fire at home will not need to burn as fast as the petrol in a car engine.

You might like to make a list of your own of the things that you think a good fuel should do. Here ar[e] some suggestions.

- **a** lights easily
- **b** does not cause any pollution
- **c** gives out plenty of heat
- **d** leaves no ash
- **e** is fairly cheap
- **f** is easy to store

There are several other properties you could list. Perhaps you can already see that our fuels are far from ideal.

Time for thought

What advantages and disadvantages does coal have compared with natural gas?

COAL

There are three main types of coal (see page 134).
Lignite is rather brown and soft. It is about 70 per cent carbon.
Bituminous coal is black and quite crumbly. It is much older than lignite. It contains about 85 per cen[t] carbon.
Anthracite is the oldest form of coal. It is hard, black and shiny. Anthracite contains about 95 per cent carbon.

All forms of coal may contain small amounts of sulphur. When the coal burns sulphur dioxide gas is formed. This is a serious pollutant. It is the gas that causes **acid rain** (see page 196).

As well as being a good fuel, coal is a very important source of chemicals. These are obtained from the coa[l] by a process called **destructive distillation**.

The main products of the destructive distillation of coal (see investigation 15.1) are

- **a coke** – this can be used as a fuel but, it is used mainly in the extraction of iron in the blast furnace (see page 206).
- **b coal tar** – this is a valuable source of chemicals. Th[e] tar is further distilled to produce benzene, phenol and other useful substances. These products can be used to make dyes, drugs and insecticides. It is even possible to make a form of petrol from the coal tar. Do you think that this is a likely source of petrol when oil runs out?
- **c ammoniacal liquor** – although most ammonia is now produced by the Haber process (see page 203) the ammonia obtained from coal can still be made into useful fertilizers.
- **d coal gas** – until the early 1960s this was how all our gas was made. Now local industries may use the gas, otherwise it is just burned away in the air.

INVESTIGATION 15.1

THE DESTRUCTIVE DISTILLATION OF COAL

Before North Sea gas was discovered destructive distillation was used to produce gas for domestic use. Today it is more useful for other products.

a Carefully heat the coal (see figure 15.4). When smoke comes out of the glass tube, see if you can light the gas.

b Continue heating until there seems to be no further change to the coal. As soon as you stop heating make sure that you disconnect the tube containing the water. Otherwise the water will run back into the hot test tube.

Questions

1 What does the coal look like at the end of the experiment?

2 How would you test the pH of the water?

3 What are some of the disadvantages of producing gas in this way compared with collecting natural gas?

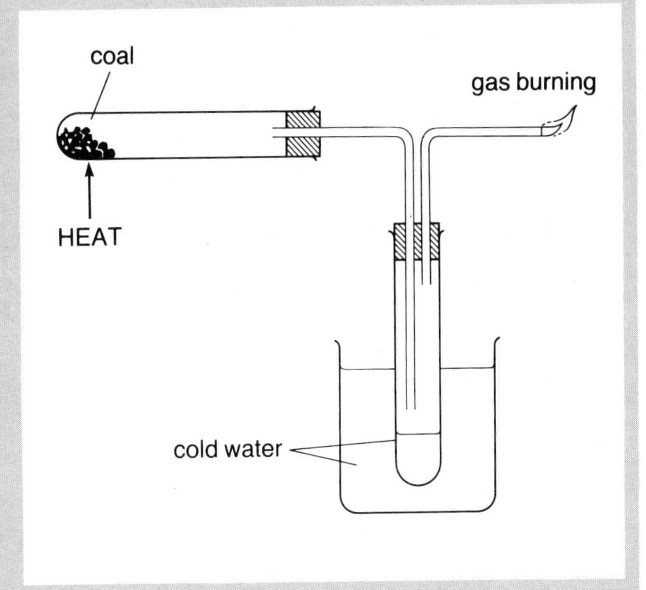

figure 15.4 Destructive distillation of coal

OIL

Processing oil

Crude oil, as it comes out of the ground, is not much use as a fuel. The oil is a complicated mixture of hydrocarbons. Some of the molecules contain only a few atoms, while others contain long chains of carbon atoms. The work of an **oil refinery** is to separate the oil into simpler mixtures and to increase the proportion of the more useful hydrocarbons.

figure 15.5 An oil rig is used to bring oil to the surface

Time for thought

The oil and gas from the North Sea will not last much more than another 30 years. What effect do you think that this will have on the coal industry?

figure 15.6 An oil refinery where fractional distillation takes place

The separation of the oil is done by **fractional distillation**. This is a way of separating a mixture of liquids with different boiling points (see page 168). The different hydrocarbons in the oil all have different boiling points. In general the smaller the molecule the lower its boiling point.

UNIT 15

INVESTIGATION 15.2

FRACTIONAL DISTILLATION OF CRUDE OIL

This investigation is best carried out in a fume cupboard or a very well ventilated laboratory.

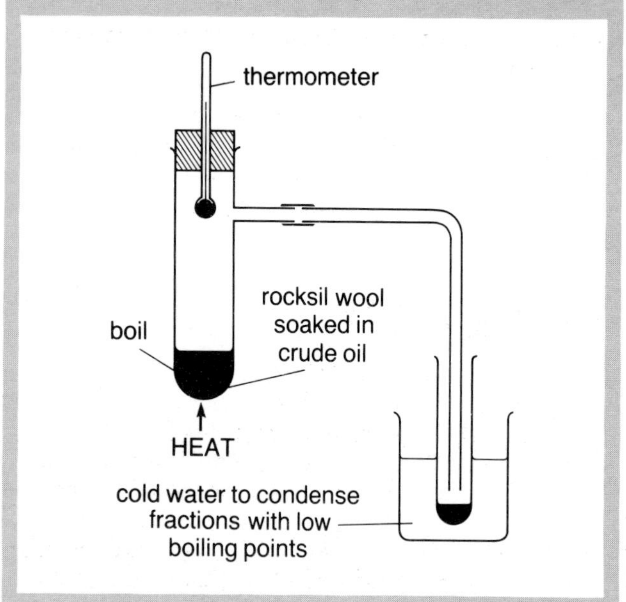

figure 15.7 *Laboratory distillation of oil*

a Place about 2 cm depth of rocksil wool in the bottom of a boiling tube. Add crude oil to the boiling tube so that the oil level is just above the rocksil wool.
b Connect up the apparatus as shown in figure 15.7, making sure that the thermometer scale is facing you and that you are using a 0–360 °C thermometer.
c Gently heat the oil using a small Bunsen flame. Collect all the products given off in test tube 1 until the temperature reaches 70 °C.
d Change the collection test tube and put a rubber stopper in test tube 1.
e Continue heating until the temperature reaches 120 °C. Change the collection test tube again and stopper test tube 2.
f In test tube 3 collect the fraction that boils between 120 °C and 170 °C.
g In test tube 4 collect the fraction that boils between 170 °C and 220 °C.
h In test tube 5 collect the fraction that boils between 220 °C and 270 °C.
i In test tube 6 collect the fraction that boils between 270 °C and 330 °C. *Take extra care at these high temperatures.*
j Examine the fractions and note their colour, runniness, smell, ease of lighting, colour of flame and how they burn.

Questions

1 Which fraction was there least of? Can you explain why?
2 Which fraction do you think is most like petrol?
3 Why would the fraction in test tube 6 not be very good for use in an oil heater in the home?

An oil refinery

Fractional distillation is one of the main processes that take place at an oil refinery. Unlike the laborator situation, however, the people working at an oil refinery do not want to have to keep changing the collection vessels. They would much prefer to have a continuous process. This is done using a special **fractionating column** (see figure 15.8).

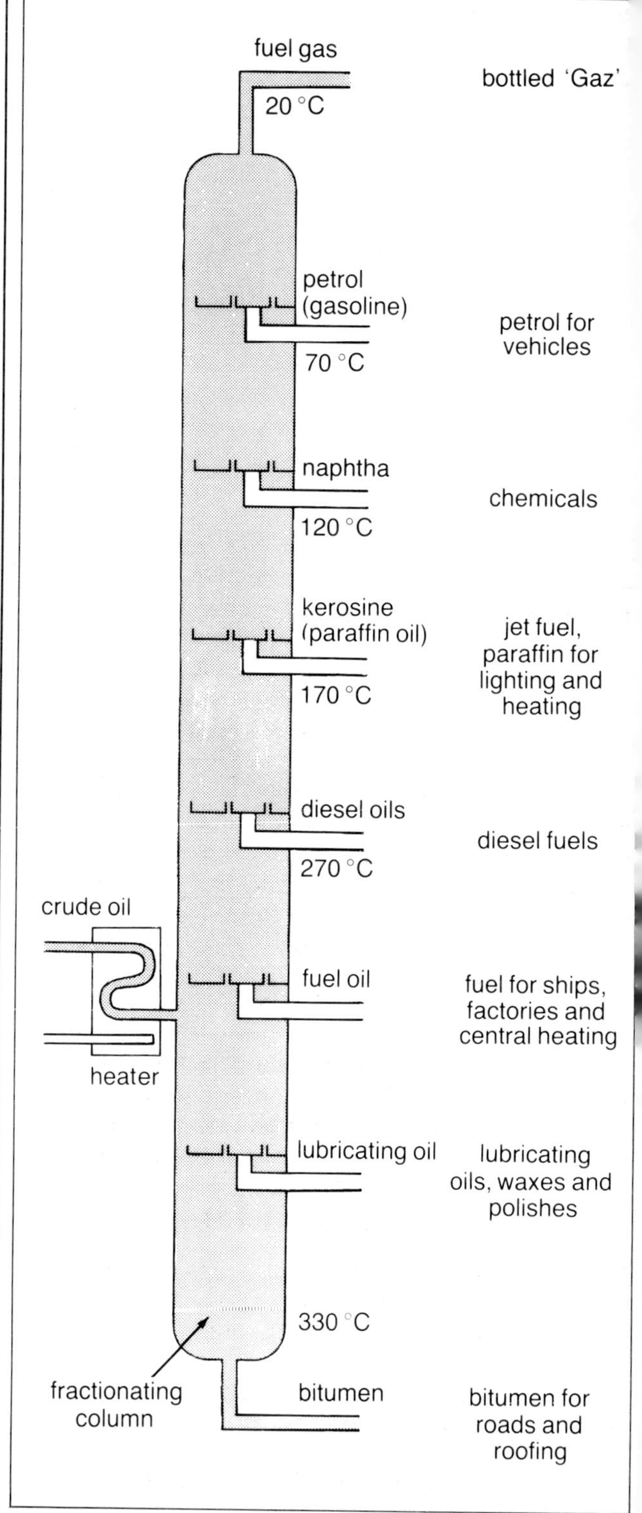

figure 15.8 *Fractional distillation of oil*

The hot crude oil is fed into the bottom of the column. The liquids (or fractions) with the higher boiling points stay at the bottom but the ones with lower boiling points progress up the column. As the temperature gets cooler up the column, more of the liquids condense. At various levels up the column there are shelves where the different fractions can collect. Before a substance can move up the column it has to bubble through the different layers by means of a special structure called a **bubble cap**. This ensures a better separation of the fractions (see figure 15.9).

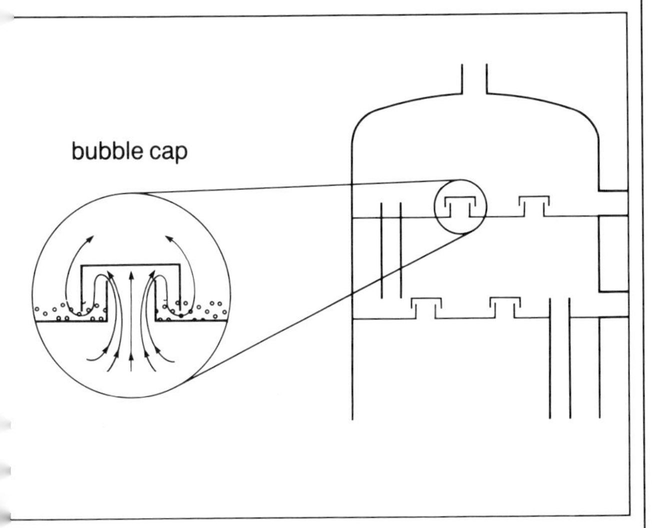

figure 15.9 A bubble cap improves separation

Cracking oil

You can see from figure 15.8 that it is the smaller, lower boiling point fractions that are the most useful ones. Unfortunately, British crude oil does not contain a very high percentage of these compounds. Oil from some parts of the world contains higher proportions of these more useful chemicals.

The second important job of an oil refinery is to try and break down some of the larger, not so useful compounds into the smaller, more useful ones. This process is called **cracking**. The higher boiling point fractions are heated very strongly, often in the presence of a catalyst. This causes the large molecules to split up and make smaller ones. The new mixture produced can be fractionally distilled as before.

Sometimes an additional process called **reforming** is carried out. In this process the arrangement of the atoms in a molecule is altered in order to produce a more useful substance.

Time for thought

Where would you expect an oil refinery to be sited? Check your reasoning by looking at an atlas of the British Isles.

ELECTRICAL ENERGY FROM CHEMICALS

Fuels are burned to produce heat and light energy. At home we can also use **electrical energy** to produce both heat and light (see units 8 and 9). It is also possible to produce electrical energy from chemical reactions.

An electric current is a flow of electrons (see unit 8), and we know that chemical reactions are concerned with the electrons in the outer shells of atoms. So it should not be too surprising to find a connection between the two.

INVESTIGATION 15.3

ELECTRICITY FROM CHEMICALS

- **a** Half fill a 250 cm^3 beaker with salt solution.
- **b** Choose two different metals and connect them to a voltmeter. (A general purpose Avo type meter set to measure a low voltage d.c. is suitable.) Choose from copper, iron, zinc, magnesium, lead and carbon.
- **c** Dip the two metals into the beaker of salt solution (see figure 15.10). Make sure that the two metals do not touch.
- **d** Repeat the experiment with several different pairs of metals. Notice whether it makes any difference which terminal of the meter the metals are connected to.
- **e** Choose two pieces of the same metal and see what results you get.

Questions

1. Which pair gives the highest voltage?
2. Which pair gives no voltage?
3. Out of each pair, which metal needs to be connected to the negative terminal of the meter?
4. How could you use this experiment to list the metals in order of reactivity?

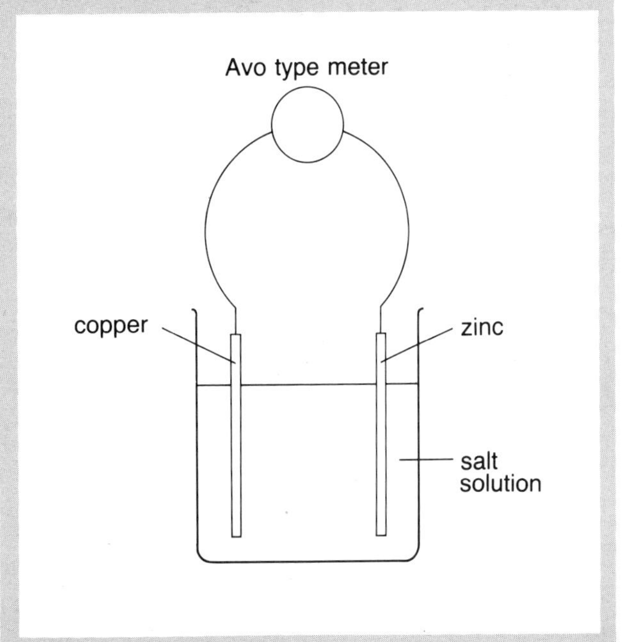

figure 15.10 A simple cell

You probably found in investigation 15.3 that any two metals connected to a meter and then dipped in a conducting solution will produce an electric current. The greater the difference in reactivity the greater the voltage. Electrons are released by the more reactive metal (see unit 13). As electricity is produced, the more reactive metal corrodes away.

Application 15.1

THE DRY CELL

A dry cell, like the one used in a torch or transistor radio, is an example of a **primary cell**. This means that it makes electricity directly from a chemical reaction. In a dry cell, zinc and carbon are used as the electrodes. The carbon behaves like an unreactive metal. A conducting paste connects the two electrodes. The zinc corrodes away, freeing electrons when the appliance is switched on. Eventually, the zinc will corrode away completely and the cell will no longer work.

Hydrogen can build up round the carbon rod and so powdered manganese(IV) oxide is used to oxidize the hydrogen to water. (Figure 15.11 shows a cell and figure 15.12 the inside of a cell.) A number of cells connected together is known as a **battery**.

figure 15.11 *Dry cells*

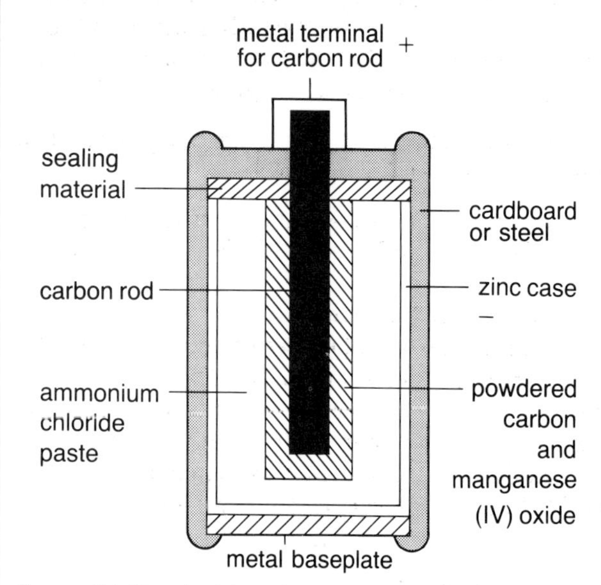

figure 15.12 *Inside a cell*

Storing electricity

When the chemicals in a primary cell have been used up they cannot be re-used and the cell cannot be **recharged**. These cells are often found in personal stereos etc. A second set of cells, known as **secondary cells**, work by a chemical reaction which can be reversed. These cells can be recharged and re-used. A particular example of a secondary cell is the **lead-acid accumulator** which is used as a car battery.

Cells and batteries store chemicals which produce electricity as part of a chemical reaction. They do not store electricity.

INVESTIGATION 15.4

MAKING A SIMPLE ACCUMULATOR

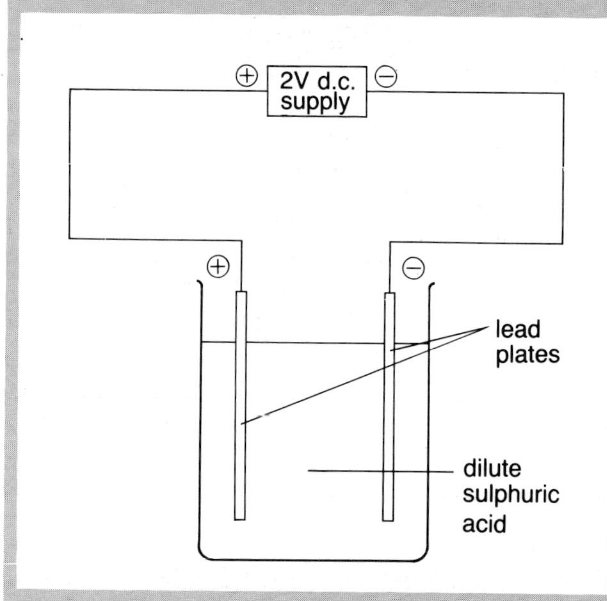

figure 15.13 *A simple accumulator*

a Half fill a 100 cm^3 beaker with dilute sulphuric acid (*care*).
b Fix two strips of lead foil to the sides of the beaker using crocodile clips (see figure 15.13). Make sure that the strips do not touch.
c Connect the crocodile clips to a 2 V d.c. supply and leave it on for about three minutes.
d Disconnect the wires from the voltage supply and connect them instead to a 2 V bulb.

Questions

1. How long did the bulb stay lit?
2. Did the pieces of lead change their appearance while the battery was charging?
3. Batteries of this sort can be used to power electric milk floats and electric cars. What are the main disadvantages of using this type of battery in this way?
4. Design an experiment to test other metal combinations to see whether they act as accumulators.

Application 15.2

FUEL CELLS

We know that hydrogen and oxygen react readily together. The usual test for hydrogen is that it pops when a lighted spill is put into it. In larger quantities, explosions can occur. Hydrogen could be burned as a fuel, but the reaction would be more efficient if hydrogen and oxygen could be made to combine and give out the energy formed as electrical energy. This is the aim of a fuel cell. This type of cell is still being developed. Fuel cells are used on most space rockets as a means of producing electricity. Figure 15.14 shows the fuel cells in an Apollo space rocket.

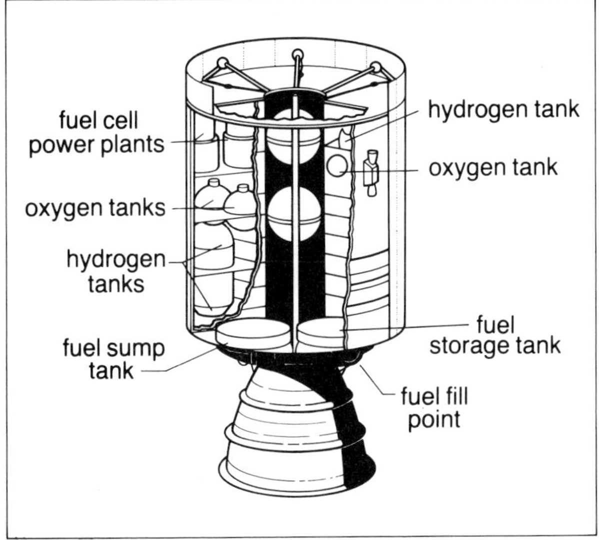

figure 15.14 *Fuel cells in an Apollo service module*

CHEMICALS FROM NATURE

There is a whole branch of chemistry devoted to the study of chemicals found in nature, having their origin in living things. It is called **organic chemistry**. All organic compounds contain the element **carbon**. There are thousands of different organic chemicals so they are divided up into smaller groups whose members have similar properties.

The hydrocarbons

Hydrocarbons are compounds that are made from the elements hydrogen and carbon. They are all covalent compounds. Carbon has four electrons in its outer shell and so lacks four. It obtains these by sharing with other atoms. A carbon atom always has four covalent bonds linking it with other atoms. There is a very large number of organic compounds, because of the ability of carbon atoms to form covalent bonds with other carbon atoms, and so form chains. Within the hydrocarbon group compounds are divided into even smaller groups called **homologous series**. The simplest of these is called the **alkanes**. Table 15.1 shows the simple alkanes.

<table>
<tr><th>name</th><th>formula</th><th>structural formula</th></tr>
<tr><td>methane</td><td>CH_4</td><td>H
|
H—C—H
|
H</td></tr>
<tr><td>ethane</td><td>C_2H_6</td><td>H H
| |
H—C—C—H
| |
H H</td></tr>
<tr><td>propane</td><td>C_3H_8</td><td>H H H
| | |
H—C—C—C—H
| | |
H H H</td></tr>
<tr><td>butane</td><td>C_4H_{10}</td><td>H H H H
| | | |
H—C—C—C—C—H
| | | |
H H H H</td></tr>
</table>

table 15.1

You will probably have heard of some of these compounds. **Methane** is the main gas in North Sea gas; **propane** is used as camping gas and **butane** is used as the gas in some cigarette lighters.

The names of the alkanes are very important because the names of all other organic compounds are based on them. The next member of the series is called pentane. Can you see why? What do you think for formula of pentane is?

A **structural formula** is a way of representing how the atoms are joined together. It is a flat representation and so does not really show the actual shape of the molecule. In the structural formula all the bonds appear to be at right angles. In fact, the four bonds are equally spaced around the carbon atom. In the case of methane, the four hydrogen atoms are at the corners of a tetrahedron (a triangular pyramid).

Alkanes are useful fuels. They burn easily easily in air to form carbon dioxide and water.

$$\text{propane} + \text{oxygen} \rightarrow \text{carbon dioxide} + \text{water}$$
$$C_3H_8 + 5O_2 \rightarrow 3CO_2 + 4H_2O$$

Other simple hydrocarbons

There are several other families of hydrocarbons, the **alkenes** being the most important. The simplest member of the alkene family is **ethene**, C_2H_4. Ethene is used in making polythene (poly-ethene). The structural formula of ethene shows that it contains a **double bond** between two carbon atoms.

ethene C_2H_4

```
H   H
|   |
C = C
|   |
H   H
```

UNIT 15

All the alkenes contain a double bond between two carbon atoms. A hydrocarbon that contains a multiple bond between two carbon atoms is said to be **unsaturated**. You will have seen the word 'polyunsaturates' on a margarine tub. It means that the fat contains molecules with many of these double bonds. These molecules make the fat softer and easier to spread. Also, fats containing polyunsaturates contain less **cholesterol**. Cholesterol is a fat that can build up on the inside of artery walls and increase the risk of heart attack (see pages 41 and 55).

figure 15.15 Polyunsaturated margarine

Ethyne, C_2H_2, contains a triple bond between two carbon atoms. Ethyne (more commonly called **acetylene**) is used to weld and cut metals with oxy-acetylene torches (see page 215).

Isomerism

When the length of the carbon chain in the hydrocarbons is greater than three, it is possible to have compounds with the same formula but different structures. These different structures are called **isomers**. The isomer is always named after the longest straight chain of carbon atoms.

The isomers of C_4H_{10} are:

```
    H   H   H   H             H   H   H
    |   |   |   |             |   |   |
H — C — C — C — C — H     H — C — C — C — H
    |   |   |   |             |   |   |
    H   H   H   H             H   |   H
                              H — C — H
                                  |
                                  H
     butane               2-methyl propane
```

(2-methyl propane gets its name from the methyl group attached to the second carbon atom in the longer propane chain.)

See if you can work out the isomers and their names for the compound C_5H_{12}.

Fermentation and alcohols

Fermentation is a process that has been known for a very long time. If fruits like grapes or apples are left to rot away in a bucket of water they start to **ferment** and produce alcohol. Microorganisms react with the sugars in the fruits. Today fermentation is a major industrial process, producing alcohol for drinks and chemical use. The enzymes needed for the breakdown of the sugars are found in yeast.

figure 15.16 Whisky distilleries

INVESTIGATION 15.5

FERMENTATION

a Place about 30 cm^3 distilled water in a 100 cm^3 conical flask.
b Dissolve about a teaspoonful of glucose in the water.
c Add a spatula measure of yeast to the flask.
d Connect up the delivery tube as shown in figure 15.17 and then leave the apparatus in a warm place for a few days.
e Decant (pour off) the liquid from the flask into a larger distillation flask.
f Ask your teacher to fractionally distil the mixture to obtain a sample of alcohol (see figure 15.18)

Questions

1 What happened to the lime water? What does this show?
2 How could you show that the liquid produced was alcohol?
3 Why do you think that it is illegal to distil a fermentation mixture at home?
4 It is the enzymes in yeast that change the sugar to alcohol. What would happen if yeast was added to boiling sugar solution?
5 Find out what you can about the 'fruit fly'/'vinegar fly'.

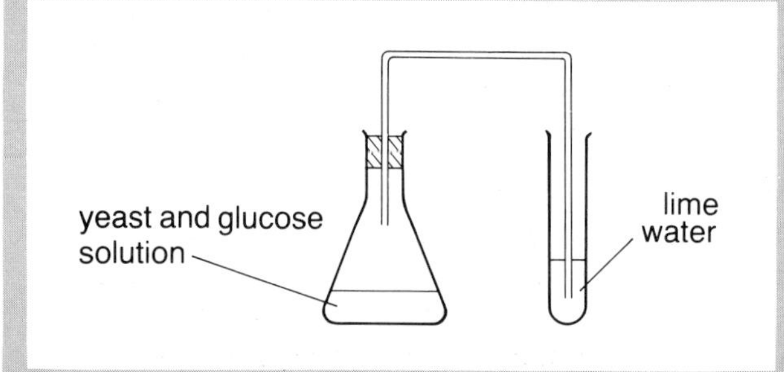

figure 15.17 Laboratory fermentation

Enzymes present in yeast break down glucose to carbon dioxide and **ethanol** (ordinary alcohol).

$$\text{glucose} \rightarrow \text{carbon dioxide} + \text{ethanol}$$
$$C_6H_{12}O_6 \rightarrow 2CO_2 + 2C_2H_5OH$$

The solution produced contains about 5 per cent ethanol. This concentration kills the yeast cells. In order to concentrate the alcohol, the mixture has to be distilled. Ethanol has a boiling point of 78 °C and water a boiling point of 100 °C. This means that they can be separated by fractional distillation (see page 168). The first few cm^3 of distillate will be about 95 per cent pure ethanol. It should be concentrated enough to burn easily.

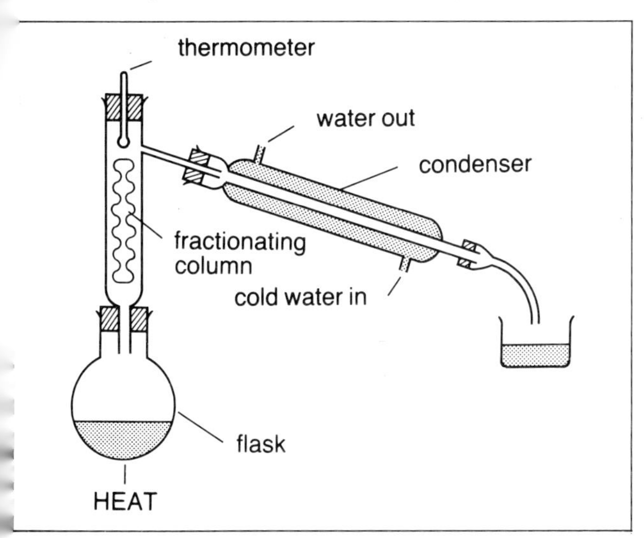

figure 15.18 *Fractional distillation of alcohol and water*

Table 15.2 shows the strengths of some common alcoholic drinks.

drink	approximate percentage alcohol
beer	5
wine	12
sherry	25
whisky	40

table 15.2

The drinks industry sometimes expresses the strength of the alcohol in terms of **proof spirit**.

Proof spirit was the name given to alcohol which when poured onto gunpowder would still allow it to burn. If the alcohol was too weak the gunpowder would not burn.

This definition was not very accurate and so a better description was needed. Proof spirit is now determined by measuring the density of the liquid (see pages 8 – 9). Proof spirit contains 49.3 per cent alcohol by weight or 57 per cent alcohol by volume. (In America, proof spirit is 50 per cent alcohol by volume.)

Pure alcohol is 175 per cent proof. A bottle of whisky containing 40 per cent alcohol by volume will be about 70 per cent proof (see figure 15.19).

figure 15.19 *Alcoholic drinks labels tell you the percentage of alcohol present*

Alcohols form another homologous series. All alcohols contain an OH group. The simplest member of the series is **methanol**, CH_3OH.

$$\begin{array}{c} \text{H} \\ | \\ \text{H}-\text{C}-\text{O}-\text{H} \\ | \\ \text{H} \end{array} \quad \text{methanol}$$

Table 15.3 shows some of the simple alcohols with their physical properties. Notice how the names of the alcohols are based on the names of the alkanes having the same number of carbon atoms. Notice also the gradual change in properties, especially the boiling point: the larger the liquid's molecules, the higher its boiling point.

alcohol	formula	molecular mass	boiling point (° C)	density (g/ cm³)
methanol	CH_3OH	32	65	0.79
ethanol	C_2H_5OH	46	78	0.79
propanol	C_3H_7OH	60	97	0.80
butanol	C_4H_9OH	74	118	0.81

table 15.3

Alcohols all burn and can be used as fuels. Once again, the products are carbon dioxide and water.

$$C_2H_5OH + 3O_2 \rightarrow 2CO_2 + 3H_2O$$

Breaking down starch

Starch is a complex carbohydrate (see page 42) produced by plants during photosynthesis (see page 219). Carbohydrates, including starch, are a valuable food source for animals. Foods we eat that contain large amounts of starch include bread and potatoes.

In the body the starch is broken down by enzyme action. This takes place in the mouth by the action of saliva and in the intestine by the action of other enzymes called **amylases**.

INVESTIGATION 15.6

THE BREAKDOWN OF STARCH

By enzyme action

a Make about 20 cm^3 of starch solution by dissolving a spatula measure of soluble starch in water at about 35 °C.

b Add a few drops of a solution of the enzyme in saliva to a test tube half full of starch solution. Leave the test tube in a beaker of warm water for about 30 minutes.

c Test a small sample of the original starch solution for the presence of a simple sugar using Benedict's test. This involves adding Benedict's solution and then carefully boiling the mixture. An orange/red precipitate indicates that a simple sugar is present (see page 43).

d Repeat Benedict's test on a sample of the starch solution that has been left with the enzyme.

Questions

1 Was any simple sugar present in the original starch solution?

2 Was a simple sugar formed by the action of the enzyme?

By the action of an acid

a Pour above 1 cm depth of starch solution into a boiling tube.

b Add an equal volume of dilute hydrochloric acid to the starch solution and boil the mixture for a few minutes.

c Test the resulting mixture with Benedict's solution to see whether a simple sugar is present.

Chromatography of sugars

Investigation 15.6 identifies the presence of sugar, although does not identify the actual sugars produced.

We have already come across **chromatography** as a way of separating mixtures of coloured substances (see page 168). Here the substances are colourless and so a **locating agent** has to be used to show where the sugar spots are.

The chromatography paper is spotted with solutions containing

a sucrose(s)
b glucose(g)
c maltose(m)
d starch solution after reaction with acid (s/a)
e starch solution after reaction with enzyme (s/e)
(see figure 15.20).

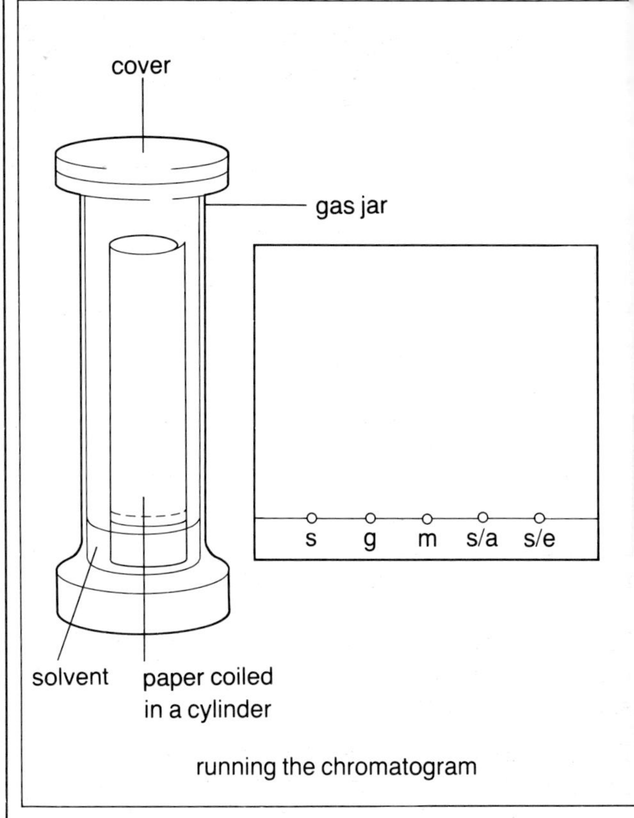

figure 15.20 Chromatography of sugars

The chromatogram is allowed to run using a solvent containing 1 part water, 1 part acetic acid (ethanoic acid) and 3 parts propan-2-ol. The solvent may take several hours to travel up the paper.

When the paper is dry it is dipped in a locating agent. (This is a solution containing 5 parts 2% aniline in propanone, 5 parts 2% diphenylamine in propanone and 1 part 85% phosphoric acid.)

The paper must then be dried in an oven at 100 °C. The different spots should then be visible (see figure 15.21).

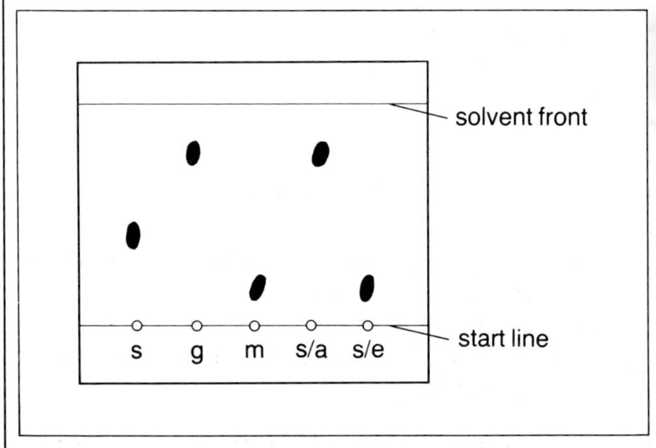

figure 15.21 The developed chromatogram

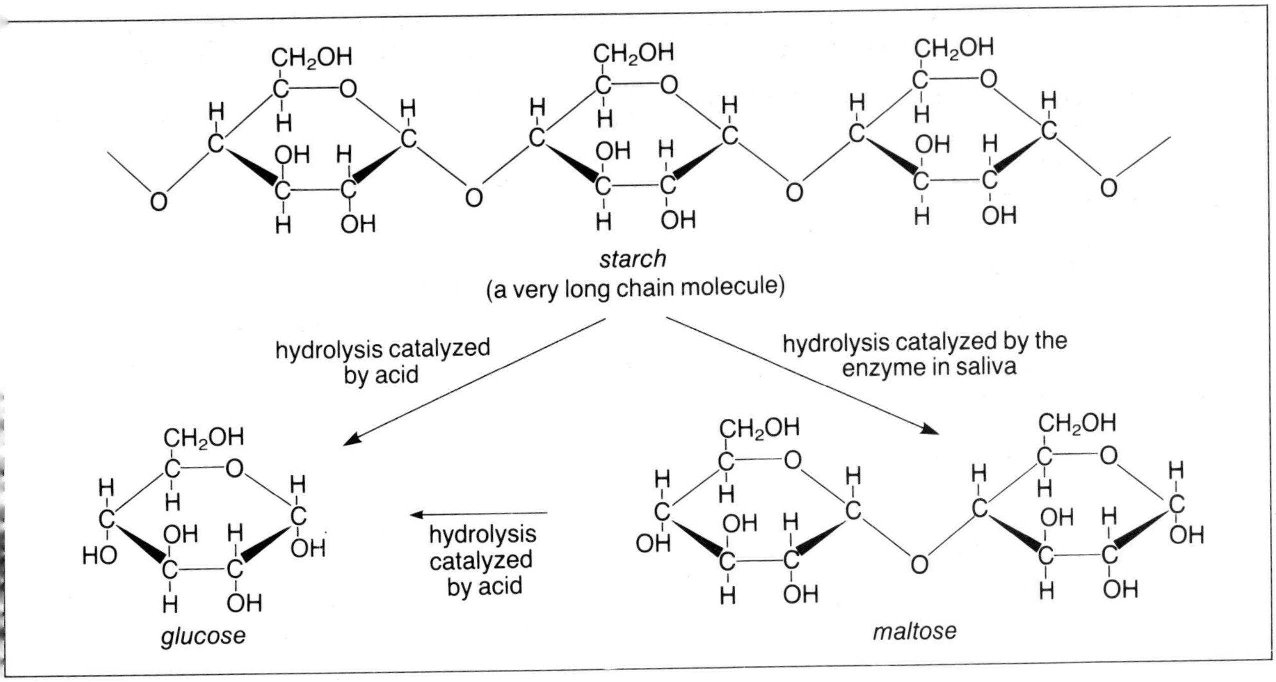

starch
(a very long chain molecule)

hydrolysis catalyzed by acid

hydrolysis catalyzed by the enzyme in saliva

hydrolysis catalyzed by acid

glucose

maltose

figure 15.22 *The breakdown of starch*

The chromatogram shows that the enzymes in saliva break down the starch to **maltose**. The acid breaks the starch down to **glucose**.

Figure 15.22 shows that starch is a large molecule made up of many repeating units, i.e. it is a **polymer**. The repeating units in the starch are glucose molecules. The enzymes break starch into molecules containing two glucose units (maltose) but acid breaks starch up into individual glucose molecules.

Try chewing a piece of bread for a few minutes and see if you can detect the sweet taste of glucose.

MODERN MATERIALS

Many of the household materials we use today had not even been heard of 50 years ago. These include plastics, some dyes, some medicines and synthetic fibres. Think of all the things you use every day that fit into these groups. How could you manage without them?

Plastics and polymers

The word **polymer** describes a molecule made up of many repeating units. The repeating unit is sometimes called the **monomer**. Starch is a natural polymer made up from large numbers of glucose molecules linked together.

All plastics and synthetic fibres are polymers. They are produced by making relatively simple molecules join together to form very big molecules. The relative molecular mass of a polymer may well be as high as 150 000 – with thousands of molecules linked together.

Polythene
This is one of the commonest and simplest polymers. It is made from ethene, C_2H_4. Ethene is an unsaturated hydrocarbon – it contains a double bond. Polythene is a shortened form of the word poly-ethene and the polymer is made by making the ethene molecules **polymerize** (join together).

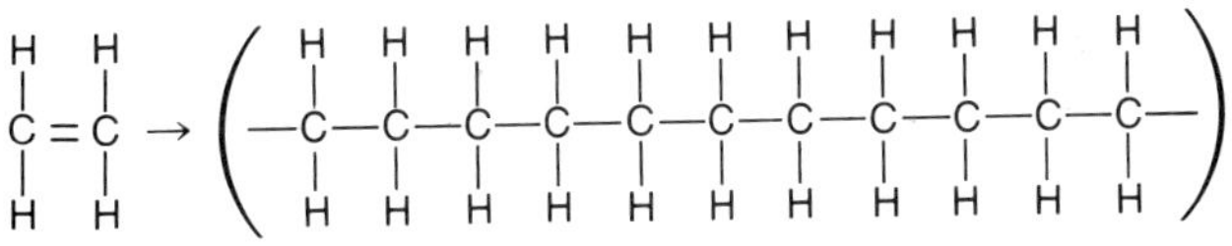

$$H_2C{=}CH_2 \rightarrow (-CH_2-CH_2-CH_2-CH_2-CH_2-CH_2-CH_2-CH_2-CH_2-CH_2-CH_2-)$$

The ethene is made to polymerize using complex catalysts. By using different conditions of temperature and pressure the size of the polymer molecules can be controlled.

figure 15.23 *Containers made from polythene*

monomer	polymer			
formula/name	name	trade name	formula	uses
$\begin{matrix}H & & H\\ & C=C & \\ Cl & & H\end{matrix}$ vinyl chloride (chloroethene)	polyvinylchloride	PVC	$\left(\begin{matrix} & H & H & \\ - & C & C & - \\ & H & Cl & \end{matrix}\right)_n$	records, clothes, electrical wire insulators
$\begin{matrix}H & & H\\ & C=C & \\ H & & C_6H_5\end{matrix}$ styrene (phenylethene)	polystyrene		$\left(\begin{matrix} & H & H & \\ - & C & C & - \\ & H & C_6H_5 & \end{matrix}\right)_n$	packaging materials, ceiling tiles, plastic model kits
$\begin{matrix}H & & H\\ & C=C & \\ H & & CN\end{matrix}$ acrylonitrile	polyacrylonitrile	Orlon, Courtelle, Acrilan	$\left(\begin{matrix} & H & H & \\ - & C & C & - \\ & H & CN & \end{matrix}\right)_n$	synthetic fibre
$\begin{matrix}F & & F\\ & C=C & \\ F & & F\end{matrix}$ tetrafluoroethene	polytetrafluoethene	Teflon PTFE	$\left(\begin{matrix} & F & F & \\ - & C & C & - \\ & F & F & \end{matrix}\right)_n$	coating for non-stick saucepans; bridge bearings

table 15.4

Polymers made by adding together similar simple molecules are called **addition polymers**. Table 15.4 shows some common addition polymers and their uses.

> **Time for thought**
>
> *One of the common uses of polythene is in making polythene bags. What disadvantage do they have?*

Nylon

Other polymers can be made by reacting two different types of molecule together. Each of the molecules involved has reactive groups at each end so that the reaction can take place many times. This again results in the formation of very large molecules. When these types of polymer are formed a simple substance, such as water, is also produced as a by-product of the reaction. Polymers formed in this way are called **condensation polymers**. Nylon is a condensation polymer.

INVESTIGATION 15.7

MAKING NYLON

This investigation is best carried out in a fume cupboard or a well-ventilated laboratory.

The name nylon refers to a group of polymers. In this investigation you will make nylon 6.6.

a Add about 10 cm^3 of a solution of hexane-1, 6-dioyl chloride to a 100 cm^3 glass beaker (a 5% solution in cyclohexane is needed).

b Carefully add an equal volume of a solution of 1, 6-diaminohexane (a 5% solution in water). This will form a separate layer on top of the organic solvent.

Do not be put off by the names of the chemicals. What do you think that the 'hexa' part means in each name?

The reaction takes place at the interface, i.e. where the two solutions meet.

c Pick out the cloudy film that forms between the two solutions with a pair of forceps. As the film is removed a new film forms. If you are careful you should be able to wind the nylon round a glass rod (see figure 15.24).

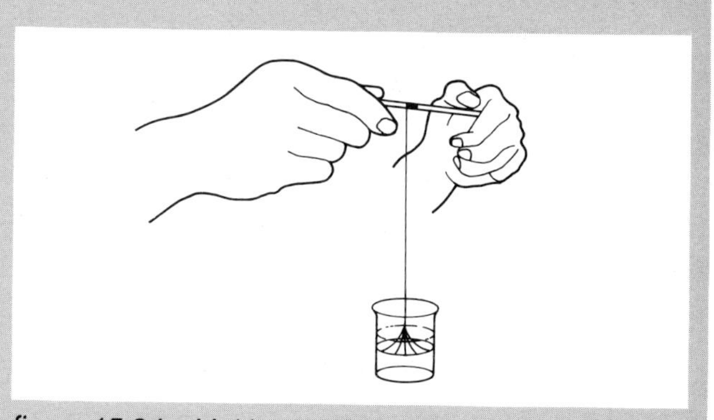

figure 15.24 Making nylon

Questions

1 What name is given to two liquids that do not mix?

2 When does the reaction stop?

3 Why do you think that this type of nylon is called nylon 6.6?

4 In this reaction what do you think is produced as a by-product? How would you test this idea?

otice how the reacting molecules in investigation .7 have reactive groups at each end.

$$H_2N-CH_2-CH_2-CH_2-CH_2-CH_2-CH_2-NH_2 + ClOC-CH_2-CH_2-CH_2-CH_2-COCl$$

$$H_2N-CH_2-CH_2-CH_2-CH_2-CH_2-CH_2-NH-CO-CH_2-CH_2-CH_2-CH_2-COCl$$

he process can be repeated many times to produce a ng chain molecule. The different types of nylon ave different numbers of carbon atoms between ach link.

Application 15.3

USES OF NYLON

Nylon has many different uses. It can be made into a fine thread or moulded into tough blocks. Uses include: making fabric for shirts, blouses and underwear; making stockings and tights; making fishing line; making strong ropes; making machinery parts, especially tough gear wheels.

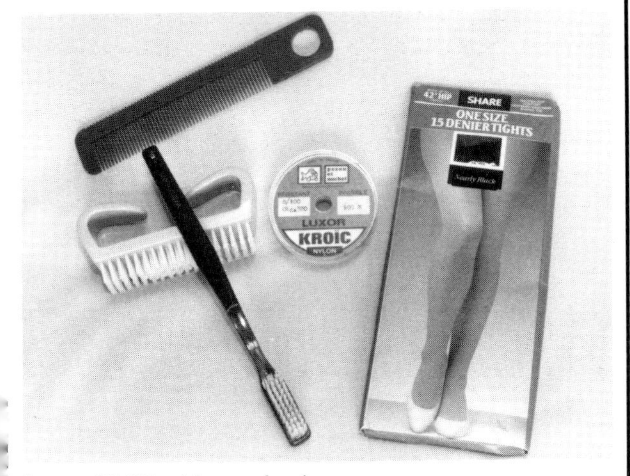

figure 15.25 Uses of nylon

Polyesters form another group of condensation polymers.

Properties and structure of polymers

The general physical properties of a substance are determined by the way the particles are arranged. In polymers that can easily be made into fibres the molecules of the polymer contain long chains of atoms. In these cases there is little attraction between the chains. The technical term for this is that there is little **cross-linking**.

In some polymers, however, there is considerable attraction between the polymer molecules and a great deal of cross-linking takes place.In these cases a much stronger three-dimensional framework is produced (see figure 15.26).

In plastics, the arrangement of the polymer molecules can be altered by heating the plastic. Some plastics become soft on heating and then harden again when they cool down. Polythene and nylon are examples of this type of plastic. They can be heated and moulded several times. Plastics like this are called **thermoplastics**. Other plastics like Bakelite and Melamine become hard and brittle on cooling when strong cross-linking takes place. Plastics of this kind are called **thermosetting plastics**. These are the types of plastics used for making electrical sockets (see figure 15.27).

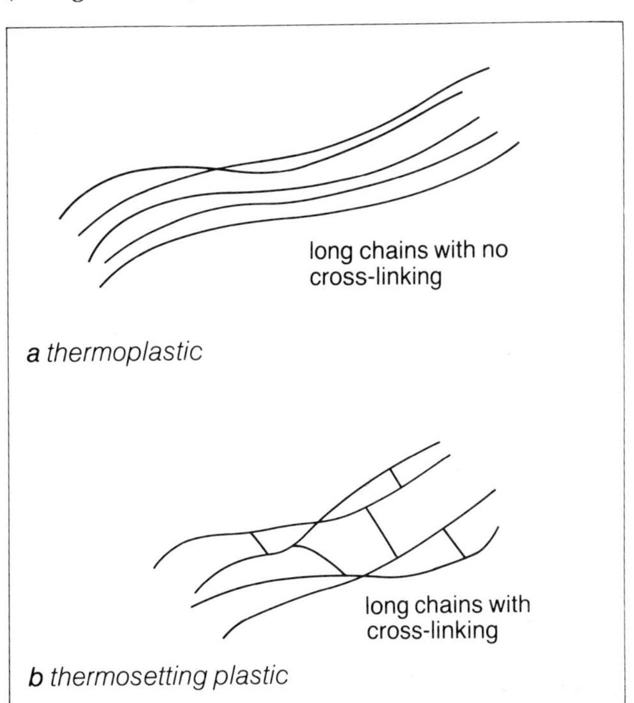

figure 15.26 Structure of polymers

figure 15.27 Electrical sockets are made of thermosetting plastics

Time for thought

Make a list of everyday objects that are made from polymers and in each case suggest an alternative material that could be used.

What advantages do synthetic fibres have over natural fibres?

Giant structures for giant structures

Concrete and **glass** cannot really be classed as modern materials. Glass in particular has been known for thousands of years. However, when we look at a modern building we can see the extensive use made of these two materials. Again it is the arrangement of the atoms in the materials that give them their properties.

Cement

Cement is made by heating a mixture of limestone (calcium carbonate) and clay (mainly aluminium silicate). The resulting mixture is ground to a very fine powder.

When cement is mixed with water and allowed to dry, it sets to a hard solid. This is because the structure in the cement changes and a cross-linked crystal pattern is formed.

If the cement is mixed with sand and gravel **concrete** is formed. This is even stronger than cement because the sand and gravel particles give extra support to the crystal structure.

Time for thought

Where would you expect to find a cement works? Is there one anywhere near you?

What problems, if any, would you have if you lived near to a cement works?

INVESTIGATION 15.8

MAKING CEMENT AND CONCRETE BLOCKS

Make a wooden framework about 10 cm long and 2 cm wide from pieces of wood that will allow you to cast a block 1 cm thick.

Make a block using only cement and water. Then make several more blocks using concrete, and concrete containing strips of metal.

Devise a way of testing which block is strongest.

You can then investigate whether the amount of sand in the concrete alters its strength, and whether the pebble size of the gravel makes any difference.

It is developments in the manufacture of reinforced concrete that have enabled us to build the huge skyscrapers we see today in all big cities (see figure 15.28).

figure 15.28 Concrete and glass make up modern office blocks

Glass

In most modern buildings the parts that we can see from the outside are made from either concrete or glass.

Like cement, glass contains silicon compounds. The main ingredient for making glass is sand. Sand is basically silicon dioxide (silica). Sand itself is a crystalline substance, i.e. it is made up from a giant regular arrangement of particles. When sand is heated with sodium carbonate it cools to give this rather unusual substance we call glass. It is unusual because it is actually **supercooled liquid** and the particles are not able to get into a regular pattern. When the glass is warmed again it can be easily shaped into all sorts of objects (see figure 15.29).

figure 15.29 Blowing glass

> ## Application 15.4
> TYPES OF GLASS
>
> There are several different types of glass. Ordinary glass is sometimes called **soda glass** and contains sodium and calcium silicates. **Crystal glass** used for making cut-glass ornaments and glasses contains potassium and lead silicates as well as sodium and calcium silicates. **Pyrex glass** used for laboratory glassware and kitchenware contains silicates of boron and aluminium. This 'boro-silicate' glass can withstand higher temperatures and a more sudden change of temperature.
>
> Colours can be added to glass by adding various metal oxides to the molten glass, e.g. cobalt oxide produces a blue colouration.

Detergents

A **detergent** is the name given to a substance that aids cleaning. It is sometimes described as a **cleaning agent**.

A detergent has three main jobs to do.

a It helps the water to wet the material.
b It helps to dislodge dirt and grease.
c It helps to disperse the dirt and grease.

All detergent molecules have two basic parts, a long covalently bonded chain of carbon atoms and an ionic group at one end of the molecule. The molecule can be thought of as being a bit like a tadpole with an ionic head and covalent tail (see figure 15.30). The covalent tail part of the molecule does not like being in water, it prefers to be in grease. The ionic head prefers to be in water and does not like grease. Figure 15.31 shows a detergent in action.

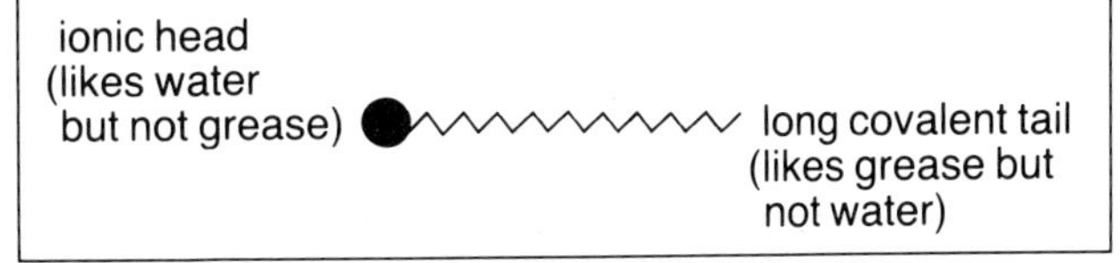

figure 15.30 A detergent molecule

Helping the water wet the material

Water is not very good at wetting things. This might seem a strange statement but you can quite easily demonstate it. If you get a brand new handkerchief or tea towel, stretch it tightly and add a drop of water to its surface, you will find that the water will not soak in very easily. If you now add a drop of detergent solution to the drop of water, it will soak in straight away. The reason for this is that the detergent molecules gather at the surface of the water so reducing the **surface tension** of the water, allowing it to wet the material.

Helping dislodge dirt and grease

During washing, the material is agitated to dislodge the dirt. As soon as the dirt comes away from the material it is surrounded by detergent molecules which stop it joining the material again.

Helping to disperse the dirt and grease

Once the dirt is in the water it is completely surrounded by detergent molecules. These have their ionic heads on the outside and because like charges repel, the bits of dirt are forced apart. This also prevents the particles of dirt and grease sticking back on to the material.

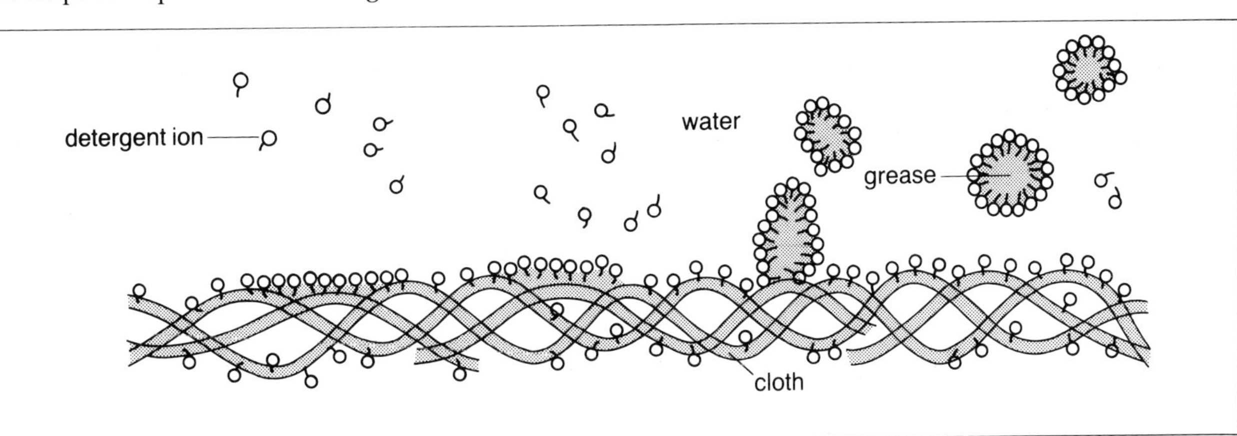

figure 15.31 How a detergent works

INVESTIGATION 15.9

MAKING SOAP

Soap is the name given to a particular type of detergent. Soaps are produced when fats are boiled with alkalis. The technical word for this reaction is **saponification**.

Soap is the sodium or potassium salt of a fatty acid. The length of the carbon chain in the tail of the soap molecule is about 20–30 atoms.

Goggles must be worn during this investigation.

a Place about 3 cm^3 of castor oil in a 100 cm^3 glass beaker.
b Add about 25 cm^3 of dilute (2M) sodium hydroxide to the beaker.
c Heat the mixture gently until it boils **stirring all the time**. If you heat it too quickly or do not stir the mixture it sometimes 'jumps' out of the beaker.
d Boil the mixture for about 5 minutes and then add 25 cm^3 of saturated sodium chloride (salt) solution. Bring the mixture back to the boil.
e Allow the mixture to cool. The soap should collect as a solid on the surface of the liquid.
f Test a small piece of the soap by shaking it in a test tube half full of distilled water.

Questions

1 What effect did the soap have on the distilled water?
2 What would have happened if the soap had been shaken with a sample of hard water?
3 Why would it not be sensible to try and wash yourself with a sample of the soap that you have made?

Soapless detergents

Many people seem to think that it is rather strange to talk of a soapless detergent. Detergents are cleaning agents and the word 'soap' refers only to the type of compound that we came across in the last section. **Soapless detergents** are cleaning agents other than soaps. They can be made from vegetable oils or mineral oil (petroleum oil that comes from under the ground).

Soapless detergents have the great advantage of not being affected by hard water. They are, however, usually more expensive to make. They can be solids or liquids. Washing-up liquids are soapless detergents. Both soaps and soapless detergents work in the same way.

INVESTIGATION 15.10

MAKING A SOAPLESS DETERGENT

Soapless detergents are made by reacting an oil with concentrated sulphuric acid, so great care is needed in this investigation.

Goggles must be worn during this investigation.

a Place about 1 cm^3 of castor oil in a Pyrex test tube.
b Using a dropping pipette, carefully add about 2 cm^3 of concentrated sulphuric acid a drop at a time. Make sure that no drops go anywhere else. Stir the mixture with a glass rod. The reaction mixture is likely to become very hot.
c Carefully pour the liquid into a 100 cm^3 beaker containing about 25 cm^3 distilled water. Stir the mixture gently. The detergent should form a solid.
d Pour off the water and wash the solid detergent two or three times with fresh distilled water.
e Test some of the detergent by shaking it with distilled water.

Questions

1 Does the detergent form a good lather?
2 Some detergents are **biodegradable**. What does this mean and what advantage do these sorts of detergents have (see pages 228 – 9).

Modern soap powders

A great deal of chemical research has resulted in today's washing powders. These powders are usually mixtures of several active chemicals.

A typical powder might contain:
a a soap
b a soapless detergent
c a water softener
d an optical brightener
e a fabric conditioner
f a dye and a scent.

Some powders are also 'biologically active'. These powders contain enzymes (see page 202) which help to breakdown 'natural' stains, e.g. food, blood. Enzmes are destroyed by high temperatures and so to get the best out of a 'biological' powder the clothes should be soaked in cold or warm water before washing. Many modern washing machines have programmes that allow you to do this.

ALL YOUR OWN WORK

The three main ________ ________ are oil, gas and coal. They are being used up quite quickly. At the present rate of use ________ and ________ will run out during the next 50 years but ________ should last much longer.
These fuels all contain the element ________ and so when they burn ________ ________ is formed. If there is not enough air some ________ ________ may be formed. This is dangerous because it is very ________.

Oil is a complex mixture of ________. It can be separated into different fractions because the different compounds have different ________ ________. This separation is done at an oil ________. Another process that takes place is ________. This involves breaking the larger, not so useful molecules into smaller ones.

To make a simple cell two different ________ must be placed in a ________ solution. The negative terminal will always be the ________ ________ metal. The size of the voltage depends on the ________ in reactivity of the two metals.

5 What is the main advantage of a soapless detergent? What problems can detergents cause?

6 Coal is an important fuel in many parts of the world. When coal is burned, however, it creates pollution. Figure 15.32 shows some of the pollutant from burning coal.

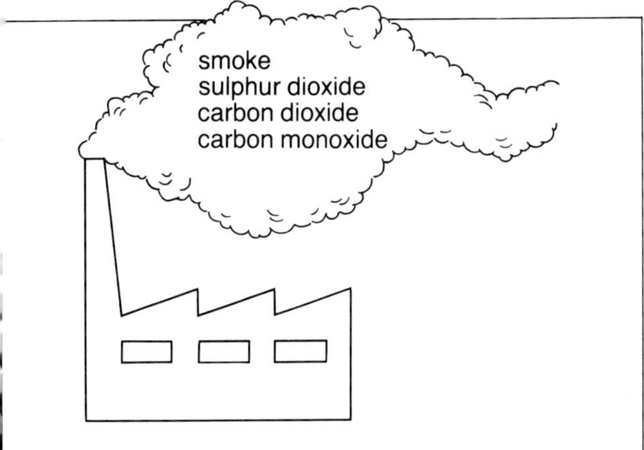

figure 15.32

a Below are four problems caused by burning coal. Copy them and alongside each problem write in a pollutant most responsible for it. (The same pollutant may be used more than once.)

problem	**pollutant**
reduction of light	________
damage to stone buildings	________
clogging of leaf stomata	________
damage to lungs	________

b Burning coal is not the only source of pollutants. Give **two** other examples of pollutants that can effect the health of a human being.

c How might **one** of these be controlled?

7 What are the main properties of a good fuel? List the good and bad points about using oil as a fuel.

8 What are the three parts of the fire triangle? How does water help put fires out?

9 Supplies of petrol will one day run out and so alternative fuels will have to be found. One suggestion is to use electric cars. What are the main problems that have so far prevented the electric car becoming popular?

10 A family of organic compounds is called an homologous series. Members of an homologous series have:
a similar ________ properties
b a similar ending to the ________
c a general chemical ________
d a gradual change in ________ properties as the ________ ________ increases.

11 What are isomers? Give the names of the two isomers of C_4H_{10}.

12 Fermentation is a way of producing ________ from sugar or other ________ . ________ in yeast break down the large sugar ________ . Bubbles of ________ ________ are produced during the fermentation. The ________ produced can be concentrated by ________ distillation because ________ and water have different ________.

13 What is a polymer? Make a list of common plastics or fibres that begin with 'poly'.

14 Although Pyrex type glass is much more expensive than soda glass, most of the glass apparatus used in schools is Pyrex. Why do you think this is so?

CHEMICAL AMOUNTS

HOW MUCH?

This unit is concerned with how much. We often need to know how much of each chemical is needed for a particular reaction, or how much of each element a compound contains.

Symbols and formulae

We saw in unit 11 that each chemical element has its own symbol. The symbol consists of one or two letters, usually taken from the name of the element. We also know that the chemist often uses a 'shorthand' form using these symbols to represent compounds, for instance, we are all familiar with the formula for water, H_2O. This tells us that in one molecule of water there are two atoms of hydrogen and one atom of oxygen.

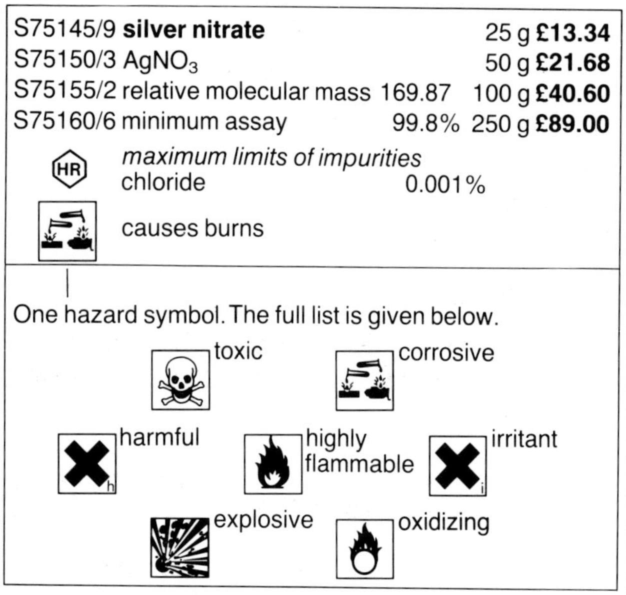

S75145/9 **silver nitrate** 25 g **£13.34**
S75150/3 $AgNO_3$ 50 g **£21.68**
S75155/2 relative molecular mass 169.87 100 g **£40.60**
S75160/6 minimum assay 99.8% 250 g **£89.00**
HR *maximum limits of impurities*
chloride 0.001%
causes burns

One hazard symbol. The full list is given below.

toxic
corrosive
harmful
highly flammable
irritant
explosive
oxidizing

figure 16.1 *Chemical labels give information to help in working out how much to use, and also precautions to take when using the chemical*

In unit 11 the covalent diagram for water explained why H_2O is the correct formula. It is not possible, however, to work out the formula of **ionic** compounds in this way. So how can we do it?

modern name	old name	symbol
gold	Sol (sun)	⊙
iron	Mars	♂
silver	Luna (moon)	☾

figure 16.2 *Some ancient symbols used to represent elements*

Valency

Each atom or group of atoms has a **valency**. This is the combining power of the atom or group. The valency is just a number. In the case of ions, the valency is the same number as the number of charges on the ion. In the case of covalent compounds, the valency is the number of electrons shared by an individual atom in the compound.

Table 16.1 shows the valencies of some metals and of hydrogen.

valency 1		valency 2		valency 3	
hydrogen	H	magnesium	Mg	aluminium	Al
lithium	Li	calcium	Ca	iron (III)	fe
sodium	Na	barium	Ba		
potassium	K	zinc	Zn		
silver	Ag	copper (II)	Cu		
ammonium	NH_4	iron (II)	Fe		
copper (I)	Cu	lead	Pb		

table 16.1 *Valencies of some metals*

Table 16.2 shows the valencies of non-metals and groups.

From these two tables, it should now be possible to work out the formulae of common compounds.

valency 1		valency 2		valency 3		valency 4	
fluorine	F	oxygen	O	phosphate	PO_4	Carbon	C
chlorine	Cl	sulphur	S	nitrogen	N		
bromine	Br	sulphate	SO_4	phosphorus	P		
iodine	I	carbonate	CO_3				
hydroxide	OH						
nitrate	NO_3						
hydrogencarbonate	HCO_3						
hydrogen sulphate	HSO_4						

table 16.2 *Valencies of some non-metals*

Formulae

In order to work out the formula of a compound, we need to know the valencies of the groups involved. In a compound all the valencies must cancel out.

e.g. for sodium chloride

sodium Na – valency 1
chloride Cl – valency 1

If we have one atom of each, the valencies cancel out and so the formula for sodium chloride is **NaCl**.

If we look at copper chloride, we see a difference in valencies.

copper (II) Cu – valency 2
chlorine Cl – valency 1

We need two lots of chlorine to cancel out the valency of copper, and so the formula of copper(II) chloride is $\mathbf{CuCl_2}$.

You might find it easier to think of the valencies as 'arms' which have to be linked together. If you have the correct number of atoms, there will be no spare arms.

e.g. sodium sulphate

sodium – valency 1 Na (1 arm)
sulphate – valency 2 SO_4 (2 arms)

sodium sulphate Na > SO_4 < Na

formula $\mathbf{Na_2SO_4}$

Notice how numbers in the formula are small numbers and come after the atom to which they refer.

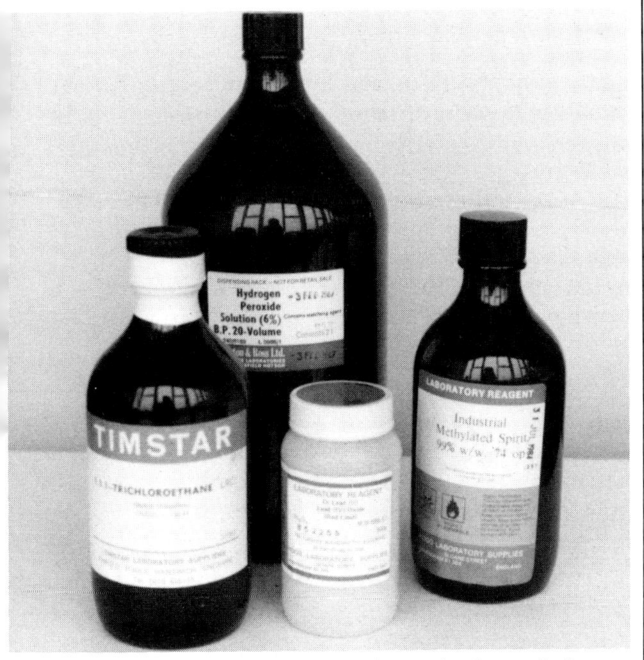

figure 16.3 It is possible to work out the formula for any compound

See if you can work out the formulae of the following.

a potassium iodide
b sodium hydroxide
c ammonium chloride
d copper(II) oxide
e lead carbonate
f aluminium chloride
g silver nitrate
h sodium sulphide
i potassium hydrogencarbonate
j aluminium oxide

Using brackets

Sometimes you will find that you need more than one lot of a group of atoms, e.g. a nitrate group. The group of atoms is put in brackets and the small number put after the brackets.

e.g. lead nitrate

lead Pb – valency 2
nitrate NO_3 – valency 1

Pb < NO_3, NO_3

formula $\mathbf{Pb(NO_3)_2}$

Try these compounds involving brackets.

a magnesium hydroxide
b zinc nitrate
c iron(III) hydroxide
d ammonium sulphate
e iron(III) sulphate

Remember that you need brackets only if you need more than one lot of a symbol containing more than one element.

EQUATIONS

Chemical equations are a way of representing the overall change that has taken place in a chemical reaction. The equation may use words or symbols. The big advantage of a symbol equation is that it tells us not only what substances have reacted and been formed, but also how much of each substance is involved.

If you follow the rules set out below, it is not difficult to write symbol equations.

a Write out the equation in words. i.e. write out the chemicals reacting together on the left-hand side and the products of the reaction on the right-hand side.

magnesium oxide + dilute sulphuric acid → magnesium sulphate + water

b Write down the correct formulae for all the chemicals involved.

magnesium oxide MgO
dilute sulphuric acid H_2SO_4
magnesium sulphate $MgSO_4$
water H_2O

c Substitute the formulae into the word equation.

$$MgO + H_2SO_4 \rightarrow MgSO_4 + H_2O$$

d Check to see that the equation is **balanced**, i.e. there is an equal number of atoms of each element on each side of the equation. This is the most important step.

$$MgO + H_2SO_4 \rightarrow MgSO_4 + H_2O$$

This is already balanced.

e Add the **state symbols**. State symbols tell us what form or state each chemical is in. The symbols go in small brackets, after the chemical.

(s) for solid
(l) for liquid
(g) for gas
(aq) for solution in water (aqueous)

Our complete equation is then

$$MgO(s) + H_2SO_4(aq) \rightarrow MgSO_4(aq) + H_2O(l)$$

Look at the reaction between magnesium oxide and dilute hydrochloric acid.

a magnesium oxide + dilute hydrochloric acid → magnesium chloride + water

b

magnesium oxide	MgO
dilute hydrochloric acid	HCl
magnesium chloride	$MgCl_2$
water	H_2O

c $MgO + HCl \rightarrow MgCl_2 + H_2O$

This is not balanced.

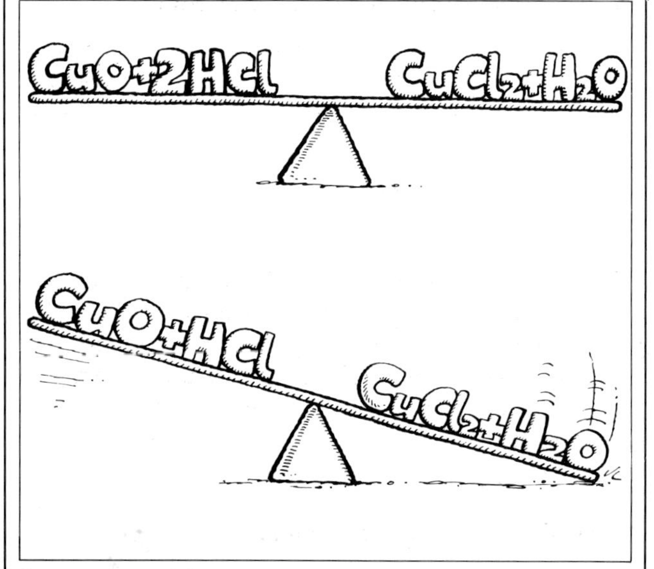

figure 16.4 A balanced equation

d We can see that there is more chlorine and more hydrogen on the right-hand side than there is on the left-hand side. In order to balance the equation we **cannot alter any formulae**, we can only alter the amounts of each chemical. What we need in this case is two lots of dilute hydrochloric acid.

$$MgO + 2HCl \rightarrow MgCl_2 + H_2O$$

Notice how the large 2 goes in front of the hydrochloric acid.

e $MgO(s) + 2HCl(aq) \rightarrow MgCl_2(aq) + H_2O\,(l)$

See if you can balance the following equations.

1 $CaCO_3 + HCl \rightarrow CaCl_2 + H_2O + CO_2$
2 $Na + H_2O \rightarrow NaOH + H_2$
3 $CH_4 + O_2 \rightarrow CO_2 + H_2O$

Write a fully balanced equation for the reaction of copper(II) carbonate and dilute nitric acid.

CHANGES IN MASS

You might already have experimented to see if there is any change in mass when substances are heated. Common substances to heat are magnesium and red lead.

If you weigh some magnesium in a crucible, burn it and re-weigh the crucible and ash, it will have increased in mass. This is because the magnesium has combined with oxygen from the air.

If you heat red lead it loses mass because it gives off oxygen to the air.

In either of these examples, has there been any **overall** change in mass?

Carry out investigation 16.1.

You probably confirmed the **law of conservation of mass**. In any chemical reaction, there can never be any overall change in mass. This is because matter cannot be created or destroyed in a chemical reaction.

The remainder of this unit contains the more mathematical quantitative chemistry. Check whether your syllabus requires these sections.

THE MOLE

The **mole** is a name given to a specific amount of a chemical substance. In some of our investigations we need to be able to relate a mass we can weigh out on the balance to an actual number of atoms or molecules. The mole enables us to do that. Atoms are too small to be weighed individually, but it is possible to compare the masses of atoms using a machine called a **mass spectrometer** (see figure 16.6). The results obtained give us the **relative atomic mass** of ach atom. These are shown on the periodic table of lements (see page 177). The relative atomic mass ells us how heavy an atom is compared with ydrogen.

INVESTIGATION 16.1

IS THERE ANY CHANGE IN MASS IN A CHEMICAL REACTION?

a Carefully place about 20 cm^3 sodium chloride solution in a 100 cm^3 conical flask.
b Tie a piece of cotton round a small test tube and three-quarters fill the tube with silver nitrate solution.
c Carefully lower the test tube into the conical flask, making sure that you do not spill any of the silver nitrate solution (see figure 16.5).
d Trap the cotton with a rubber bung.

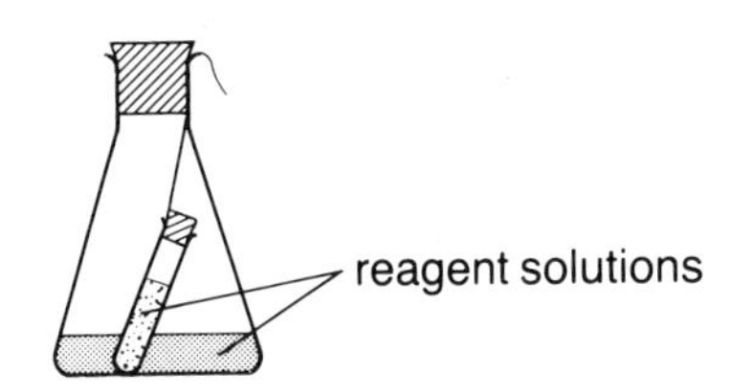

figure 16.5

e Weigh the flask and contents and record its mass.
f Tip the flask up to mix the two solutions.
g Re-weigh the flask and contents after the reaction has finished.

Questions
1 What evidence did you see of a reaction?
2 Was there any change in mass?
3 Write a balanced equation for the reaction.

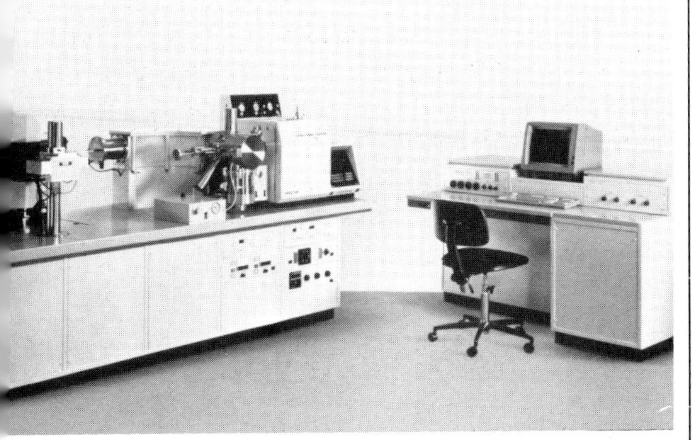

figure 16.6 A mass spectrometer tells us the relative atomic mass of a compound

The relative atomic mass expressed in grams is called **one mole** of that element. **One mole of any element contains the same number of atoms**, i.e. 1 g hydrogen contains the same number of atoms as 24 g magnesium or 32 g sulphur.

figure 16.7 Avogadro

As you can imagine, the actual number of atoms in one mole of an element is very large. The actual number has been shown to be about 6×10^{23}, i.e. 600 000 000 000 000 000 000 000. This number is called **Avogadro's number**. You can see how a mass we can weigh out can be directly related to an actual number of atoms. The size of Avogadro's number means that we do not often use the actual number of atoms, but prefer to work in moles and fractions of moles.

One mole of a **compound** is the **relative formula mass** in grams. The relative formula mass is found by adding together the individual relative atomic masses.

How to calculate one mole for compounds

water	H_2O	$2 \times 1 + 16$	= 18 g
carbon dioxide	CO_2	$12 + 2 \times 16$	= 44 g
calcium carbonate	$CaCO_3$	$40 + 12 + 3 \times 16$	= 100 g

Numbers of moles

To work out the number of moles in a mass of a substance we simply divide the mass we have by the mass of one mole. Using magnesium as an example: Mg = 24, therefore 1 mole of magnesium has a mass of 24 g. If 48 g magnesium were used this would represent

48 g is $\frac{48}{24} = 2$ moles

A smaller mass, say 6 g, would be

6 g of magnesium is $\frac{6}{24} = 0.25$ mole

UNIT 16

Work out the following numbers of moles. (You will need to look up the relative atomic masses).

a 64 g oxygen
b 3 g carbon
c 9 g water
d 200 g calcium carbonate
e 1 g calcium

What mass do the following have?

a 2 moles sulphur
b 1 mole iron
c 0.5 mole carbon
d 0.01 mole magnesium
e 5 moles water

Using moles to find formulae

Earlier in the unit we used valency tables to work out the correct formulae for different compounds. By using the mole it is possible to work out formulae from experimental data. See investigation 16.2.

Calculating formulae

Example

Find the formula of an oxide of copper where 0.64 g copper combines with 0.16 g oxygen (Cu = 64, O = 16).

Convert the masses to moles.

$\frac{0.64}{64}$ moles copper join with $\frac{0.16}{16}$ moles oxygen

i.e. 0.01 moles copper join with 0.01 moles of oxygen

one mole of copper would join with one mole of oxygen (one mole of copper contains the same number of atoms as one mole of oxygen)

1 **atom** of copper joins with 1 **atom** of oxygen

formula = CuO

Work out the formula of an oxide of carbon where 0.3 g carbon join with 0.4 g oxygen (C = 12, O = 16).

Percentage composition

The formulae of compounds can be worked out from their percentage composition in exactly the same way. All you have to do is assume you have 100 g of the compound.

Example

Find the formula of an oxide of sulphur that contains 50 per cent of each element.

In 100 g the oxide there would be 50 g sulphur joined with 50 g oxygen. Change the mass to moles (S = 32, O = 16)

$\frac{50}{32}$ moles sulphur join with $\frac{50}{16}$ moles oxygen

1.5625 moles sulphur join with 3.125 moles oxygen

1 mole sulphur joins with $\frac{3.125}{1.5625}$ moles oxygen

1 mole of sulphur joins with 2 moles of oxygen

1 **atom** of sulphur joins with 2 atoms of oxygen

formula = SO_2 (sulphur dioxide)

Calculations from equations

A chemical equation tells us how much of each substance is reacting, and how much is formed.

Calcium carbonate decomposes to calcium oxide and carbon dioxide when it is strongly heated.

INVESTIGATION 16.2

TO FIND THE FORMULA OF MAGNESIUM OXIDE

a Accurately weigh a crucible and lid.
b Accurately weigh out 0.12 g magnesium ribbon. (This should be a strip about 12 cm long.)
c Heat the magnesium strongly in the crucible. Remove the lid from time to time to allow more air in. Try not to let any smoke out.

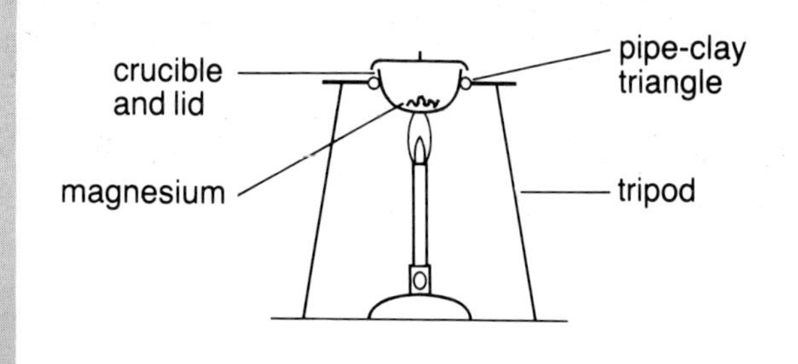

figure 16.8 *Heating magnesium*

d When all the magnesium appears to have burned, remove the lid and continue strong heating for about 1 minute.
e When the crucible is cool, re-weigh it. *Don't forget to re-weigh the lid as well.*

Questions

1 What mass of magnesium oxide was formed?
2 What mass of oxygen was there in the magnesium oxide?
3 How many moles of magnesium is 0.12 g magnesium?
4 How many moles of oxygen did this combine with?
5 How many moles of oxygen would combine with one mole of magnesium?
6 Remember that one mole of any element contains the same number of atoms. How many atoms of oxygen combine with one atom of magnesium?
7 Does your formula agree with the one you worked out from the valancy table? If not, can you suggest any reasons why not?

$$CaCO_3 \rightarrow CaO + CO_2$$
$$1 \text{ mole} \rightarrow 1 \text{ mole} + 1 \text{ mole}$$
$$\text{i.e. } 100\text{ g} \rightarrow 56\text{ g} + 44\text{ g}$$
$$(40 + 12 + 3 \times 16)\text{ g} \rightarrow (40 + 16)\text{ g} + (12 + 2 \times 16)\text{ g}$$

This tells us that the maximum amount of calcium oxide we can get from 100 g calcium carbonate is 56 g, and the maximum amount of carbon dioxide is 44 g. All calculations concerning equations are done in a similar way.

Example

How much nitric acid is needed to dissolve 8 g copper(II) oxide?

$$CuO + 2HNO_3 \rightarrow Cu(NO_3)_2 + H_2O$$
$$1 \text{ mole} + 2 \text{ moles} \rightarrow 1 \text{ mole} + 1 \text{ mole}$$
$$80\text{ g} + 126\text{ g} \rightarrow 188\text{ g} + 18\text{ g}$$
$$(64 + 16)\text{g} + 2(1 + 14 + 3 \times 16)\text{g} \rightarrow (64 + (14 + 3 \times 16)_2)\text{ g} + (2 \times 1 + 16)\text{g}$$

The equation tells us that 80 g copper(II) oxide will react with 126 g nitric acid, so 8 g copper(II) oxide will react with 12.6 g nitric acid.

Molar solutions

You may have noticed on reagent bottles, especially dilute acids, the symbol 2 M. This refers to the **concentration** of the acid. 2 M means a 2 molar solution.

A molar solution is one in which 1 dm^3 (1000 cm^3) of the solution contains 1 mole of the solute, i.e. 2 M sulphuric acid means that each dm^3 of the acid contains 2 moles of sulphuric acid.

Apart from being a good way of describing the concentration of solutions, this is also useful when considering the reactions involving solutions. It is easier to work out the volume of the solution required, rather than converting it back to the mass of solution it contains.

Examples

a How much sulphuric acid is needed to make 10 dm^3 of 2M concentration?

to make 1 dm^3 of 1 M solution we need 1 mole
1 mole of sulphuric acid (H_2SO_4) = $(2 + 32 + 4 \times 16)$g = 98 g

to make 1 dm^3 of 1 M sulphuric acid we need 98 g
to make 1 dm^3 of 2 M sulphuric acid we need 196 g
to make 10 dm^3 of 2 M sulphuric acid we need 1960 g

b What volume of 2M sulphuric acid is needed to react with 6 g magnesium?

$$Mg + H_2SO_4 \rightarrow MgSO_4 + H_2$$
$$1 \text{ mole} + 1 \text{ mole} \rightarrow 1 \text{ mole} + 1 \text{ mole}$$
$$24\text{ g} + 98\text{ g} \rightarrow 120\text{ g} + 2\text{ g}$$

Instead of writing 98 g for 1 mole of sulphuric acid, we can write 1 dm^3 of 1 M acid.

24 g magnesium react with 1 dm^3 of 1 M acid
or 24 g magnesium react with 500 cm^3 of 2 M acid (if the acid is twice as concentrated we only need half as much)

6 g of magnesium would then react with $\frac{500}{4}$ dm^3 ($\frac{1}{4}$ of 24) of 2 M acid

125 cm^3 of 2 M sulphuric acid react with 6 g magnesium

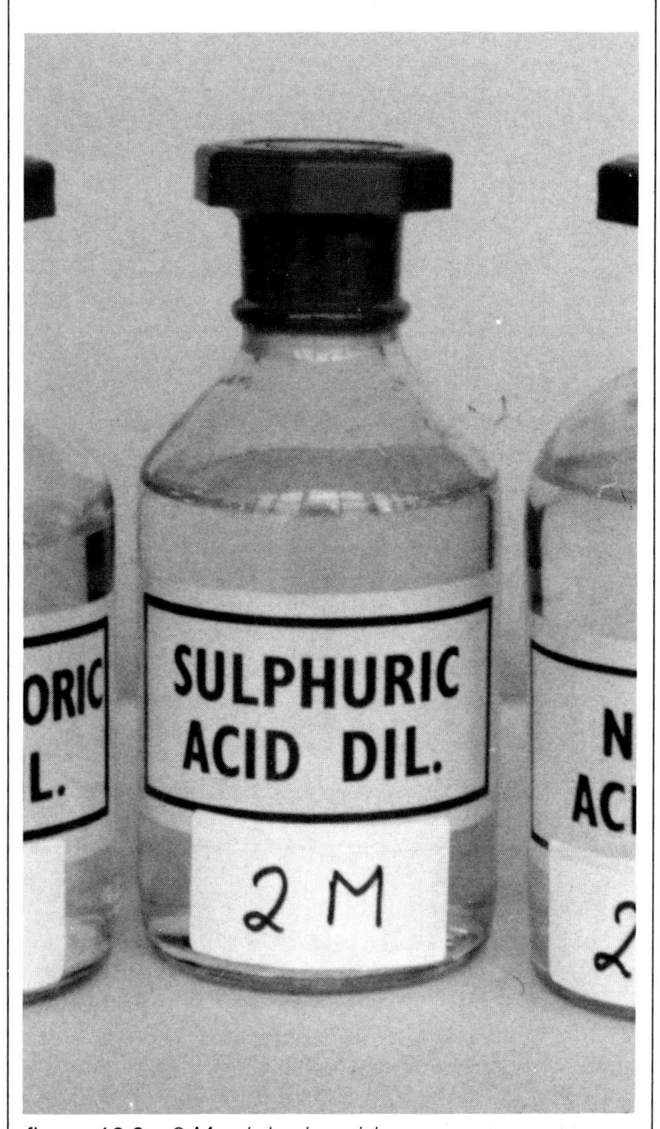

figure 16.9 2 M sulphuric acid

ALL YOUR OWN WORK

1 The formula of a compound tells us how many ________ of each ________ there are in one molecule of the compound.

2 The valency of a metal ion is the same as the ________ on the ion.

3 In a balanced equation there must be an ________ ________ of ________ of each ________ on either side of the equation.

4 The number of atoms in one mole of any element is called ________ ________

5 The ________ ________ ________ of an atom tells us how heavy the atom is compared with hydrogen.

6 A molar solution is one in which ________ of solution contains ________ of solute.

7 Write down the correct formulae for
a lead sulphate
b iron(II) chloride
c aluminium hydroxide
d silver carbonate
e calcium phosphate.

8 What are the total numbers of atoms in one molecule of
a CO_2
b $Cu(NO_3)_2$
c $FeSO_4$
d $(NH_4)_2SO_4$
e $CuSO_4.5H_2O$?

9 Natural gas is mainly methane (CH_4).
a Write a word equation for the burning of methane in oxygen.
b Write a balanced equation for this reaction using chemical symbols.

10 Calculate the mass of one mole of
a $CuSO_4$
b CO_2
c NH_3
d $Fe_2(SO_4)_3$
e $KHCO_3$

11 Work out the formula of an oxide of lead where 2.07 g lead join with 0.32 g oxygen.

12 Calculate the percentage of nitrogen in the following (percentage is $\frac{\text{mass of nitrogen}}{\text{mass of 1 mole}} \times 100\%$)
a NH_4NO_3
b KNO_3
c $(NH_4)_2SO_4$
d NH_3

13 What mass of carbon dioxide is given off when 1 g calcium carbonate is heated?

$$CaCO_3 \rightarrow CaO + CO_2$$

14 Iron reacts with hot water to produce the oxide Fe_3O_4 and hydrogen gas.
a Construct an equation for this reaction.
b How many moles of the oxide can be made from 1 mole of iron?

15 What mass of sodium hydroxide is needed to make
a 1 dm^3 of 2 M solution
b 20 cm^3 of 1 M solution
c 20 cm^3 of 4 M solution?

16 Balance the following equations.
a $H_2SO_4 + NaOH \rightarrow Na_2SO_4 + H_2O$
b $CuO + HCl \rightarrow CuCl_2 + H_2O$
c $MgCl_2 + AgNO_3 \rightarrow Mg(NO_3)_2 + AgCl$
d $Fe + Cl_2 \rightarrow FeCl_3$

17 Which of the following contains the greatest percentage of oxygen?
a SO_2
b $NaNO_3$
c $Mg(NO_3)_2$
d $CaCO_3$
e $Fe_2(SO_4)_3$

18 Copper can be displaced from copper (II) sulphate solution by iron. Iron (II) sulphate is also formed
a Write a word equation for the reaction.
b How would you show that iron (II) ions were present in the resulting solution?
c Write the symbol equation.
d How much copper would be displaced by 7 g of iron?

19 When ammonia gas is passed over heated copper (II) oxide, copper is formed.

$$2NH_3 + 3CuO \rightarrow 3Cu + N_2 + 3H_2O$$

a What is the ammonia doing to the copper (II) oxide, and what type of reaction is this?
b What mass of copper could be obtained from 2.4 g of copper (II) oxide?
c What volume of nitrogen would be produced when 2.4 g of copper (II) oxide react with ammonia? (1 mole of any gas has a volume of 24 dm^3 at room temperature and pressure.)

20 The equation below represents the neutralization reaction between dilute hydrochloric acid and sodium hydroxide solution.

$$HCl + NaOH \rightarrow NaCl + H_2O$$

It was found by experiment that 25 cm^3 of 2 M sodium hydroxide solution was exactly neutralized by 20 cm^3 of dilute hydrochloric acid.
a What does 2 M mean?
b In this experiment which is the more concentrated solution, the acid or the alkali?
c How many moles of sodium hydroxide are present in 25 cm^3 of 2 M solution?
d How many moles of hydrochloric acid must be present in the 20 cm^3 of acid?
e What is the concentration of the dilute hydrochloric acid used in this experiment?

KEEPING THINGS STEADY

figure 17.1

Keeping steady is vital for this tightrope walker. How does she do it? Her brain uses information from her sense organs to calculate her stability on the tightrope. As she becomes off-balance her brain sends messages to her muscles which contract or relax to return her to a balanced position. Tiny adjustments like this occur many times each second. You can investigate for yourself how you manage to keep upright (see investigation 17.1).

Stability is important in many situations. Think about any **system**: a mechanical system such as a gas oven; a biological system such as your body or an ecosystem such as a pond. Systems consist of a large number of factors that vary over time – these are **variables**. The tightrope walker's position or the temperature of an oven are examples of variables. Keeping these variables under control is known as **homeostasis**.

CONTROLLING ENERGY

When energy is used for a specific purpose we usually want to control it. Energy out of control can be disastrous. The heat energy released in an oven must be carefully controlled to avoid a burnt dinner (see figure 17.2).

INVESTIGATION 17.1

KEEPING YOUR BALANCE

a Stand with your feet together and your arms by your sides look straight ahead. What is happening to the position of your body? Try to stop this movement. Is it possible?
b Now close your eyes. How does this affect your body movement? How is your sight involved in keeping your balance?
c Open your eyes. Concentrate on the soles of your feet. How does the pressure on your soles vary?
d Keep your eyes open. Now concentrate on your calf muscles. Lean forward. Which muscles feel tense? What movements stop you falling over?

Any system under control has an **input** and an **output**. The input in this case is the amount of gas being burnt in the oven. The output required is a steady oven temperature. We can use information about the output (the oven temperature) in order to control the input (the gas flame). We could use the information and control the oven ourselves. By using a thermometer, we could keep a constant watch on the oven temperature. When it got too high we could turn down the gas flame. As the temperature fell too far we could turn up the flame and so bring the oven temperature back to the level required. We would need continually to alter the gas flame to keep a steady temperature. You can explore this method of control in investigation 17.2, overleaf.

This way of controlling the oven temperature would be very time-consuming for a busy cook. Fortunately, the oven has a control system that does the job automatically. It works like this: a special switch, called a **thermostat**, can be set at a desired temperature level (the **set point**). The thermostat detects when the temperature rises above the set point and turns down the gas flame. The oven temperature begins to fall. When the temperature falls below the set point the thermostat turns up the gas flame automatically.

figure 17.2 Energy out of control can be disastrous

UNIT 17

Any control system relies on a device called a **sensor** that can detect the level of the output. The actual output level is compared with the desired output level. This task is carried out by the **comparator**. In the example of the oven, the thermostat acts as both comparator and sensor. The system also needs an **actuator**. This is the part of the system that can bring the level of the output back to the set point. The gas flame is the actuator. It maintains the oven temperature at the required level. The actuator needs to be linked to the sensor so that the correct action is taken. Information about the oven temperature is linked to the size of the gas flame. This information is known as **feedback**.

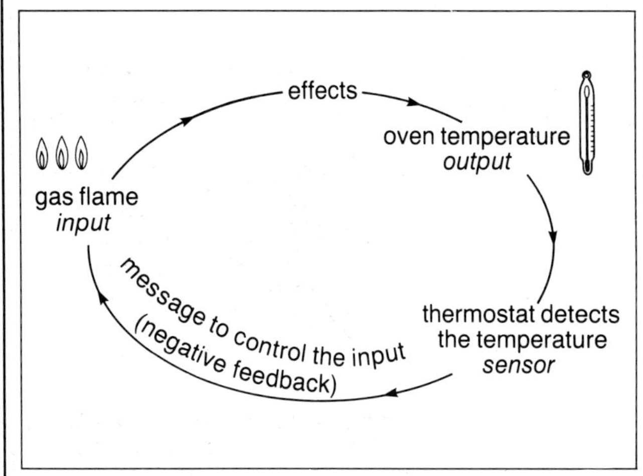

figure 17.3 *How feedback carries information to control the input*

Positive and negative feedback

Feedback carries information in the form of simple messages. Very often these messages are in code. The coded messages in the example of the oven were *'The temperature is too high. Turn down the gas flame'* and *'The temperature is too low. Turn up the gas flame'*. Feedback in this example reversed the direction of the temperature change and is known as **negative feedback.** Negative feedback allows systems to maintain homeostasis.

Positive feedback leads to a system going out of control. A positive feedback message would say *'The temperature is too high. Increase the gas flame'*. Clearly, positive feedback cannot provide homeostatic control!

KEEPING YOUR BODY TEMPERATURE STEADY

Controlling body temperature is similar to controlling the temperature of an oven. A pea-sized structure on the underside of the brain called the **hypothalamus** acts as a thermostat. The hypothalamus is continually checking the temperature of the blood. If it varies

INVESTIGATION 17.2

INVESTIGATING HOMEOSTASIS

The aim of this experiment is to keep a beaker of water at a constant temperature of 60 °C.

Work in pairs, one partner making a record of the temperature readings.

a Place a thermometer in a beaker of water. Record its temperature.
b Heat the beaker of water gently over a low Bunsen flame, taking temperature readings every minute.
c Holding the base of the Bunsen burner in your hand, try to keep a constant water temperature of 60 °C by removing the flame from under the beaker and returning it at intervals. Your partner should continue to take readings for 15 minutes.
d Plot the temperature readings on a line graph with time on the horizontal axis. Draw a horizontal line on the graph representing an 'ideal' constant temperature of 60 °C.

Questions

1 What difficulties did you find in trying to keep a constant temperature?
2 Did the graph show any improvement in your ability to maintain a constant temperature?
3 In this system, what is the input, the output and the sensor (see figure 17.3)?
4 Would a thermostat be more reliable in maintaining a constant temperature? If so, explain why.

from the set point of 37 °C, messages are sent to all parts of the body. Changes occur which return the blood temperature to normal. These changes are as follows. (See figure 17.4.)

Skin capillaries
As warm blood flows through capillaries near the skin surface it loses heat. In warm conditions these capillaries widen (**dilate**) and allow more blood to flow. The blood loses heat more quickly and becomes cooler. The widening of blood capillaries is called **vasodilation**. The opposite occurs when we are too cold. The blood vessels narrow (**vasoconstriction**), less blood flows to the surface and less heat is lost.

You can observe these changes in light-skinned people. When they are hot their skin becomes pinker. In the cold, they will look pale. Vasoconstriction at sub-zero temperatures can completely stop the flow of blood to the fingers and toes. Without a blood supply the cells begin to die. This is called **frostbite**. In severe cases the affected part may have to be amputated.

Sweating
In cool weather we do not produce much sweat. If we get too hot, our sweat glands become more active. Sweat is poured onto the surface of the skin where it

evaporates and cools the body (see investigation 17.3). A person who exercises vigorously in hot weather may lose up to a litre of water every hour.

Body metabolism
Chemical reactions in our cells generate a lot of heat. The muscles and liver are particularly important. The heat they produce is distributed around the body by the blood circulation. In cold weather the muscles make small but rapid contractions. We describe this as **shivering**. When we become chilled, shivering generates extra warmth.

Hair
In cold conditions we may find that our skin is covered in tiny bumps. These are **goose pimples** caused by the contraction of very small muscles. These muscles make each body hair stand on end rather than lie flat. Although we do not benefit from this reaction, other mammals do. Their fur stands out from the skin and gives them a thick layer of insulation.

INVESTIGATION 17.3

CHANGING TEMPERATURES DURING HOMEOSTASIS
An electrical immersion heater is used to heat the water in a tank. The temperature of the water is controlled by a thermostat. After switching on a heater the water temperature was measured at one-minute intervals. The readings are shown in table 17.1.

time after heater switched on (min)	water temperature (°C)
0	25
1	27
2	30
3	33
4	36
5	39
6	42
7	41
8	40
9	39
10	40
11	41
12	42

table 17.1

Questions
1 Plot these data on a line graph.
2 Label these points on the graph
  a when the thermostat turns the heater off
  b when the thermostat turns the heater on.
3 What temperature has the thermostat been set at?
4 Draw a line on the graph to show how you think the water temperature would have changed if the thermostat had been set at 30 °C when the heater was first switched on.

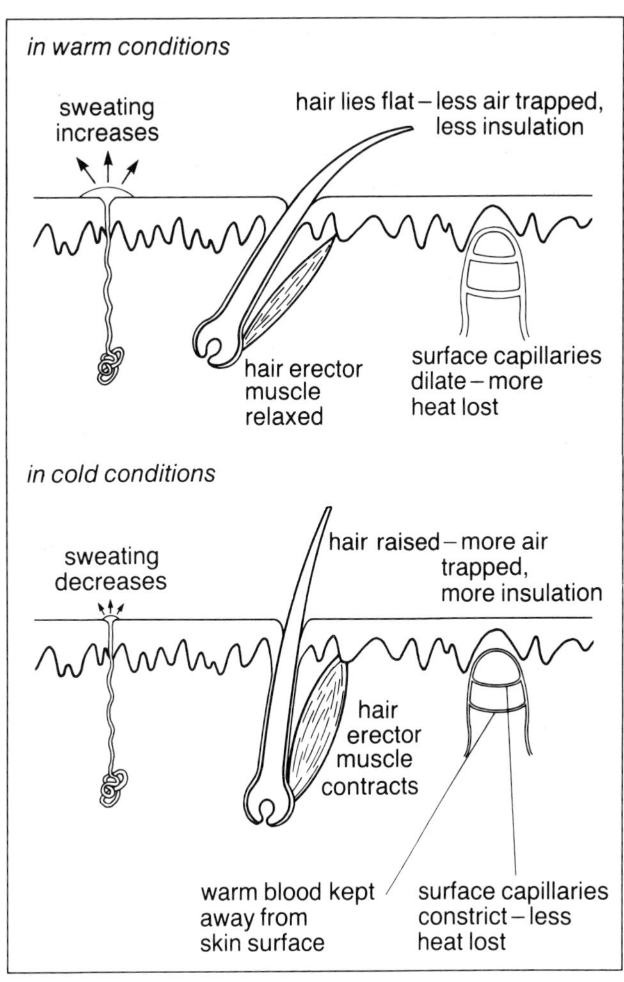

figure 17.4 How body temperature is controlled

INVESTIGATION 17.4

CONTROLLING HEAT LOSS
a Use a clamp and stand to hold a round-bottomed flask a few inches above the laboratory bench.
b Fill the flask with boiling water and immediately place a rubber bung containing a thermometer into the neck of the flask. The bung should have a second hole to allow the escape of steam.
c Make an initial temperature reading and repeat at two minute intervals for 20 minutes.
d Devise an appropriate table to record your results.
e Repeat steps **a – c** with a flask covered by a layer of cotton wool. Record your results.
f Repeat steps **a – c** with an uncovered flask. Wipe the outside of the flask every two minutes with cotton wool soaked in ethanol. Record you results.
g Plot a line graph of your results for the three treatments with time on the horizontal axis.

Questions
1 Which flask lost heat most rapidly?
2 Which flask lost heat most slowly?
3 Which flask demonstrates the effect of sweating?
4 Which flask demonstrates the effect of insulation?
5 What structures in a mammal are the equivalent of the cotton wool in this investigation?

Why is small size a problem for keeping warm?

The skin surface of a small animal is relatively large in relation to its volume when compared with that of a bigger animal (see investigation 17.5).

The relatively large surface area of small mammals means that they tend to lose their body heat very rapidly. To make up for this heat loss their cells produce heat by burning up food more rapidly. They have a fast rate of metabolism. This explains why the heart beat of a mouse is about 500 beats per minute. Human babies also have to overcome the problem of a relatively large surface area for their size. They should be wrapped up in extra layers of clothing in cold weather to slow down the rapid heat loss. As they grow in size, this problem lessens.

INVESTIGATION 17.5

LOSING HEAT: A PROBLEM FOR LARGE OR SMALL ANIMALS?

The two diagrams in figure 17.5 represent the approximate body shapes of two differently sized mammals: a mouse and a tiger.

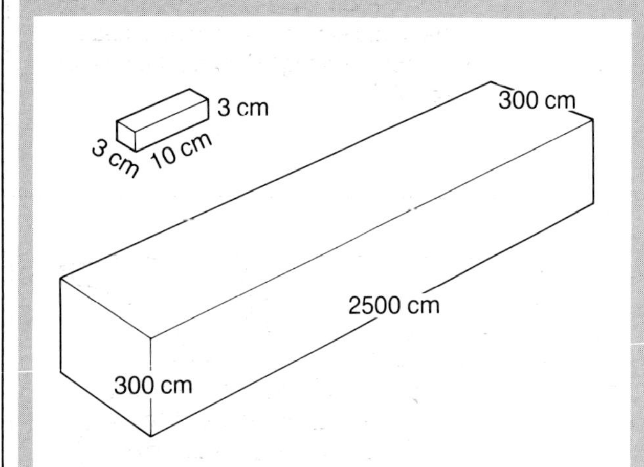

figure 17.5

Questions

1 For each mammal, calculate
  a the total skin surface area
  b the total volume of the mammal
  c the ratio $\frac{\text{surface area}}{\text{volume}}$ (a ÷ b).
2 Which of the mammals has the higher surface area to volume ratio?
3 Which mammal would lose heat more rapidly?
4 The body temperature of these two mammals is approximately the same. How do you think the mouse manages to keep its body temperature as high as the tiger's?
5 Each day the tiger needs to eat only a small fraction of its own weight in food. The mouse will only survive if it eats more than its own weight in food every day. How do you explain this difference?

HORMONES AND HOMEOSTASIS

A **hormone** is a chemical that is released from special organs known as **endocrine glands**. Hormones are carried in the bloodstream and have specific effects on particular parts of the body (see figure 17.6). Hormones work as *chemical messengers*. Sometimes they act like switches, controlling the level of a particular variable. The amount of sugar in the blood and the water content of the body are two variables controlled by hormones.

Controlling our blood sugar level

Glucose is being taken from the bloodstream and used up by our body cells all the time. After a meal, large quantities of glucose flood into the bloodstream from the small intestine, yet the concentration of glucose in the blood remains fairly steady. How is this achieved?

Glycogen is an insoluble carbohydrate that is stored in the liver. Glucose entering the liver from the small intestine can be converted to glycogen. When needed, glycogen can be converted back into glucose and released into the blood. This control system requires a sensor to check the blood glucose levels and feed back this information to the liver. In this case, the sensor consists of special tissue in the pancreas. This tissue will produce the hormone **insulin** when the glucose level rises above the set point. Insulin travels in the blood to the liver where it activates the conversion of glucose to glycogen. Glucose passing through the liver is removed from the blood and converted to glycogen. The blood glucose concentration decreases and returns to normal.

At other times the blood glucose level will decrease below normal. For example, when we exercise glucose is taken from our blood to supply energy to our muscles. If the drop in glucose levels continued our cells would be starved of energy and eventually we would fall into a coma. The release of another hormone **glucagon**, ensures that this does not happen. Glucagon is released by the pancreas when the blood glucose level drops below 100 mg/100 ml of blood. When it reaches the liver, glucagon switches on the conversion of glycogen to glucose, which is then released into the blood. This control system is summarized in figure 17.7.

Keeping our water level steady

The average adult body consists of 58 per cent water. It is important that this amount of water remains constant. If we gain too much water, our body fluids

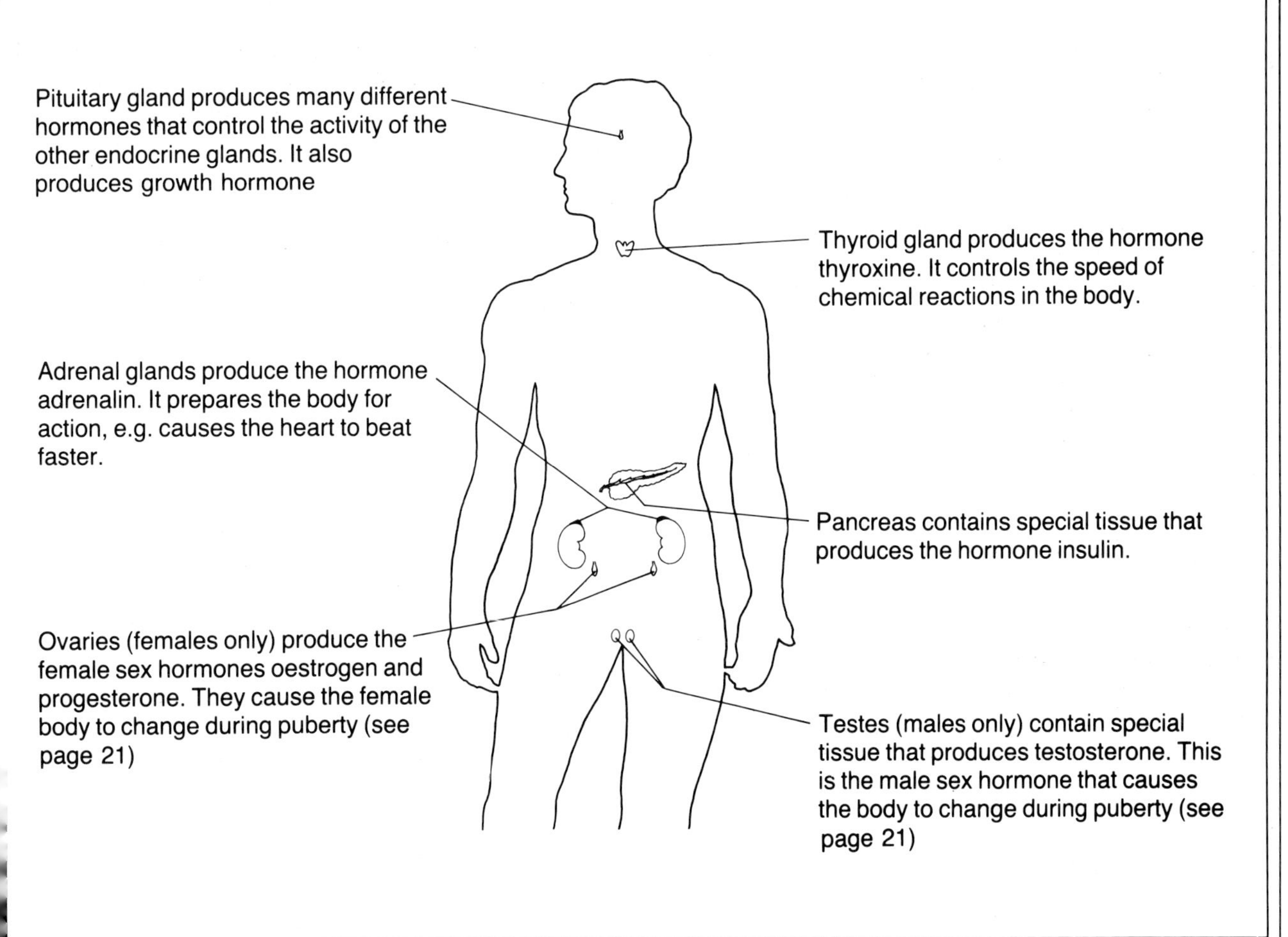

Pituitary gland produces many different hormones that control the activity of the other endocrine glands. It also produces growth hormone
Thyroid gland produces the hormone thyroxine. It controls the speed of chemical reactions in the body.
Adrenal glands produce the hormone adrenalin. It prepares the body for action, e.g. causes the heart to beat faster.
Pancreas contains special tissue that produces the hormone insulin.
Ovaries (females only) produce the female sex hormones oestrogen and progesterone. They cause the female body to change during puberty (see page 21)
Testes (males only) contain special tissue that produces testosterone. This is the male sex hormone that causes the body to change during puberty (see page 21)

figure 17.6 *The major endocrine glands and their functions*

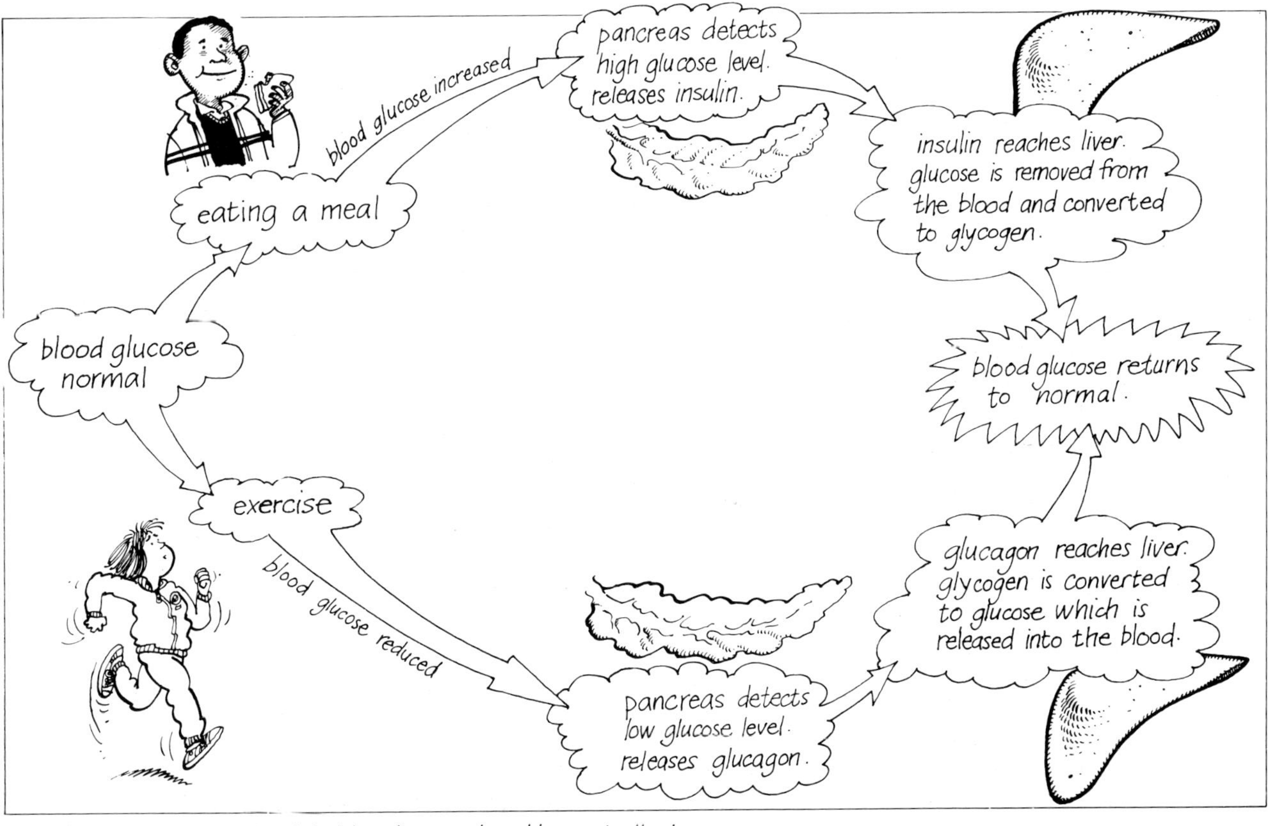

eating a meal

pancreas detects high glucose level. releases insulin.

pancreas detects low glucose level. releases glucagon.

figure 17.7 *How the body's blood sugar level is controlled*

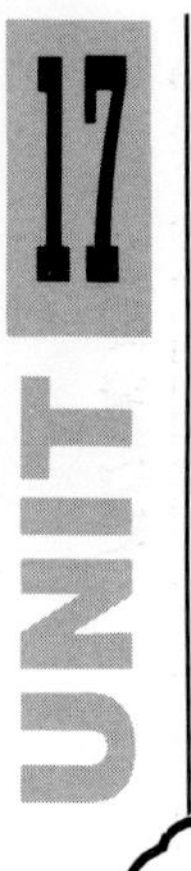

figure 17.8 Keeping a steady water content in the body

Application 17.1

DIABETES – BLOOD SUGAR OUT OF CONTROL

Diabetes is disease that affects one in 50 people in the UK. Scientists are uncertain of the causes of diabetes but once it has developed life-long treatment is necessary.

There are two types of diabetes. **Insulin dependent diabetes** develops in childhood when the pancreas stops producing insulin. The diabetic in this case is dependent on daily injections of insulin. This regulates the blood sugar levels.

Non-insulin dependent diabetes occurs during adulthood. It is more common in men than in women. It is more likely to strike the overweight as their high-sugar diet puts an extra strain on the pancreas. In this type of diabetes the pancreas will only produce a small amount of insulin which is insufficient to control the blood glucose level. Treatment with glucose-regulating pills rather than insulin injections is given.

A careful balance between diet and exercise is crucial to the diabetic. Vigorous exercise will drain the blood of sugar which must be rapidly replaced. Any sugar taken in the diet must be carefully controlled. If the blood sugar rises above 180 mg/100 ml the kidneys are unable to hold the sugar back and it floods over into the urine. Diabetics replace the homeostatic control normally carried out by the pancreas by their own conscious control of sugar intake.

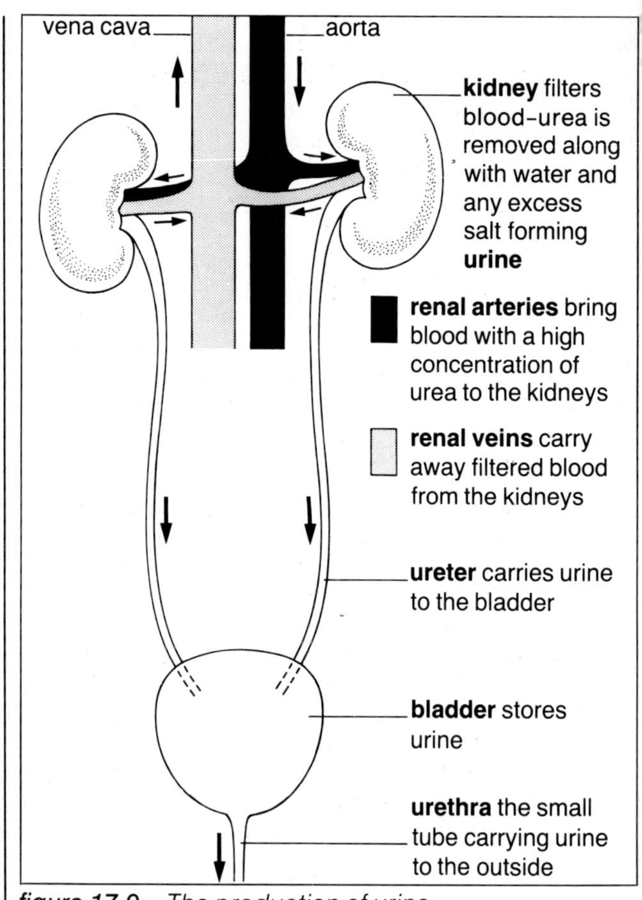

figure 17.9 The production of urine

become too dilute. If we lose too much water, our body fluids become too concentrated. In either case our body cells would be unable to work properly. Keeping the body fluids at the right concentration is known as **osmoregulation**.

The organ responsible for controlling the amount of water leaving the body is the kidney. The kidney acts like the plug in the bath. If we run in too much water we can take out the plug. If we drink a lot of water our kidney can 'take out the plug' by removing more water from the blood. This excess water forms **urine**. The opposite will happen if we are short of water, for example, after sweating profusely. The kidney will then reduce the amount of water removed from the blood.

How does the kidney know it should produce more or less urine? The hypothalamus continually checks the blood concentration. If we are short of water the hypothalamus instructs the pituitary gland to release the hormone **ADH**. ADH is carried in the bloodstream to the kidney. The kidney stops producing so much urine and more water is kept in the body. The complete control system is shown in figure 17.8.

Excretion and homeostasis

Your body must be able to cope with any waste products produced by its cells. Getting rid of these waste products is called **excretion**. This keeps the level of poisonous waste chemicals in the blood at a low and safe level. Excretion is an example of homeostasis. If the level of a waste chemical increases, the body responds by working harder to get rid of it. If the level of carbon dioxide in the blood increases, a part of brain senses this. It sends off nerve messages which cause an increase in your breathing rate. Your rapid breathing removes the carbon dioxide more quickly and the level returns to normal.

Your liver produces a poisonous waste chemical called **urea**. Urea is produced when amino acids are broken down. It is released into the bloodstream and carried around the body. If it stayed in the bloodstream it would eventually kill you. In most of us, urea is filtered out of our blood by our kidneys (see figure 17.9) and is removed from our bodies dissolved in urine. In some people the kidneys fail to do this job. For them the solution is either a kidney transplant (see application 3.8, page 60) or regular treatment on a kidney machine (see application 17.2).

SENSING CHANGES

Our survival depends on being able to sense what is happening in our environment. Our environment includes changes inside our body such as our blood temperature, as well as changes in our surroundings. Any change in the environment that we can detect is a **stimulus**. There are other changes that we cannot detect, for example radio waves and X-rays. Fortunately scientists have devised instruments that can detect the stimuli that our own senses cannot.

Specialized body cells sensitive to stimuli are called **receptors**. Receptors are usually sensitive to one type of stimulus. The eye contains a large number of receptors that are sensitive only to light, not sound or temperature.

Application 17.2

THE ARTIFICIAL KIDNEY

You can survive with one healthy kidney. If *both* kidneys fail you will die. Fortunately, the job of the kidney can be done by a machine. A patient's bloodstream is connected to the kidney machine for three to six hours, two or three times a week. During this time, the blood is filtered and the waste chemicals removed.

A tube connected to an artery carries the patient's blood into the kidney machine. Inside the machine the blood flows through tubing made from a selectively permeable membrane (see page 6) such as cellophane. This membrane is surrounded by a solution that is constantly being replaced. The solution has the same concentration of sugar and salts as the blood plasma. Urea and other waste chemicals pass from the patient's blood, across the membrane and into the solution. Proteins and blood cells cannot cross the membrane and so they remain in the blood. Sugar and salts pass across it until the concentration in the blood and the solution is equal.

figure 17.10 A patient on an artificial kidney machine

INVESTIGATION 17.6

WHICH PARTS OF YOUR TONGUE CAN DETECT DIFFERENT TASTES?

a Get your partner to stick out their tongue as far as possible. Draw four outline drawings of the tongue surface. Use these outlines to record the areas of the tongue that are sensitive to a specific taste.
b Your teacher will provide you with four small beakers containing different solutions. These solutions represent the four basic tastes: bitter, salty, sweet and sour.
c Take one beaker and using a clean glass rod place a drop of the solution on the tip of your partner's tongue.
d If your partner can taste it, shade in the tip on the outline drawing of the tongue.
e Test other parts of the tongue with the same solution. Make sure that you shade in all the areas that can detect this taste.
f Carry out this procedure with each of the four solutions making sure that you clean the glass rod in distilled water before you begin to test a new taste.

Questions

1 Compare your outline maps with others in your class. Can you draw any general conclusions about the sensitivity of different people's tongues.?
2 What does this experiment tell you about our sense of taste?

Receptors may be grouped together in sense organs such as the eye or ear, or they may be spread throughout the body. For example, receptors sensitive to touch are located beneath the entire skin surface.

Responding quickly

A driver would have to act quickly to avoid hitting the child who ran out into the road (see figure 17.11). The sight of the child running into the road acts as the stimulus. The receptors in the eye form an image of the scene which is carried along nerves to the brain. The brain makes a decision about the correct action to be taken. The brain sends instructions along nerves to the leg muscles which contract and cause the foot to press down on the brake pedal.

figure 17.11 A driver needs to respond quickly

Any reaction caused by a stimulus is called a **response**. The organ carrying out the response, in this case of the leg muscle, is called an **effector**. See figure 17.12.

Braking sharply to avoid a collision requires the brain to make a decision. Sometimes we react quickly without having to make a decision. When we prick a finger or touch a red-hot object we pull our hand away without thinking. We have no power to stop this response. This is an example of a very simple type of behaviour known as a **reflex action**. Reflex actions are so vital for our survival that we are born with them. They will always occur if the right stimulus is given. Activities such as breathing, swallowing and sweating are all reflex actions that are essential for keeping us alive. Figure 17.13 shows how the knee-jerk reflex is co-ordinated by the nervous system.

INVESTIGATION 17.7

WHICH PARTS OF YOUR SKIN ARE THE MOST SENSITIVE?

a Bend a paper clip so that the points are 1 cm apart.
b Ask your partner to look away placing the palm of his or her hand facing upwards on the table.
c Place either one *or* both points on the tip of your partner's finger. Your partner should say how many points he or she can feel. Record in a table whether the answer is correct or not by a tick or cross.
d Repeat this test 20 times. Change the number of points you use at random.
e Carry out a similar set of tests on the palm, the back of the hand, the inside of the arm and the elbow. *Keep a careful record of the results of each test.*
f Count up the number of correct answers given for each area tested.

Questions

1 Which part of the body could detect the number of points most accurately?
2 What good biological reason is there for some parts of the body being more sensitive to touch than others?

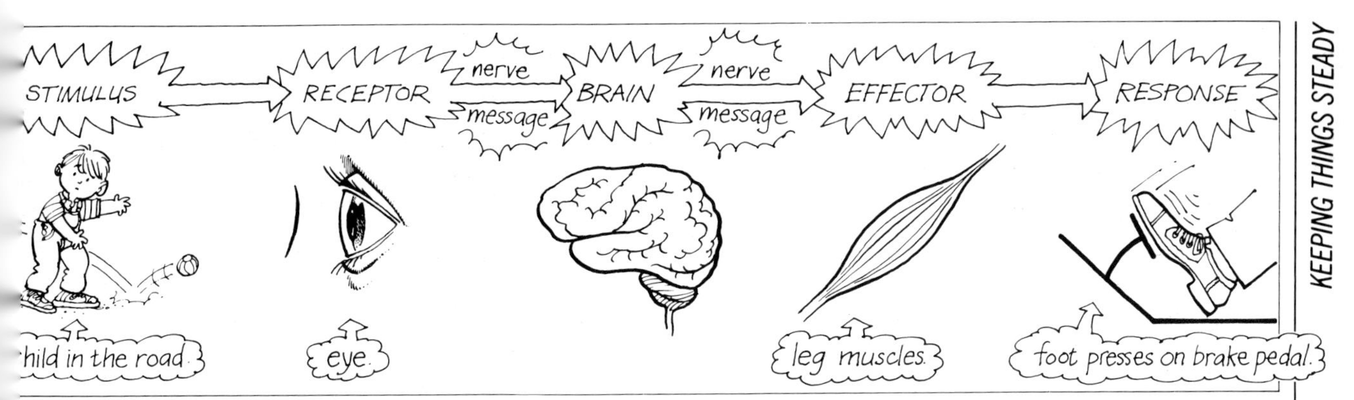
STIMULUS
RECEPTOR
nerve
message
BRAIN
nerve
message
EFFECTOR
RESPONSE
hild in the road
eye
leg muscles
foot presses on brake pedal

Figure 17.12 How the nervous system helps you to respond quickly

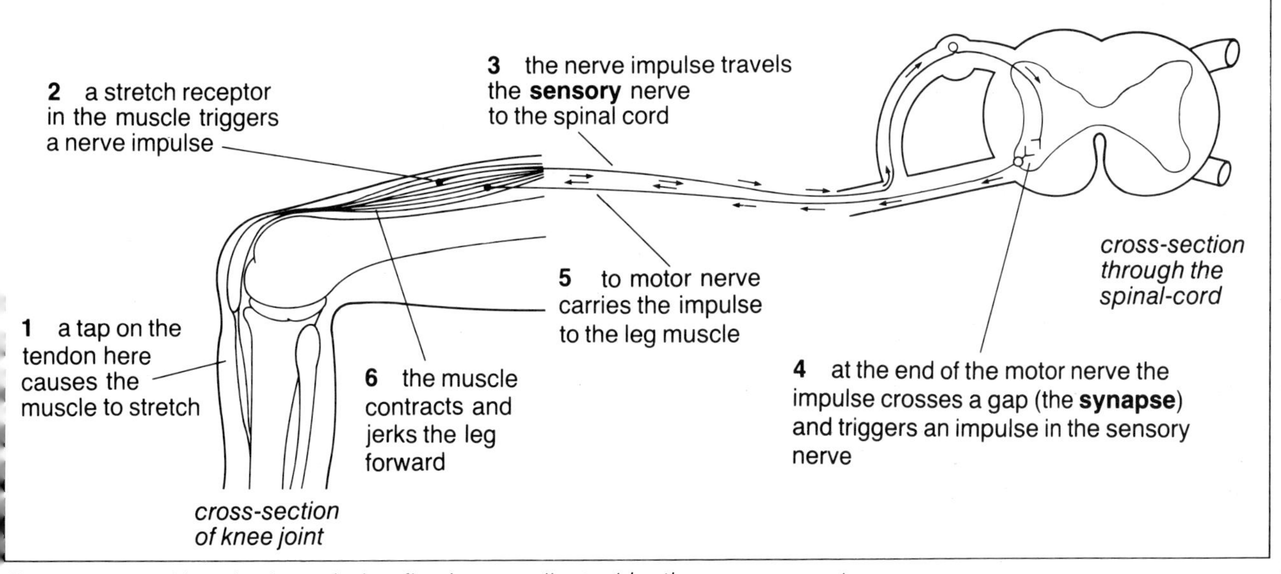
2 a stretch receptor in the muscle triggers a nerve impulse
3 the nerve impulse travels the **sensory** nerve to the spinal cord
cross-section through the spinal-cord
1 a tap on the tendon here causes the muscle to stretch
5 to motor nerve carries the impulse to the leg muscle
6 the muscle contracts and jerks the leg forward
4 at the end of the motor nerve the impulse crosses a gap (the **synapse**) and triggers an impulse in the sensory nerve
cross-section of knee joint

figure 17.13 How the knee-jerk reflex is co-ordinated by the nervous system

Fast reactions

Driving a car, saving a penalty in a football match or playing a computer video game all need fast reactions. The time between the detection of the stimulus and the response occurring is the **reaction time**. You can investigate your own reaction time in investigation 17.8. Many factors can affect our reaction time. Tiredness and lack of attention slow it down. Certain drugs also cause a serious reduction in our reaction time. This is why driving after drinking alcohol is so dangerous.

INVESTIGATION 17.8

HOW FAST ARE YOUR REACTIONS?

- a Stand on a chair and hold one end of a metre rule so that the zero mark is between your partner's thumb and forefinger, but *not* touching them.
- b Without telling your partner, release the rule. Your partner must try to catch the rule as quickly as possible.
- c Record the distance that the rule has dropped before being caught.
- d Repeat the above steps 10 times. Record your result each time.
- e Carry out this experiment again. This time hold the rule so that it just touches the inside of your partner's thumb.
- f Carry out the experiment a third time. This time your partner's eyes must be closed. Do not allow the rule to touch your partner's hand. Call out 'now' as you let go of the rule.

Questions

1. In *each* of the above experiments, which senses do you use to detect the rule dropping?
2. How does practice affect your reaction time? How do you explain this effect?
3. Does your reaction time change when you rely only on your sense of hearing? Can you suggest reasons for this change?
4. List five situations where you rely on fast reactions.
5. Write a description of a car accident that might be caused by a driver whose reaction time was delayed by alcohol.

Time for thought

Automatic control systems have replaced humans in many jobs. Computers now do the tedious accounting calculations previously carried out by enormous numbers of clerks. Motor cars can now be constructed by robots instead of assembly-line workers. This trend raises a number of questions.

1. *In the future, which jobs should be taken over by robots?*
2. *Are there some jobs you would not like to see carried out by robots?*
3. *If we continue to replace people's jobs with robots, what can we do to avoid mass unemployment?*

figure 17.14 The Fiat robot assembly line

Application 17.3

CONTROL DEVICES

Mechanical control devices are everywhere. You might think of a door handle as a very simple example of a control device. The more the handle is turned the more the catch turns. A spring-loaded handle will return to its original position when released. A tap is another simple control device. Unlike door handles, taps do not have return springs. Both types of device are examples of **direct control**. One action controls another. However, neither example *automatically* controls something else. They both have to be operated manually. In an ordinary central heating system there is a number of mechanical control devices which operate automatically.

Automatic water control: ballcocks

A **ballcock** is an automatic tap. A round plastic or copper ball, at one end of a lever, floats on the water surface in a water tank. At the other end the lever presses against a piston connected into the water supply. When water is drawn out of the tank the float lowers. As it does so the piston moves and water flows into the tank. Water flowing into the tank causes the float to rise cutting off the water supply. See figure 17.15.

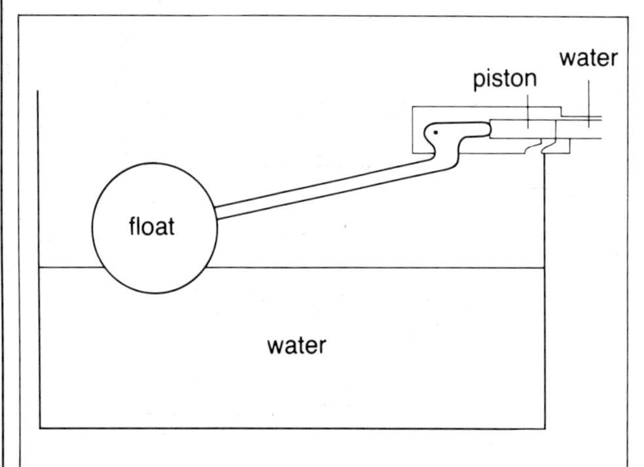

figure 17.15 Automatic water control

Automatic temperature control: expansion thermostats

When metals heat up they expand. This characteristic can be used to control the flow of hot water in a pipe network. The example shown in figure 17.16 is in the shape of a spring. As it heats up and expands it reduces the amount of water flowing past it. It may even expand to block the pipe and stop the flow altogether. When the temperature falls it contracts and hot water will again flow past. These devices are often used on individual room radiators.

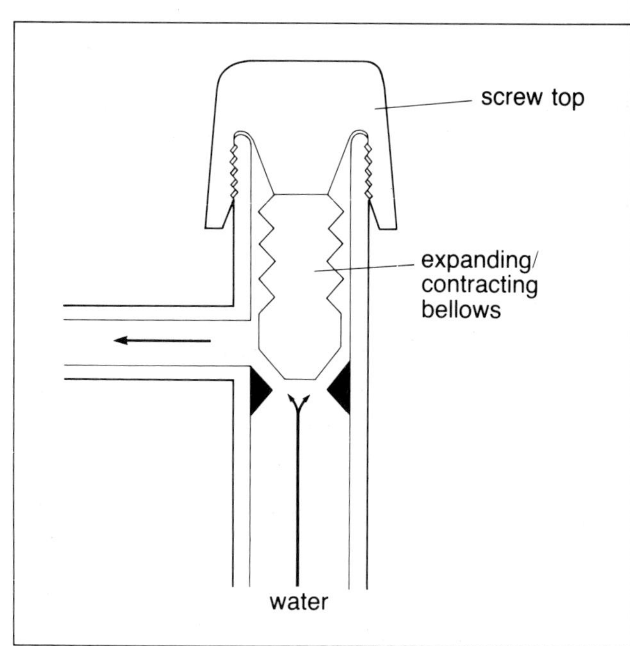

figure 17.16 Spring-activated thermostat

Automatic temperature control: bimetallic strips

A bimetallic strip is made up from two dissimilar metals joined together. When heated, they expand at different rates, causing the strip to bend. The bending movement can be used to switch various devices on or off. The one in figure 17.17 will switch on a heater when the strip becomes cold.

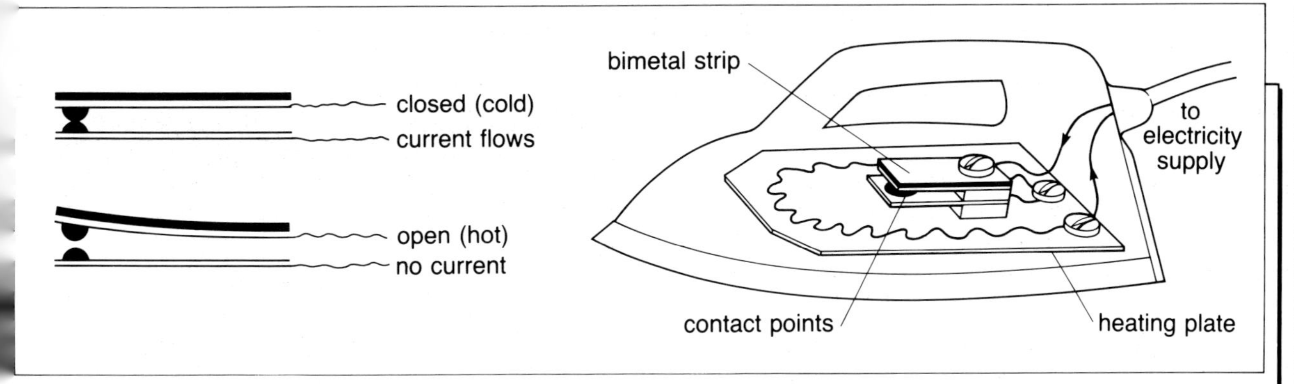

figure 17.17 Bimetallic strip-activated thermostat

Automatic 'flame-out' detector: bimetallic strips
When a pilot light in a gas boiler or cooker is on, it causes the bimetallic strip to expand and bend. This opens the valve and allows gas to flow. When the flame goes out, the strip contracts, cutting off the gas supply.

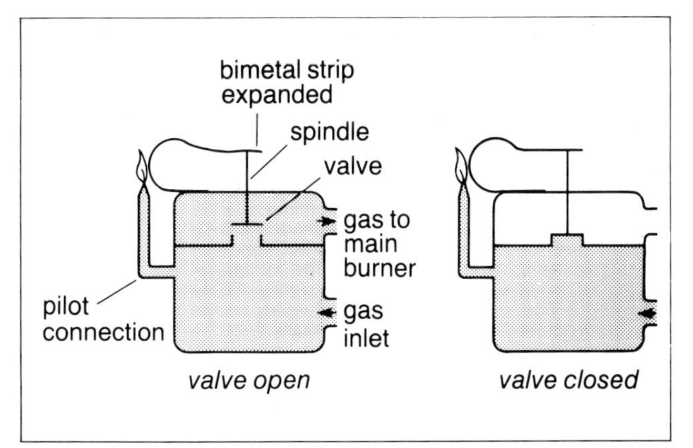

figure 17.18 Automatic 'flame-out' detector

Automatic water sprinkler for fire control: liquid expansion bulbs
A liquid-filled bulb is used as a 'plug' to seal a water pipe. When the liquid in the bulb becomes very hot (a room fire, for example) the liquid expands so much that it smashes its container. The water is then free to flow through the sprinkler, extinguishing the fire.

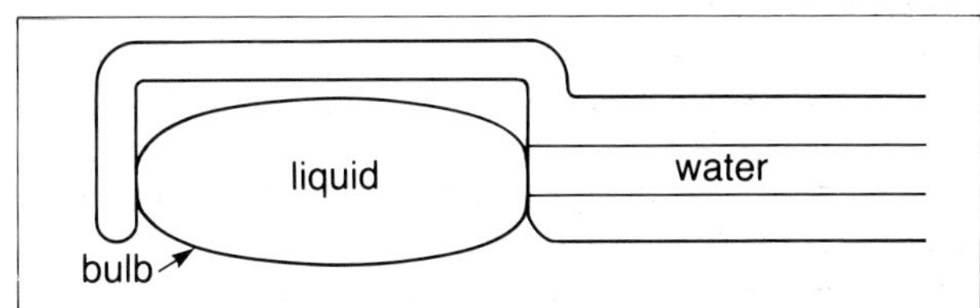

figure 17.19 Expansion-activated sprinkler system

The mechanical control system
Figure 17.20 shows a number of these mechanical control devices installed as part of an ordinary water supply system.

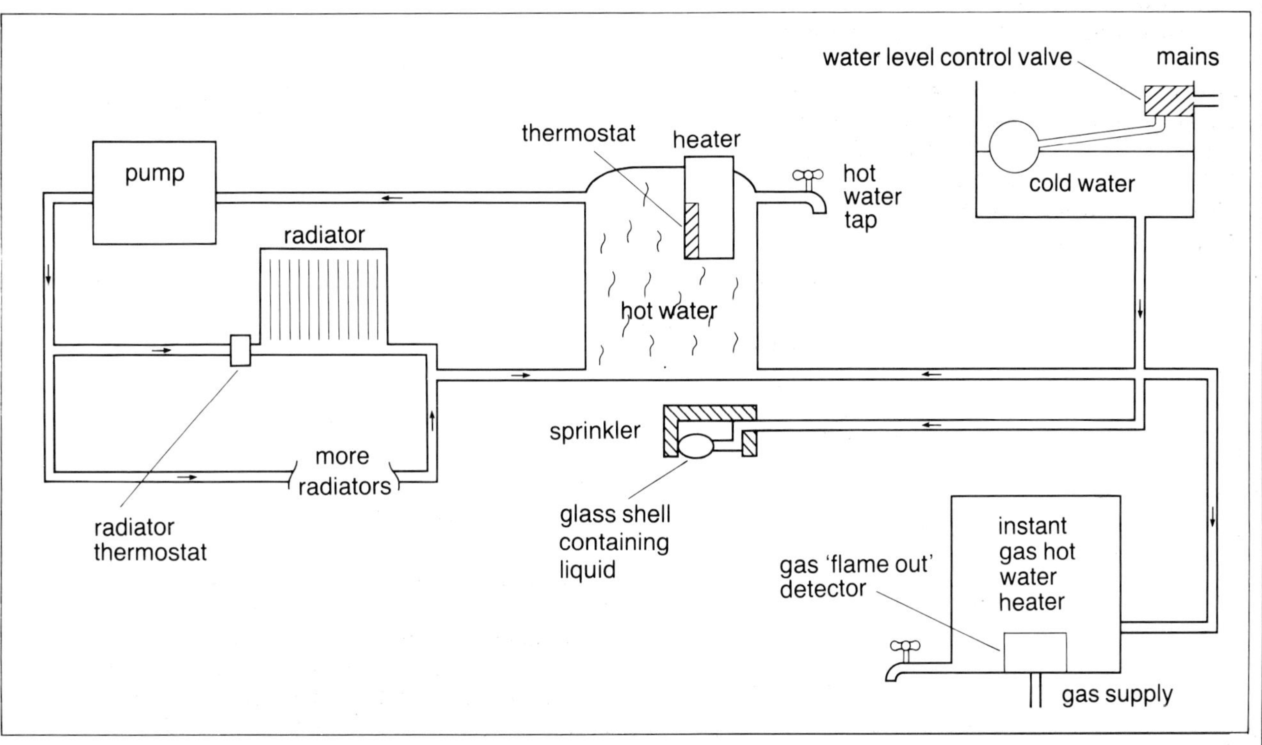

figure 17.20 How control devices are used in practice

ELECTRICAL AND ELECTRONIC CONTROL

Unlike mechanical control devices, electrical and electronic ones usually work **indirectly**. They control other devices in two stages. Sensors activate a trigger or switch, which in turn does the controlling.

Triggers and sensors

Transistors and thyristors are electronic triggers. They are switches. They need a sensor to trigger them if they are to operate other devices. Other devices include alarm bells, electric motors, electric water pumps etc. This is the two-stage process: a sensors trigger a transistor or thyristor which in turn controls the device.

Sensors as triggers

There is a number of electronic sensors which can be used as triggers in the circuit diagram in investigation 17.9. Each of the additions in application 17.4 can be used with that circuit to provide a different type of control device. (For each new circuit, resistors have been added where necessary to limit the amount of current flowing. They are included to protect other more expensive components.)

Caught in the act

Connecting a number of electronic sensors together and adding a loud bell forms the beginning of a comprehensive burglar alarm.

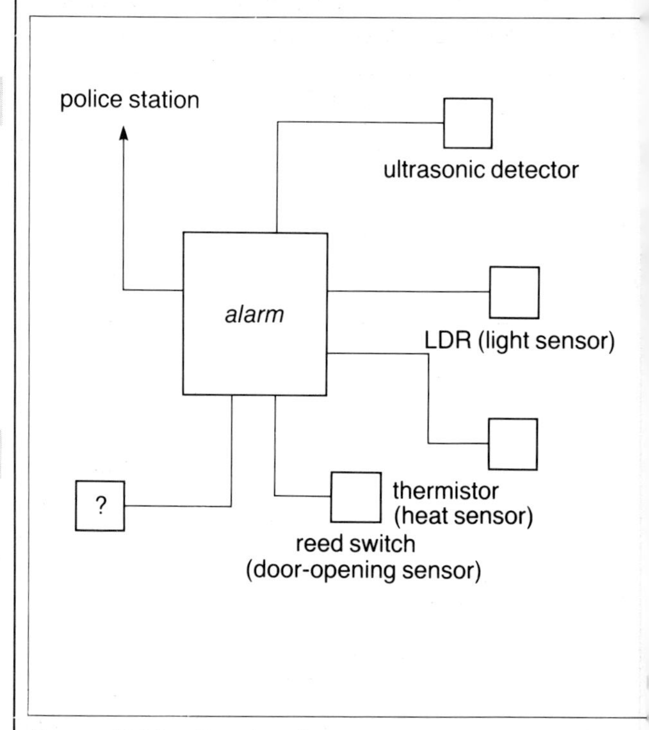

figure 17.22 Burglar alarm system

INVESTIGATION 17.9

TWO-STAGE CONTROL

This investigation looks at how to use two very simple electronic devices, and the differences between them. The small resistors included in the circuit are to protect the light emitting diode (LED) and the transistor or thyristor.

a Do not connect the battery.
b Construct the circuit shown in figure 17.21 using a transistor.
c When you are confident of the circuit, connect the battery.
d The LED should not light up.
e With a damp finger or a resistor make a direct connection between the positive rail and the base ('third') connection of the transistor.

Questions

1 What happens to the LED?
2 What happens when the 'third' connection is disconnected?
3 What happens to the LED when resistors of different value are used as the 'third' connection?
4 What does the 'third' connection do to the transistor?
5 What does the transistor do to the LED?
6 By repeating the experiment find the difference between the transistor and the thyristor.

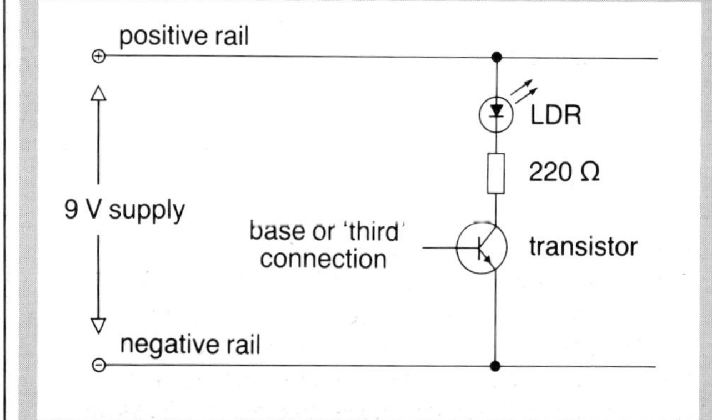

figure 17.21 Two-stage control

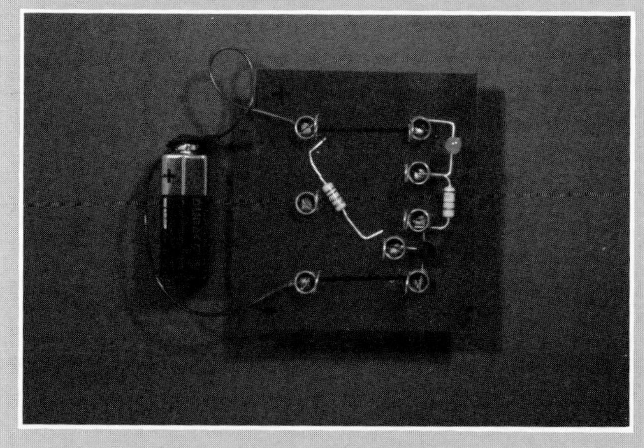

Application 17.4

CONTROL BY ELECTRONICS

Control by light: the LDR

A **light-dependent resistor (LDR)** is affected by light. It allows more electricity to flow through it when the light level is high. It can be used to switch other devices on or off when the light level changes (see figure 17.23).

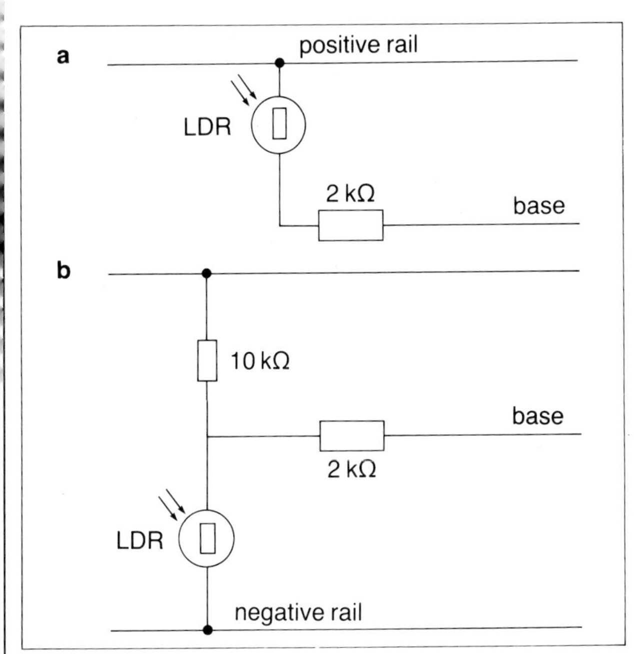

figure 17.23 Which circuit switches the LED on when the light level becomes low?

Control by heat: the thermistor

Thermistors allow more electricity to flow through them when they become warm. (See figure 17.24.)

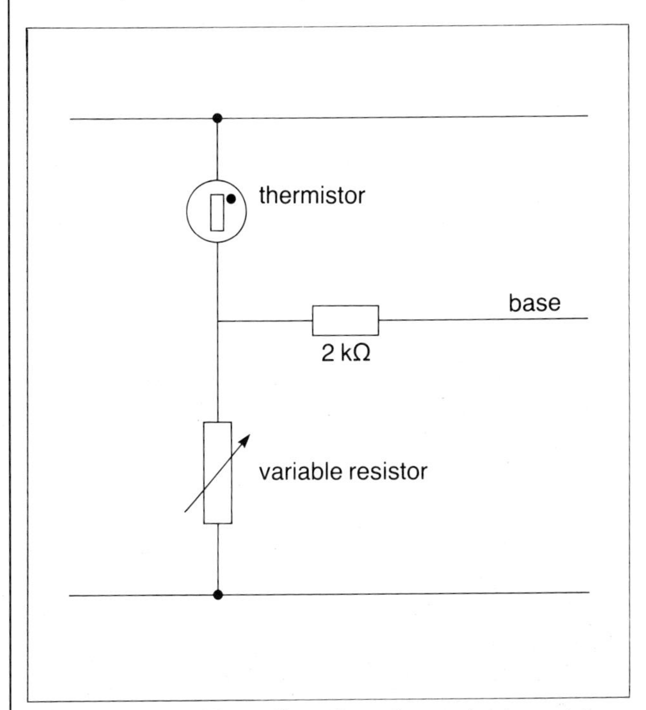

figure 17.24 What effect does the variable resistor have?

Control by magnetism: the reed switch

When a magnetic field passes through a **reed switch**, the two internal reeds or contacts touch. When this happens the switch becomes a conductor (see figure 17.25a).

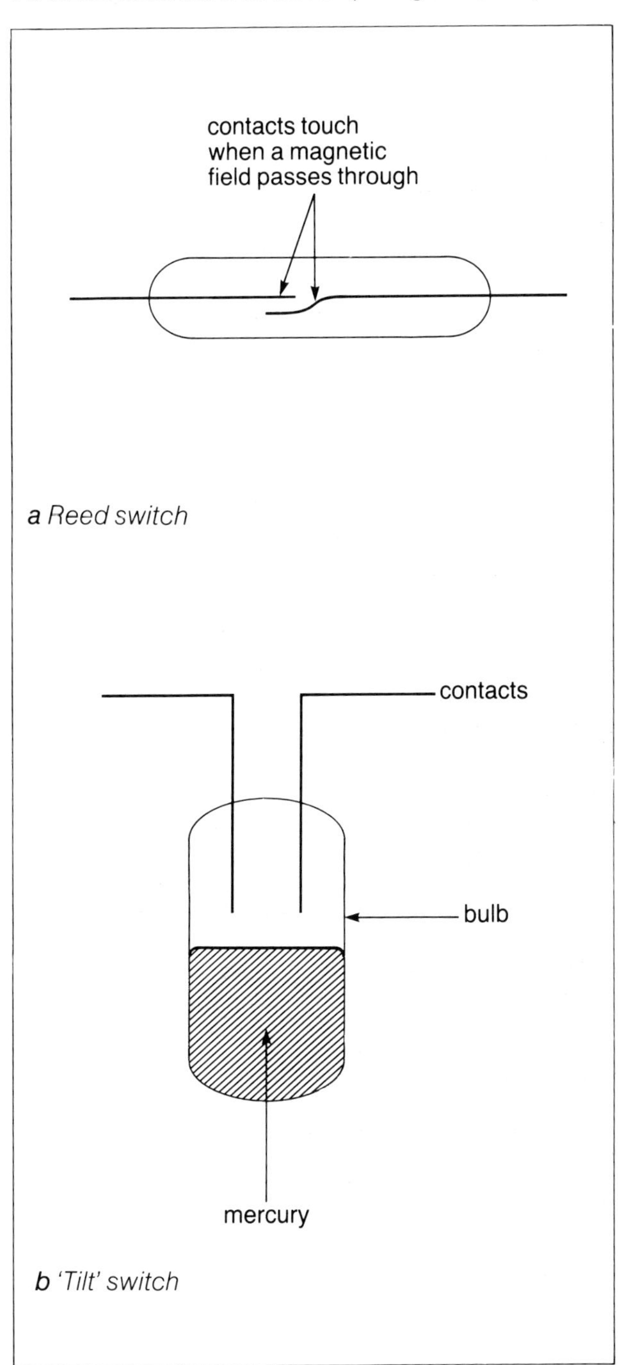

figure 17.25 Switches

Control by balance: the mercury switch

This type of switch is often made from a glass bulb containing mercury and two contacts (see figure 17.25b). The two contacts do not touch, and when the switch is held in one direction, neither of them can touch the mercury. When the switch moves the mercury flows and completes an electric pathway between the contacts.

UNIT 17

Application 17.5

HIGH LEVEL COMPUTER CONTROL

There are many processes that need to be controlled continuuously but which are too complicated to be controlled by simple mechanical, electrical or electronic devices. You may well have seen examples of computer control in school. The 'Turtle' and the 'BBC Buggy' are two well known devices that can be controlled by computer. However, both of these devices can only be programed to obey instructions that have been stored in the computer previously. They cannot think for themselves. Two important advantages of using computer control are

- the computer can store a large number of instructions and can therefore carry out very complicated processes easily.
- the speed at which computers work make them ideal for use in situations where the conditions change very rapidly.

Problem Cement is produced at high temperatures in a cement kiln. The quality of cement is affected by the noxious NC_x compounds (nitrogen oxides) contained in the waste gases. The energy used in the cement production increases if the combustion conditions in the kiln are not correct.
Solution A sensor is used to detect the nitrogen oxides in the waste gases. The information is fed to a computer and compared with the 'ideal kiln' conditions stored in the computer program. Adjustments are then automatically made to the fuel and air flows to return the kiln conditions to normal. Because the kiln is always operating correctly the cement is of high quality and the energy consumption is kept to a minimum.

Problem Staffordshire County Council set itself the task of reducing the coal consumption of a number of its school buildings.
Solution Monitoring stations (intelligent outstations) were set up in three schools to gather information such as internal and external temperatures, hot water demand, conditions within each boiler system etc. The outstations are all connected to a central control station via the telephone network and each is able to control its own school's heating system – when instructed to do so from the central station which monitors the information sent out.

The system proved so successful that it was extended to include another 18 schools. Although the installation was very expensive to install (it cost around £160 000) it paid for itself in less than four years and reduced the average annual energy consumption of the 21 schools by 13 per cent (approximately 750 tonnes coal/year).

ALL YOUR OWN WORK

1 A ________ system relies on information from a ________. This information is known as ________.

2 A thermostat is used to control the temperature in a domestic water heater. It has a preset value to come on and switch off. What would happen if it did not switch off?

3 What type of control device might be installed as part of a car engine, and how would it operate?

4 A heater is first turned on when the room is at 50 °C. Figure 17.24 shows how the temperature varies with time.

a At what point does the thermostat operate?
b What would the graph look like if the thermostat was set for 80 °C?

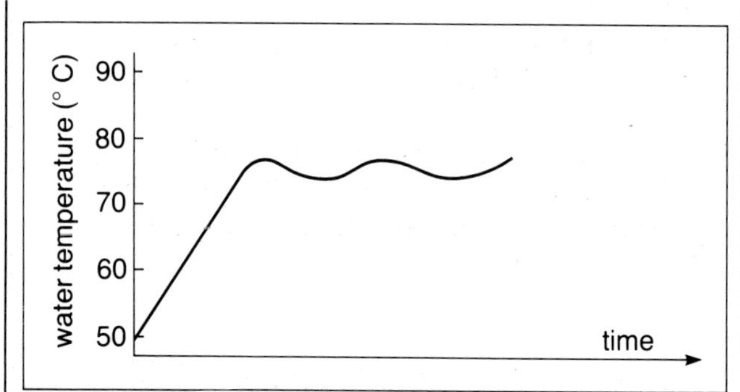

figure 17.26

5 Mammals lose heat through their skin surfaces. Research on the skin area of a type of rat gave the results in table 17.2.

	skin area	body weight
baby	1	1
young	4	8
young adult	9	27
adult	16	64

table 17.2

a What happens to the amount of skin this animal has as it grows up?
b How does its increase in weight compare with its increase in skin area?
c Use this information to explain why a baby rat finds it harder to keep warm.

ANSWERS

1 STARTING OUT

1 cells; organ; system; pancreas, small intestine
2 Plant cells have cell walls, large vacuoles and chloroplasts. Animal cells do not have these structures.
3 moves; kinetic
4 all; joule
5 Energy arrows are diagrams that show energy changes.
6 The kinetic theory is all about the movement of molecules. Whether a material is a solid, a liquid or a gas depends on how much energy the molecules have and how free they are to move.
7 **a** mixture density = 0.96 g/cm^3.
b oil is less dense than vinegar.
c oil – less than 0.96 g/cm^3.
vinegar – more than 0.96 g/cm^3.

2 CONTINUITY OF LIFE

1 zygote; gamete; sperm; tail
2 mutation; mutation; Down's
3 Co-dominance occurs when both alleles at a locus are expressed in the phenotype, as occurs with blood group AB.
4 Genetic counsellors can assess the risk of passing on the gene to the person's children. They can help them decide whether the risk is worth taking.
5 **a i** 3 **ii** 8 **iii** Brown is dominant because red-haired children are produced by two brown-haired parents in generation 2. Each parent must pass on one recessive gene.
b i P has genotype **RR**. **R** is recessive gene and will only show up if the person is homozygous.
ii Q has genotype **BR**. Some of Q's children have red hair so they must have gained an **R** gene from both parents.

3 HEALTHY BODY SYSTEMS

1 C; rickets; A
2 foetus; fat; carbohydrate
3 strengthens; drinking water
4 reduces; supple; strength
5 a virus causing influenza; a bacterium causing typhoid; a fungus causing athlete's foot; a protozoan causing malaria
6 A stimulant drug such as nicotine speeds up the activity of the nervous system. A depressant drug such as alcohol slows down the reactions of the nervous system.
7 **a** niacin
b $\frac{30}{100} \times 1485$ kJ = 445.5 kJ
c The cereal contains no fat, calcium or water.
d Milk would increase the fat, calcium and fluid intake; sugar would increase the energy and carbohydrate intake.
8 **a** The more cigarettes a person smokes the greater their chance of dying.
b No. Some non-smokers die of lung disease, especially if they live in cities.
c Whether they smoke or not, country dwellers have a smaller chance of dying from lung disease.
9 **a** food uncovered; sharp knives; rubbish uncovered
b House flies can transfer microbes onto uncovered food. Child may receive a cut from the knife. Uncovered rubbish allows house flies to pick up microbes.

4 PLANET EARTH

1 plates; earthquake
2 magma; sea; metamorphic
3 ploughing and removal of hedgerows
4 Clay has smaller particles and is more likely to become waterlogged. A sandy soil consists of larger particles that allow water to drain away more quickly.
5 water vapour
6 cumulonimbus and nimbostratus
7 clockwise; anticlockwise; hemisphere
8 to provide valued employees with work inside when work outside is difficult because of bad weather

5 ORGANISMS AND THEIR ENVIRONMENTS

1 sunlight; producers (green plants); consumers; carnivores
2 less
3 It is a method of observation requiring the unbiased selection of a small proportion of the total population
4 carbon dioxide – passes in through stomata; water – is carried up from the roots in the xylem vessels
5 **a** the Sun **b** the grass **c** The bullock is very active and uses up some energy. The bullock does not consume all the grass. Some energy is lost during respiration.

6 ON THE MOVE

1 constant; uniform
2 acceleration; m/s^2

3 It is useful for calculations and comparing journeys. Practical journeys involve changing speeds.
4 by changing the mass or velocity of the object
5 It is the reluctance of an object to move if it is still or to increase or decrease its speed if it is moving.
6 Weighing machines are usually marked in kg or g which are the units of mass. Weight is a force and so should be measured in newtons.
7 **a** 4 seconds (A speed of 20 m/s is decreased at the rate of 5 m/s every second. This process will take $\frac{20}{5}$ = 4 seconds to complete.)
b A friction force is created by applying the brakes to the wheels.
c KE = ½ mv^2 = ½ × 500 × 20 × 20 J
= 100 000 J
It is turned to heat in the brakes.
8 **a** 375 m
b 300 m
c 9.5 m (approx.)

7 OUT OF THIS WORLD

1 and **2** Craters suggest little or no atmosphere. Long shadows suggest deep craters, short shadows shallow craters. Craters inside other craters are younger. Surface lines may suggest rivers long ago (but not necessarily rivers of water).

8 ELECTRICITY

1 charge; positive; negative
2 conductor; insulator
3 current; amperes; potential difference or voltage; volts
4 magnetic; heating; chemical (see page 210)
5 **a** Current flow is measured in amperes. The higher the current value the greater the heating effect.
b If one light goes off the rest stay on.
c Electricity cannot reach the appliance until it is switched on.
6 **a** The trace would move in the Y direction and become smaller.
b The traces would become closer together.
7 2.5 kW = 2500 W $\quad I = \frac{P}{V} = \frac{2500}{240}$ A = 10.4 A
Fuse needed is the next highest value which is 12 A, however see main text for reason why it may be necessary to fit this appliance with a 13 A fuse.
8 **a** As the float moves up the magnet enters the coil which produces a voltage. When the float moves down the magnet moves up through the coil and produces a second voltage but this time in the opposite direction. The motion is continuous and so a voltage is produced continuously, alternating in direction.
b The voltage can be increased by **i** increasing the strength of the magnet **ii** increasing the number of wire coils **iii** using the generator where the waves have a larger amplitude.
c frequency = 2 Hz. Measure it on a CRO.

9 ENERGY

1 renewable; non–renewable
2 fossil
3 Refer to text.
4 PE = mgh
= 25 × 10 × 15 J
= 3750 J
If 3750 J represents 75% of the energy used then

$$100\% \text{ of the energy used} = \frac{3750 \times 100}{75} \text{ J} = 5000 \text{ J}$$

$$\text{input power} = \frac{5000 \text{ J}}{10 \text{ s}} = 500 \text{ W}$$

$$\text{output power} = \frac{3750 \text{ J}}{10 \text{ s}} = 375 \text{ W}$$

6 to **10** Refer to text.

10 WAVES

1 energy
2 electromagnetic spectrum (EMS); transverse; longitudinal
3 wavelength; frequency; hertz
4 incidence; angle; reflection
5 See text page 153.
6 They need to avoid waves travelling around corners and into the harbours or onto the beaches.
7 **a** reflects all colours, absorbs none
b absorbs all colours, reflects none
c reflects red light only, absorbs all others
d reflects yellow light (which is made up from red and green) only, absorbs remaining blue light
8 **a** distance = speed × time = 1500 × 0.1 m = 150 m
b If the echo distance is 150 m the distance to the shoal is 75 m.
The echo lasts longer because not all the fish will be at exactly the same depth, some will be in deeper water, some will be in shallower water.

11 CHEMICAL CLASSIFICATION

1 elements
2 elements; join, react
3 protons; neutrons; electrons
4 ionic; covalent
5 elements; atomic number
6 The noble gases are unreactive because their outer shells are already full.
They do not need to gain or lose any electrons.
7 Simple covalent compounds are gases at room temperature because the molecules do not carry any charge and have little attraction for each other.

3 a i 11; 17
ii neutrons
iii sodium group I, chlorine group 7
b i It loses one electron.
ii It gains one electron.
c i The ions become free to move around. The ions become further apart.
ii The oppositely charged ions attract each other and form a very strong lattice (crystal framework).

12 RADIOACTIVITY

1 At A, the activity increased and the thickness decreased, for a short time.
At B, the activity decreased and the thickness increased, for a short time.
At C, the activity has not returned to normal level, the thickness increased for a longer time.
At D, the activity increased dramatically, probably due to a break in the paper.
The remedy for the variation in thickness could involve improved control of the paper rollers.
2 a i group A
ii group B
b A higher percentage of health problems were associated with the higher radiation doses recieved.

	group A	group B	group C
i	10.0%	1.5%	1.8%
ii	17	62	106
iii	57%	91%	94%

The children may have died from cancers – a long term effect of radiation.

13 CHEMICALS IN ACTION

1 hydrogen
2 water; hydrogen ions
3 two from magnesium; zinc; iron
4 pH; hydrogen ion
5 Reactions are fastest at the beginning because there are the greatest number of reactant particles present and so most collisions.
6 There is a finite amount of each metal in the ground. If we continue to use it up at the present rate we shall soon run out. If metals are carefully recycled they will last a lot longer.
7 a i hydrogen
ii 68 cm^3
iii 53 seconds
b ii B
iii The smaller pieces have a greater surface area and so reaction B is faster.
c i A
ii The acid was in excess and so adding more magnesium would give more hydrogen.
d The risk of explosions is high at a flour mill because the air contains a very fine dust of flour particles. The small particles have a very large surface area and will burn explosively.

14 CHEMICALS AROUND US

1 mixture; nitrogen; 80%; oxygen
2 copper; syringes; copper; copper oxide; volume; 80 cm^3; 20
3 fractional distillation
4 three of: both need oxygen; both produce energy; both produce carbon dioxide; both need fuel (food)
5 lather; calcium; magnesium; temporary hard water; permanent hard water; temporary hard water; permanent hard water; washing soda; ion exchange column
6 a i carbon dioxide from the air; water taken in from the soil
ii light
iii the sun
b i carbohydrate + oxygen → carbon dioxide + water + ENERGY
ii yeast
c i heart disease – too much cholesterol in fat causes the arteries to get blocked
ii soft margarines contain polyunsaturated fats rather than cholesterol

15 CHEMICALS IN USE

1 fossil fuels; oil; natural gas; coal
2 carbon; carbon dioxide; carbon monoxide; poisonous
3 hydrocarbons; boiling points; refinery; cracking
4 metals; conducting; more reactive; difference
5 Soapless detergents are not affected by hard water. If detergents are not biodegradeable they can cause unpleasant foaming in rivers, damaging the river life.
6 a smoke
sulphur dioxide
smoke
b smoking; lead fumes from petrol
c use lead free petrol
7 A good fuel should: light easily; give out plenty of heat; burn slowly with a clean flame; produce little ash; produce no pollution; be readily obtainable; be fairly cheap and easy to store.
Oil needs a special burner and storage tank. It has to be delivered by tanker and is not particularly cheap. However, it is easy to light, burns with a fairly clean flame and leaves no ash.

8 Fuel, air (oxygen) and heat. Water reduces the temperature of the fire and also, to some extent, prevents the air from getting to the fuel.
9 The main problem with electric cars is that the electricity is not easy to store. This means that cars would have to be connected to a supply like trains and trolley buses are, or carry their own batteries. So far the batteries or accumulators that have been developed are very heavy and not really suitable for cars.
10 **a** chemical
b name
c formula
d physical; molecular mass
11 Isomers are compounds that have the same molecular formula but different structures. The two isomers are butane and 2-methyl propane (the 2 indicates that the methyl group is joined to the second carbon atom of the propane group).
12 alcohol (ethanol); carbohydrates; enzymes; molecules; carbon dioxide; alcohol; fractional; alcohol; boiling points.
13 A polymer is a chemical made up of large molecules which are themselves made up of many repeating units. Polythene, polystyrene, polyester, polyurethane, polyvinyl chloride (PVC) and others
14 Pyrex glass is much stronger. It can also withstand heating to higher temperatures and sudden changes of temperature.

16 CHEMICAL AMOUNTS

1 atoms; element
2 charge
3 equal number; atoms; element
4 Avogadro's number
5 relative atomic mass
6 1 dm^3; 1 mole
7 **a** $PbSO_4$
b $FeCl_2$
c $Al(OH)_3$
d Ag_2CO_3
e $Ca_3(PO_4)_2$
8 **a** 3
b 9
c 6
d 15
e 21
9 **a** methane + oxygen → carbon dioxide + water
b $CH_4 + 2O_2 \rightarrow CO_2 + 2H_2O$
10 **a** 160 g
b 44 g
c 17 g
d 400 g
e 100 g
11 PbO_2
12 **a** 35%
b 13.9%
c 21%
d 82%
13 0.44 g
14 **a** $3Fe + 4H_2O \rightarrow Fe_3O_4 + 4H_2$
b 0.33
15 **a** 80 g
b O.8 g
c 3.2 g
16 **a** $H_2SO_4 + 2NaOH \rightarrow Na_2SO_4 + 2H_2O$
b $CuO + 2HCL \rightarrow CuCl_2 + H_2O$
c $MgCl_2 + 2AgNO_3 \rightarrow Mg(NO_3)_2 + 2AgCl$
d $2Fe + 3Cl_2 \rightarrow 2FeCl_3$
17 **c** $Mg(NO_3)_2$, 65% oxygen
18 **a** copper (II) sulphate + iron → iron (II) sulphate + copper
b Add sodium hydroxide solution, a green precipitate of iron II hydroxide.
c $CuSO_4 + Fe \rightarrow FeSO_4 + Cu$
d 8 g
19 **a** The ammonia is taking away the oxygen from the copper oxide. This is a redox reaction.
b 1.92 g
c 0.24 dm^3
20 **a** The solution has a concentration of 2 moles/dm^3.
b The acid is more concentrated.
c 0.05 moles
d 0.05 moles
e 2.5 M

17 KEEPING THINGS STEADY

1 control; sensor; feedback
2 The water would eventually boil, and the tank might explode.
3 Expansion thermostat – it opens to allow more water to pass through to the car radiator when the water is hot. It closes down when the water is cold.
4 **a** around 75 °C
b The wavy portion of the graph would start at around 80 °C.
5 **a** It increases slowly.
b It becomes much larger much more quickly.
c With a large skin area and a small weight, it loses heat very quickly.

INDEX